Parasitic and Infectious Diseases

Parasitic and Infectious Diseases

Epidemiology and Ecology

Edited by

Marilyn E. Scott

Institute of Parasitology
Macdonald Campus, McGill University
Ste-Anne de Bellevue, Québec
Canada

Gary Smith

School of Veterinary Medicine
University of Pennsylvania
Kennett Square, Pennsylvania

Academic Press

San Diego New York Boston
London Sydney Tokyo Toronto

This book is printed on acid-free paper. ♾

Academic Press, Inc.
A Division of Harcourt Brace & Company
525 B Street, Suite 1900, San Diego, California 92101-4495

United Kingdom Edition published by
Academic Press Limited
24–28 Oval Road, London NW1 7DX

Library of Congress Cataloging-in-Publication Data

Parasitic and infectious diseases : epidemiology and ecology / edited by Marilyn E. Scott, Gary Smith,
p. cm.
Includes bibliographical references and index.
ISBN 0-12-633325-4
1. Parasitic diseases--Epidemiology. I. Scott, Marilyn E. II. Smith Gary, DATE
RA644, P18P37 1994
614.5'5--dc20 93-39576
CIP

PRINTED IN THE UNITED STATES OF AMERICA
94 95 96 97 98 99 QW 9 8 7 6 5 4 3 2 1

This book is dedicated to Joan and Bob Scott,
Bob Harvey, Caroline Smith, Catherine Smith, and James Smith
and to Anne Keymer, friend and colleague.

Contents

5

Parasite Population Density Is Regulated

Gary Smith

6

Ecological Epidemiology and Standard Measures of Disease Occurrence

Gary Smith

7

Models and Public Health Applications

Joan L. Aron & Barry A. Silverman

8

Host Populations: Genetics and Immunity

D. Wakelin

9

Nutritional Status of Host Populations Influences Parasitic Infections

Noel W. Solomons & Marilyn E. Scott

10

Host Behavior and the Flow of Parasites Through Host Populations

Victor Apanius & Gerhard A. Schad

17
Trichostrongylid Parasites of Domestic Ruminants
Michael J. Coyne & Gary Smith

18
Echinococcosis
M. G. Roberts & M. A. Gemmell

19
Population Biology of *Ixodes scapularis*, the Vector of Lyme Disease in the Eastern and North Central United States
Stephen Vail, Gary Smith, & Cynthia Lord

20
Heligmosomoides polygyrus: A Laboratory Model for Direct Life-Cycle Nematodes of Humans and Livestock
Marilyn E. Scott & Geneviève Tanquay

21
Population Biology of *Trichostrongylus tenuis* in the Red Grouse, *Lagopus lagopus scoticus*
Andrew P. Dobson & Peter J. Hudson

22

Population Biology of the Diplostomatid Trematode *Uvulifer ambloplitis*

Gerald W. Esch

23

Myxomatosis

Frank Fenner

24

The Sealworm Situation

Michael D. B. Burt

Contributors

Numbers in parentheses indicate the pages on which the authors' contributions begin.

Victor Apanius (115), Institute of Parasitology, MacDonald Campus, McGill University, 21,111 Lakeshore Road, Ste-Anne de Bellevue, Quebec, Canada H9X 3V9

Joan L. Aron (73), Department of Population Dynamics, School of Hygiene and Public Health, The Johns Hopkins University, Baltimore, Maryland 21205, USA

Ben M. Bolker (219), Department of Ecology and Evolutionary Biology, Princeton University, Princeton, New Jersey 08544, USA

Michael D. B. Burt (347), Department of Biology, Bag Service #45111, University of New Brunswick, Fredericton, New Brunswick, Canada E3B 6E1

Michael J. Coyne (235), SmithKline Beecham Animal Health, 812 Springsdale Drive, Exton, Pennsylvania 19341, USA

D. W. T. Crompton (175), WHO Collaborating Center for Soil-Transmitted Helminthiases, Department of Zoology, University of Glasgow, Glasgow G12 8QQ, Scotland, United Kingdom

Andrew P. Dobson (301), Department of Ecology and Evolutionary Biology, Princeton University, Princeton, New Jersey 08544-1003, USA

Christopher Dye (159), Department of Medical Parasitology, London School of Hygiene and Tropical Medicine, University of London, Keppel Street, London WC1E 7HT, United Kingdom

Gerald W. Esch (321), Department of Biology, Wake Forest University, Winston-Salem, North Carolina 27109, USA

Frank Fenner (337), The John Curtin School of Medical Research, The Australian National University, Canberra ACT 2603, Australia

M. A. Gemmell (249), Department of Clinical Veterinary Medicine, University of Cambridge, Madingley Road, Cambridge CB3 OES, United Kingdom

Bryan T. Grenfell (219), Department of Zoology, University of Cambridge, Downing Street, Cambridge CB2 3EJ, United Kingdom

Peter J. Hudson (301), Scottish Grouse Research Project, Game Conservancy, Crubenmore Lodge, Inverness-shire PH20 1BE, Scotland, United Kingdom

Cynthia Lord (263), Department of Biological Sciences, University of Notre Dame, Notre Dame, Indiana 46556, USA

Hamish McCallum (29), Department of Zoology, University of Queensland, St. Lucia 4072, Australia

Angela R. McLean (129), Zoology Department, University of Oxford, South Parks Road, Oxford OX1 3PS, United Kingdom

Graham F. Medley (141), Ecosystems Analysis and Management Group, Department of Biological Sciences, University of Warwick, Coventry CV4 7AL, United Kingdom

M. G. Roberts (249), AgResearch, Wallaceville Animal Research Center, P.O. Box 40063, Upper Hutt, New Zealand

Gerhard A. Schad (115), Laboratory of Parasitology, Department of Pathobiology, School of Veterinary Medicine, University of Pennsylvania, Philadelphia, Pennsylvania 19104-6050, USA

Marilyn E. Scott (1, 21, 29, 101, 279), Institute of Parasitology, Macdonald Campus, McGill University, Ste-Anne de Bellevue, 21,111 Lakeshore Road, Quebec, Canada H9X 3V9

Barry A. Silverman (73), Cambridge Consulting Corporation, Reston, Virginia 22091, USA

Gary Smith (9, 21, 47, 65, 235, 263), School of Veterinary Medicine, University of Pennsylvania, New Bolton Center, 382 West Street Road, Kennett Square, Pennsylvania 19348, USA

Noel W. Solomons (101), Center for Studies of Sensory Impairment, Aging, and Metabolism, Hospital de Ojos y Oïdos "Dr. Rodolfo Robles V.", Diagonal 21 y, 19 Calle, Zona 11, Guatemala City 01011, Guatemala

Geneviève Tanquay (279), Fonds pour la Formation de Chercheurs et l'Aide à la Recherche, 3700, rue du Campanile, Sainte-Foy, Quebec, Canada G1X 4G6

Stephen Vail (263), Department of Biology, William Patterson College, Wayne, New Jersey 07470, USA

D. Wakelin (83), Department of Life Science, University of Nottingham, Nottingham NG7 2RD, United Kingdom

Mark E. J. Woolhouse (197), Department of Zoology, University of Oxford, South Parks Road, Oxford OX1 3PS, United Kingdom

Preface

Epidemiology is the study of the occurrence of disease. Ecological epidemiology deals with the way in which the occurrence of infectious disease can be explained in terms of population processes. There are at least two populations involved in any infectious disease: the population of parasites and the population of hosts that is one of the necessary causes of the disease. The abundance of both populations is determined by a variety of factors. It is the operation of these factors that is the principal concern of the ecological epidemiologist.

Our understanding of how these processes work has increased enormously with the use of mathematical models, and many of the chapters in this book focus on how we have used such models. Unfortunately, it has been our experience as teachers that those who could make best use of the insights of ecological epidemiology are often those who are least equipped to understand how those insights were obtained. Models are a means of representing and manipulating something that would otherwise be inaccessible, but couching models in mathematical language renders them unintelligible to those who might benefit from them the most. Even highly motivated students quickly find that there seems to be no place to start. The best texts are much too difficult for the beginner, and we know of no elementary text that combines straightforward explanations of the mathematics with more detailed development in concrete case studies. We hope that this book will help to remedy this deficiency.

This book is intended for that vast army of medical students, veterinary students, and biologists who have expressed an interest in ecological epidemiology but do not know where to begin. Chapters 1–6 summarize and explain in more than the usual detail the fundamentals of how one writes deterministic mathematical models of infectious disease. These early chapters contain elementary material and are deliberately written in a rather informal style. Chapters 7–13 elaborate on some of the general principles and objectives of the discipline, focusing on factors that influence host–parasite population interactions, including the influence of certain approaches to control. Chapters 14–24 are a collection of case studies of parasitic and infectious diseases in humans, livestock, and wild animals. Each case study attempts to show how links between field observations and experimentation, laboratory experiments, and theoretical studies have helped us to better understand the specific infection. The mathematical content of these chapters varies substantially. Some chapters contain no mathematics at all but are included because they are excellent illustrations of the way in which ecological epidemiologists view infectious diseases as dynamic entities. Other chapters contain the kind of mathematics that most beginners skip in the hope of finding something that does make sense. We have no illusions that the first few chapters (Chapter 1–6) will equip the novice with all that is required to follow the mathematics in some of these case studies, but the reader should find enough that is familiar and intelligible to be persuaded that further study will be rewarded. We hope that the combination of stimulating material (the case studies) and straightforward explanations of basic mathematical techniques will encourage persistence.

A great many people have assisted in the genesis of this book. Anyone who knows us will understand the particular debt we owe to Roy Anderson, but we have been fortunate also to benefit from the wit and wisdom of a growing cadre of modelers and population biologists on both sides of the Atlantic. We express our heartfelt gratitude to the friends and colleagues who graciously agreed to write the chapters that form the bulk of this book. Mick Roberts, Stephen Vail, Bryan Grenfell, David Gordon, and Bob Harvey read early drafts of some of the chapters, and Cindy Plyler, Ruth Owen, Mike MacKay, and Jenny Anderson diligently and uncomplainingly assisted us in the day-to-day tasks of putting the book together. We also thank our ever-patient editors at Academic Press, Charlotte Brabants and Michelle Walker. We also acknowledge the determined contribution of colleagues at a recent Epidemic Modelling Program at the Isaac Newton Institute in Cambridge who persuaded us that it was mathematically more precise (and grammatically more defensible) to call the basic reproductive rate the basic reproduction ratio. As a final note, American spelling is used throughout the book in accordance with the publisher's style. This does not necessarily reflect the preference of individual contributors.

Marilyn E. Scott
Gary Smith

Marilyn E. Scott

Populations Are Dynamic

1

I. So You Understand Dynamics

In this book, the single most important unifying principle is that of population dynamics. Therefore, at the outset, it is critically important that you have a clear idea of what is meant by dynamics, whether host population dynamics, parasite population dynamics or host–parasite population dynamics. When we think of dynamics, we think of change, movement, addition or removal of individuals. When we try to determine whether a dynamic process is occurring, our first step is often counting the actual numbers of individuals in the population at various points in time. Possible results are shown in Fig. 1. Which of these graphs represent the result of dynamic processes? Many investigators would consider Fig. 1b, c, and d, but not Fig. 1a, to depict the results of dynamic processes. This belief reflects the first common misconception of population dynamics: If the number of individuals in the population remains constant, the population is not dynamic. To dispel this idea, consider the analogy of a clinical microbiology laboratory. A laboratory technician has a shelf on which specimens wait for processing. Each day, samples are processed and new ones arrive. Is the number of samples sitting on the shelf dynamic? Yes. If the laboratory technician is sick one day, presumably the samples will accumulate (rate of input exceeds rate at which they are processed). Under normal circumstances, however, the laboratory technician tries to process samples at the same rate they are received. In this situation, the "standing crop" (number of samples sitting on the shelf) remains constant. Despite the fact that the number of samples is constant, some samples are leaving the shelf and others are being added.

II. Reviewing the Basics of Population Dynamics

When we think of a "host population," we think of a group of animals living in a defined area (a village, a pasture, a lake). Four basic biological processes determine how the numbers of individuals in this population change through time: birth, death, immigration, and emigration. Throughout this chapter, and throughout the book, these processes will be represented diagrammatically using a "flow chart." In the flow chart, the box represents the number of individuals in the population. Arrows entering the box represent biological processes that lead to increases in the numbers of individuals.

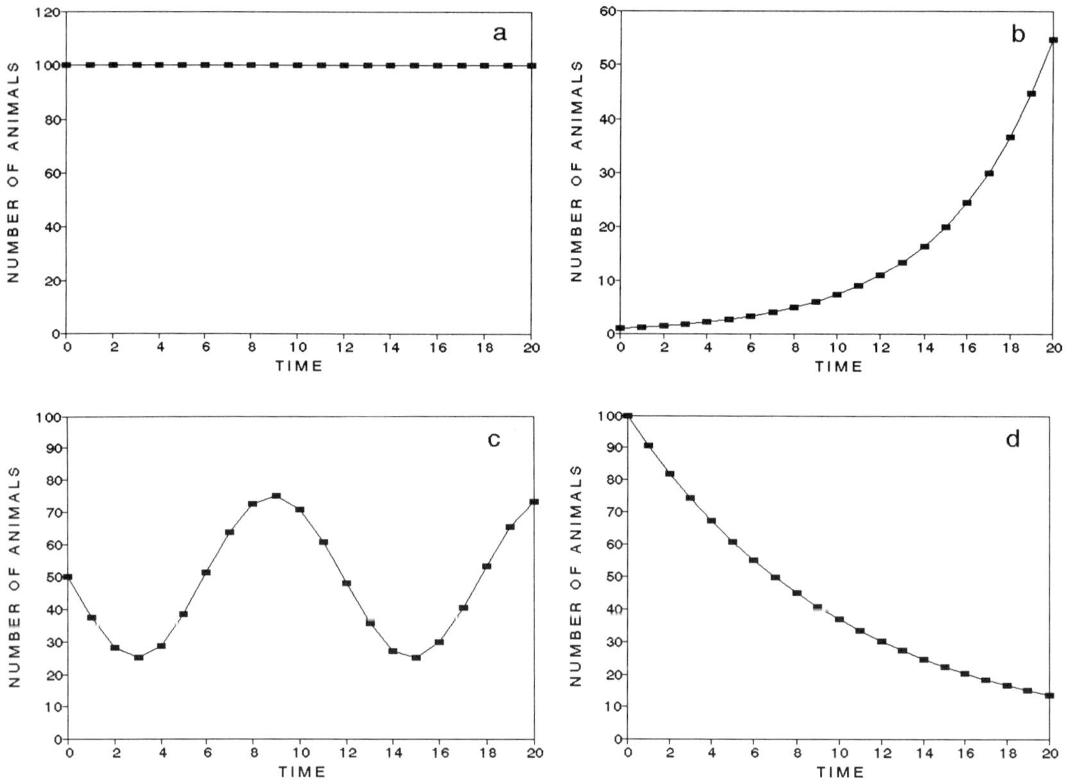

Figure 1 Patterns of change in the number of individuals with time. Which of these graphs reflect the results of a dynamic process?

Arrows leaving the box represent biological processes that lead to decreases in the number of individuals in the population. For example, we could represent the number of animals in the population by the symbol H (see Fig. 2). The number of animals can increase in two ways (birth and immigration), each of which is represented by an arrow coming into the H box. The number of animals can decrease in two ways (death and emigration), each of which is represented by an arrow leaving the H box. Intuitively, if the rate at which animals enter the population exceeds the rate at which animals leave the population, the actual number of animals will increase.

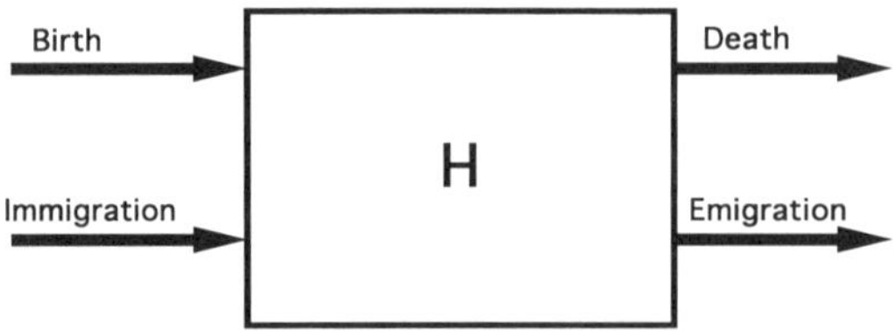

Figure 2 Flow chart representing population dynamics of an animal population. H, Number of animals in the population.

In a very simple example, no immigration or emigration occurs. On average, each individual produces one offspring every 6 mo. The average birth rate per individual is 2/year. This rate is called the *per capita* birth rate, b. If the population contains 10 individuals, 20 offspring would be produced in the year. However, if the population contains 100 individuals, 200 offspring would be born in a year. When we multiply the *per capita* birth rate, b, by the number of individuals in the population, H, we obtain the net birth rate *(bH)*. If, on average, each individual lives for 20 yr, the average death rate per individual is 0.05/yr (1/20). This rate is the *per capita* death rate, *d*. The net loss due to death depends on the *per capita* death rate, but also depends on how many individuals are in the population. The net death rate, therefore, is *dH*.

A key step in really understanding population dynamics is understanding that these processes occur simultaneously, yet they are independent of each other. The easiest way to visualize this is by imagining a game in which one person represents "birth" and another person represents "death." The person representing "birth" determines how many individuals are alive today, and calculates the number of new births over the next time interval. Meanwhile, the person representing "death" determines how many individuals are alive today, and calculates the number of deaths over the next time interval. This process continues over several time intervals and a tally of the actual number of individuals in the population at each time point is obtained.

Figure 3 shows the results that are obtained from this exercise, using a time interval of 1 yr. Note that the actual number of births increases over time although the *per capita* birth rate is constant. Also note that the actual number of deaths increases over time, although the *per capita* death rate is constant. In this most simple example, the

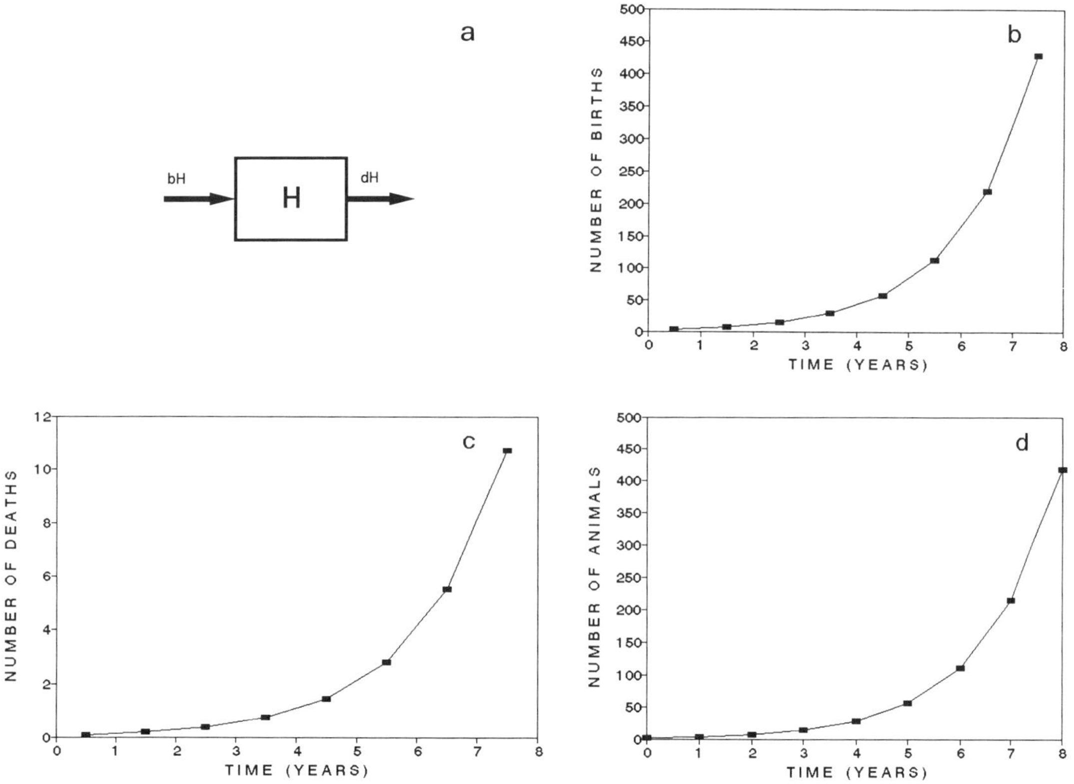

Figure 3 Changes in the number of individuals (H) through time as a result of two biological processes: birth and death. The *per capita* birth rate (b) is 2/year. The *per capita* death rate (d) is 0.05/yr. The initial population size is 2. Calculations were done on yearly intervals. (a) Flow chart representing the biology. (b) Number of individuals born in each year. (c) Number of individuals that die in each year. (d) Total number of individuals in the population.

population will continue to grow indefinitely, because the birth rate (2/yr) exceeds the death rate (0.05/yr).

Although this exercise may seem trivial, it illustrates very well the independence of these processes and how their combined effects determine the size of the population. Now add a third biological process, emigration, to the system so, on average, each individual stays in the population for 5 yr before emigrating. Therefore, the *per capita* emigration rate (e) is 1/5 or 0.2/yr and the *net* emigration rate is eH. Add a new player to the game who represents "emigration." The

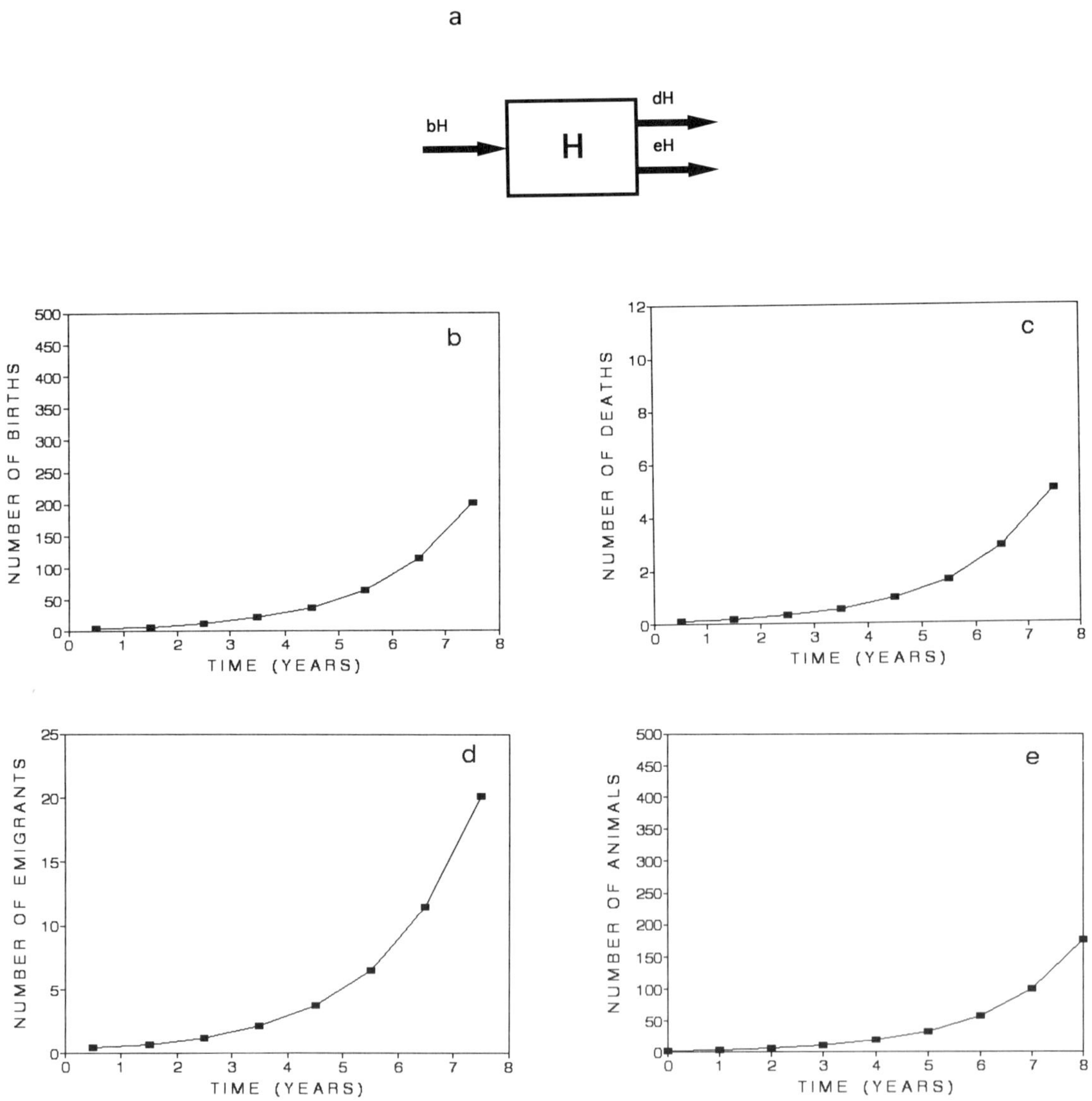

Figure 4 Changes in the number of individuals (H) through time as a result of three biological processes: birth, death, and emigration. The *per capita* birth (b) and death (d) rates and the initial population size are the same as in Fig. 3. The *per capita* emigration rate (e) is 0.2/yr. (a) Flow chart representing the biology. (b) Number of individuals born in each year. (c) Number of individuals that die in each year. (d) Number of individuals emigrating from the population in each year. (e) Total number of individuals in the population.

people representing "birth" and "death" continue playing as before, oblivious to the fact that a third person has joined the game. At each time interval, the original players count the number of individuals in the population and perform their calculations. Similarly, the person representing "emigration" is oblivious to the actions of the other players. Each person only counts the actual number of individuals alive in the population at each point in time. Increasing or decreasing any of these *per capita* rates, or adding new biological processes, has absolutely no effect on the fundamental biology of the other processes. What changes is the impact these processes have on the population. The total number of individuals born or the number that die is influenced, simply because the population size is affected. This result can be seen most clearly by comparing Fig. 4 (including emigration) with Fig. 3. Addition of a new biological process results in fewer total births in each time interval and fewer deaths in each time interval, despite the fact that the *per capita* rates have remained the same.

Clearly this situation is very artificial. We know that populations do not grow unchecked forever. A second key step in understanding population dynamics is understanding the phenomenon of density dependence. Density-dependent processes enable a population to increase in size quickly when the population is small, but to increase in size more slowly as the population grows. We can very easily incorporate the concept of density dependence into our game. Return to the situation in which only two players participate: birth and death. Now assume that the rate at which each individual dies increases as the population density increases (see Fig. 5). You could imagine this change to result from limited food resources, for example. Now, each time the person representing "death" calculates the number of new deaths, he or she first must determine the *per capita* death rate that is relevant for that population density. Then the net loss due to death is calculated. In our simple example, the direct consequence of adding a density-dependent death rate is that the total number of individuals dying at any point in time is higher than in the absence of density dependence (Fig. 6).

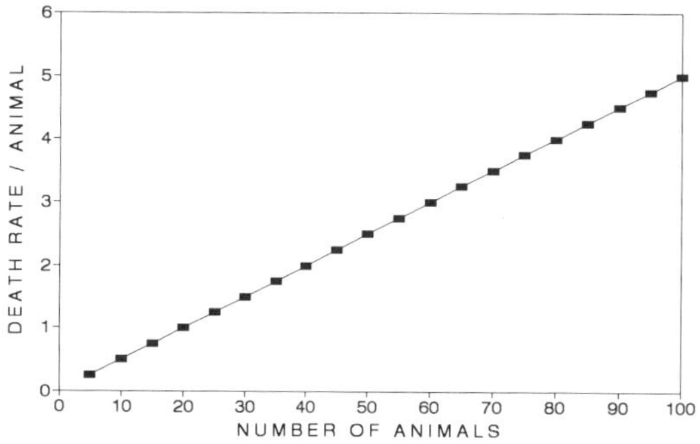

Figure 5 Effect of changes in the number of animals (H) on the *per capita* death rate (d). Note that, as the number of animals increases, the death rate of each animal increases.

Many investigators believe they understand density dependence, saying that, obviously, when more individuals are present in the population, more individuals will die. This reflects the second common misconception in population dynamics. This statement is true whether or not density dependence occurs (compare Figs. 3 and 6). Understanding that density dependence refers to the rate at which a single individual is affected by a process is fundamentally important. When a biological process is density dependent, the rate at which the process acts on an individual changes as the density of the population changes.

III. Host–Parasite Population Dynamics

The study of host–parasite population dynamics can be defined as the study of the processes that affect the numbers of hosts and parasites in the population. Many parasites exist as at least two distinct populations, one within the host and one in the environment. These populations can increase as a result of reproduction or trans-

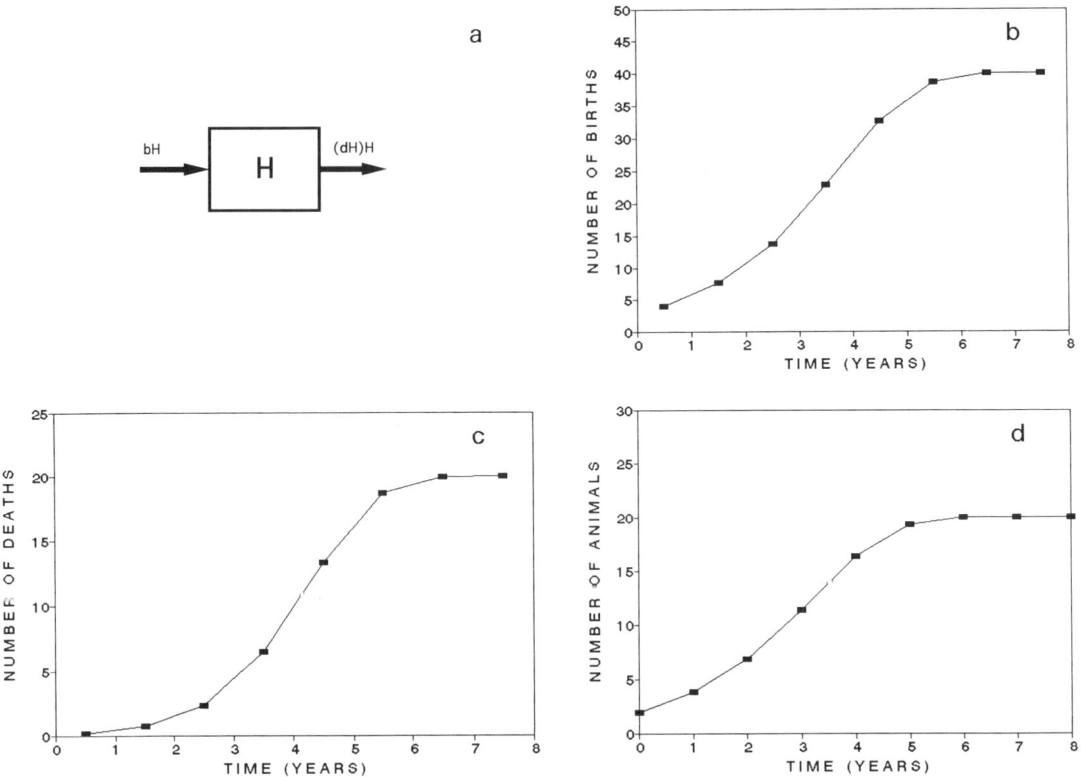

Figure 6 Changes in the number of individuals (H) through time as a result of two biological processes: birth and density-dependent death. The *per capita* birth rate and the initial population size are as in Fig. 3. The *per capita* death rate (dH) is 0.05 H/yr (see Fig. 5). Calculations were done on using a time interval of 1 yr. (a) Flow chart representing the biology. (b) Number of individuals born in each year. (c) Number of individuals that die in each year. (d) Total number of individuals in the population.

mission and can decrease as a result of death or transmission. Death of free-living stages in the environment may not be affected in any major way by the host population, whereas transmission of the free-living stages to the host population is influenced very much by the host population. In many cases, the presence of the parasite in the population alters the survival or reproduction of the host. Because of these connections, whereby the parasite affects the processes relevant to the host population or the host population affects the processes relevant to the parasite population, we consider host–parasite population dynamics to be linked. Simply understanding the population biology of the parasite or of the host without understanding the other is not sufficient.

To illustrate the importance of the linkage between the two populations and to illustrate how changes in the dynamics of one of the populations can affect the dynamics of the other population, consider a direct life cycle nematode with the following general biology: adult parasites (A) live in the intestine of the host; eggs (E) are passed in the feces, survive in the soil, and are ingested by the host. This simple life cycle can be summarized to consist of four biological processes: transmission (β), reproduction of adult parasites (λ), death of adult parasites (μ_1), and death of larvae (μ_2). This biology can be represented as a flow chart (Fig. 7). First imagine a situation in which the host population consists of a single individual (one host), for example, a pony that is kept in a small pasture. Therefore, the total adult

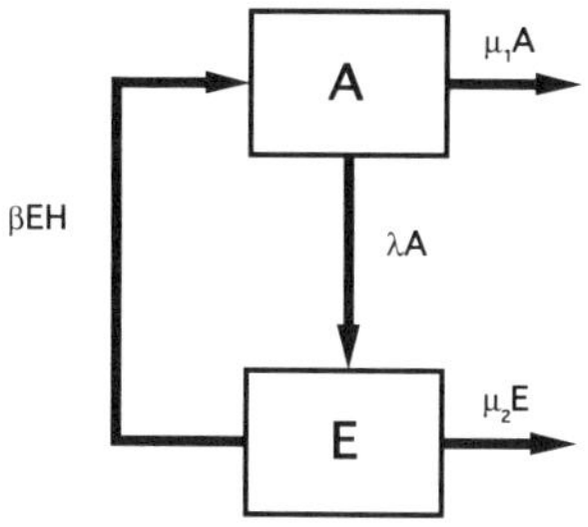

Figure 7 Flow chart representing the biology of a simple direct life cycle nematode with adult parasites (A) in a pony (H) and eggs (E) in the environment. μ_1, *Per capita* death rate of adult parasites; μ_2, *per capita* death rate of eggs; λ, *per capita* rate of egg production; β, *per capita* rate of transmission.

parasite population at any point in time is contained within that pony.

If we focus on the transmission process, first we notice that when transmission of an egg to the pony occurs, there is loss of an egg but concomitant gain of an adult parasite occurs. Transmission depends on the pony contacting the egg and the egg being infective to the pony. These two factors can be combined to form a single *per capita* transmission rate (β). The net rate of transmission (βE) depends on the *per capita* transmission rate and on the number of eggs. Interestingly, however, if a second pony is added to the pasture, the net rate of uptake of eggs from the pasture is likely to double (βEH). Therefore, movement from the egg population to the adult population is dependent on the number of hosts. In the extreme condition, with no ponies on the pasture, no transmission would occur. It is not possible for us to discuss even the simplest population dynamics of the parasite unless we take into account the number of hosts.

Another example consists of a sheep population in which lambs are born and adult sheep are culled (Fig. 8). As the number of hosts present in the population must be known to understand the population dynamics of the parasite, how these processes of birth and culling affect the host population dynamics must be understood. In addition, a third biological process may be involved. In many situations, direct life cycle nematode parasites kill sheep. Therefore, the number of hosts available for transmission of the parasite will be determined by not only the birth rate and the rate of culling of the sheep, but also the rate at which parasites cause the host to die. This effect is linked directly to the number of adult parasites that exist in the host. Therefore, we cannot fully understand the host population dynamics unless the number of parasites in the population is known, yet we cannot fully understand the parasite population dynamics unless the number of hosts in the population is known. This linkage leads to the term "host–parasite population dynamics."

IV. When Numbers Are Not Enough

The final point that should be considered in this chapter is the extent to which "numbers" are sufficient. It is important to know how many individuals (parasites or hosts) are in the population to determine the extent of infection in the host population, the risk of a new disease spreading through the population, or the likely impact of some intervention. However, all hosts are not the same.

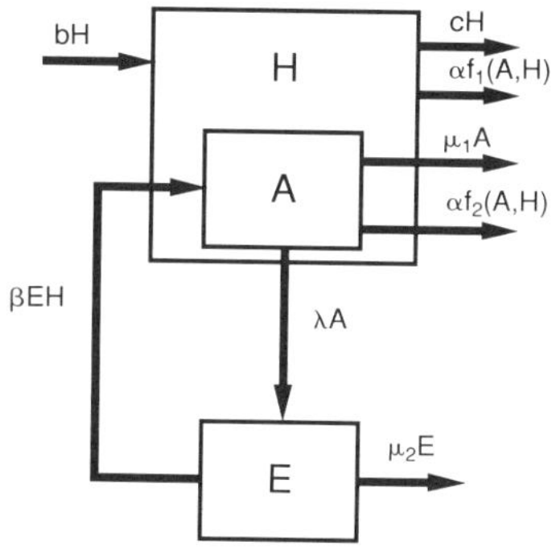

Figure 8 Flow chart representing the biology of a simple direct life cycle nematode with adult parasites (A) in a sheep population (H) and eggs (E) in the environment. μ_1, *Per capita* death rate of adult parasites; μ_2, *per capita* death rate of eggs; λ, *per capita* rate of egg production; β, *per capita* rate of transmission; *b*, *per capita* birth rate of sheep; *c*, *per capita* culling rate of sheep; α, *per capita* death rate of sheep resulting from the parasite.

We know, for example, that host age can be an important determinant of susceptibility to parasitic infections. Therefore, a population that is aging, yet in which the number of individuals remains constant, presents a changing "target" for the parasite population. Even if the age-structure of the population remains stable, changes in which individuals are present (perhaps because of immigration and emigration) can be important in the host–parasite association. Individuals that leave are likely to take their parasites with them. Those that arrive may be uninfected and, therefore, more susceptible to infection because they lack protective immunity that would have developed through previous exposure to the parasite. Even if the same individuals remain in the population, changes that are relevant for host–parasite population dynamics may occur from one month to another, for example, changes in nutritional status of the population related to seasonal availability of food. Once again, although the same number of individuals is present, the "quality" of these individuals as potential hosts may change.

Similarly, we can ask whether "all the parasites are alike." Although the number of individual parasites in the population is of fundamental importance, the characteristics of individual parasites may affect the dynamic processes. The number of eggs or larvae present in the environment may be constant from one month to the next. However, continuing use of a drug to which the parasite population is becoming resistant may lead to a situation in which a greater proportion of the larvae present in the environment is resistant to that drug. This change will have a direct effect on the net ability of that parasite population to infect the treated host population. Changes in the gene pool of both host and parasite are thus important.

I introduce this concept of the "quality" as well as the "quantity" of host and parasite to reinforce the perspective that the entire interaction between host and parasite populations is dynamic. Not only how many individuals, but also which individuals, are present must be known because this population distribution can affect the transmission rate, the survival, or the rate of reproduction of both the host and the parasite. As you can imagine, it is very complicated to include all of these sorts of characteristics when studying the population biology. Therefore, the first important step in any study is determining which of the various processes is most important in the system. This decision-making process is addressed in more detail in the next chapter.

V. Summary

- A range of biological processes influences the number of individuals in a population.
- Each process acts independently of the others.
- The combined result of these processes leads to a dynamic, constantly changing set of individuals, even if the actual number of individuals remains constant.
- The rate of growth of the population can be affected if one of the biological processes is influenced directly by the density of the population.
- Parasite population dynamics are tied intimately to host population dynamics.
- Although the number of parasites or hosts is important, understanding that their "quality" also can affect host–parasite population dynamics is also important.
- *Per capita* and net rates have been distinguished and the concept of a flow chart has been introduced.

Gary Smith

So You Want to Write a Model

2

I. Introduction

A model is a means of representing and manipulating a process or system that otherwise would not be accessible. Models are extremely useful. One of the aims of this book is to illustrate how useful they can be. Unfortunately, many investigators who could make the best use of the results of studies involving models never read the papers describing them because of the impenetrable nature of modeling jargon. Arcane shorthand is a necessary consequence of editorial constraints on the length of scientific papers and is not limited to modeling studies, but articles containing large quantities of mathematics seem to be especially daunting. Achieving a foothold in the discipline seems impossible because of the lack of a starting point. The prospect of writing a model oneself appears simply ludicrous. A single chapter cannot teach you all the elements of modeling, but we hope it may convince you that you could easily learn enough to understand the operations involved. This chapter is intended to build your confidence, and to provide that necessary starting point.

II. Essential Steps

Models are built for some purpose; different kinds of models solve different kinds of problems. Still, the basic steps in the modeling process are always the same.

A. Choosing an Appropriate Model

Choosing an appropriate model for the problem at hand is the first step. How to make that choice is beyond the scope of this chapter, but experience, tradition, and personal preference are all important factors. This discussion concentrates on differential equation models because this is the kind of model you will encounter most often in the later chapters of this book.

B. Drawing a Flow Chart

The second step in developing a model is establishing the various elements of the model and their connectivities; in other words, you construct a flow chart. The notion of flow charts was introduced in Chapter 1. The golden rule of modeling is "We model what we can count," so each compartment (box) in a flow chart represents an entity that can be counted. In this context, distinguishing between infections caused by microparasites (viruses, bacteria, protozoa) and infections caused by macroparasites (helminths and arthropods) is convenient. Clearly, although we cannot count the num-

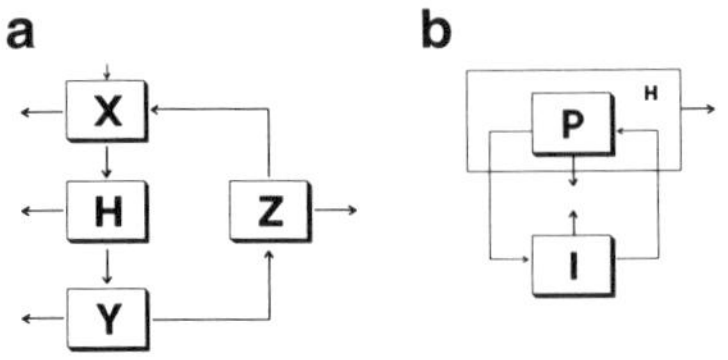

Figure 1 (a) Flow chart for a microparasitic infection. X, Susceptible hosts; H, infected but not yet infectious hosts; Y, infectious hosts; Z, recovered/immune hosts. (b) Flow chart for a macroparasitic infection. P, Parasites; I, free-living infectious stages; H, hosts.

ber of virions in an infected host, we can count the number of hosts that are seropositive. Thus, in the case of microparasitic infections, each compartment in the flow chart represents a particular class of host (e.g., susceptible, infected, immune). An example of a flow chart for a microparasitic infection is given in Fig. 1a. In contrast, estimating the number of nematodes in any given host is a much easier task, so flow charts for models of macroparasitic infections generally consist of idealized life cycles in which each compartment represents the total number of parasites in a particular stage of development. An example is given in Fig. 1b. The arrows connecting the boxes in Fig. 1 indicate the direction of particular transitions, from infected to immune status, for example, or from free-living infective stage to parasitic stage. Constructing flow charts may seem fairly simple, but at this stage one often realizes how little is known about the system under study. A student of mine tried to write a model for a common respiratory disease of chickens. She wondered whether one of the boxes should represent a class of newborn chicks that was protected from disease by maternal antibodies. A search of the literature revealed that whether the maternal antibodies were protective was not known, nor was it known how much time passed before infected birds became infectious to others or how long infectious birds remained so. In this case, a series of experiments was done to measure these parameters. This example illustrates an important aspect of modeling: that it guides and stimulates further experimental work.

C. Writing the Equations and Investigating Model Behavior

The flow chart is the basic blueprint for the model. The next step in developing the model is translating the chart into a series of mathematical equations that will express the dynamics implicit in the arrows that connect one compartment with another. We use the language of mathematics to describe the system because this language is more adequate than any other language. The usual method of conversion begins with writing one equation for each compartment in the flow chart. Sometimes, the model indicated by the chart is determined to be unwieldy or unwritable at this stage. Often one makes some modifications to the boxes and arrows before returning to the mathematical representation.

Finally, having translated the flow chart into a system of equations, investigating model behavior remains as the last task. The modeler assumes that the behavior of the model is the same as the behavior of the system being studied. Investigating model behavior and posting appropriate "what if" questions is no easy task. Successful modelers are individuals who ask the right questions in the right manner. The essentials of the technique, as well as details on how to write the equations, are addressed in the final sections of this chapter and in Chapter 3.

III. Differential Equations

Most of the models in this book are written using differential equations. Other mathematical formats for models (e.g. difference equations and Leslie matrices) exist but we shall not address these. Nor shall we address in any formal way the theory and development of differential equation sys-

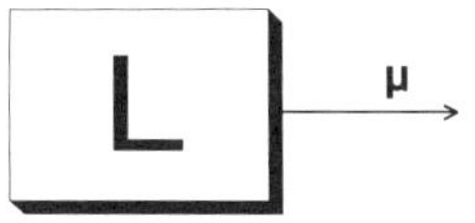

Figure 2 Flow chart for a free-living larval population (L) on pasture. μ, Instantaneous *per capita* death rate.

Table 1 Number of Infective Nematode Larvae Remaining on a Pasture at Monthly Intervals[a]

Month	Larvae (L)	Slope (dL/dt)	μ (dL/dtL)
0	100.00	91.63	0.92
1	40.00	36.84	0.92
2	16.00	14.55	0.92
3	6.40	5.86	0.92
4	2.56	2.34	0.92
5	1.06	0.97	0.92

[a] The proportional mortality each month is 60%.

tems. Instead we shall appeal to your intuition and introduce a simple empirical method that will allow you to translate simple flow charts into the corresponding models.

A. Losses

1. Constant Death Rate

We begin with the very simplest of situations. Imagine a pasture previously grazed by calves infected with a gastrointestinal nematode parasite. The pasture surface is now covered by thousands of infective third-stage larvae. These larvae have two potential fates: they can be ingested by grazing calves or they can die. Since no calves graze on our imaginary pasture, the larvae die. We can represent the number of larvae per m^2 (L) by the single compartment shown in Fig. 2. We can represent the rate at which the larvae die (move out of the box) by a single arrow (labeled μ). Suppose 60% of the larvae die every month, representing a constant death rate. Assuming an initial average of 100 larvae per m^2, an average of 40 larvae per m^2 will persist at the end of the first month, an average of 16 larvae per m^2 at the end of the second, and so on. In fact, we can calculate how many larvae there will be at the beginning of every month by applying Eq. 1 over and over again.

$$L_t = L_{t-1}(1 - 0.60). \tag{1}$$

Here, L_t is the density of larvae at the beginning of month t and L_{t-1} is the density of larvae at the beginning of the previous month. Since 60% of the larvae die each month, 40% survive. The calculations have been done for you and are set out in Table 1 and Fig. 3a. Figure 3a plots the number of larvae still alive at each successive time interval. This plot is called a survivorship curve. First, note that a constant death rate does not result in a linear survivorship curve (a common misconception). Next, note that the slope of the survivorship curve becomes smaller as L_t decreases. This slope represents the overall rate of change with respect to time in the density of larvae on the pasture which, in mathematical shorthand, is

$$\text{Slope} = \frac{dL}{dt}. \tag{2}$$

The values of dL/dt (slope) for various values of L also are given in Table 1. Now divide dL/dt by L (the solutions are given in Table 1 and Fig. 3b). Note that the result is always the same, no matter what value of L is chosen. Thus, if dL/dt (slope) is divided by the corresponding value of L, a constant μ, is obtained. This constant, μ, is given by:

$$\frac{dL}{dt\,L} = \mu. \tag{3}$$

This equation can be simplified by multiplying both sides by L:

$$\frac{dL}{dt} = \mu L. \tag{4}$$

Equation 4 states that dL/dt (slope) equals a constant multiplied by the corresponding value of L. However, Eq. 4 does not reveal

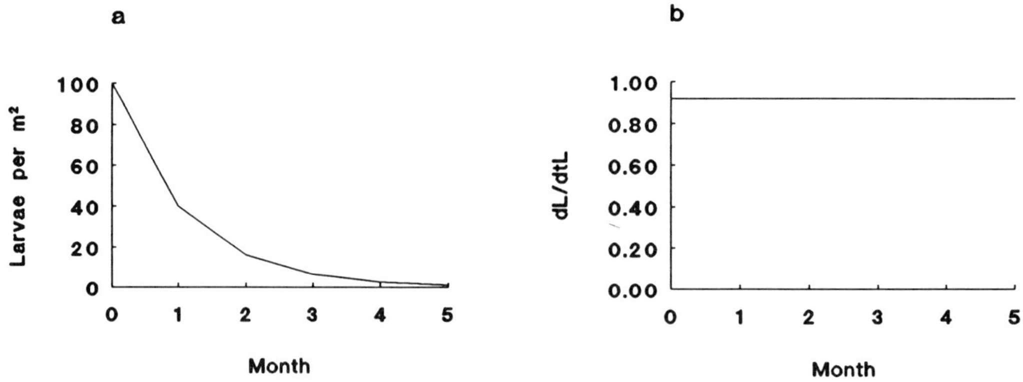

Figure 3 (a) Larval survivorship curve given 60% mortality each month. (b) Instantaneous *per capita* death rate/larva/month ($dL/dt \cdot L$).

whether the rate of change with respect to time (dL/dt) in the density of L is such that the value of L is decreasing or increasing. We can overcome this by adopting a simple convention. If L is increasing, dL/dt will be deemed positive; if L is decreasing, dL/dt will be deemed negative. Alternatively, the minus sign could be placed in front of μ to denote that mortality causes a loss. Whichever convention is used, Eq. 4 now is

$$\frac{dL}{dt} = -\mu L. \tag{5}$$

This differential equation model corresponds to the box and arrow flow chart in Fig. 2.

The word "differential" refers to the slope (dL/dt) or the rate of change in the value of L. Thus, differential equations give information about the rate at which a quantity changes, which usually is not very useful. Usually, what we want is an equation that gives information about how many larvae there are at any given time, that is, an equation similar to Eq. 1. Differential equations can be manipulated so they can be rewritten in a form similar to Eq. 1. This process is called solving the equation. For example, our differential equation can be solved to give the formula for L:

$$L_t = L_0 e^{-\mu t}, \tag{6}$$

where L_t is the number of larvae at time t and L_0 is the number of larvae at the start (time 0). Equations 1 and 6 are identical if the value of t in Eq. 6 is set equal to 1 mo:

$$L_1 = L_0(1 - 0.60) = L_0 e^{-\mu 1}, \tag{7}$$

in which case

$$(1 - 0.60) = e^{-\mu}. \tag{8}$$

Remember that μ is dL/dt divided by L. In other words, μ is the *per capita* rate of change with respect to time in the density of larvae. Note that the *per capita* rate of change (dL/dtL) is constant, whereas the overall rate of change (dL/dt) declines with larval density. This distinction is important, it was mentioned in Chapter 1 and will come up again later. The change in larval density is caused by mortality, so μ is the instantaneous *per capita* mortality rate. The units, in this case, would be "per larva per month." You should be very careful to distinguish between the instantaneous death rate (μ) and the finite death rate (0.60 per larva per month); Eq. 8 shows that the relationship between the two is quite complicated. Instantaneous rates are what we use in differential equations. As we shall see, the convenient thing about them is that we can add them together. This cannot be done with finite rates and that is why differential equation systems are so much more convenient to work with (in case you were wondering).

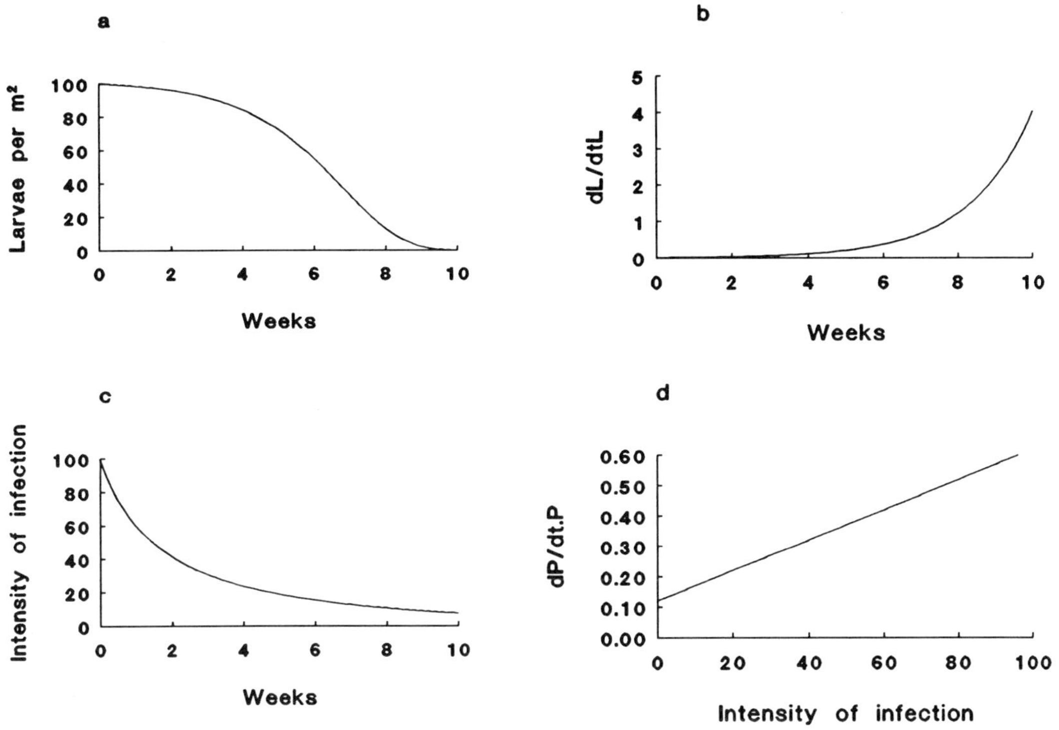

Figure 4 (a) Larval survivorship curve given an age-dependent death rate. (b) Instantaneous *per capita* death rate/larva/week ($dL/dt \cdot L$), calculated from a. (c) Parasite survivorship curve given a density-dependent death rate. (d) Instantaneous *per capita* death rate/parasite/week ($dP/dt \cdot P$), calculated from c.

2. Age-Dependent Death Rate

In the example just considered, the death rate was constant, which need not always be the case. Often the rate at which an individual dies is greatest in older individuals (age dependent) or in high density populations (density dependent). When mortality is age dependent, the resulting survivorship curve takes the form shown in Fig. 4a. Proceeding as before, dividing the slope (dL/dt) by L gives a value for μ that increases as the larvae get older. Many functions have the shape shown in Fig. 4b, but the most convenient is

$$\mu = ae^{bt}, \tag{9}$$

where a is the death rate of newly molted larvae and b is a constant that determines how quickly μ increases with larval age. Substituting this expression for μ,

$$\frac{dL}{dt} = -ae^{bt}L, \tag{10}$$

which is a simple differential equation model for a population of larvae with an age-dependent death rate.

3. Density-Dependent Death Rate

We have to proceed a little differently when mortality is density dependent (Fig. 4c). The flow chart shown in Fig. 2 can be used but L should be changed to P to signify parasites, not free-living larvae. We begin as before and divide dP/dt by P. However, instead of plotting this result against elapsed time (t), as done previously, dP/dtP is plotted against P (Fig. 4d). The relationship between μ and P is linear, and can be modeled using the equation for a straight line:

$$\mu = a + bP, \tag{11}$$

where $a + b$ is the mortality suffered by a population consisting of just one individual and b is a constant that determines how quickly mortality rises as population density increases. We now substitute this expression for μ and obtain

$$\frac{dP}{dt} = (a + bP)P, \tag{12}$$

which is a simple differential equation model for a population with a density-dependent death rate.

Of course, neither Eq. 10 nor 12 is of much use unless the values of a and b can be estimated. Parameter estimation is addressed in Chapter 4. For the moment let us just concentrate on deriving the appropriate model for each flow chart.

4. Development, Migration, and Infection

Most of the processes that result in a decrease in the number of individuals represented by the single box in the flow chart can be modeled in the same way in which mortality was modeled. The important condition is that the overall rate of change in the number of individuals as a result of the process decreases with the number of individuals, as it did for mortality. For example, if the probability is constant that any given individual will leave the compartment (box) as the result of some development process, the equation

$$\frac{dL}{dt} = -\sigma L, \tag{13}$$

where σ is the instantaneous *per capita* rate of development, is appropriate. Equally, if a constant probability exists that any given individual will leave the compartment (box) as a result of moving to another location (migration), the equation

$$\frac{dL}{dt} = -\delta L, \tag{14}$$

where δ is the instantaneous *per capita* rate of migration, is appropriate. Note that, although σ and δ are constants, the overall rate of change (dL/dt) in each case changes with L (i.e., the rate of change in L gets smaller as L gets smaller).

One more example that should be discussed before proceeding concerns the infection process itself. Suppose we now re-introduce calves onto the imaginary pasture. Some of the larvae that disappear from the pasture will do so because they have been eaten by calves, which ingest additional parasites each time they swallow contaminated herbage. Imagine also that the climate is so favorable to larval survival that none of the larvae die during the short period in which the calves graze the pasture. Thus, the only losses are due to the infection process and are shown in Fig. 5a. When dL/dt is divided by L the result is constant over all values of L but changes with the stocking density of calves (Fig. 5b). This result is reasonable since the rate of loss of larvae from the pasture will depend on the volume of grass eaten, and that changes with the number of calves. When the calf stocking density is 1 we define the instantaneous *per capita* rate of loss of larvae due to ingestion as β. Thus,

$$\frac{dL}{dtL} = -\beta. \tag{15}$$

If two calves are introduced, the rate of loss doubles; if three calves are added, the rate of loss triples; and so on. Representing the stocking density by H,

$$\frac{dL}{dtL} = -\beta H. \tag{16}$$

Multiplying both sides by L gives

$$\frac{dL}{dt} = -\beta HL, \tag{17}$$

which is a differential equation model for larval losses due to ingestion by grazing calves.

For those of you more interested in microparasites, the infection process can be

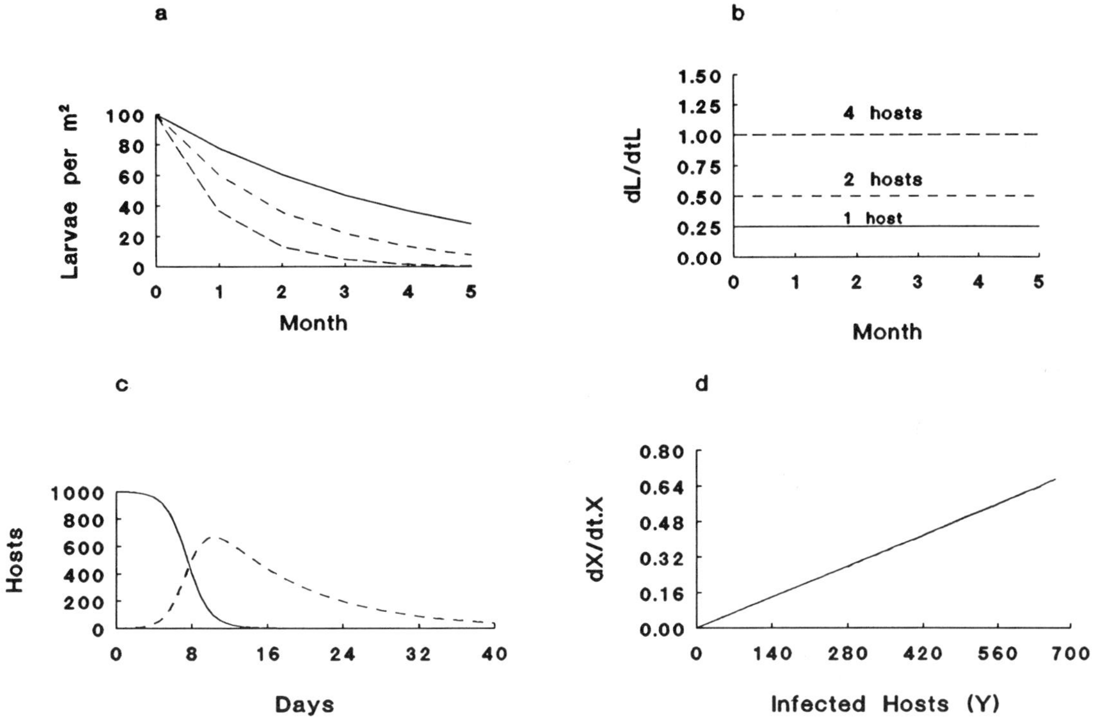

Figure 5 (a) Loss of larvae from pasture due to ingestion by grazing hosts. (b) Instantaneous *per capita* rate of loss of larvae due to ingestion by grazing hosts. One host, ——; 2 hosts, - -; 4 hosts, — —. (c) Number of susceptible and infected hosts during a simple epidemic (microparasitic infection). Susceptible, —; infected, - -. (d) Force of infection during a simple epidemic (microparasitic infection).

represented in the same way, except that the model deals with the rate of loss of susceptible hosts (X) to an infected group (Y). Changes in the numbers of susceptible and infected individuals are shown in Fig. 5c. The susceptible hosts become infected when they come into contact with an infectious host. As might be expected, when dX/dt is divided by X, the result tracks the number of infected hosts (Fig. 5d). Clearly, the likelihood of any given individual becoming infected increases with the number of infectious hosts with which that individual might come into contact. The instantaneous *per capita* rate of change in the number of susceptible hosts (dX/dtX) is called the "force of infection." (Classical epidemiologists call this quantity the "incidence" of disease; see Chapter 6). Since this value appears to be directly proportional to the number of infectious hosts (Y),

$$\frac{dX}{dtX} = -\beta Y, \tag{18}$$

where β is the force of infection when one infectious host is present. Multiplying both sides of Eq. 18 by X yields

$$\frac{dX}{dt} = -\beta XY, \tag{19}$$

which is a differential equation model for the changes in the number of susceptible hosts during a simple epidemic.

B. Gains

1. Development and Migration

The rather detailed exposition in the previous section was designed to present the basis for a simple empirical method of translating flow charts into differential equations. Stripped of their explanatory verbiage, the steps are these: First, assess whether the overall rate of change in the compartment varies with the number of individuals in that compartment. If so, write a differential equation of the form

$$\frac{dN}{dt} = \alpha N. \tag{20}$$

If the nature of the change is such that the number of individuals in the compartment is decreasing, place a minus sign in front of the generic parameter a (otherwise, insert a plus sign). Finally, determine whether a is a constant or some function of quantities such as elapsed time, age, population density, temperature, or humidity. If so, substitute an appropriate functional form. All the examples we have addressed so far have dealt with losses from a compartment, but the system works for gains as well. As an exercise, demonstrate that Eq. 20 is a reasonable model for a population, the density of which is increasing due to births.

The first step in the empirical method entails an assessment of whether the overall rate of change in the compartment varies with the number of individuals in that compartment. However, under some circumstances, the overall rate of change in a compartment does not vary with the number of individuals in that compartment. Usually, this is observed in the context of processes that lead to increases in population density. For example, suppose the changes were due to recruitment of individuals that had moved from another compartment because of development, migration, or infection. This situation is represented by the two-compartment model shown in Fig. 6a and is expressed in differential equation form as

$$\frac{dL}{dt} = -\sigma L$$

$$\frac{dP}{dt} = \sigma L. \tag{21}$$

Equations 21 and 22 show individuals mov-

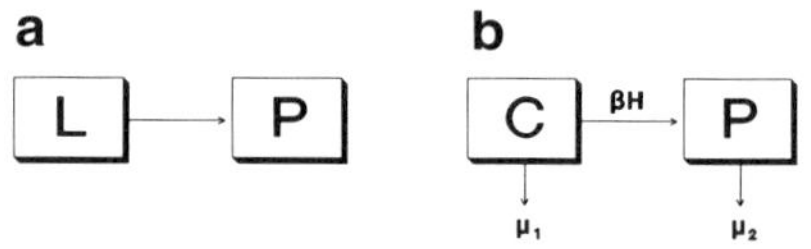

Figure 6 (a) Flow chart for a development or migration process. L, Larvae; P, adult parasites. (b) Flow chart for a *Fasciola* model. C, Infective cysts, P, parasitic flukes in sheep; H, hosts; βH, rate of loss due to infection process; μ_1, instantaneous *per capita* death rate of cysts; μ_2, instantaneous *per capita* death rate of parasites.

ing from compartment L (larvae) to compartment P (adult parasites) because of development and migration, respectively.

$$\frac{dL}{dt} = -\delta L$$

$$\frac{dP}{dt} = \delta L. \tag{22}$$

The rule for constructing the equations for P is simple and logical. Consider Eq. 22: If $-\delta L$ is the overall rate of change with respect to time in compartment L due to movement of individuals to compartment P, the overall rate of change in compartment P due to the same cause must be δL. The only difference is that the sign changes from negative to positive, which indicates that the overall rate of change in compartment P is such that the number of individuals is increasing. In this model, although the rate of change in compartment L varies with L, the rate of change in compartment P is unrelated to P. Note that circumstances involving migration exist under which the number of individuals moving from one compartment to another is unrelated to the number in either compartment. Experimental infections constitute a good example. In such cases, hosts are infected with a known number of parasites per day, the number being determined only by the experimenter. The overall daily rate of change in the number of parasites (P) per host is simply the number of parasites in each infecting dose. If the experimenter administers 100 infective stages each day, the overall rate of change is simply

$$\frac{dP}{dt} = 100. \tag{23}$$

2. Reproduction

Finally, one kind of recruitment does not involve the movement of individuals from one compartment to another. This occurs in any reproductive process when the newborns are assigned to a different compartment from their parents. Suppose a helminth (P) laid eggs (E) at the rate of λ per female per day. The overall rate of change in the number of eggs would be represented by

$$\frac{dE}{dt} = \lambda \frac{P}{2}. \tag{24}$$

The number of helminths is divided by 2 because we assume that only half the population (P) is assumed to be female.

IV. Putting the Pieces Together

In Chapter 1, we introduced the idea that all the events and processes that influence population size could be thought of as exerting their effect quite independently of one another. That idea provides a foundation from which to define each element of a model separately and then show how these elements can be put together.

We begin with a two-compartment model (Fig. 6b). The first compartment represents an infective encysted stage (C) in a macroparasitic life cycle; the second compartment represents the parasites (P) in the population of hosts. Imagine C_0 infective stages at the start. Those infective stages have two fates: they can die or they can infect one of the hosts. Consider each of these processes in turn. I hope it will be obvious by now that the overall rate of change in the number of infective stages due to mortality (μ_1) is

$$\frac{d'C}{dt} = -\mu_1 C, \tag{25}$$

and the overall rate of change due to the infection process is

$$\frac{d''C}{dt} = -\beta HC, \tag{26}$$

where H is the density of hosts. The flow chart indicates that no recruitment to the infective stage compartment, so Eqs. 25 and 26 completely define the rate of change in infective stage density. The problem now is to put the two equations together. This process is simple. The net rate of change due to mortality and the infection process is merely the sum of the individual rates of change defined by Eqs. 25 and 26:

$$\frac{dC}{dt} = \frac{d'C}{dt} + \frac{d''C}{dt} = -\mu_1 C - \beta HC. \tag{27}$$

Earlier in the chapter, one of the convenient features of instantaneous rates was noted to be that they could be added. This method of assembling the separate elements of a model into a single equation is an example of the usefulness of that property. The complete translation of the flow diagram in Fig. 6b is

$$\frac{dC}{dt} = -(\mu_1 + \beta H)C$$

$$\frac{dP}{dt} = \beta HC - \mu_2 P. \tag{28}$$

V. Building a Useful Model

We are now in a position to model more complicated systems, but first, a word of warning must be given. Beginners often make the mistake of constructing models that are too complicated to be useful. Models provide a way to address the complexity found in natural systems. The model is useless if it is so complicated that its behavior is as difficult to understand as that of the natural system it mimics. Experienced modelers know that one of the most useful things about the modeling process is that it points out what can be left out, that is, it reveals which processes are important and which are not. Of course, in some contexts, the order of importance of different model elements will change depending on the problem being studied. As I remarked earlier, models are built for some purpose and their structure reflects that purpose. This section illustrates how to simplify the modeling task by concentrating on only those aspects of the natural system that are relevant to the job at hand, and by deliberately restricting the circumstances under which the model is applicable. Specifically, we shall show that the model represented by Eq. 28 is an adequate model for the common liver fluke, *Fasciola hepatica.*

The liver fluke is a helminth parasite of the bile ducts of cattle and sheep, and undergoes a phase of asexual multiplication in an intermediate snail host. At first, the life cycle appears to be impossibly complicated (Fig. 7). The number of individuals in each stage of the life cycle is determined by a plethora of climatic and immunological factors, to which are added all the possible variations in farm management practices, resulting in what seems to be an utterly intractable system from the modeling point of view.

The fallacy in this line of reasoning, of course, is assuming that you need to know all there is to know about a system to answer worthwhile questions. There are many worthwhile questions that can be answered quite adequately, even without all the facts, provided the pertinent facts are known. This distinction between pertinent facts and all facts is at the root of many of the misconceptions about modeling. Modeling this entire system is not necessary to answer many of the questions that may be raised. In fact, a system may not necessarily be modeled correctly, yet our understanding of it is enhanced. Often, the process of realizing that the model is wrong and attempting to correct it leads to the greatest leaps in understanding the system. The next section discusses the simplest possible liver fluke model that does not violate the underlying biology.

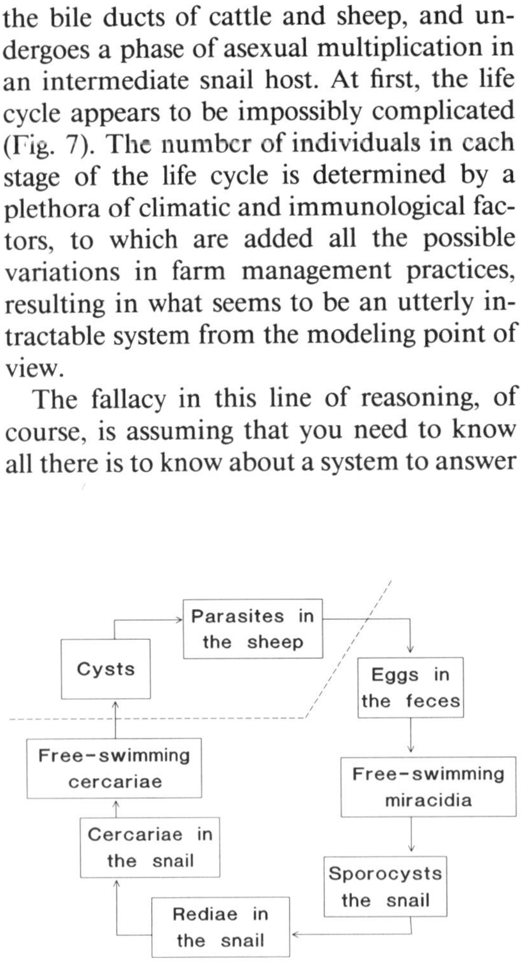

Figure 7 Life cycle of the common liver fluke, *Fasciola hepatica,* in sheep. All the life-cycle stages below the dashed line were omitted from the final fluke model described in the text.

A. Simplifying the *Fasciola* Model

We can simplify the model immediately if we *ignore the intermediate host.* In temperate regions, most of the infective metacercariae are shed from the snail during a period of 2–3 wk. In Europe, for example, cercarial shedding happens in late summer or early fall. From that point on, until the same time next year, the intermediate host has no role to play, so it can be ignored (Smith, 1982). By making that simplification, we have implicitly defined the starting point of the model as a cohort of metacercariae recently shed from snails in the intermediate host population.

Those metacercariae have two fates: they can die as their energy reserves become exhausted or they can be eaten by grazing cattle or sheep. In the laboratory, the survivorship curve of *Fasciola* metacercariae is a simple exponential decline, suggesting that the death rate is constant irrespective of the age of the organism. Surely the weather will cause variations in metacercarial death rate in the field! Fortunately, the variations are not as great as you might think, and, provided we use one death rate for the summer, and a different (lower) rate for the winter, a simple exponential decline will suffice as a model for the survivorship of metacercariae in the field also (Smith, 1982). The model can be simplified further if we consider only the period between cercarial shedding (fall) and the end of winter. In that case, a single value for metacercarial mortality (μ_1) is sufficient. What happens to the ingested metacercariae? First, determine whether the metacercariae were ingested by cattle or sheep. Cattle are known to develop a very effective acquired resistance to infection with *Fasciola,* whereas sheep do not (Smith, 1984). In sheep, the death rate of the parasite is unaffected by the host's experience of infection. The parasites simply accumulate in the sheep; this accumulation is balanced only by the intrinsic death rate of the parasites. Thus, if we choose sheep as our hosts, we do not have to model an immune response and parasite mortality can be held constant. Finally, we can restrict our study to well-managed flocks in which parasite burdens never reach levels sufficient to kill the host. Under these circumstances, flock size (H) remains constant throughout the period of interest.

A review of the assumptions made re-emphasizes the circumstances to which the model can be properly applied. The crucial assumptions are (1) European context, (2) fall to late winter, (3) sheep, and (4) constant flock size. Assumptions 1 and 2 allow us to dispense with the intermediate host and permit a constant parameter for metacercarial mortality (μ_1). Assumption 3 eliminates the effect of the host immune system, and permits a constant parameter for parasite mortality (μ_2). Assumption 4 eliminates consideration of parasite-induced host deaths or host deaths due to any other cause, so parameter H is a constant. The resulting model (Eq. 29) is a very good mimic of the population biology of the common liver fluke in Britain, for example, and was used to devise and compare various parasite control strategies (Smith, 1982).

VI. Conclusions

The basic rule of model building is that we model what we can count. The first step is constructing a flow chart of the system you are interested in and being prepared to simplify the diagram several times over before achieving a tractable and useful basis for the model. Differential equation analogs of the diagram can be assembled easily by constructing submodels for each component in term and simply adding them. Never be misled by individuals who criticize your model because it is "too simple." You have to start somewhere and it is surprising how many complex systems can be explained in terms of very simple models.

References

Smith, G. (1982). An analysis of variations in the age structure of *Fasciola hepatica* populations in sheep. *Parasitology* **84,** 49–61.

Smith, G. (1984). Density-dependent mechanisms in the regulation of *Fasciola hepatica* populations in sheep. *Parasitology* **88,** 449–461.

Gary Smith & Marilyn E. Scott

3

Model Behavior and the Basic Reproduction Ratio

I. Asking the Right Questions

This chapter assumes that you have reviewed the model building scheme outlined in Chapter 2, drawn a flow chart and translated that chart into a series of equations. You are now at the point in modeling where science gives way to art in determining the right questions to ask about a system. The right questions address important issues and are suited uniquely to the modeling context. Questions that can be answered better by other means are not right questions, nor are questions that exceed the capacity of the model. Beginners often expect more of a model than it actually can deliver. At best, models are aids to thought, or frameworks for decision making. Good models are essentially algorithms for making the most rational use of the available information. We made the point in Chapter 2, that you do not need to know all there is to know about a system to write a useful model (i.e., one that can answer worthwhile questions). On the other hand, some decisions require more information than others. The art of modeling lies in knowing whether your model is adequate to the task at hand and, if so, how it might be used to best advantage.

Posing the right questions is difficult but, once you have developed the right model, the mechanics of obtaining a useful answer are relatively straightforward. These mechanics amount to investigating model behavior, which can be done in several ways. The most obvious method is solving the equations and calculating how the number of parasites or infected hosts changes with time. Curiously, this method is often not the most efficient for determining what the model will do. As we shall see, there are indirect methods of investigating model behavior that allow us to make useful generalizations about what will happen in any given set of circumstances. In particular, we discuss a number called the basic reproduction ratio (R_0). If an expression for the basic reproduction ratio can be derived in terms of the parameters of the model, we have a very powerful inferential tool at our disposal. Although we shall address this topic in more detail later in this and other chapters, one remark is worth making: an infection can be eradicated if we alter one or more of the model parameters sufficiently to cause the basic reproduction ratio to fall below 1. Thus, the expression that defines the basic reproduction ratio for your particular model provides a great deal of informa-

Parasitic and Infectious Diseases

tion about steps required to bring the infection under control.

II. Analytical and Numerical Solutions

Differential equations explicitly describe the rate at which some quantity is changing. Usually, the actual value of the quantity at any given instant is of greater interest. We often know the starting value (the initial conditions) and wish to discover various subsequent values. Some of these values can be obtained quickly and easily by determining whether the model has any equilibrium points and, if so, how many individuals are in each compartment at those points. An equilibrium is said to exist if the rate of change in the number of individuals in each compartment of the model is zero. At equilibrium, the number of individuals in each compartment stays the same. Equilibria are important because they represent an end point toward which the model is moving constantly. Information about the nature of the equilibrium provides an important indication of model behavior. In some models, the end point is zero in all compartments; in others, the number of individuals increases indefinitely. Zero for all compartments is a perfectly respectable equilibrium (especially if the goal is eradicating an infection). Infinitely large equilibrium values usually mean that something important is missing from the model. Many models, however, have equilibria that are somewhere in between zero and infinity. For an equilibrium to exist, the inputs to each compartment must be balanced exactly by the outputs. Consider the liver fluke model introduced in Chapter 2:

$$\frac{dC}{dt} = -(\mu_1 + \beta H)C$$

$$\frac{dP}{dt} = \beta HC - \mu_2 P. \tag{1}$$

An equilibrium, if it exists, is the point at which dC/dt and dP/dt are both zero:

$$\frac{dC}{dt} = -(\mu_1 + \beta H)C = 0$$

$$\frac{dP}{dt} = \beta HC - \mu_2 P = 0. \tag{2}$$

The first of these equations can be true only when C is zero. The second is true when $P = \beta HC/\mu_2$, but C is already known to be zero at equilibrium, so P also must be zero at equilibrium. These arguments lead to the trivial conclusion that, if enough time elapses and no other factors intervene, the number of metacercariae on the pasture and the number of flukes in the sheep gradually will decline until none are left. So far, a starting value and a potential final value have been determined for each compartment. Unfortunately, we do not know what happens in between. One way to find out is to solve the equations. Solutions to Eq. 1 would have the form

$$C = f_1(t)$$

$$P = f_2(t), \tag{3}$$

where $f_1(t)$ and $f_2(t)$ are expressions involving t. These solutions give the value of C or P at any time t. To progress from Eq. 1 to their corresponding solutions, integration is necessary. Integration is basically a matter of rearranging the elements of the differential equation until a "standard form" is recognized for which the answer (the integral) already has been determined. The integrals of Eq. 1 are

$$C_t = C_0 e^{-(\mu + \beta H)t}$$

$$P_t = \frac{\beta HC_0}{(\mu_1 + \beta H - \mu_2)}[1 - e^{-(\mu_1 + \beta H - \mu_2)t}]. \tag{4}$$

Given Eq. 4, the number of individuals in each compartment can be determined using a simple calculator. The solutions shown in Eq. 4 are called analytical solutions because they were arrived at by an analysis of the original differential equations. However, very few useful models have convenient analytical solutions such as these. In most cases, numerical solutions are required. Numerical solutions are obtained using computers. A large number of computer algorithms will give answers very close to those obtained by analytical solutions. The computer solves the equations; the process is rapid and the inaccuracies that do occur usually are so small that they are not troublesome. Why are other approaches ever necessary?

The answer lies in the nature of the questions we generally ask of our models. Rules or generalizations about model behavior are desired. We cannot generalize about model behavior using numerical solution methods unless dozens of solutions have been generated over a sufficiently wide range of parameter values. This process is very time consuming and very tedious. We need a method that will lead to generalizations about model behavior in a much shorter time frame. The remainder of this chapter discusses methods used to uncover generalizations about model behavior that do not involve solving the equations.

III. Basic Reproduction Ratio

We can learn a great deal about an infection if an expression for its basic reproduction ratio (R_0) can be derived. Some authors call this parameter the contact number (Hethcote, 1989). It is a measure of the maximum reproductive potential of a parasite between one generation and the next for a given host population in a given environment. The definition is worded differently for microparasites and macroparasites because of the different ways in which infections are "counted."

A. Macroparasitic Infections

For macroparasites, R_0 is defined as "the average number of offspring (or female offspring in the case of a dioecious parasite) produced through the reproductive life-span of a mature parasite that themselves survive to maturity in the absence of density-dependent constraints to population growth" (Anderson and May, 1991). Density-dependent constraints are mechanisms that regulate the abundance of parasite populations (see Chapter 5), the effects of which are least obvious when an infection first is introduced into a host population and parasite density is very low. At this point the reproductive potential of the parasite is greatest. The basic reproduction ratio is a hypothetical index of this maximum reproductive potential; it is hypothetical because it assumes that absolutely no constraints on population growth exist, which is never the case. Even when the infection first is introduced, density-dependent processes will exert an effect, albeit very small. Thus, the actual or effective reproduction ratio when the infection first is introduced is less than the basic reproduction ratio, although at this point the two are closest in value. The differences between basic and effective reproduction ratios are considered further later in this chapter. For the moment, note that the basic reproduction ratio represents the maximum reproductive potential. Thus, if R_0 is greater than 1, the parasite population will increase in size, at least initially. However, if R_0 is less than 1, the parasite population can only decrease in size (Hethcote, 1989).

B. Microparasitic Infections

For microparasites, R_0 is defined as "average number of secondary infections produced when one infected individual is introduced into a host population in which every host is susceptible" (Anderson and May, 1991). Imagine that a child becomes infected with measles, and goes to visit relatives in another community before the infection has reached the infectious stage. In that community, no one has ever been infected with measles. From the time that measles becomes infectious to the time that the child recovers, the child comes in con-

Table 1 Examples of Estimates for the Basic Reproduction Ratio for Rubella, *Ascaris lumbricoides,* and *Onchocerca volvulus*[a]

Infection	Geographical location	Time period	R_0
Rubella	England and Wales	1960–1970	6–7
	West Germany	1970–1977	6–7
	Czechoslovakia	1970–1977	8–9
	Poland	1970–1977	11–12
	The Gambia	1976	15–16
Ascaris lumbricoides	Iran		4–5
	Burma		1–3
	Bangladesh		1–2
Onchocerca volvulus	Cameroon		50
	Ivory Coast		9–74

[a] Reproduced with permission from Anderson and May (1991) and by permission of Oxford University Press.

tact with 20 people who become infected directly as a result of their contact with this child. The R_0 for measles in this community would be 20. If the R_0 were 1, we would expect the child to be able to infect only one other person before he or she recovered and lost infectiousness. If the R_0 were less than 1, no one else would be expected to become infected.

This example shows clearly that an R_0 of 1 defines a threshold between persistence and extinction of an infection. This threshold assumes great significance when planning control programs. If the intent is eradication of the infection, then the basic reproduction ratio must be reduced and maintained below 1.

IV. Factors that Influence the Value of the Basic Reproduction Ratio

Table 1 provides estimates of the basic reproduction ratio for rubella (German measles), for *Ascaris lumbricoides,* and for *Onchocerca volvulus.* Note that the R_0 for rubella is 6–7 in England and Wales and 15–16 in the Gambia. This difference is important because it illustrates the fact that R_0 of any given infection is defined for a given environment and a given host community. One of many explanations for the difference is the nutritional status of the two populations. Nutritional status is an important determinant of the duration of infectiousness. The more effective the immune response, the faster an individual will recover from infection. If the nutritional status is poorer in the Gambia than in England and Wales, individuals infected with rubella may remain infectious for a longer period of time, and thus have the potential to transmit the infection to more people (see Chapter 9 for other ways in which nutritional status may affect parasite population dynamics). Thus, the R_0 would be higher in the Gambia than in England and Wales. We can use the same arguments for *Ascaris* and *Onchocerca* infections. For example, the maximum potential for *Ascaris* transmission depends on environment characteristics that affect the survival of the eggs, and on behavioral characteristics of the human population related to contact with eggs (see Chapter 14). Factors such as use of night soil as fertilizer or ritualized cleansing habits may increase or decrease risk of infection. For *O. volvulus,* geographically associated differences in the species or strain of blackfly involved with the transmission of the parasite might account for differences in R_0. For example, the differences in the life-span or biting behavior of the blackfly will affect markedly the maximum number of individuals to whom the infection can be transmitted.

V. Effective or Actual Reproduction Ratio

The basic reproduction ratio represents the maximum reproductive potential of the infection. The actual reproductive rate that pertains throughout most of the course of an infection is less than this maximum potential and is called the effective reproduction ratio (R_e). The effective reproduction ratio is less than the basic reproduction ratio because parasite populations are regulated, that is, natural constraints on their growth exist. The basic reproduction ratio, unlike the effective reproduction ratio, is defined in the absence of any regulation. In this section, we describe some simple algebraic methods for obtaining expressions for both the basic and the effective reproduction ratios and use these expressions to highlight the differences between the two numbers.

A. Macroparasitic Infections

Imagine a single pony maintained on the same pasture all year. This pony is infected with cyathostome nematode parasites. The total number of parasites in the pony is P. The total number of infective larvae on the pasture is L. The pony ingests these larvae at a net instantaneous rate βL and new infective stages appear on the pasture at a net instantaneous rate λP. Net larval mortality is μL and net parasite mortality is $(a + bP)P$. Thus,

$$\frac{dP}{dt} = \beta L - (a + bP)P$$

$$\frac{dL}{dt} = \lambda P - (\mu + \beta)L. \tag{5}$$

The mortality of the parasites $(a + bP)$ is density dependent; it increases as the number of parasites increases. This increase is a consequence of the response of the host's immune system to increases in parasite antigen. The first information that can be derived from this model is the behavior at equilibrium. Proceeding as before, set $dP/dt = dL/dt = 0$ and determine values of P^* and L^* that are consistent with this condition. The easiest equilibrium to find is $P^* = L^* = 0$, in other words, the infection has disappeared. From the equation for L, it follows that another possible value for L at equilibrium (L^*) is

$$L^* = \frac{\lambda P^*}{(\mu + \beta)}. \tag{6}$$

Use this equation to derive an expression for P at equilibrium (P^*). Set the equation for P to zero and substitute for L^*:

$$\frac{\beta\lambda P^*}{(\mu = \beta)} - (a + bP^*)P^* = 0. \tag{7}$$

Divide both sides by P^*:

$$\frac{\beta\lambda}{(\mu + \beta)} - a - bP^* = 0 \tag{8}$$

and rearrange:

$$P^* = \frac{\beta\lambda}{(\mu + \beta)b} - \frac{a}{b}. \tag{9}$$

Equations 6 and 9 are expressions for L and P at equilibrium. We have already seen that one possible value for P^* is zero. Equation 9 shows that the equilibrium value of P can be greater than zero only if

$$\frac{\beta\lambda}{(\mu + \beta)b} > \frac{a}{b}. \tag{10}$$

This criterion can be expressed another way by dividing the left-hand side of Eq. 10 by the right hand side of Eq. 10.

$$\frac{\beta\lambda}{(\mu + \beta)a} > 1. \tag{11}$$

More formal arguments (i.e., stability analysis) reveal this ratio to be the basic reproduction ratio (R_0), the ratio whose value must exceed 1 if the infection is to persist. However, in simpler models, sophisticated methods often are not required to obtain an expression for R_0.

Next, obtain an expression for R_e, the effective reproduction ratio. Begin by deriving an alternative expression for L^*, using Eq. 5. At equilibrium, $dP/dt = 0$. If the equation for P is set to zero, rearranged to get

$$L^* = \frac{(a + bP^*)P^*}{\beta}. \tag{12}$$

Now, the right hand side of Eq. 12 equals

the right hand side of Eq. 6:

$$\frac{\lambda P^*}{(\mu+\beta)} = \frac{(a+bP^*)P^*}{\beta}. \quad (13)$$

Rearrange Eq. 13:

$$\frac{\lambda\beta}{(\mu+\beta)(a+bP^*)} = 1 = R_e. \quad (14)$$

The ratio displayed in Eq. 14 is the effective reproduction ratio. At equilibrium, the effective reproduction ratio is 1 because each parasite is replaced with exactly one offspring in one generation.

This forbidding combination of parameters can be clarified by considering each component in turn. Equation 14 has a simple interpretation. The number of progeny produced per day per parasite is λ. The average life-span of an individual parasite is given by the reciprocal of parasite mortality $(1/(a+bP))$. Thus, the total number of progeny produced per parasite over the average life-span of a parasite is

$$\frac{\lambda}{(a+bP^*)}. \quad (15)$$

Of these, only a fraction $[\beta/(\mu+\beta)]$ survive to be ingested. Thus, the number of progeny produced during the lifetime of a single parasite that survive to become mature parasites is

$$\frac{\lambda\beta}{(\mu+\beta)(a+bP^*)} = R_e, \quad (16)$$

which is the effective reproduction ratio. At the outset, just before the pony in the model becomes infected with cyathostomes, the value of P is zero and Eq. 16 simplifies to

$$\frac{\beta\lambda}{(\mu+\beta)a} = R_0, \quad (17)$$

which is the basic reproduction ratio. The part missing from Eq. 17 deals with the density-dependent regulation of parasite abundance (bP). Note again that R_0 is defined in the absence of any reference to regulatory processes and does not change unless changes occur in the characteristics of the infection in that particular situation. In contrast, R_e changes as the number of parasites (P) changes (see Eq. 16). In one sense, R_0 is the unattainable maximum value of R_e. Since R_e must be ≥ 1 for the infection to persist, and since $R_0 > R_e$ by definition, R_0 must be greater than 1 for the infection to persist.

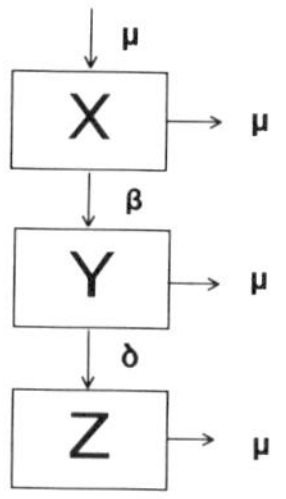

Figure 1 Flow diagram for a simple measles model. X, Susceptible hosts; Y, infected/infectious hosts; Z, immune hosts; β, transmission parameter; $1/\delta$, mean time spent as an infectious host; μ, birth rate and death rate.

B. Microparasitic Infections

The theoretical framework for defining the basic reproduction ratio for microparasites follows logic similar to that used for macroparasites. For example, use the simple flow chart for measles (Fig. 1). Here, the host population is composed of susceptible (X), infectious (Y), and immune (Z) individuals. Movement from the susceptible to the infectious state depends on the rate of transmission (β), the density of susceptible individuals (X), and the density of infectious individuals (Y). Movement from the infectious state to the immune state depends on the rate of recovery from infection (δ), which is the inverse of the duration of the infectious period. In this example, assume that immunity is lifelong, that the host population has a constant size $(N = X + Y + Z$: birth rate equals death rate, $\mu)$, and that measles does not cause additional host mortality.

The differential equations describing this

system are

$$\frac{dX}{dt} = \mu N - \mu X - \beta XY \quad (18)$$

$$\frac{dY}{dt} = \beta XY - \mu Y - \delta Y \quad (19)$$

$$\frac{dZ}{dt} = \delta Y - \mu Z. \quad (20)$$

Consider Eq. 19. The number of individuals infected with measles will decline if dY/dt is negative and will remain constant or increase if dY/dt is positive. From Eq. 19, it follows that dY/dt will be positive provided

$$\beta XY \geq (\mu + \delta)Y. \quad (21)$$

Rewrite this condition as

$$\frac{\beta X}{(\mu + \delta)} \geq 1 = R_e. \quad (22)$$

The ratio on the left hand side of Eq. 22 is the effective reproduction ratio. Like the effective reproduction ratio of macroparasitic infections, the effective reproduction ratio of microparasitic infections changes as the infection spreads through the population. However, in the case of microparasitic infections, R_e changes with the number of susceptible hosts (X). At the outset, just prior to introduction of the infection, all individuals in the population are susceptible to measles, so, under these circumstances, $X = N$. Substituting for X in Eq. 22,

$$\frac{\beta N}{(\mu + \delta)} = R_0, \quad (23)$$

which is the basic reproduction ratio.

We can use Eq. 23 to derive another very useful quantity. If measles is to persist in this population, R_0 must be greater than 1, that is,

$$\frac{\beta N}{(\mu + \delta)} > 1 \quad (24)$$

or, alternatively,

$$N > \frac{(\mu + \delta)}{\beta}. \quad (25)$$

Equation 25 shows that the number of susceptible individuals at the start must be greater than a certain fixed amount if the infection is to spread and persist in the population. This certain fixed amount is called the critical threshold density of susceptible hosts, and is an important parameter often designated X_T.

We can summarize these relationships as follows. The number of susceptibles at the outset (N) must exceed a critical threshold density (X_T) if the disease is to persist. We can modify Eq. 23 to express the same idea:

$$R_0 = \frac{N}{X_T}. \quad (26)$$

VI. Local Stability Analysis

Local stability analysis has two goals: (1) to determine whether an equilibrium point is stable to small perturbations and (2) to determine the manner and time course of the return of a system to equilibrium. This point can be illustrated with a case in which a virus is re-introduced into a population from which it previously had been eliminated. If the virus spreads through the population and persists, the previously virus-free state would not be considered stable; if, on the other hand, the virus does not persist, the virus-free state would be considered stable. The next question concerns the manner in which the system returns to equilibrium. Suppose we had treated a long-standing disease problem in a given population and the individuals in the population were now being observed for signs of re-infection. If the infection previously had been at equilibrium and that equilibrium had been stable (i.e., resistant to change), the population soon should revert to the pretreatment state. Figure 2 shows two of the possible trajectories the population might follow as it becomes re-infected. Damped oscillations to some stable point (Fig. 2b) might occur, or a smooth transition with no oscillations might be made (Fig. 2a). Other trajectories also are possible. Local stability analysis facilitates

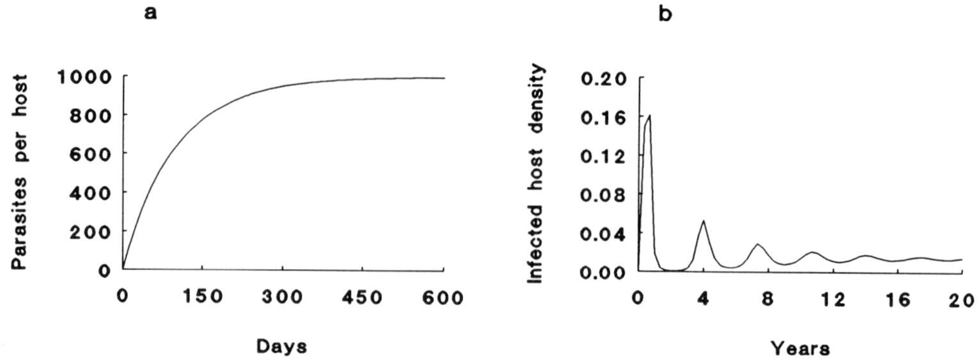

Figure 2 Possible trajectories for a simplified viral infection. (a) Monotonic return to equilibrium. (b) Damped oscillatory return to equilibrium.

prediction of the most likely course of events.

The basic method of analysis is to consider the behavior of the model when the numbers of individuals in each compartment are very close to those that would pertain at equilibrium. Under these circumstances, the equations can be rewritten in a simpler, more tractable form. A set of rules (the Routh–Hurwitz criteria) has been established that enables the investigator to determine the behavior of these simpler equations by straightforward (if laborious) algebraic methods. The interference is that the behavior of the simpler equations is representative of the behavior of the original equations from which they were derived. Thus, local stability analysis allows the modeler to state what model behavior will be provided certain parameter values pertain. Examples of this approach can be found in Chapter 21. The methodology of this very powerful method of analysis is beyond the scope of this chapter, but the interested reader is referred to May (1975, 1986), and Smith and Grenfell (1990) for an introductory exposition.

VII. Conclusions

This chapter has summarized methods for investigating model behavior. Sometimes, in the modeling literature, how the author arrived at the stated conclusions is not always apparent to the novice. As a rule of thumb, any sentence beginning "The model is stable if and only if . . ." indicates that stability analysis has been used. In practice, however, most papers consist of a mixture of stability analysis and numerical (or, less frequently, analytical) solutions. The latter often are included to illustrate the generalities revealed by the former and include examples generated widely differing parameter values to indicate how sensitive the model is to changing circumstances.

References

Anderson, R. M., and May, R. M. (1991). "Infectious Diseases of Humans: Dynamics and Control." Oxford University Press, Oxford.

Hethcote, H. B. (1989). Three basic epidemiological models. *In* "Applied Mathematical Ecology" (S. A. Levin, T. G. Hallam, and L. J. Gross, eds.), pp. 119–144. Springer-Verlag, Berlin.

May, R. M. (1975). "Stability and Complexity in Model Ecosystems," 2d Ed. Princeton University Press, Princeton, New Jersey.

May, R. M. (1986). Population biology of microparasitic infections. *In* "Biomathematics" (T. G. Hallam and S. A. Levin, eds.), pp. 87–103. Springer-Verlag, Berlin.

Smith, G., and Grenfell, B. T. (1990). The population biology of pseudorabies in swine. *Am. J. Vet. Res.* **51**, 148–155.

Hamish McCallum & Marilyn E. Scott

Quantifying Population Processes: Experimental and Theoretical Approaches

4

I. Introduction

In principle, the dynamics of any population are affected by four factors: the rates of birth, immigration, death, and emigration, often called the vital rates. These factors lead to the most fundamental equation in population dynamics,

$$\Delta N = \text{births} + \text{immigration} - \text{deaths} - \text{emigration}, \quad (1)$$

where ΔN is the change in population size over some time period. Even for single populations, conditions are more complicated in practice because each rate is likely to depend on population size and structure and to be variable in time.

Host and parasite populations are, like all others, subject to Eq. 1, but the vital rates of each are linked with those of the other. By most definitions of parasitism (Price, 1980; Schmidt and Roberts, 1989; Begon *et al.*, 1990), a parasite must either decrease the fecundity of its host or increase its mortality. Similarly, the vital rates of parasite populations are affected by the host population: transmission is dependent on host density and the death of a host has consequences for the parasites being carried. Additional complications arise from host immune responses, vectors, and intermediate hosts.

Mathematical models provide a formal language in which these interactions can be written explicitly and understood. From the perspective of a modeler, two main broad classes of parasite exist: macroparasites and microparasites. As the names suggest, this division is made mainly on the basis of size: large multicellular parasites (primarily helminths) compared with small unicellular parasites (viruses, bacteria, and protozoans). However, size *per se* is not as important as the fact that (in general) microparasites have the ability to reproduce rapidly within their host, whereas macroparasites do not. Hence, as a first approximation, all hosts infected with microparasites can be treated as equivalent. In contrast, for macroparasites, it is essential to keep track of the parasite burden of hosts, which is determined substantially by the number of infective stages encountered. Within both of these classes of parasite, the way that parasites are transmitted has a central influence on population dynamics. In the simplest case, transmission may be direct, from host-to-host, or equivalently via a short-lived infective stage. This chapter concentrates on direct transmission. Frequently, complications arise from long-lived infective stages, latent periods, vectors, intermediate hosts, vertical (parent–offspring), or sexual transmission. Many of these conditions will be considered in later chapters.

A. Microparasites

The simplest epidemiological models of microparasites divide the host population into three classes: susceptible, infected, and resistant hosts ("SIR" models). Such models have a long history, being developed initially by Bernoulli (1760) and applied extensively in malaria modeling early this century (Ross, 1916; Ross and Hudson, 1917). Anderson and May (1979) provide a summary of this basic model framework. Letting X_t be the number of susceptible hosts, Y_t be the number of infected hosts, and Z_t be the number of resistant hosts at time t, the population dynamics of a direct life cycle microparasite can be represented with the following equations:

$$\frac{dX}{dt} = a(X + Y + Z) - \mu X - \beta XY + \gamma Z \quad (2)$$

$$\frac{dY}{dt} = \beta XY - (\alpha + \mu + \sigma)Y \quad (3)$$

$$\frac{dZ}{dt} = \sigma Y - (\mu + \gamma)Z. \quad (4)$$

Here, the basic population parameters are a, the birth rate of hosts; μ, their disease-free death rate; β, the infectivity of the disease; α, the increment in death rate caused by infection; σ, the recovery rate from the disease; and γ, the rate at which immunity is lost.

B. Macroparasites

For macroparasites, we need a rather different approach. Among infected members of a host population, infection levels can (and usually do) vary greatly. Therefore, describing a host as infected is not sufficient; knowledge of the degree of infection is required. Hence, it is necessary to keep track of the host population size, the parasite population size (or the mean parasite burden per host), and, finally, the distribution of parasites within the host population (Crofton, 1971; Anderson and May, 1978; May and Anderson, 1978, 1979). As shown subsequently, this distribution may have a major effect on the behavior of a host–parasite interaction. In this case, the simplest set of equations capable of representing the interaction is

$$\frac{dI}{dt} = \lambda P - \mu_2 I - \beta IH \quad (5)$$

$$\frac{dH}{dt} = (a - \mu)H - \alpha P \quad (6)$$

$$\frac{dP}{dt} = \beta IH - (\mu + \mu_1)P - \alpha HE(i^2). \quad (7)$$

Here, H is the host population, P is the parasite population, and I is the population of infective stages. As in the microparasite models, α represents the pathogenicity of the parasite and β the infectivity. However, important differences include α as the increment in the death rate of an infected host *per parasite* and β as the rate of contact between infective stages and hosts, rather than between infected and susceptible hosts. The death rate of parasites in hosts is represented by μ_1 and that of free-living infective stages by μ_2. The rate at which parasites produce infective stages is λ. An entirely new complication is contained in the term $E(i^2)$. In most parasitic infections, the death of the host results in the death of all parasites harbored in it. If a host carrying i parasites has an increased death rate of αi, then the total rate of loss due to death of infected hosts will be αi^2. The overall loss rate of parasites from the system as a whole depends on characteristics of the distribution of parasites within the host population, represented by $E(i)$ (the expectation of finding i parasites in a host). Therefore, the mean or expected value of the overall loss rate due to death of infected hosts, over all hosts, is $\alpha E(i^2)$. As a result, the rate at which parasite-induced death causes loss from the host population depends not just on the mean parasite burden, but also on the way in which the parasites are distributed between hosts; the more aggregated they are for a given mean, the greater $E(i^2)$ will be. If the effect of parasite burden on host death is nonlinear, the distribution of parasites among hosts still determines the loss rate from the host population due to parasite-induced death, but the relationship is more complex.

C. Why Quantify Population Processes?

Being clear about *why* quantitative data are required is essential. We shall assume that the intention is to use the information in a model; however, remember the wide variety of models and purposes for which these can be used. At one extreme are differential equation models similar to those pioneered by Anderson and May (1978,1979), as outlined in the previous section. These models are designed as "strategic" tools to formalize the understanding of parasite–host interactions in a very general way. These models are also valuable for providing rapid qualitative predictions, particularly on the possible consequences of parasite introductions into novel systems (for which detailed information suitable for accurate parameter estimation is unlikely to be available). A good example of this approach is provided by Pech and Hone (1988), who used a simple microparasite model to evaluate possible control strategies for foot and mouth disease, should it infect feral pig populations in Australia.

Simple differential equation models are not suitable for making detailed predictions about particular systems. Therefore, putting a large effort into statistically rigorous and elaborate sampling designs to quantify parameters for such models is pointless. The correct functional form for relationships and "order of magnitude" estimates is necessary. In any event, a major use of such models is identifying where experimental effort is directed best. If the models cannot be applied until extensive experimental work is complete, they defeat their own purpose.

Simulation models lie at the other extreme. These models may be necessary in highly specific cases, to assess the impact of a parasite in a particular situation. Since the results of simulation models are only as reliable as their weakest parameter estimate, effort put into detailed estimation procedures will be repaid.

II. Host Population Parameters

Numerous publications cover the estimation of the abundance of free-living animals (Southwood, 1978; Seber, 1982; Krebs, 1990; Pollock *et al.,* 1990). The two methods most widely used are direct counts (either quadrat or transect based) and mark–recapture methods. Estimating appropriate values for host birth rate a and mortality μ is more difficult. Given extensive mark–recapture data, methods exist for estimating birth and death rates (see, for example, Cormack, 1989, or Pollock *et al.,* 1990). In practice, data of sufficient quality are rarely available. In any event, the birth and death rates a and μ required for models such as those in Eqs. 2–4 or Eqs. 5–7 are the rates occurring in the absence of density dependence and parasite-induced effects. Field estimates may include either or both of these factors.

As a result, various less direct estimation methods generally are used. Information on fecundity often will include time until first breeding, interval between breeding events, and litter or clutch size. Reliable data on mortality in most vertebrate populations are harder to obtain. One useful approach, when the death rate is constant and independent of age, is estimating the instantaneous death rate μ as the inverse of the life expectancy. Provided migration can be neglected and individuals can be aged, the size of cohorts in successive years also can be used to estimate survival.

Vital rates in vertebrate hosts are usually age dependent. These rates can be used directly in simulation models or in stage-structured models (e.g., McCallum and Singleton, 1989), but manipulation is needed to derive appropriate estimates for the age-independent rates a and μ required for use in Eqs. 2–4 or Eqs. 5–7. Any method of condensing age-structured rates into age-independent rates will involve approximations and loss of information. However, the best strategy is using the Euler equation (Keyfitz, 1977), which relates the intrinsic rate of population growth r (an age-independent quantity) to m_x, the age-specific fecundity (the number of females born per female in age class x), and l_x, the age-specific survival (the number of individuals surviving from birth to the beginning of age class x):

$$\sum e^{-rx} m_x l_x = 1. \tag{8}$$

The equation is implicit, which means that it can be manipulated only rarely into a form $r = n$, where n is some mathematical expression. Usually, this equation must be solved by some process of successive approximation. Technically, the Euler equation applies only to a population with a stable age structure. (Whatever the initial age distribution of a population, if m_x and l_x remain constant, the proportion of individuals in each age class eventually will settle to levels determined by l_x and m_x alone.) This is the stable age distribution. In reality, confirming that a stable age distribution has been reached will be impossible, but this condition is a reasonable first approximation. Additional details can be found in any text on age-structured populations (e.g., Caswell, 1989; Krebs, 1990).

The following example, from McCallum and Singleton (1989), illustrates the use of the Euler equation to derive estimates of a and μ for the house mouse *Mus domesticus* under ideal conditions. Life expectancy is perhaps 4 mo (122 days), the inverse of which gives $\mu = 0.0082$ per day (l_x is thus $e^{-\mu x}$). Mature females can produce a litter of 8 every 21 days, but mice do not produce their first litter until at least 49 days of age. m_x is, therefore, 0 for $x < 49$ and 0.19 (8/21 × 0.5) for $x > 49$ (assuming half the offspring are female). If these values are substituted into the Euler equation, the left and right sides become equal when r is 0.036. Since $r = a - \mu$, the age-independent birth rate then is estimated to be $a = 0.044$. This result should be contrasted with the estimate of $a = 0.19$ (hence, $r = 0.1823$) which is obtained if the prereproductive period is ignored.

III. Parasite Population Parameters

As with all organisms, parasite populations will persist only if, on average, each individual at least replaces itself in the next generation. More particularly, a parasite can invade a host population only if, at low parasite densities, the number of descendants each parasite leaves in a new host (R_0) exceeds 1. For macroparasites, R_0 can be interpreted as the number of second generation parasites per first generation parasite. In the case of microparasites, R_0 usually is interpreted as the number of secondary infections per primary infection. For a parasite to replace itself in the next generation in a new host, two distinct processes are required: replication and transmission. This statement is equally true of microparasites and macroparasites, but for simple microparasite models (Eqs. 2–4) it is not necessary to quantify replication explicitly.

A. Quantifying Replication

Replication in most helminths involves the production of eggs by the adult parasitic stages, which usually are shed via the host gut into the external environment. In principle, then, replication can be quantified by measurement of the rate of egg production in conjunction with the parasite burden of the host. An example of this approach is provided by Elkins *et al.* (1991), who counted the eggs of *Opisthorchis viverrini* per gram in the feces of humans in Thailand and found a high correlation with the worm burden subsequently expelled after treatment with an anthelmintic. As in many studies of this type (e.g., Bundy *et al.*, 1987), a complication was encountered because of a density-dependent reduction in fecundity; fewer eggs were produced per worm in individuals with heavy infections than in individuals with light infections.

Although the objective of the study by Elkins *et al.* (1991) was validating the use of eggs per gram of feces as an index of parasite burden, we can use their data to quantify replication. A regression of log (eggs per g feces) versus log (worms expelled) should yield a slope close to 1. If the slope is significantly less than 1, evidence exists of density dependence. Provided the slope is 1, the antilog of the Y intercept, multiplied by the g feces per day, estimates the egg production per worm per day. Figure 1 shows an example of such an analysis based on the data of Elkins *et al.* (1991). Clear evidence of density dependence in egg production exists. (Note that a regression on logged data, constrained to pass through the origin, as used by Elkins *et al.*, 1991, is not appropriate when quantifying replication rates). Despite the highly significant regression, a large amount of scatter occurs around the line of best fit. Detection of density dependence in parasite egg production, especially in field studies, is complicated by this large variability. For further discussion, see Shostak and Scott (1993).

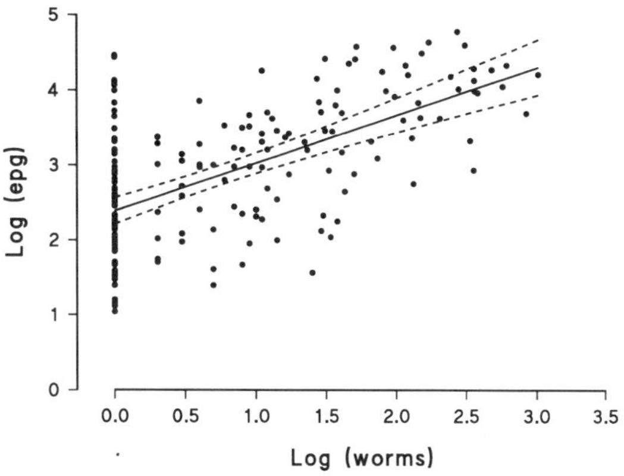

Figure 1 Regression of $\log_{10}$(eggs per gram of feces) versus $\log_{10}$(worms expelled after chemotherapy) for *Opisthorchis viverrini* in a Thai village community. Data were obtained from M. Haswell-Elkins and were published previously by Elkins *et al.* (1991). Data points at which either eggs per gram (epg) or worms expelled was zero have been omitted. The solid line is the least squares regression line, $\log_{10}(\text{epg}) = a + b \cdot \log_{10}(\text{worms})$; the dashed lines represent 95% confidence limits for the line. The least squares estimate for a is 2.389 with a standard error of 0.0719; the least-squares estimate for b is 0.6373 with a standard error of 0.0617 (191 degrees of freedom). Since the estimate for the slope differs significantly from 1, clear evidence suggests density dependence in egg production. Hence, $\text{epg} = 10^a \times (\text{worms})^b$ or $\text{epg} = 244.9 \times (\text{worms})^{0.6373}$. Assuming a daily fecal production of approximately 180 g (M. Haswell-Elkins, *personal communication*), total egg production can be estimated. However, evidence for other parasites (for example, *Ascaris*; Elkins, 1986) suggests that stool mass and worm burden may be correlated. This possibility obviously complicates estimates of parasite fecundity.

B. Quantifying Transmission Rates

Transmission is always difficult to quantify adequately under field conditions, although it is one of the most crucial factors in determining R_0 and the threshold host population for disease maintenance, H_T. In fact, estimating (or guessing) H_T itself, and then using this number as a means of estimating the transmission rate often will prove easier (e.g., Dobson and Hudson, 1992). However, several methods can be used to measure transmission rates directly. Under controlled conditions, transmission in macroparasites can be quantified by exposing a fixed number of hosts in an arena to a known number of infective stages for a set period of time. The infective stages have three possible fates: they can die before they encounter a host, they can encounter a host but fail to develop, or they can encounter a host and develop into a detectable parasite. Given the simplest assumptions, infection occurs as a "binary collision" between infective stage and host; the death rate of the infective stages is constant. The infection process then can be described by the differential equations (McCallum, 1982):

$$\frac{dI_t}{dt} = -\mu_2 I_t - \beta H I_t \quad (9)$$

$$\frac{dP_t}{dt} = \beta H I_t s. \quad (10)$$

Here, I_t is the number of infective stages present at time t, μ_2 is their death rate, and β is the infection rate. (Since host populations usually are measured as densities per unit area, the units of the infection rate are area per unit time. Thus, the infection rate can be thought of as the area traversed by each host per unit time, from which all infective stages present would be picked up.) H is the number of hosts present and P_t is the number of parasites establishing by time t. The proportion of infections producing detectable parasites is s. The time-dependent solution of Eqs. 9–10 is

$$P_t = \frac{HI_0 s}{\mu_2/(\beta + H)}(1 - e^{-(\mu_2 + \beta H)t}). \quad (11)$$

Here, I_0 is the number of infective stages present at the start of the experiment. Equation 11 predicts a linear relationship between the number of infective stages present and the parasite burden that subsequently develops. This kind of relationship has been observed in most experiments with infective stages that directly infect the host through contact (e.g., Anderson, 1978; Anderson *et al.*, 1978; McCallum, 1982; Smith, 1987). Deviations from a linear relationship may occur because of density-dependent constraints on parasite development after infection or, if infective stages are ingested, through a host-feeding functional response (Keymer and Anderson, 1979).

Note that simple infection experiments in which hosts are left exposed to infective stages for an extended period cannot measure the infection rate explicitly; it is confounded in Eq. 11 with both the proportion of infections that develop successfully, s, and the infective stage death rate, μ_2. To measure β directly, several infection periods must be used and an independent estimate

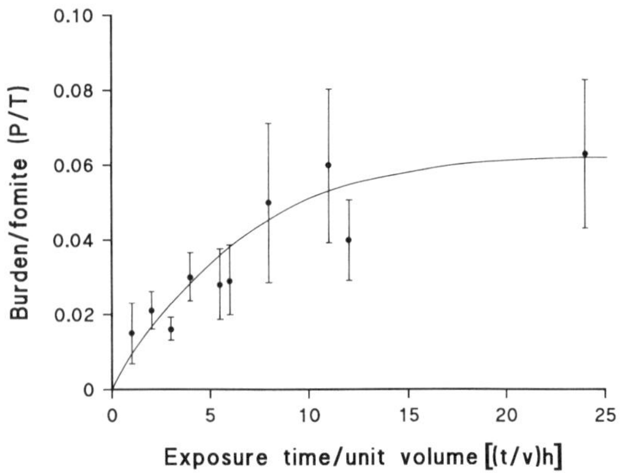

Figure 2 The number of *Ichthyophthirius multifiliis* trophozoites established per infective stage (P/T), as a function of exposure time (hr) per liter of water. The curve shown is Eq. 11 (with $\mu_2 = 0$, since infective stage death over the 6 hr of the experiment was found to be negligible). The infection rate b is estimated to be 0.14 liter/hr [From McCallum (1982). Reprinted with permission from Cambridge Univ. Press.]

of the infective stage death rate must be made. An example of the results of such an experiment on the ciliate *Ichthyophthirius multifiliis* (from McCallum, 1982) is shown in Fig. 2.

Whether the results of such experiments provide quantitative guidance about the transmission rate of the parasite in natural conditions is subject to debate. The principal problem is that direct infection experiments can be carried out only on a fairly small scale; otherwise obtaining sufficient infective stages to undertake the experiments would be logistically impossible. Under natural conditions, habitat use by the host is likely to be very nonrandom, and infective stage distributions also are likely to be patchy. Straightforward application of an infection rate derived from a small uniform arena to the larger heterogeneous natural environment is likely to be inappropriate.

Under natural or seminatural conditions, attempts may be made to estimate infectivity by introduction of "sentinel" or "tracer" animals, parasite-free animals introduced into the study area for a short period, after which they are removed and their parasite burden determined (Scott, 1987b; Quinnell, 1992; Chapter 20). Tracers can provide a direct estimate of the force of infection Λ, the rate of acquisition of new parasites per host, which is related to the infection rate β by

$$\Lambda = \beta I s. \qquad (12)$$

Here, I is the infective stage density (which must be assumed to be constant over the exposure period) and s is the proportion of infections that develops into detectable parasites.

An approach conceptually similar to the use of tracers is estimating the rate of acquisition of infection using the relationship between host age and parasite intensity. Newborns, thus, are used as "natural tracers." If the infection rate and loss rate of parasites can be considered constant and independent of age, the rate of parasite acquisition with age can be represented as a simple immigration–death process:

$$\frac{dM_x}{dx} = \Lambda - \mu_1 M_x, \qquad (13)$$

where Λ is the force of infection, M is the mean parasite burden at age x, and μ_1 is the death rate of the parasites within hosts. Including more realistic features such as an age-dependent infection rate is also possible but an appropriate form for the age dependence may prove difficult to determine (Anderson and May, 1985). The solution of Eq. 13 is

$$M_x = \frac{\Lambda}{\mu_1}(1 - e^{-\mu_1 x}), \qquad (14)$$

which generates a curve of the form shown in Fig. 3a. Provided an estimate of the death rate of parasites is available (see Section III,C), Λ can be estimated from the asymptote of the curve. For many helminth infections of humans, the mean worm burden increases initially with age, but then decreases (Fig. 3b), probably as a result of age-dependent exposure, although host-age-related parasite survival could generate the same pattern. In such cases, an approximate estimate of Λ in young hosts can be obtained using a nonlinear regression procedure to fit Eq. 14 to the increasing part of the curve (Bundy *et al.*, 1987). However, given that the decrease in mean burden in older individuals probably is generated by Λ varying with age, Λ is unlikely to be constant for young individuals.

For microparasites, an equivalent method is the use of age at first infection (Grenfell and Anderson, 1985) or age-dependent serology. An excellent overview of methods for estimating transmission rates in microparasites is provided by Hone *et al.* (1992).

C. Quantifying Parasite Death Rates

In a direct life cycle macroparasite, two additional "parasite" parameters are the loss or death rate of parasites from hosts and the death rate of free-living infective stages. Of these, the death rate of the free-living stages in the most straightforward to measure. Observing motility of survival of free-living stages may be possible under experimental conditions, or a "bioassay" approach may be used, in which infective stages of various ages are used to infect hosts in a standard way. In much of the parasitological literature (Anderson and Whitfield, 1975; McCallum, 1982), a curve is fitted to the proportion of infective stages surviving as a function of time. This approach probably provides a reasonable description of the survival curve. However, the method is dubious statistically because successive points include the same individuals and are, therefore, not independent. A better approach is the use of survival analysis (Cox and Oakes, 1984; Grenfell *et al.*, 1986), which uses the time until death of each individual, rather than the proportion of individuals surviving, as the response variable. The mathematical theory supporting such approaches is difficult. However, standard statistical packages such as GLIM (Crawley, 1993) or SAS (PROC LIFEREG; SAS, 1988) can perform the analysis quite easily. The mor-

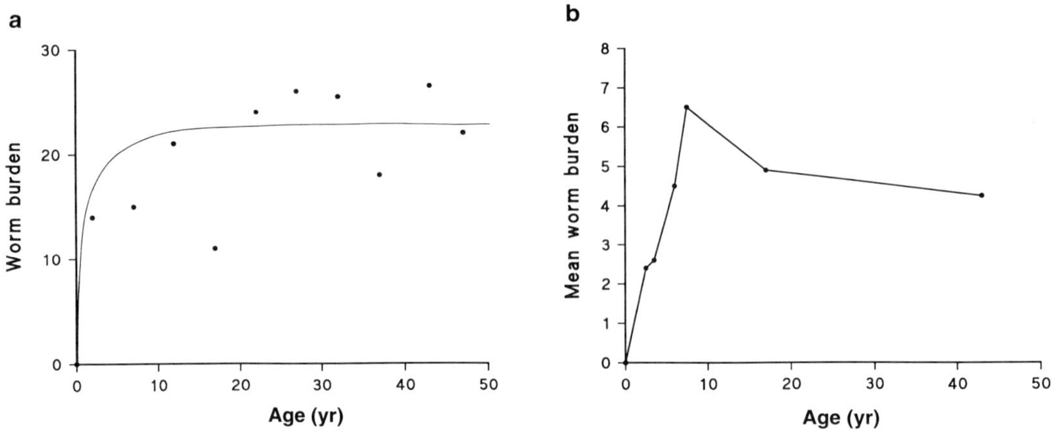

Figure 3 Age–intensity curves for human helminth infections. (a) *Ascaris lumbricoides* within a rural Iranian community. Reprinted with permission from Anderson and May (1985). The solid line shown is Eq. 14. (b) *Ascaris lumbricoides* within a Carribean village [Reprinted with permission from Bundy *et al.* (1987)].

tality rate of infective stages is usually age dependent, because of depletion of finite energy resources in the nonfeeding stages (Anderson and Whitfield, 1975), producing an S-shaped survival curve (Fig. 4). Such curves can be described well by assuming that the death rate increases exponentially with time. In the parasitological literature, this relationship often is described as a Gompertz survival function. In the statistical literature, the same relationship is called a Weibull hazard function.

The death rate of parasites in hosts obviously is harder to observe than that of free-living stages. However, this rate can be determined by following the decay in mean parasite burden in a population no longer exposed to infection (Anderson and May, 1985).

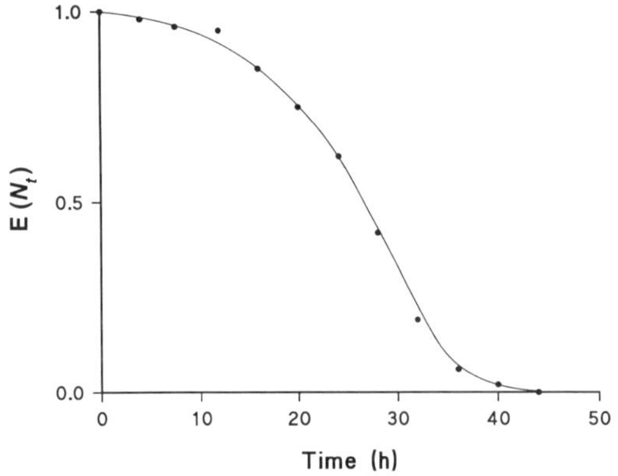

Figure 4 Survival of cercariae of *Transversotrema patialensis*. The vertical axis is the proportion of cercariae originally present that survived to the time shown. The solid line is a Gompertz survival function. [From Anderson and Whitfield (1975). Reprinted with permission from Cambridge University Press.]

IV. Effects of Parasites on Hosts

Parasites may increase the death rate of their hosts or decrease their fecundity. In some cases, these effects are dramatic. A well-known example of the effect of parasites on fecundity is provided by the crustacean parasite *Saculina,* infection with a single individual of which sterilizes male crabs (see Baudoin, 1975). In many other cases, the effects of parasites are more subtle and may occur through increased susceptibility to predation (Dobson, 1988; Hudson *et al.*, 1992; Chapter 21) or decreased foraging efficiency (see Chapter 10). Some authors have suggested that parasite-induced death is compensatory, killing hosts that would die anyway, rather than being additive, increasing the total death rate (Holmes, 1982). At times, whether parasites have any effect at all on their host is debatable (Munger and Karasov, 1991).

In this section, two questions are addressed. First, how can the effects of parasites on hosts be quantified in experimental situations in which parasite burdens can be manipulated or hosts monitored over time? Second, what, if anything, can be inferred regarding the influence of parasites on hosts from the frequency distribution of parasites in hosts under natural conditions?

A. Experimental Manipulation

By treating hosts to remove or reduce parasite burdens, investigators can demonstrate that parasites have an impact on host reproduction or mortality. For example, Brown and Brown (1987) fumigated cliff swallow *(Hirundo pyrrhonata)* nests to remove hematophagous swallow bugs *(Oeciacus vicarius)* and showed that, in the largest swallow colonies (which had the highest bug densities), fledgling success was increased twofold compared with untreated control colonies. Similarly, Hudson (1986) treated red grouse with an anthelmintic and found increased production of young. However, to quantify the impact of parasites on hosts fully, investigators also must measure the parasite burden in both treated and control populations. Surprisingly few studies exist in which this has been done. The most common approach has been to expose or dose hosts with varying numbers of infective stages, and then to monitor their survival. For example, McCallum (1985) exposed black mollies, *Poecilia latipinna,* to infective stages of the ciliate *I. multifiliis* and found an approximately linear relationship between median survival time and parasite burden. The survivorship curves for infected fish, however, were very different from the simple exponential decay curves implicitly assumed by the differential equation models of Anderson and May (1978).

Scott (1990) dosed laboratory mice with up to 1000 larvae of the nematode *Heligmosomoides polygyrus* and found a relationship of the form

$$P = 1 - ae^{bL_4}, \tag{15}$$

where P is the proportion of mice surviving to 14 days, L_4 is the number of stage 4 larvae the mouse is carrying, and a and b are constants. In this case, the relationship between infection level and death rate is quite different from the linear relationship assumed by Eq. 6.

As a guide to the effect of parasites on wild hosts, most such experiments may be criticized on two counts. First, they measure only acute mortality. Parasite burdens at levels insufficient to kill a captive animal fed *ad libitum* are likely to have major effects on the ability of the host to locate food (see Chapter 9) and to escape predators. Thus, the true deleterious effect of the parasite on the host may be underestimated considerably. Second, a single pulse of infection may have different effects on hosts than a continual trickle infection, which is more likely to occur in the wild.

B. Inferring Mortality from Frequency Distributions

Clearly mortality among the most infected members of a host population will have the effect of truncating the upper end of the frequency distribution of parasites among hosts, decreasing both the mean parasite burden and the degree of aggregation of the parasites among the host population. Unfortunately, detection of such truncation is difficult, and factors other than host mortality also may truncate parasite distributions. Anderson and Gordon (1982) conducted extensive model simulations to examine this question. These researchers concluded that a peak in parasite distribution in hosts of intermediate age, coupled with a decline in the degree of dispersion with age, may provide evidence of parasite-induced mortality. This result was, however, subject to qualifications: age-related changes in the rate of infections, or age-related changes in the strength of host resistance to infection, also may generate this same pattern.

The suggestion that parasite-induced mortality may be inferred from truncation of the negative binomial distribution first was made by Crofton (1971). This approach has been used several times. Lanciani and Boyett (1980) found a poor fit of the untruncated negative binomial to the distribution of mites on *Anopheles crucians* larvae (on which mites are known, from laboratory experiments, to have a deleterious effect) and a good fit of the negative binomial to the distribution of mites on the aquatic insect *Sigara ornata* (on which mites were thought to have a minimal effect). These investigators believed that this result demonstrated the validity of the truncation approach. Royce and Rossignol (1990) used the method to infer mortality caused by tracheal mites infecting honey bees. A negative binomial was fitted to the first few terms of the distribution, and parasite-induced mortality was inferred from the presence of fewer heavily infected hosts than predicted by the negative binomial. Adjei *et al.* (1986) used a more sophisticated version of the same approach. These researchers fitted a negative binomial distribution to the first three terms of the parasite frequency distribution (the number of fish with 0, 1, and 2 parasites) and used the parameters obtained to infer the tail of the distribution that should exist. Comparison of this expected tail with the observed distribution (which was considerably more truncated) suggested that parasite-induced death could be detected.

All applications of this approach have their problems. First, the method assumes that the negative binomial distribution accurately represents the distribution before the action of parasite-induced mortality, and the method is particularly sensitive to this assumption. Second, factors other than parasite-induced mortality (as discussed by Anderson and Gordon, 1982) also may truncate parasite distributions. In the particular case examined (a cestode on a marine fish), Adjei *et al.* (1986) considered this second problem not to be serious, since all fish apparently were infected over a short period in the life cycle (thus, age-related changes in infection were not a problem) and no evidence was found of encapsulated parasites that would indicate a host immune response.

C. Collecting Data on Effects of Parasites on Hosts under Natural Conditions

Although the prospect of inferring mortality directly from frequency distributions is tantalizing, no currently published method can be recommended unequivocally. As discussed previously, treating a randomly chosen fraction of the host population with an antiparasitic agent and comparing either fecundity or mortality with that of untreated controls does not quantify the effect of parasites on the host population fully, unless the parasite burden of both controls and treated animals is estimated. Under some circumstances, this estimation may be possible. A second approach that does not appear to have been used is to mark individually animals with parasite burdens that have been estimated and to follow survivorship. With the availability of mortality-sensing radio transmitters, this approach may be feasible. Alternatively, the parasite distribution on hosts that have died naturally could be compared with the parasite distribution on hosts sampled from the living population. Although such experiments would be invaluable, they would suffer from the problem that the direction of causality is unknown: did hosts die because of high parasite burdens or did otherwise stressed hosts acquire heavy parasite burdens? Only experiments in which parasite burdens are manipulated can clarify this issue.

V. Distribution of Parasites among Hosts

The distribution of parasites within the host population is of crucial importance to the outcome of host–parasite interactions. In almost all cases, parasites are highly overdispersed, meaning a majority of the parasites is harbored by a minority of the hosts (Crofton, 1971; Anderson and May, 1978). Overdispersion is defined mathematically to occur when

$$s^2 > \bar{x}, \tag{16}$$

where s^2 is the variance in parasites per host and $\bar{x}$ is the mean number of parasites per host. The actions of density-dependent constraints on parasite population growth, such as parasite-induced host death or host resistance, are intensified by overdispersion, leading to increased stability of the host–parasite interaction (Anderson and May, 1978).

Many processes can generate overdispersion (Anderson and Gordon, 1982), which can be represented by many probability distributions. In many models of host–parasite interactions, the negative binomial distribution is used to describe overdispersion:

$$P(X = x) = \frac{\Gamma(k + x)}{x!\Gamma(k)} \left(\frac{\mu}{\mu + k}\right)^x \left(1 + \frac{\mu}{k}\right)^{-k}, \tag{17}$$

where, $\Gamma(x)$ is the gamma function, μ is the mean parasite burden, and k is a parameter that inversely measures the extent of overdispersion. In a negative binomial distribution, k, the population mean μ, and variance σ^2 are related by

$$k = \frac{\mu^2}{\sigma^2 - \mu}. \tag{18}$$

Thus, a Poisson (random) distribution with $\sigma^2 = \mu$ has $k = \infty$; aggregated distributions have small values of k. Negative binomial distributions frequently provide adequate representations of parasite frequency distributions, but considerable debate exists over whether the negative binomial is the most suitable description of an overdispersed distribution and, more particularly, whether its parameter k is an appropriate index of the degree of overdispersion (Taylor *et al.*, 1979).

A. Quantifying Overdispersion

Parasitologists frequently require an index of overdispersion. Ideally, they would be able to measure aggregation or overdispersion independent of the mean parasite burden, so overdispersion on different populations or at different times could be compared validly. For example, Bundy *et al.* (1987) and Haswell-Elkins *et al.* (1989) have used k of the negative binomial for this purpose. This goal is probably unachievable in general with a single index, although particular indices may be appropriate for this purpose under particular circumstances. In the following section, some of the commonly used indices are discussed.

1. *k* of the Negative Binomial Distribution

As an index of overdispersion, k has considerable theoretical advantages. Chief among these advantages is that k is invariant under random thinning (Pielou, 1977), meaning a population of parasites that has a negative binomial distribution (with a given value of k) prior to the action of some mortality factor still will have a negative binomial distribution (with the same value of k) after that mortality factor has acted, assuming that the mortality factor acts independently on each individual parasite in such a way that each parasite has the same probability p of surviving.

Whether k does, in fact, describe aggregation independent of mean density is debatable. Haswell-Elkins *et al.* (1989) computed k for *Ascaris lumbricoides* in South Indian villagers, both before and 8 mo after treatment with an anthelmintic. Although the parasite burden established on re-infection was approximately 50% of the original parasite burden, a remarkable consistency

($k = 0.81$ before treatment, $k = 0.85$ afterward) was seen in the parameter k. This result contrasted with a change in the variance : mean ratio from 11.4 to 6.9. Bundy *et al.* (1987) calculated k for *A. lumbricoides* and *Trichuris trichuria* in the same community and commented on the "remarkable similarity" of the k values, suggesting that this result indicated that the differences in susceptibility or exposure to infection that generated the overdispersion were related for both parasites. Other workers have found a systematic relationship between k and density within parasite populations (Scott, 1987a). Rather than being evidence that k is an unsuitable index of aggregation, these results may indicate that some interesting density-dependent processes are at work.

Two main methods for estimating k from a frequency distribution are possible. The easiest is the moment estimate:

$$\hat{k} = \frac{\bar{x}^2}{s^2 - \bar{x}}, \tag{19}$$

where $\bar{x}$ is the mean number of parasites per host and s^2 is the variance in parasites per host. This estimate is obtained from Eq. 18. Unfortunately, it has rather poor statistical properties, although its inverse is better behaved (Pielou, 1977). An alternative is the use of a maximum likelihood estimator (Pielou, 1977; Krebs, 1990). A disadvantage of this approach, in addition to computational difficulty, is that the result has meaning only if the distribution is negative binomial, whereas the moment estimate is an index of overdispersion regardless of the underlying distribution. Sometimes, only data on infected individuals are available; the zero class is unobservable. If this is the case, Grenfell *et al.* (1990) provide a possible maximum likelihood approach.

2. Other Indices of Overdispersion

Existing mathematical models almost exclusively use k as the index of overdispersion. Other indices are, however, often used in the parasitological literature. Following the definition of overdispersion (Eq. 16), the ratio of variance to mean, $s^2/\bar{x}$, provides perhaps the most obvious measure of overdispersion.

A standard method of scaling the variability in measurements (as distinct from counts) is using the coefficient of variation:

$$\text{CV} = \frac{s}{\bar{x}}, \tag{20}$$

or, equivalently, the square of coefficient of variation. For measurements, this value has the desirable property of being independent of the units used; the CV of a given sample of animal lengths is the same whether length is measured in millimeters or meters. As a means of describing aggregation in counts, CV is less appropriate, particularly for weakly overdispersed distributions. Given that, in a Poisson distribution, the variance is equal to the mean, clearly the CV of a Poisson distribution with a large mean is lower than that of a Poisson distribution with a small mean. The variance of log-transformed measurement data is, like the CV, scale independent. Also, as does CV, the variance of log(x) will decline with the mean if the underlying distribution is Poisson. An additional problem is that most parasite distribution data include large numbers of zeros, requiring the use of log($x + 1$). Thus requirement complicates interpretation of the variance. The variance of log(x) is, however, the most appropriate way of describing variability in log-normally distributed data, which some parasite distributions approach. This measure should not be used unless (1) the distributions being compared are all highly overdispersed and (2) few zeros are found in the data set.

VI. Estimating the Basic Reproduction Ratio

A. Macroparasites

Estimating the basic reproduction ratio (R_0) for macroparasites is no simple task. Anderson and May (1991) considered a simple two-equation model for a human helminth infection in which worm abundance was regulated by a density-dependent constraint on egg production. The constraint assumed that parasites were distributed in a manner that could be mimicked using the negative binomial probability distribution. In a single host with i worms, fecundity (λ) was given by

$$\lambda = \lambda_0 e^{-\gamma(i-1)} = \lambda_0 z^{(i-1)}. \quad (21)$$

In the community of hosts, average fecundity was given by

$$\lambda = \lambda_0 \left[1 + \frac{M(1-z)}{k}\right]^{-(k+1)} = \lambda_0 f(M). \quad (22)$$

Given these assumptions, Anderson and May (1991) were able to show that, if the basic reproduction ratio is greater than 1, the system settles down to an equilibrium mean worm burden (M^*) given by

$$R_0 f(M^*) = 1. \quad (23)$$

Equation 23 shows how the effective reproduction ratio of the parasite is reduced to 1 at equilibrium under the influence of the regulatory density-dependent process embodied by $f(M^*)$. This equation also provides an opportunity to estimate R_0 provided k, the exponent of the negative binomial distribution; z, an inverse measure of the severity of density dependence; and M^*, the mean worm burden at equilibrium can be estimated.

First, let's consider how we might get a value for M^*. Two points are particularly important. Estimates for M^* usually are taken from age–intensity profiles obtained from cross-sectional surveys of host populations. If this profile increases to a plateau, the value for M^* is taken as the worm burden at the plateau. However, in many human helminth infections, the intensity increases rapidly initially, then declines to a low plateau. In such a situation, M^* refers to the equilibrium worm burden in the adult population. The second important point is that M^* is defined as "worms," not as eggs per gram or eggs per day. This distinction is particularly important in the context of human parasites, for which detailed age–intensity data on the actual numbers of parasites are rare. Usually egg count data are available. Therefore, to obtain estimated worm numbers from data on egg output, assumptions about the relationship between eggs in stool and numbers of adult parasites become critical.

Some methods for estimating k already have been presented (Section V). In addition to these methods, a crude estimate of k can be obtained from age–prevalence and age–intensity profiles:

$$p_a = 1 - (1 + M_a/k)^{-k}, \quad (24)$$

where p_a is the prevalence in age class a and M_a is the average worm numbers in age class a. k is assumed to be independent of host age, that is, the degree of aggregation of the parasites in each age class of the host population is assumed to be comparable. Solution of Eq. 24 for k involves iterative procedures requiring a computer.

Finally, z is the parameter that deals with the density-dependent constraint. Before methods for estimating z are presented, warning is in order. Remember that the two-equation Anderson and May model (1991) assumes that the constraint on fecundity is the only density-dependent constraint operating. Others may exist but, because of limitations of working with parasite infections in human communities, this parameter is the only one known.

Given data on egg output (per gram of feces or per day), and data on the actual number of worms releasing these eggs (obtained by counting gut parasites expelled with anthelmintics), nonlinear curve-fitting procedures can be used to estimate the parameters of Eq. 21.

By now, a few general points should be obvious. First, obtaining estimates of R_0 without some information on actual numbers of worms (rather than solely egg count data) will be difficult. Second, if the estimates of R_0 must be taken from field data, much overlap exists between equations for λ, k, and R_0. To some extent, the process becomes circular; this confounding should be recognized.

B. Microparasites

Obtaining estimates of the R_0 for microparasites seems somewhat easier than for macroparasites. From Eqs. 2–4 it follows that

$$R_0 = \frac{\beta N}{(\alpha + \mu + \sigma)}. \tag{25}$$

In this simple example, only five values are required: β, N, α, μ, and σ. The host population density (N) usually can be estimated quite accurately. In human populations, demographic data usually are available to give an indication of the death rate of the population (μ). This rate may be more difficult to obtain in wild animal populations. The duration of infectiousness ($1/\sigma$) usually is available as well, from hospital records or epidemiological studies, as is the mortality due to the disease (α). The value that presents the most difficulty is the transmission rate, β. In fact, obtaining good values for β is often impossible. Therefore, workers have sought other methods of estimating R_0. If we assume that births balance deaths in our infected population, and the force of infection (λ) is constant through time and for all ages of host, it is possible to show that

$$R_0 = 1 + \lambda L, \tag{26}$$

where L is the mean expected life-span of individuals in the host population (Dietz, 1975,1976). In Chapter 6, it is shown that the force of infection is the reciprocal of the average time between the introduction of the virus or bacterium into the population and a host becoming infected. In an epidemic situation, this value is merely the reciprocal of the average age at infection (A). Thus,

$$R_0 = 1 + \frac{L}{A}. \tag{27}$$

The parameters L and A are measured relatively easily. In human populations, demographic data give a clear idea of the average life expectancy (from birth) in the population; the average age at infection can be estimated from age–prevalence or age–seroprevalence profiles at the age at which 50% of the population has experienced the infection. Nevertheless, remember that Eq. 27, useful though it is, applies only under a very particular set of circumstances: the host population must not be growing, host mortality must be constant (i.e., independent of age), and the force of infection also must be constant (independent of age).

VII. Overview

Anderson and May (1978) demonstrated the theoretical capability of parasites to regulate the population density of their hosts, and used the simple models represented by Eqs. 2–4 and Eqs. 5–7 to illustrate the point. Since then, considerable progress has been made in estimating the parameters of these models and in examining the assumptions on which they are based. This progress has, however, been restricted largely to laboratory populations or populations in small-scale controlled environments (Keymer, 1981; McCallum, 1985; Scott, 1987a,b; Gregory, 1991). There are remarkably few field host-parasite systems for which all the parameters necessary for use of the simple models described here have been estimated.

The two factors most critically in need of further study are (1) the effect of parasites on host fecundity and mortality under field conditions and (2) the transmission dynamics of parasites. These factors have not been overlooked; the lack of good field data occurs because these processes are not easy to investigate in free-ranging populations. Until such data are collected, parasite regulation of natural host populations will remain a hypothesis (Scott and Dobson, 1989) rather than a demonstrated reality.

However, despite the theoretical interest in the possibility that parasites can regulate the abundance of their host population, other reasons exist for attempting to obtain good estimates of various model parameters. Chapter 3 showed that the basic reproduction ratio is a parameter of fundamental interest to individuals responsible for disease control. Good estimates of R_0 indicate the extent of the effort that may be required to control or eliminate an infection (see also Chapters 11 and 12). Further, if estimating the impact of a given control program on the value of basic reproduction ratio is possible, we have a method or ranking strategies in their likely order of efficacy prior to their full scale implementation in the field. Unfortunately, estimating R_0 is a challenging and difficult task under all but the very simplest of circumstances.

References

Adjei, E. L., Barnes, A., and Lester, R. J. G. (1986). A method for estimating possible parasite-related host mortality, illustrated using data for *Callitetrarhynchus gracilis* (Cestoda: Trypanorhyncha) in lizardfish (*Saurida* spp.). *Parasitology* **92,** 227-243.

Anderson, R. M. (1978). Population dynamics of snail infection by miracidia. *Parasitology* **77,** 201-224.

Anderson, R. M., and Gordon, D. M. (1982). Processes influencing the distribution of parasite numbers within host populations, with special emphasis on parasite-induced host mortalities. *Parasitology* **85,** 373-398.

Anderson, R. M., and May, R. M. (1978). Regulation and stability of host-parasite interactions. I. Regulatory processes. *J. Anim. Ecol.* **47,** 219-247.

Anderson, R. M., and May, R. M. (1979). Population biology of infectious diseases. Part I. *Nature (London)* **280,** 361-367.

Anderson, R. M., and May, R. M. (1985). Helminth infections of humans: Mathematical models, population dynamics, and control. *Adv. Parasitol.* **24,** 1-101.

Anderson, R. M., and May, R. M. (1991). "Infectious Diseases of Humans." Oxford University Press, Oxford.

Anderson, R. M., and Whitfield, P. J. (1975). Survival characteristics of the free-living cercarial population of the ectoparasitic digenean *Transversotrema patialensis* (Soparker, 1924). *Parasitology* **70,** 295-310.

Anderson, R. M., Whitfield, P. J., and Dobson, A. P. (1978). Experimental studies of infection dynamics: Infection of the definitive host by the cercariae of *Transversotrema patialensis*. *Parasitology* **77,** 189-200.

Baudoin, M. (1975). Host castration as a parasitic strategy. *Evolution* **29,** 335-352.

Begon, M., Harper, J. L., and Townsend, C. R. (1990). "Ecology," 2d Ed. Blackwell Scientific Publications, Oxford.

Bernoulli, D. (1760). Essai d'une nouvelle analyze de la mortalité causée par la petite variole et des advantages de l'inoculation pour la prevenir. *Mem. Math. Phys. Tires Regist. Acad. R. Sci. (Paris) 1760,* 1-45.

Brown, C. R., and Brown, M. B. (1987). Ectoparasitism as a cost of coloniality in cliff swallows *(Hirundo pyrrhonata)*. *Ecology* **67,** 1206-1218.

Bundy, D. A. P., Cooper, E. S., Thompson, D. E., Didier, J. M., and Simmons, I. (1987). Epidemiology and population dynamics of *Ascaris lumbricoides* and *Trichuris trichuria* infection in the same community. *Trans. R. Soc. Trop. Med. Hyg.* **81,** 987-993.

Caswell, H. (1989). "Matrix Population Models." Sinauer Associates, Sunderland, Massachussetts.

Cormack, R. M. (1989). Log-linear models for capture-recapture. *Biometrics* **45,** 395-413.

Cox, D. R., and Oakes, D. (1984). "Analysis of Survival Data," Chapman and Hall, London.

Crawley, M. J. (1993). "GLIM for Ecologists." Blackwell, Oxford.

Crofton, H. D. (1971). A quantitative approach to parasitism. *Parasitology* **62,** 179-194.

Dietz, K. (1975). Transmission and control of arbovirus diseases. *In* "Epidemiology" (D. Ludwig and K. L. Cooke, eds.), pp. 104-121. Society of Applied and Industrial Mathematics, Philadelphia.

Dietz, K. (1976). The incidence of infectious diseases under the influence of seasonal fluctuations. *In* "Mathematical Models in Medicine. Lecture Notes in Biomathematics" (J. Berger, W. Buhlen, and R. Regges, eds.), pp. 1-15. Springer-Verlag, Berlin.

Dobson, A. P. (1988). The population biology of parasite-induced changes in host behaviour. *Q. Rev. Biol.* **63,** 139-165.

Dobson, A. P., and Hudson, P. J. (1992). Regulation and stability of a free-living host-parasite system: *Trichostrongylus tenuis* in red grouse. II. Population models. *J. Anim. Ecol.* **61,** 487-498.

Elkins, D. B. (1986). "The Epidemiology and Control of *Ascaris lumbricoides* in a South Indian Fishing Community." Unpublished PhD. thesis, University of London.

Elkins, D. B., Sithithaworn, P., Haswell-Elkins, M., Kaewkes, S., Awacharagan, P., and Wongratanacheewin, S. (1991). *Opistorchis viverrini*: Relationships between egg counts, worms recovered, and antibody levels within an endemic community in Northeast Thailand. *Parasitology* **102,** 283-288.

Gregory, R. D. (1991). Parasite epidemiology and host population growth: *Heligmosomoides polygyrus* (Nematoda) in enclosed wood mouse populations. *J. Anim. Ecol.* **60,** 805-821.

Grenfell, B. T., and Anderson, R. M. (1985). The estimation of age-related rates of infection from case notifications and serological data. *J. Hyg. (Cambridge)* **95,** 419–436.

Grenfell, B. T., Smith, G., and Anderson, R. M. (1986). Maximum likelihood estimates of the mortality and migration rates of the infective larvae of *Ostertagia ostertagi* and *Cooperia oncophora. Parasitology* **92,** 643–652.

Grenfell, B. T., Das, P. K., Rajagopalan, P. K., and Bundy, D. A. P. (1990). Frequency distributions of lymphatic filiariasis microfilariae in human populations: Population processes and statistical estimation. *Parasitology* **101,** 417–427.

Haswell-Elkins, M., Elkins, D., and Anderson, R. M. (1989). The influence of individual, social group, and household factors on the distribution of *Ascaris lumbricoides* within a community and implications for control strategies. *Parasitology* **98,** 125–134.

Holmes, J. C. (1982). Impact of infectious disease agents on the population growth and geological distribution of animals. *In* "Population Biology of Infectious Diseases" (R. M. Anderson and R. M. May, eds.), pp. 37–51. Springer-Verlag, Berlin.

Hone, J., Pech, R., and Yip, P. (1992). Estimation of the dynamics and rate of transmission of classical swine fever (hog cholera) in wild pigs. *Epidemiol. Infect.* **108,** 377–386.

Hudson, P. J. (1986). The effect of a parasitic nematode on the breeding production of red grouse. *J. Anim. Ecol.* **55,** 85–92.

Hudson, P. J., Dobson, A. P., and Newborn, D. (1992). Do parasites make prey vulnerable to predation? Red grouse and parasites. *J. Anim. Ecol.* **61,** 681–692.

Keyfitz, N. (1977). "Applied Mathematical Demography." Wiley, New York.

Keymer, A. (1981). Population dynamics of *Hymenolepis diminuta* in the intermediate host. *J. Anim. Ecol.* **50,** 941–950.

Keymer, A. E., and Anderson, R. M. (1979). The dynamics of infection of *Tribolium confusum* by *Hymenolepis diminuta*: The influence of infective-stage density and spatial distribution. *Parasitology* **79,** 195–207.

Krebs, C. J. (1990). "Ecological Methodology." Harper and Row, Cambridge.

Lanciani, C. A., and Boyett, J. M. (1980). Demonstrating parasitic water mite-induced mortality in natural host populations. *Parasitology* **81,** 465–475.

May, R. M., and Anderson, R. M. (1978). Regulation and stability of host–parasite interactions. II. Destabilizing processes. *J. Anim. Ecol.* 47, 249–267.

May, R. M., and Anderson, R. M. (1979). Population biology of infectious diseases. Part II. *Nature (London)* **280,** 455–461.

McCallum, H. I. (1982). Infection dynamics of *Ichthyophthirius multifiliis. Parasitology* **85,** 475–488.

McCallum, H. I. (1985). Population effects of parasite survival of host death: experimental studies of the interaction of *Ichthyophthirius multifiliis* and its fish host. *Parasitology* **90,** 529–547.

McCallum, H. I., and Singleton, G. R. (1989). Models to assess the potential of *Capillaria hepatica* to control population outbreaks of house mice. *Parasitology* **98,** 425–437.

Munger, J. C., and Karasov, W. H. (1991). Sublethal parasites in white-footed mice: Impact on survival and reproduction. *Can. J. Zool.* **69,** 398–404.

Pech, R. P., and Hone, J. (1988). A model of the dynamics and control of an outbreak of foot and mouth disease in feral pigs in Australia. *J. Appl. Ecol.* **25,** 63–77.

Pielou, E. C. (1977). "Mathematical Ecology." Wiley, New York.

Pollock, K. H., Nichols, J. D., Brownie, C., and Hines, J. E. (1990). Statistical inference for capture–recapture experiments. *J. Wildlife M.* **54,***(Suppl.)* 1–97.

Price, P. W. (1980). "Evolutionary Biology of Parasites." Princeton University Press, Princeton, New Jersey.

Quinnell, R. J. (1992). The population dynamics of *Heligmosomoides polygyrus. J. Anim. Ecol.* **61,** 669–679.

Ross, R. (1916). An application of the theory of probabilities to the study of *a priori* pathometry. Part I. *Proc. R. Soc. London A* **92,** 204–230.

Ross, R., and Hudson, H. P. (1917). An application of the theory of probabilities to the study of *a priori* pathometry. Parts II and III. *Proc. R. Soc. London A* **93,** 212–240.

Royce, L. A., and Rossignol, P. A. (1990). Honey bee mortality due to tracheal mite parasitism. *Parasitology* **100,** 147–151.

SAS Institute, Inc. (1988). "SAS/STAT™ User's Guide Release 6.03 Edition." SAS Institute, Inc., Cary, N.C.

Schmidt, G. D., and Roberts, L. S. (1989). "Foundations of Parasitology," 4th Ed. Times Mirror/Mosby, St. Louis, Missouri.

Scott, M. E. (1987a). Temporal changes in aggregation: A laboratory study. *Parasitology* **94,** 583–595.

Scott, M. E. (1987b). Regulation of mouse colony abundance by *Heligmosomoides polygyrus. Parasitology* **95,** 111–124.

Scott, M. E. (1990). An experimental and theoretical study of the dynamics of a mouse nematode *Heligmosomoides polygyrus)* interaction. *Parasitology* **101,** 75–92.

Scott, M. E., and Dobson, A. P. (1989). The role of parasites in regulating host abundance. *Parasitol. Today* **5,** 176–183.

Seber, G. A. F. (1982). "The estimation of Animal Abundance and Related Parameters" 2d Ed. Griffin, London.

Shostak, A. W., and Scott, M. E. (1993). Detection of

density-dependent growth and fecundity of helminths in natural infections. *Parasitology* **106,** 527–539.

Smith, G. (1987). The relationship between the density of *Fasciola hepatica* micacidia and the net rate of miracidial infection in *Lymnaea truncatula. Parasitology* **95,** 159–163.

Southwood, T. R. E. (1978). "Ecological Methods: with Particular Reference to the Study of Insect Populations," 2d ed. Chapman and Hall, London.

Taylor, L. R., Woiwood, I. P., and Perry, J. N. (1979). The negative binomial as a dynamic ecological model for aggregation and the density dependence of *k. J. Anim. Ecol.* **48,** 289–304.

Gary Smith

Parasite Population Density Is Regulated

5

I. Introduction

A. What Is Regulation?

The density of a regulated population fluctuates considerably less than would be expected from a "random walk" over the period of study (Hassell, 1982). If the abundance of a regulated population is perturbed by some transient extrinsic agency (for example, weather), it will tend to return to its previous value. Thus, we expect a regulated population to persist at some consistent level for long periods of time and to return to previous values if perturbed.

B. Regulation in Microparasitic and Macroparasitic Infections

As you might expect, we think about the regulation of macroparasitic and microparasitic infections rather differently. Macroparasites are large enough to count as individuals; the consequences of regulation in this group can be described simply in terms of changes in the number of parasites per host. Microparasites are too small to count, so it is convenient as well as conventional to think of regulation in microparasitic populations in terms of the effect of these parasites on the proportion of the host population that is currently susceptible, infectious, or immune.

C. Evidence that Regulation Occurs

The evidence that regulation actually occurs is of three kinds. First, long-term surveys show that population densities fluctuate between very restricted limits for long periods of time unless environmental conditions change. Second, similar surveys show a return to previous densities following some radical change in population numbers. Third, we have been able to identify processes that constrain population growth and whose effects are proportionately more severe at higher population densities (Lack, 1967; Quinnell *et al.*, 1990).

For viruses and bacteria, long-term case notification records of human disease provide evidence of the stability of microparasitic infections in humans (Anderson and May, 1991). Similar data for microparasitic infections of nonhuman hosts are harder to find. Figure 1 illustrates that the prevalence and incidence of a variety of microparasitic infections of nonhuman species vary over a

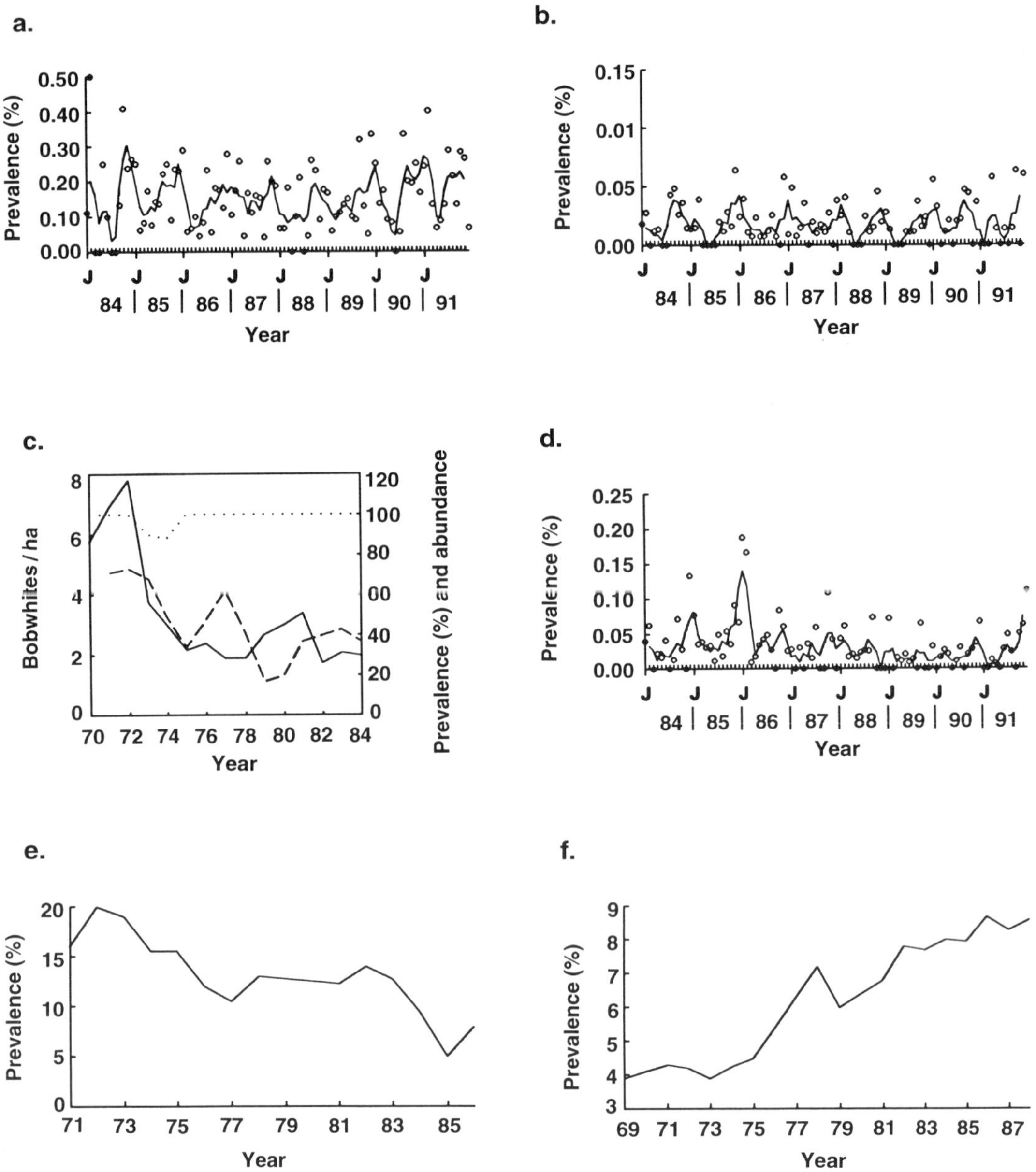

Figure 1 Prevalence of infection for various macroparasites. (a) Gut-dwelling nematodes (ascarids) in cats admitted to the University of Pennsylvania Small Animal Hospital. (b) Tapeworms in dogs admitted to the University of Pennsylvania Small Animal Hospital. (c) A caecal nematode (*Heterakis isolonche*) in North American Quail. Bobwhite density, ———; prevalence of infection · · ·; mean intensity of infection, – – –. (d) Nematode parasite of the circulatory system (heartworm) in dogs admitted to the University of Pennsylvania Small Animal Hospital. (e) Liver condemnations due to *Fasciola hepatica* infections in sheep in Northern Ireland. (f) Liver condemnations due to lesions caused by migrating juvenile stages of *Ascaris suum* in pigs in Northern Ireland. Data for a, b, and d provided by T. Nolan and used here with permission. Other data from Davidson *et al.* (1991) and McIlroy *et al.* (1990a,b).

rather narrow range. Very few studies have been done of the long-term stability of the prevalence and intensity of macroparasitic infections (Quinnell *et al.,* 1990). A striking feature of the data shown in Fig. 1 is that most do not provide a very powerful argument for regulation. Certainly, the prevalence of ascarids in cats and tapeworms in dogs fluctuates over a very narrow range (Fig. 1a,b) but under some circumstances the intensity of infection can change even if prevalence remains constant (Fig. 1c). The remaining examples in Fig. 1 represent instances in which a significant increase or decrease in prevalence occurs over the relatively short time-span studied (Fig. 1d,e,f). That prevalence or intensity changes in this way, even in a regulated population, should not be surprising. The equilibrium level to which the population density continuously tends changes with environmental variables. The seasonal nature of these fluctuations is apparent in Fig. 2a,b, but some environmental changes extend over several years. For example, a decline in the abundance of caecal worms infecting bobwhites (Fig. 1c) was a consequence of changes in bobwhite density, which itself was a consequence of the change in habitat quality. The decline in heartworm infections in dogs in the United States (Fig. 1d) is coincident with the introduction and widespread use of a very powerful anthelmintic agent.

Better evidence for regulation probably is provided by perturbation studies. Indications of regulation are seen in instances in which vaccination coverage has declined and a resurgence in disease prevalence has occurred (e.g., pertussis in the United Kingdom; Anderson and May, 1982). Similar evidence of regulation is provided by perturbation studies involving macroparasitic infections (Quinnell *et al.*, 1990). In these studies, nematode populations (usually) are depleted by anthelmintic treatment and then followed as they recover to pretreatment levels.

Finally, investigators have been able to identify a substantial number of candidate regulatory processes for both micro- and macroparasitic infections (for example, see Keymer, 1982; Smith, 1984a; Tizard, 1987; Quinnell *et al.,* 1990). These processes will be considered in detail later in this chapter.

II. Basic Principles

A regulated population is maintained at some equilibrium value by processes that constrain population growth and the effect of which is proportionately more severe at higher population densities. Sometimes the equilibrium is some fixed value; sometimes it is a cyclically repeated set of values. Note that even a cycling population is considered a regulated population because the idea of regulation refers only to the tendency to return to some previous state. This concept implies nothing about the exact nature of that state except its consistence.

A. General Single Species Model without Time Delays

The processes that are responsible for regulation all can be described as functions of population density. For example, a frequently discussed model for a general single species population is

$$\frac{dN_t}{dt} = aN_t - f(N_t)N_t, \qquad (1)$$

where N_t is population density at time t, a is the birth rate, and $f(N_t)$ is an unspecified function of the current population density representing the instantaneous *per capita* mortality. If the value of $f(N_t)$ increases with N_t, the ensuing changes in mortality will regulate population density. Suppose, for example, that the loss rate is a simple linear function $f(N_t) = b + cN_t$, where b and c are constants. Then

$$\frac{dN_t}{dt} = aN_t - (b + cN_t))N_t. \qquad (2)$$

Regardless of the initial value of N, the population will move toward an equilibrium

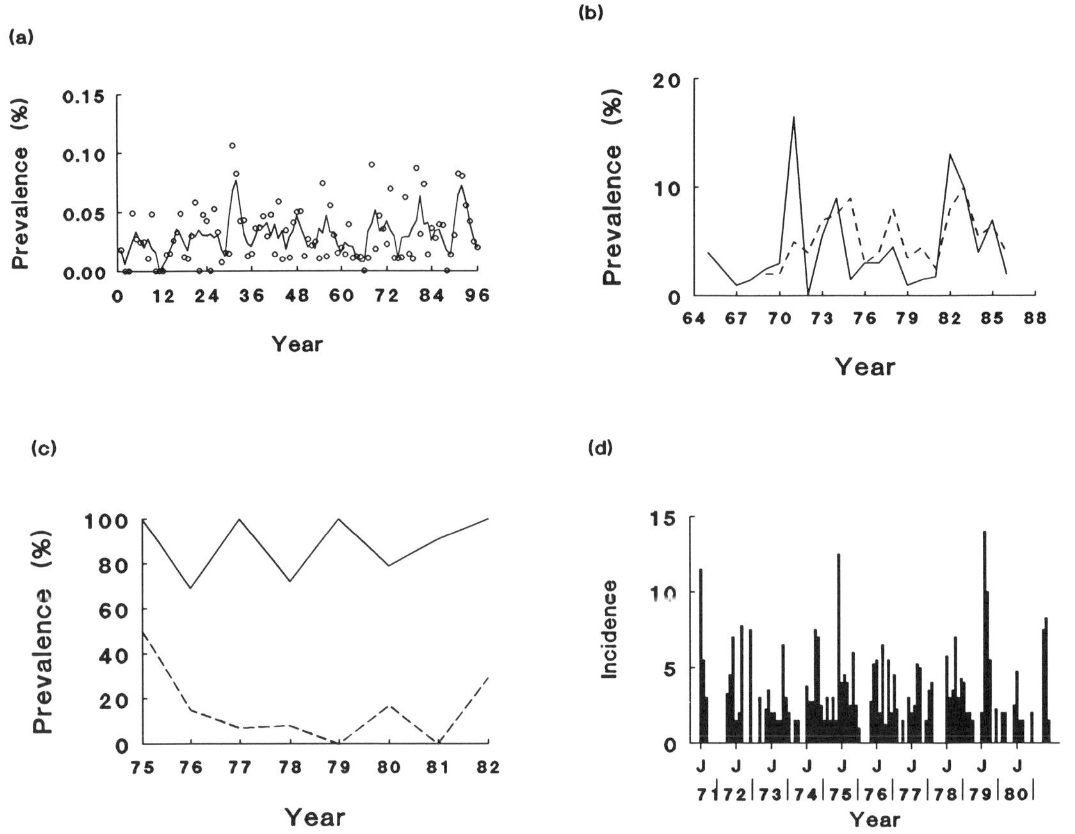

Figure 2 Prevalence of various microparasitic infections. (a) Protozoan infections (coccidia) in dogs admitted to the University of Pennsylvania Small Animal Hospital. (b) Rabies in bats assessed from two overlapping data sets (Illinois). (c) Serum antibody prevalence for infectious canine hepatitis virus (solid line) and canine distemper virus (dashed line) in wolves (Alaska). (d) Incidence of salmonellosis in horses admitted to Large Animal Clinic of the University of California, Davis. Incidence was measured as the number of new cases each month divided by the corresponding in-patient population, expressed as a percentage. Data from Carter *et al.* (1986), Zarnke and Ballard (1987), and Burnett (1989). Data in a provided by T. Nolan and used here with permission.

value N^* defined by

$$N^* = \frac{(a-b)}{c}. \tag{3}$$

This equilibrium is calculated easily by setting the right hand side of Eq. 1 equal to zero and solving for N. This method works because Eq. 1 represents the net rate of change in the value of N. At equilibrium, this net rate of change is zero. The behavior of the model is illustrated in Fig. 3. Note that a similarly regulated population could have been obtained if mortality had been held constant and the birth rate, a, had been some *decreasing* function of N_t.

B. General Single Species Models with Time Delays

The assumption implicit in Eq. 1 is that mortality is immediately responsive to changes in population density. A more realistic model might be

$$\frac{dN_t}{dt} = aN_t - f(N_{(t-T)})N_t, \tag{4}$$

where some delay (T) occurs before mortal-

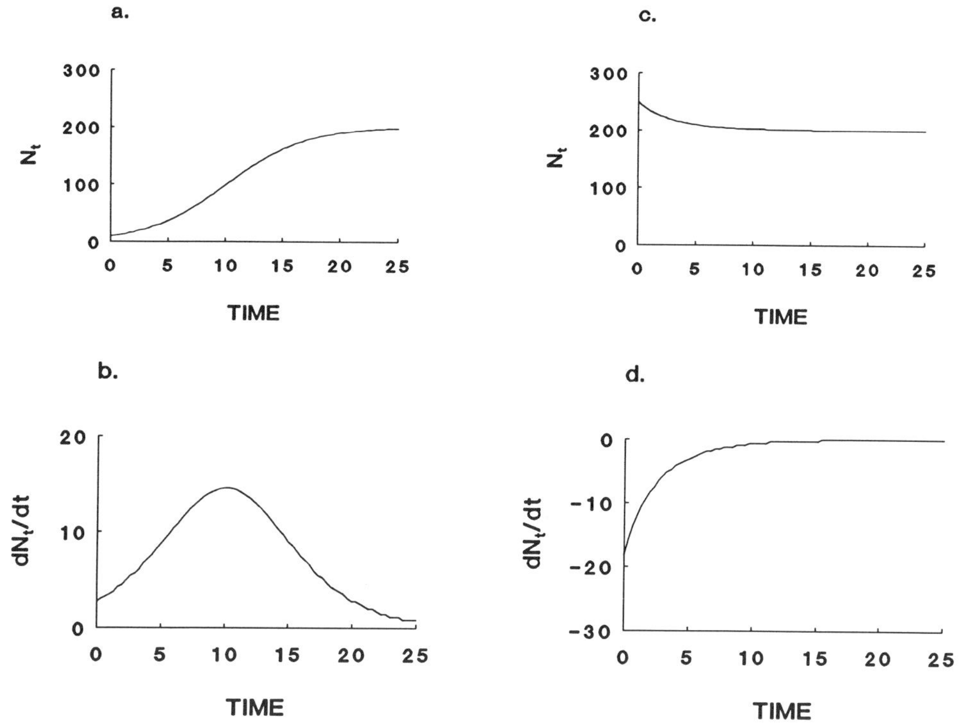

Figure 3 Behavior of a single species model without time delays (Eq. 2). In these examples, the final equilibrium (N^*) is 200 individuals. In a and b, the initial value of N was 10. In c and d, the initial value of N was 250.

ity is adjusted. If T is large relative to the natural response time of the system, population density will not approach its final equilibrium position in a smooth (monotonic) way but will oscillate like a pendulum. As T increases, the tendency to oscillate changes into a stable limit cycle in which population density perpetually oscillates with an amplitude and frequency determined by the parameters of the model.

C. Regulation and the Potential for Population Increase

Considering regulation to have meaning only when applied to a population that has the potential for growth simplifies matters conceptually. In some cases, regulatory processes are operating but the population simply "runs down" for lack of input. For example, when calves are infected only once with the nematode *Ostertagia ostertagi,* the resulting parasite survivorship curve is consistent with the hypothesis that the mortality of the worms is density dependent. Nevertheless, the final parasite burden is still zero (Fig. 4; Anderson and Michel, 1977). The model that describes this experiment is identical to Eq. 2, but lacks an input term (the magnitude of the original infecting dose is dealt with as an "initial condition"):

$$\frac{dN_t}{dt} = -(b + cN_t)N_t. \tag{5}$$

Lacking an input term, the right hand side of Eq. 5 is always negative, so N_t will decrease continuously until it reaches zero. The reader should not assume, however, that in other similar cases the end point will always be zero. For example, in the simple epidemic model, one of the defining features is that its time course is sufficiently rapid

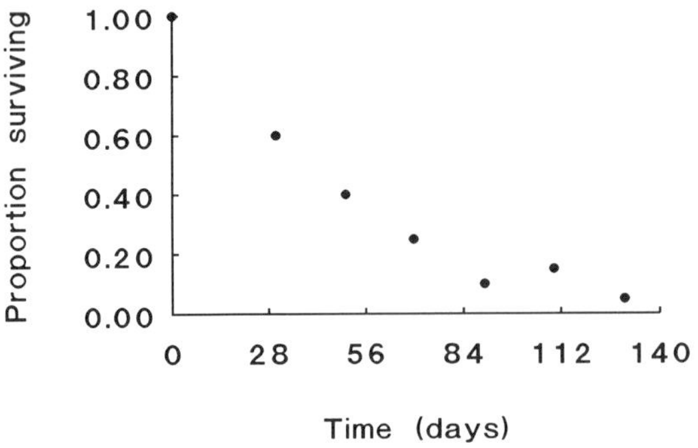

Figure 4 Survivorship of *Ostertagia ostertagi* in calves following a single infecting dose. The mortality of the worms is density dependent but the population nevertheless declines for want of further input. Data from Anderson and Michel (1977).

that births and deaths in the host population can be ignored. No input to the susceptible class exists, so it "runs down" in the sense alluded to earlier, despite (as we shall show) the operation of a potentially regulatory process. Nevertheless, as shown in Chapter 3, whether x, the proportion of susceptible hosts, effectively goes to zero or not is entirely dependent on the value of the basic reproduction ratio (R_0). To examine this phenomenon in more detail, the basic epidemic model presented in Chapter 3 can be rewritten in terms of the *proportion* of hosts in each category. Conventionally, the host population is divided into susceptible hosts (x), infected/infectious hosts (y), and recovered/immune hosts (z). The lower case designations (x, y, z) signify that the proportion of the host population in each class (i.e., $x + y + z = 1$), not the number, is considered:

$$\frac{dx}{dt} = -\beta xy$$

$$\frac{dy}{dt} = \beta xy - \delta y$$

$$\frac{dz}{dt} = \delta y. \qquad (6)$$

Here β is the transmission coefficient, βy is the disease incidence (instantaneous force of infection), and $1/\delta$ is the average time an individual host remains infectious. The right hand side of the equation for x is always negative, so the proportion of susceptible hosts in the population will decrease until the rate of change in x (dx/dt) finally declines to zero. Clearly, one way in which dx/dt could become zero is if x itself is zero, as in the nematode example just given. However, dx/dt also could be zero if y were zero. Further, if y falls to zero before x, the equilibrium value for x will be some number greater than zero. Under some circumstances, y can reach zero before x (i.e., when $R_0 < 3$); some completely susceptible hosts escape infection merely for want of infectious individuals.

The final equilibrium in this model is attained when the rate of change in *all* host classes is zero. The rate of change in y, the proportion of infected hosts, is also zero at the peak of the epidemic curve (Fig. 5) but x and z are still changing at this stage and so y continues to change also until the final equilibrium is reached.

III. Regulation in Microparasitic Infections

A. Equilibrium Behavior of a Simple Endemic Model

A model for endemic disease can be constructed from Eq. 6 by including terms representing births and deaths in the host population. We make the net birth rate (a) equal to the net death rate [$\mu(x + y + z)$] so total population size does not increase. We assume that infectious individuals contribute to births irrespective of their disease status:

$$\frac{dx}{dt} = a - \beta xy - \mu x$$

$$\frac{dy}{dt} = \beta xy - (\mu + \delta)\, y$$

$$\frac{dz}{dt} = \delta y - \mu z\,. \qquad (7)$$

As before, x, y, and z represent the fractions

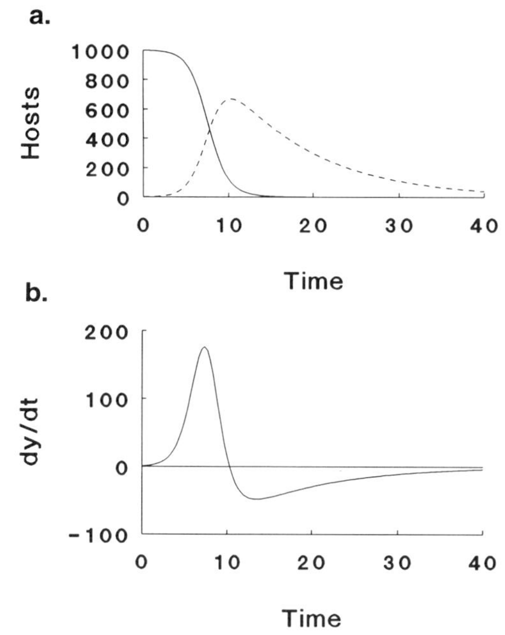

Figure 5 Behavior of the simple epidemic model (Eq. 6). (a) Changes in the number of susceptible (X; ——) and infected (Y;--) hosts. (b) Changes in dy/dt. Note that $dy/dt = 0$ at peak levels of Y as well as at equilibrium (Y^*).

of the host population in each class ($x + y + z = 1$) and x^*, y^*, and z^* represent the equilibrium values. Two sets of equilibrium values are possible: in one, the infection fails to persist ($y^* = 0$); in the other, the infection becomes endemic ($y^* > 0$). We can predict the final outcome by reference to the basic reproduction ratio (R_0) of the infection:

$$R_0 = \frac{\beta}{(\mu + \delta)}. \quad (8)$$

Reference to previous chapters might lead the reader to ask why no parameter for total population density (N) is included in the numerator of Eq. 8. The difference between this expression for R_0 and previous formulations in this book arises because now the proportion of hosts, not the actual number of hosts, in each class is considered. In this case, $N = 1$, so $\beta N = \beta$, which appears in the numerator. Provided $R_0 > 1$, the infection will persist. If $R_0 < 1$, the infection will not persist. This relationship is derived from the expression for dy/dt in Eq. 7. Dividing both sides by y, the *per capita* rate of change in y is shown to depend only on the value of x, the proportion of susceptible hosts, and the relative values of the model parameters:

$$\frac{dy}{dt \cdot y} = \beta x - (\mu + \delta). \quad (9)$$

If $\beta/(\mu + \delta) < 1$, the right side of Eq. 9 is negative for all possible values of x ($0 \leq x \leq 1$) and y declines until it reaches zero. This solution represents the infection-free state. On the other hand, if $\beta/(\mu + \delta) > 1$, a range of values of x exists for which dy/dt is positive and a range of values of x exists for which dy/dt is negative (Fig. 6a). Thus, since the original threshold criterion ($R_0 > 1$) is satisfied, y may increase or decrease depending on the current value of x and eventually will settle to some value greater than zero. This solution represents the endemic state. To estimate the equilibrium values of x, y, and z in the endemic state, the left side of Eq. 7 is set equal to zero. The manipulations that follow are often troublesome, so they are presented in some detail. The equation for y yields

$$x^* = \frac{(\mu + \delta)}{\beta} = \frac{1}{R_0}. \quad (10)$$

The equation for x yields

$$\beta x^* y^* = \mu - \mu x^*$$

$$\frac{\beta (\mu + \delta) y^*}{\beta} = \mu \left(1 - \frac{1}{R_0}\right)$$

$$y^* = \frac{\mu}{(\mu + \delta)} \left(1 - \frac{1}{R_0}\right). \quad (11)$$

The equation for z yields

$$\mu z^* = \delta y^*$$

$$\mu z^* = \frac{\delta \mu}{(\mu + \delta)} \left(1 - \frac{1}{R_0}\right)$$

$$z^* = \frac{\delta}{(\mu + \delta)} \left(1 - \frac{1}{R_0}\right). \quad (12)$$

In the next section, we shall examine the

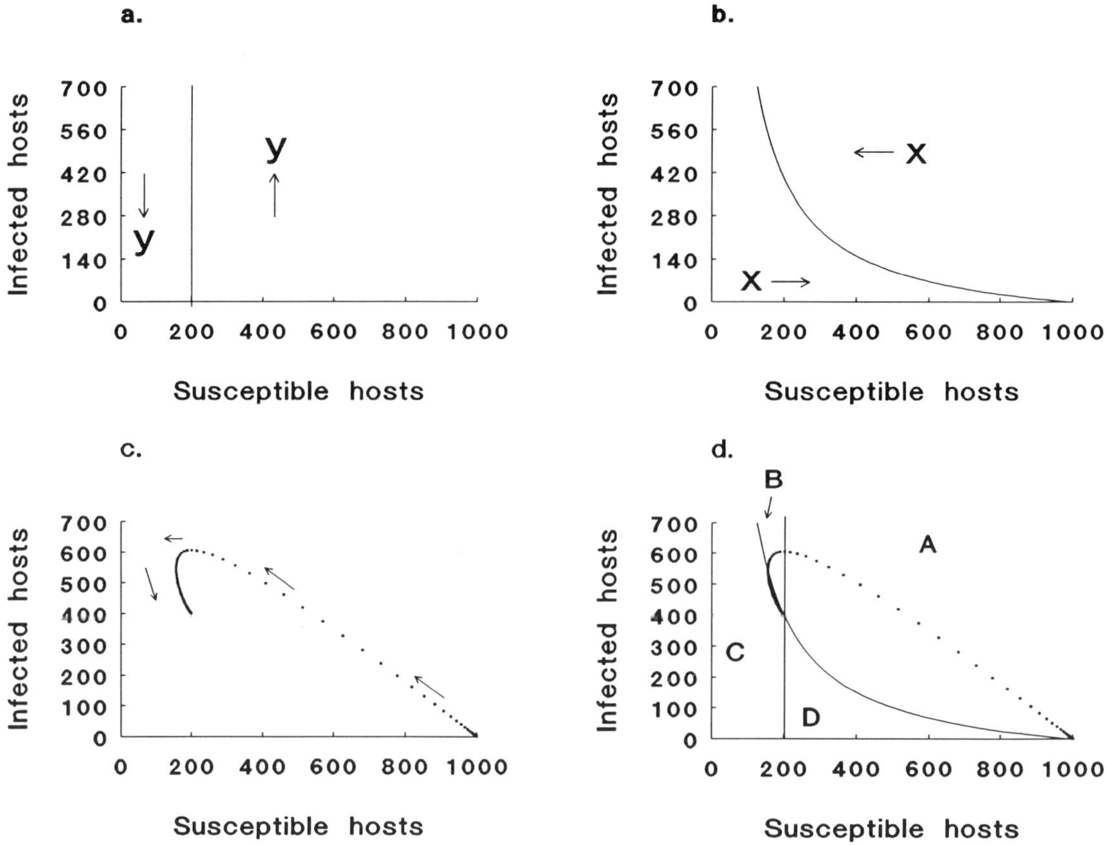

Figure 6 Phase plane method for revealing the qualitative behavior of the simple endemic model (Eq. 7). (a) The *X* isocline. (b) The *Y* isocline. (c) An example of one possible trajectory described by corresponding values of *X* (susceptible hosts) and *Y* (infected hosts) set out in Fig. 5a. (d) The complete phase plane diagram.

processes that cause the population to move toward the equilibrium represented by Eqs. 10, 11, and 12.

B. Regulation of Endemic Microparasitic Infections

As just demonstrated, the simple endemic model differs from the simple epidemic model because it includes births and deaths in the host population. In this context, regulation is reasonable since the susceptible hosts that become infected can be replaced by births. Equilibrium can be maintained by the density-dependent adjustments of the rates at which these two processes occur. In the general model for a single species population, regulation was a consequence of the density-dependent nature of one or more of the components of the instantaneous *per capita* rate of change. The same principle is true of the model for endemic microparasitic infections. Consider the equation for susceptible hosts (x). The net rate of increase in the proportion of susceptible hosts due to births is constant (a), so it can be ignored. Similarly, the instantaneous rate of loss to mortality is also constant (μ), so it, too, is ignored. However, the instantaneous *per capita* rate of loss in susceptible hosts that is attributable to infection is not constant:

$$\frac{dx}{dt \cdot x} = -\beta y. \tag{13}$$

Both sides of the equation have been divided by x, so both sides represent the *per capita* change in the instantaneous rate of loss. The *per capita* rate of loss to infection is clearly a function of y, the proportion of infectious hosts. Insofar as y is itself a function of x, the *per capita* rate of loss of susceptible hosts to infection also is a function of x. Thus, susceptible hosts are regulated by their density-dependent removal consequent to infection. This demonstration assumes, of course, that the possible range of adjustment to this rate is sufficient to achieve the required balance, which will always be true if $R_0 > 1$. However, if $R_0 < 1$, the removal rate due to infection can never be sufficient to offset recruitment due to births, so the proportion of susceptible hosts will increase until all the hosts are susceptible.

C. Oscillatory Behavior and the Simple Endemic Model

Endemic microparasitic infections often are associated with repeating cycles of high prevalence (y) followed by low prevalence. In some cases, the driving forces are extrinsic seasonal factors such as weather. In other cases, for example, the cycles of fox rabies in Europe, the causes are less clear. In the single species model with time delays, oscillatory behavior is a consequence of regulation, so the question arises whether this behavior also might be a consequence of regulation in the simple endemic model. To answer this question, we can focus on the changes that occur in the susceptible (x) and infectious (y) hosts. What happens when we introduce a single infectious individual into an otherwise wholly susceptible population?

The method requires the establishment of criteria for deciding whether x and y are increasing or decreasing. These criteria are determined simply since the values of x and y will increase if their respective differentials (dy/dt and dx/dt) are positive and will decrease if their respective differentials are negative. Thus, if current values of x and y are plotted on a graph divided according to whether the differentials are positive or negative, how the next pair of values for x and y will differ from the current values can be predicted. The graph will be divided into four regions (both differentials positive, both differentials negative, and two regions in which the signs of the differentials will be opposite), and begins with the locus of points for which $dy/dt = 0$. This line divides the graph into two regions: one in which dy/dt is positive and one in which dy/dt is negative. This division consists of a vertical line in the x, y plane so $x = (\mu + \delta)/\beta$ for all values of y (Fig. 6a). Showing that when $x = (\mu + \delta)/\beta$, no change in y occurs, that is, $dy/dt = 0$, is not difficult:

$$\frac{dy}{dt} = \beta xy - (\mu + \delta)y = \beta \frac{(\mu + \delta)}{\beta} y - (\mu + \delta)y = 0. \quad (14)$$

However, if $x > (\mu + \delta)/\beta$, dy/dt is positive, so y is increasing. Similarly, if $x < (\mu + \delta)/\beta$, dy/dt is negative and y is decreasing. With these criteria, we can determine whether y is increasing or decreasing based on the current value of x. We can follow exactly the same procedure for x. First, we define the locus of points for which $dx/dt = 0$. Using Eq. 15, we can calculate a pair of points (x, y) for which $dx/dt = 0$.

$$\frac{dx}{dt} = \mu - \beta xy - \mu x = 0$$

$$\mu - (\beta y + \mu)x = 0$$

$$x = \frac{\mu}{(\beta y + \mu)}. \quad (15)$$

The resulting line crosses the x, y plane from top left to bottom right (Fig. 6b). From Eq. 15 it follows that, if the current value of y is below this line, dx/dt is positive and x is increasing; if the current value of y is above this line, dx/dt is negative and x is decreasing. When we combine the two lines (isoclines) as in Fig. 6d, we obtain a complete set of criteria for x and y to determine the behavior of any given pair of current values. Instead of plotting x or y against time (t), Fig. 6c and d plots y against the corresponding value of x. The isoclines divide Fig. 6d into four regions, A, B, C, and D. The x,y

trajectory begins in region B. If paired values of x and y occur anywhere in this area, x must be decreasing and y must be increasing. Accordingly, the x,y trajectory moves from the lower right to the upper left of the figure. As this trajectory crosses the y isocline, the condition of $dy/dt = 0$ is satisfied momentarily. This point represents a peak value for y; the number of infected hosts is greatest at this point. However, x is still decreasing and the x,y trajectory moves into region A in which both x and y decrease. This behavior continues until the trajectory crosses the x isocline at which $dx/dt = 0$ is satisfied momentarily. This point represents the lowest value of x; the number of susceptible hosts is lowest at this point. Thereafter, the x,y trajectory moves into region C. Here, y continues to decrease but x begins to increase. All movement stops when the trajectory reaches the intersection of the x and y isoclines. At this point, both dy/dt and dx/dt are zero. This point is the final equilibrium point. In two instances in this example, either dx/dt or dy/dt was zero, but this final point is the only point at which both differentials are zero; once this point is reached, no further change in the value of x or y occurs.

In more concrete terms, equilibrium is reached only when the losses to the susceptible class due to infection and death are balanced exactly by gains due to births and the gains to the infectious class are balanced exactly by the losses due to recovery and death. Whether or not this state will be reached on the first pass depends entirely on the parameter values of the model (Fig. 7). When the loss of susceptible hosts due to infection occurs too rapidly, the system initially will balance the losses and gains in y without doing so in x. In that case, the continuing change in x will upset the precarious balance in y because both host classes are yoked by the infection process (βxy). This overcorrection leads to oscillations in the values of x and y. Sometimes the oscillations are severe (Fig. 7b), and further corrections are necessary.

The important point is that the simple endemic model has an inherent tendency to oscillate. Thus, invoking extrinsic cyclical phenomena to explain every instance of repeated episodic changes in observed prevalence (y) may not be necessary. This cyclical behavior may be the result of density-dependent regulation of infection.

IV. Regulation of Macroparasitic Infections

A. Mechanisms of Density-Dependent Regulation in Macroparasitic Infections

Density-dependent regulation of macroparasites usually is attributed to one or more of the following mechanisms: intraspecific

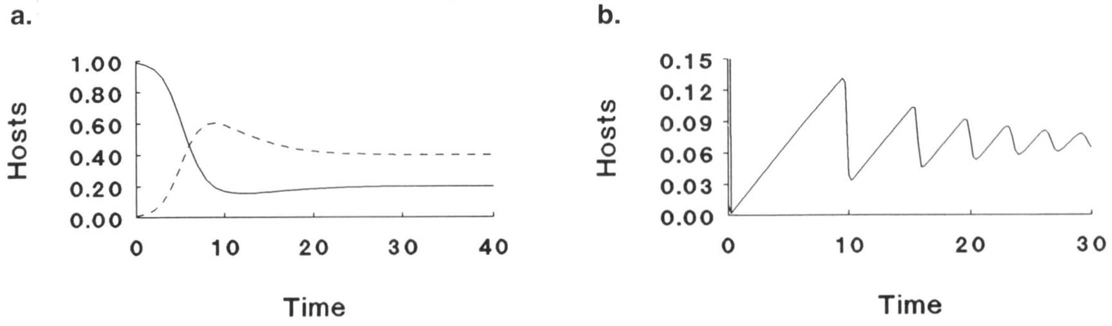

Figure 7 Oscillatory tendency inherent in the simple endemic model. All plots were obtained from numerical solutions of Eq. 7. (a) $\mu = 0.1$, $\beta = 1.0$, $\delta = 0.1$. (b) $\mu = 0.015$, $\beta = 375.0$, $\delta = 26.0$. Susceptible hosts, ——; infectious hosts, - - -.

competition for limited resources, the effects of the immune response of the host, the effects of parasite-induced pathology, and the effects of parasite-induced host mortality (Smith, 1987; Quinnell *et al.* 1990). The regulation of the common liver fluke, *Fasciola hepatica,* provides a fairly comprehensive example (Fig. 8; Smith, 1984a, 1987). In sheep, juvenile flukes illicit a nonspecific fibrotic response in the liver parenchyma as they migrate to the bile ducts (Dargie *et al.,* 1974). Eventually, the fibrosis is sufficient to impede the passage of the most recently ingested flukes, so the number of flukes reaching their predilection site diminishes because of a density-dependent increase in fluke mortality. The density-dependent nature of this mechanism is revealed in single infection experiments in which the "percentage take" is inversely proportiona to the magnitude of the infecting dose (Fig. 8a; Boray, 1967; Sykes *et al.,* 1980). These migrating juvenile flukes can cause extensive liver damage. If sheep ingest large numbers of metacercariae, the resulting damage can lead to the death of the host. When an infected host dies, the parasite burden dies with it. Since the probability of host death increases with the intensity of infection, parasite mortalities from that cause are themselves density dependent. Thus, mean parasite burdens are reduced by the parasite-induced deaths of the most heavily infected hosts (Fig. 8b). However,

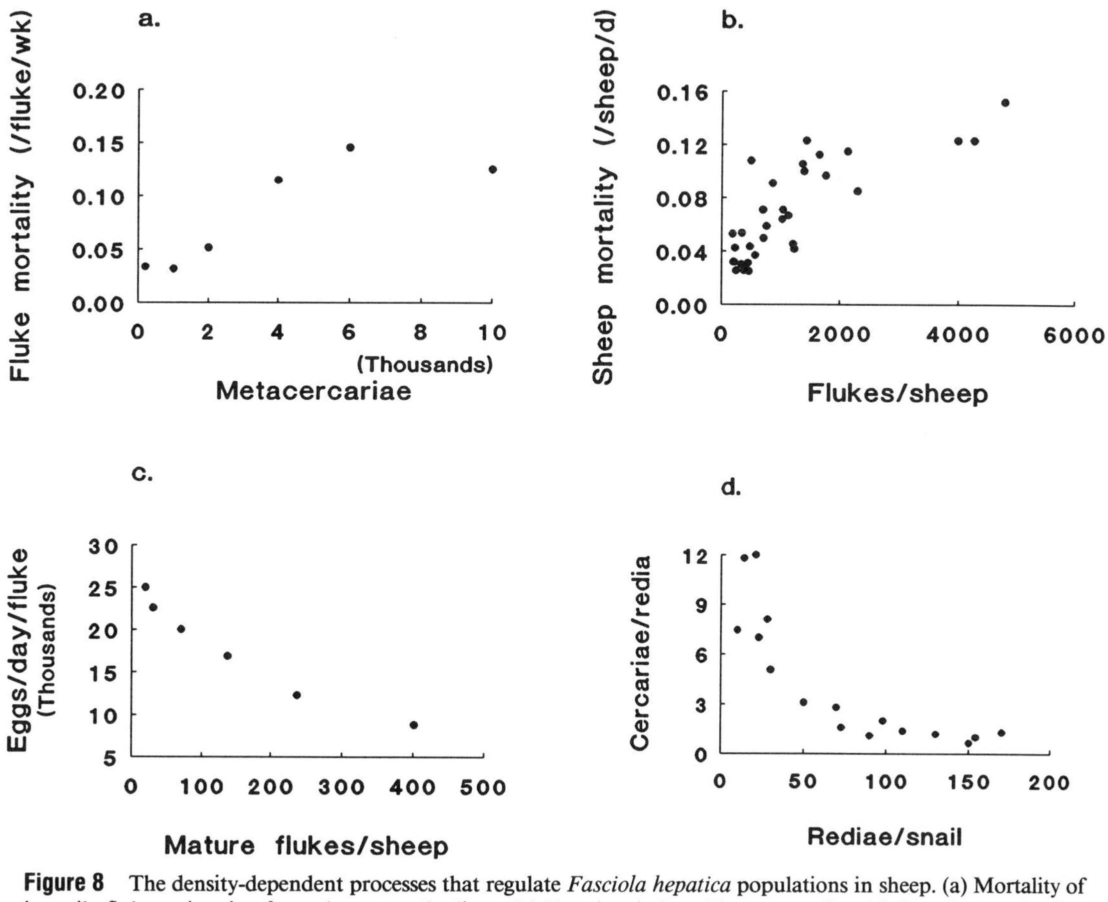

Figure 8 The density-dependent processes that regulate *Fasciola hepatica* populations in sheep. (a) Mortality of juvenile flukes migrating from the gut to the liver. (b) Parasite-induced host mortality. (c) Sexual reproduction (fecundity). (d) Asexual multiplication (cercarial production). Data from Boray (1967) and Smith (1984a,b).

not all density-dependent mechanisms operate via parasite mortality. Mature flukes can lay eggs at an astonishing rate (24,000/fluke/day, or 15–20/fluke/min), but this rate rapidly diminishes as the intensity of infection increases. Whether the effect on the fecundity of mature flukes seen in sheep is an immune-mediated mechanism or a consequence of competition between the flukes is not known (Fig. 8c). The mechanism of the rather similar constraint on asexual multiplication in the intermediate snail host is better understood (Fig. 8d). The number of mature cercariae found in any given redia is a decreasing function of redial density. Kendall (1949) and Kendall and Ollerenshaw (1963) attributed this phenomenon to intraredial competition for a finite food resource (the hepatopancreas). The one element missing from this example is an unequivocal role for an acquired immune response. Corba *et al.* (1971) were able to transfer resistance to infection with *F. hepatica* from one calf to its monozygotic twin with lymphoid cells but, in all other contexts, distinguishing between changes in establishment caused by physical barriers imposed by fibrotic livers and fluke mortality due to specific host immune responses has proved impossible.

It is relatively easy to devise empirical models for each of these processes as they occur in a single host (Smith, 1984a). However, modeling their impact on parasites summed over a population of hosts is not so easy, because parasites are not distributed uniformly among their hosts. The problem of incorporating parasite frequency distributions in mathematical models is discussed in the next section.

B. Significance of Parasite Frequency Distributions

Models of macroparasitic infections must account for parasite frequency distributions for two reasons. First, the morbidity and mortality associated with macroparasitic infections generally are supposed to be some function of the intensity of infection. Assuming this to be the case, determining the overall impact of parasitism on a population requires more information than the mean number of parasites per host. The morbidity and mortality experienced by a host population in which the parasites are shared more or less equally among hosts will be very different from the morbidity and mortality experienced by a host population in which most parasites are found in a small number of hosts. Second, the average values of density-dependent rates of parasite mortality and fecundity summed across all hosts are explicit functions of the way in which the parasites are distributed among the host population. The classical papers on the development of macroparasitic models are not especially helpful in their explanation of why this last point is true, but the interested reader can inspect the accounts in the appendices to May (1977) and Anderson and May (1978). The following explanation has been written with the novice in mind.

Our example concerns a macroparasite that is regulated by density-dependent changes in its fecundity. First, a model is formulated that describes events that occur when fecundity is unregulated. Imagine that the fecundity (λ) of a hermaphrodite trematode parasite is constant for all mature parasites. In a single host, the net rate of egg production is

$$\frac{dE}{dt} = \lambda i, \tag{16}$$

where i is the intensity of infection. Clearly, we could write a similar equation for every host in the population. Adding up all these equations would give the net rate of egg production summed over the entire host population, which could be represented as

$$\frac{dE}{dt} = \sum_{i=0}^{\infty} Hp(i)i\lambda, \tag{17}$$

where H is the total number of hosts in the population and $p(i)$ is the probability of any given host containing i mature parasites (i can be any number from zero to infinity).

The term $Hp(i)$ is, therefore, a way of calculating the total number of hosts that contain i mature parasites. Because fecundity is constant, we can simplify Eq. 17 by placing λ in front of the summation sign. If host population size is held constant, H also can be brought forward:

$$\frac{dE}{dt} = \lambda H \sum_{i+0}^{\infty} p(i)i. \tag{18}$$

By definition, the sum of all the respective values of $p(i)$ multiplied by their corresponding values of i is simply the mean value of i. Thus,

$$\sum_{i=0}^{\infty} p(i)i = m, \tag{19}$$

where m is the mean number of mature parasites per host (I/H). I represents the total number of mature parasites in all hosts, whereas i represents the number in an individual host. Equation 18 can be rewritten as:

$$\frac{dE}{dt} = \lambda Hm = \lambda I. \tag{20}$$

Equation 20 is a direct and simple analog of Eq. 16.

Next, the same process can be followed for a population of parasites in which fecundity is regulated, specifically in the case in which fecundity is a declining exponential function of the intensity of infection. A concrete example is given in Fig. 8c. In a single host, the net rate of egg production is

$$\frac{dE}{dt} = \lambda_0 e^{-\delta i} i. \tag{21}$$

Here, λ has been replaced by $\lambda_0 e^{-\delta i}$ in which λ_0 and δ are constants. By analogy with Eq. 20, one might anticipate that the net rate of egg production summed over all hosts will be simply

$$\frac{dE}{dt} = \lambda_0 e^{-\delta m} mH, \tag{22}$$

but this is wrong! To determine why this expression is incorrect, consider writing Eq. 21 for every host in the population and, proceeding as before, adding up all the equations. The result is

$$\frac{dE}{dt} = \sum_{i=0}^{\infty} Hp(i)i\lambda_0 e^{-\delta i}. \tag{23}$$

The constant terms can be placed in front of the summation sign:

$$\frac{dE}{dt} = \lambda_0 H \sum_{i=0}^{\infty} p(i)ie^{-\delta i} \tag{24}$$

but, unfortunately, no further progress can be made unless the values of $p(i)$, which depend on the parasite frequency distribution, are known. $p(i)$ could be eliminated when fecundity was constant because the summation term in Eq. 18 could be reinterpreted as the mean number of parasites per host. In Eq. 24, this simplification is not possible. We shall postpone a description of the way we address this problem, until we have considered the nature of parasite frequency distributions.

C. Negative Binomial Distribution

To describe parasite frequency distributions, investigators resort to theoretical distributions with known properties and characteristics that match those of the population under study. Parasite populations tend to be aggregated. Most parasites are found within only a relatively small proportion of susceptible hosts. Aggregated or clumped distributions such as this are described most conveniently using the negative binomial frequency distribution. This distribution is most convenient because it is defined completely by only two parameters, the mean of the distribution and the exponent k. For any given mean, the smaller the value of k, the greater the degree of aggregation (Elliot, 1977). Empirical observations suggest that, when the mean intensity of infection is small (<50), the value of k is typically less than 1 (Smith, 1984a; Guyatt *et al.*, 1990). However, for many macroparasitic infections the mean intensity of infection is measured in hundreds or even hundreds of thousands. In these cases, the value of k is frequently greater than 1 although the parasite population still clearly is aggregated (Smith, 1984a; Smith and Guerrero, 1993). Compare, for example, the estimated values of k for a trematode infection of wild ducks and a trichostrongylid nematode infection of sheep (Fig. 9). The difference in the value of k has a number of implications for the way in which models are written that are illustrated most easily in the explanation of the way in which models must take explicit account of parasite frequency distributions.

You will remember that Eq. 24 represents the net rate of change in egg production for a population of hermaphroditic trematode parasites summed over the entire population of hosts:

$$\frac{dE}{dt} = \lambda_0 H \sum_{i=0}^{\infty} p(i) i e^{-\delta i}. \quad (25)$$

We can simplify the right side of this equation by setting $e^{-\delta i} = z^i$ where $z = e^{-\delta}$.

$$\frac{dE}{dt} = \lambda_0 H \sum_{i=0}^{\infty} p(i) i z^i. \quad (26)$$

The next task is replacing $p(i)$ with more easily measurable parameters [$p(i)$ is the probability of hosts containing i parasites]. A probability generating function is required for this step. As its name implies, a probability generating function is a convenient means of calculating all possible values of $p(i)$. The probability generating function for the negative binomial distribution is $(q - pz)^{-k}$. When parasite distributions approximate the negative binomial, the values of $p(i)$ are given by the coefficients of z^i in the expansion of $(q - pz)^{-k}$ (May, 1977). Here, z is a dummy variable that can have any value. In this case, we have already chosen $z = e^{-\delta}$. For the other parameters, $q = (1 + p)$ and $p = m/k$, where m is the mean intensity of infection. By definition,

$$(q - pz)^{-k} = \left(1 + \frac{m}{k} - \frac{mz}{k}\right)^{-k}$$

$$= \sum_{i=0}^{\infty} p(i) z^i. \quad (27)$$

Note that the right side of Eq. 27 is very similar to the right side of Eq. 26. That similarity must be exploited to transform Eq. 27 so its right side is identical to the right side

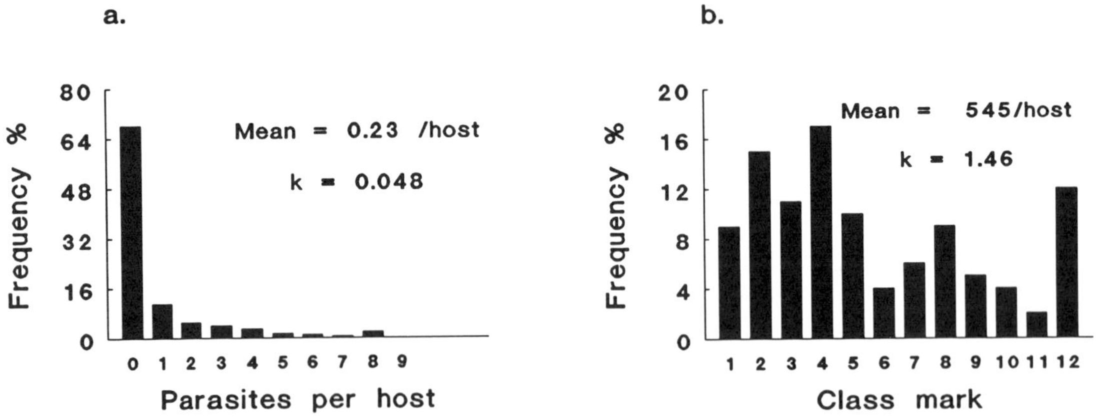

Figure 9 (a) Distribution of *Typhlocoelum cucumerinum* (Digenea) in ducks. (b) Distribution of *Teladorsagia circumcincta* (Nematoda) in grazing sheep. The class marks in b denote increments of 50 in the number of parasites per host. (k is the exponent term of the negative binomial distribution). Data from Scott *et al.* (1979) and Barger (1985).

of Eq. 26. Trial and error (and a little luck) show that this manipulation can be done by differentiating both sides of the equation and multiplying through by $zH\lambda_0$ (May, 1977; Smith, 1984a), yielding

$$zH\lambda_0 m\left(1+\frac{m}{k}-\frac{mz}{k}\right)^{-(k+1)} = H\lambda_0\sum_{i=0}^{\infty} p(i)iz^i. \quad (28)$$

The right side of Eq. 28 is identical to the right side of Eq. 26. Thus, the left side of Eq. 28 can be substituted for the right side of Eq. 26:

$$\frac{dE}{dt} = e^{-\delta}H\lambda_0 m\left(1+\frac{m}{k}-\frac{me^{-\delta}}{k}\right)^{-(k+1)}. \quad (29)$$

This expression represents the net rate of change in egg production summed over the entire population of hosts in terms of the mean intensity of infection (m), the number of hosts (H), and the exponent term of the negative binomial distribution (k). How to measure H and m seems obvious, but how is k determined? This exponent term can be calculated using the definition of the variance (v) of a negative binomial distribution (Elliot, 1977):

$$v = m + \frac{m^2}{k} \therefore k = \frac{m^2}{v-m}. \quad (30)$$

Much more efficient ways of estimating k exist (Bliss and Fisher, 1953; see also Chapter 4), but Eq. 30 illustrates one other point. When the variance is zero, all hosts contain exactly the same number of parasites. Equation 30 shows that this extremely unlikely condition arises when $k = -m$. Substituting $-m$ for k in Eq. 29 yields

$$\frac{dE}{dt} = \lambda_0 e^{-\delta m} mH, \quad (31)$$

which is identical to Eq. 22. Note now, that Eq. 22 is incorrect because this equation applies only when each host contains exactly the same number of parasites.

We have discussed in detail this method because it illustrates the complexities introduced by the incorporation of parasite frequency distributions in macroparasitic models (see Smith and Guerrero, 1993, for additional examples). Unfortunately, the functions linking parasite birth or death rates to the intensity of infection frequently are not amenable to the treatment just outlined. A best approximation is a model like the one represented by Eq. 22. To what extent does ignoring parasite frequency distributions compromise model performance? This issue is addressed in the next section.

D. When Parasite Frequency Distributions Can Be Ignored

Remember that the exponent k of the negative binomial distribution is an inverse measure of the degree of aggregation. If we have a model that incorporates the assumption of a negative binomial distribution, we can examine the impact of differing degrees of aggregation by varying the value of k over a plausible range of values. The simplest comparison (because it involves no integration) is the examination of the effect of varying k on the rate of change with respect to time in the number of individuals in some particular stage of the life cycle. For example, consider Eq. 29. The parameter values used were those obtained by Smith (1984a) for the common liver fluke, *F. hepatica*. The results are plotted in Fig. 10. When k takes values between 1 and 5, relatively little change occurs in the net rate of egg production. In other words, when the degree of aggregation is such that $k > 1$, the parasite frequency distribution is of relatively little relevance with respect to the net rate of egg production over the entire plausible range of mean parasite burdens. However, when $k < 1$, changes in its value lead to very large changes in the net rate of egg production. Furthermore when the parasite

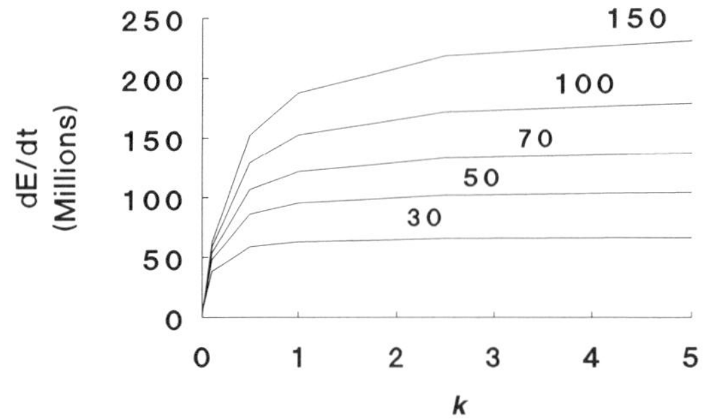

Figure 10 The effect on dE/dt of varying the value of k (the exponent term of the negative binomial distribution). Highly aggregated populations are described by low values of k. The results were obtained from Eq. 29. The figures placed above each line denote the mean intensity of infection (m).

population is highly aggregated ($k < 0.1$), the respective values of dE/dt converge. This example illustrates one of the principal effects of aggregated distributions: they enhance the ability of density-dependent processes to regulate the parasite population. In the example shown in Fig. 10, when $k < 0.1$, the net rate of change in egg production summed over the entire population is approximately the same, irrespective of the actual mean intensity of infection. Under these conditions, total daily egg output will change relatively little from one host population to another.

At this point, we should return to the question of whether model performance compromised by ignoring parasite frequency distributions? Clearly the answer depends on the actual prevailing value of k. Measured values of k for *F. hepatica* tend to increase with the prevalence and intensity of infection (Smith, 1984a) and values greater than 1 frequently occur. Under these circumstances, a model that ignores parasite frequency distributions may provide an adequate description of events.

The reader might like to use Eq. 22 and the values given in Fig. 9 to confirm that the net rate of egg output predicted by a model that takes no account of parasite distributions is very similar to a model that incorporates the assumption of aggregated distributions provided $k > 1$. Under all other circumstances, models of macroparasitic infections that ignore parasite frequency distributions are likely to be rather poor mimics of the system under study.

V. Conclusions

Demonstrating that parasite population density is regulated is by no means easy. Nevertheless, the preponderance of evidence suggests that regulation occurs. As every modeler knows, mimicking a real parasite–host system is very difficult unless one regulatory process (or more) is invoked. Indeed, one of the most convincing demonstrations of the biotic potential of most macroparasitic systems is a model that has no constraint on parasite population growth. One surprising consequence of the operation of regulatory processes may be the frequent appearance of oscillatory behavior in both micro- and macroparasitic systems. This phenomenon is of particular importance in understanding the so-called cyclical epidemic behavior of many viral infections. Finally, incorporating regulation into models of macroparasitic systems is rendered complicated by the requirement to accommodate the impact of different parasite frequency distributions.

References

Anderson, R. M., and May, R. M. (1978). Regulation and stability of host–parasite interactions. I. Regulatory processes. *J. Anim. Ecol.* **47,** 219–247.

Anderson, R., and May, R. (1982). The logic of vaccination. *New Scientist* **96,** 410–415.

Anderson, R. M., and May, R. M. (1991). "Infectious Diseases of Humans: Dynamics and Control." Oxford University Press, Oxford.

Anderson, R. M., and Michel, J. F. (1977). Density dependent survival in populations of *Ostertagia ostertagi*. *Int. J. Parasitol.* **7,** 321–329.

Barger, I. A. (1985). The statistical distribution of trichostrongylid nematodes in grazing lambs. *Int. J. Parasitol.* **15,** 645-649.

Bliss, C. I., and Fisher, R. A. (1953). Fitting the negative binomial distribution to biological data and a note on the efficient fitting of the negative binomial. *Biometrics* **9,** 176-200.

Boray, J. C. (1967). Studies on experimental infections with *Fasciola hepatica* with particular reference to acute fascioliasis in sheep. *Ann. Trop. Med. Parasitol.* **61,** 439-450.

Burnett, C. D. (1989). Bat rabies in Illinois: 1965-1986. *J. Wildlife Dis.* **25,** 10-19.

Carter, J. D., Hird, D. W., Farver, T. B., and Hjerpe, C. A. (1986). Salmonellosis in hospitalized horses: Seasonality and case fatality rates. *J. Am. Vet. Med. Assoc.* **188,** 163-167.

Corba, J., Armour, J., Roberts, R. J., and Urquhart, G. M. (1971). Transfer of immunity to *Fasciola hepatica* infection by lymphoid cells. *Res. Vet. Sci.* **12,** 292-295.

Dargie, J. D., Armour, J., Rushton, B., and Murray M. (1974). Immune mechanisms and hepatic fibrosis in fascioliasis. *In* "Parasitic Zoonoses. Clinical and Experimental Studies" (E. J. L. Soulsby, ed.) pp. 249-271. Academic Press, New York.

Davidson, W. R., Kellogg, F. E., Doster, G. L., and Moore C. T. (1991). Ecology of helminth parasitism in bobwhites from northern Florida. *J. Wildlife Dis.* **27,** 185-205.

Elliot, J. M. (1977). "Some Methods for the Statistical Analysis of Samples of Benthic Invertebrates," Scientific Publication No. 25, Freshwater Biological Association, 2d Ed. Windermere, Cumbria, UK.

Guyatt, H. L., Bundy, D. A. P., Medley, G. F., and Grenfell, B. T. (1990). The relationship between the frequency distribution of *Ascaris lumbricoides* and the prevalence and intensity of infection in human communities. *Parasitology* **101,** 139-143.

Hassell, M. P. (1982). Impact of infectious diseases on host populations-Group report. *In* "Population Biology of Infectious Diseases" (R. M. Anderson and R. M. May, eds.), pp. 15-35. Springer-Verlag, Berlin.

Kendall, S. B. (1949). Nutritional factors affecting the rate of development of *Fasciola hepatica* in *Limnaea truncatula. J. Helminthol.* **23,** 179-190.

Kendall, S. B., and Ollerenshaw, C. B. (1963). Effect of nutrition on the growth of *Fasciola hepatica. Proc. Nutr. Soc.* **22,** 41-46.

Keymer, A. (1982). Density-dependent mechanisms in the regulation of intestinal helminth populations. *Parasitology* **84,** 573-587.

Lack, D. L. (1967). "The Natural Regulation of Animal Numbers." Clarendon Press, Oxford.

May, R. M. (1977). Dynamical aspects of host-parasite associations: Crofton's model revisited. *Parasitology* **75,** 259-276.

McIlroy, S. G., Goodall, E. A., Stewart, D. A., Taylor, S. M., and McCracken, R. M. (1990a). A computerized system for the accurate forecasting of the annual prevalence of fascioliasis. *Prev. Vet. Med.* **9,** 27-35.

McIlroy, S. G., Goodall, E. A., Taylor, S. M., and McCracken, R. M. (1990b). Investigating the epidemiology of disease using a computerized system for slaughterhouse condemnation data. *Irish Vet. J.* **43,** 36-39.

Quinnell, R. M., Medley, G. F., and Keymer, A. E. (1990). The regulation of gastrointestinal helminth populations. *Phil. Trans. R. Soc. London B* **330,** 191-201.

Scott, M. E., McLaughlin, J. D., and Rau, M. E. (1979). *Typhlocoelum cucumerinum* (Digenea: Cyclocoelidae): Detailed analysis of distribution in wild ducks of southern Manitoba. *Can. J. Zool.* **57,** 2128-2135.

Smith, G. (1984a). Density-dependent mechanisms in the regulation of *Fasciola hepatica* populations in sheep. *Parasitology* **88,** 449-461.

Smith, G. (1984b). The relationship between the size of *Limnaea truncatula* naturally infected with *Fasciola hepatica* and the intensity and maturity of the redial infection. J. Helminthol. **58,** 123-127.

Smith, G. (1987). The regulation of trematode populations by density-dependent processes. *Int. J. Parasitol.* **17,** 541-548.

Smith, G., and Guerrero, J. (1993). Mathematical models for the population biology of *Ostertagia ostertagi* and the significance of aggregated parasite distributions. *Vet. Parasitol.* **46,** 243-258.

Sykes, A. R., Coop, R. L., and Rushton, B. (1980). Chronic subclinical fascioliasis in sheep: Effects on food intake, food utilization, and blood constituents. *Res. Vet. Sci.* **28,** 63-70.

Tizard, I. (1987). "Veterinary Immunology : An Introduction," 3d Ed. Saunders, Philadelphia.

Zarnke, R. L., and Ballard, W. B. (1987). Serologic survey for selected microbial pathogens of wolves in Alaska, 1975-1982. *J. Wildlife Dis.* **23,** 77-85.

Gary Smith

6

Ecological Epidemiology and Standard Measures of Disease Occurrence

I. Introduction

Norman Levine suggested that "epidemiology is nothing more than ecology with a medical and mathematical flavor" (Smith, 1991), which is certainly the impression given by the chapters in this book. Nevertheless, many individuals would disagree, and some would deny that any part of this book constitutes an epidemiological study. Part of the reason, I think, is the ascendance of some extremely powerful tools for testing hypotheses concerning associations between disease occurrence and putative causal factors. The increasingly sophisticated use of case–control studies, for example, has been described as the most outstanding development of modern epidemiology (Rothman, 1986). Such methodologies consume growing amounts of space in standard epidemiological textbooks. Indeed, one excellent textbook contains the statement," The specific objectives of an epidemiological study should always involve obtaining estimates of one of the measures of disease occurrence or effect" (Rothman, 1986), in which "effect" refers to the difference in disease incidence in two populations that differ only with respect to their exposure to some putative causal factor. This view seems too narrow and certainly does not follow automatically from the widely accepted opinion that epidemiology is the study of disease occurrence.

Fortunately, not all authors adopt so restrictive a view, and many subdivide epidemiology into various subdisciplines. First is descriptive epidemiology, which encompasses all the activities of the local or national surveillance agencies that monitor disease activity. Second is analytical epidemiology, which uses increasingly powerful statistical tools to draw inferences about causal associations between putative risk factors and disease incidence. Third is outbreak investigation and fourth is ecological epidemiology, which focuses on understanding the factors and processes that affect the transmission and maintenance of disease agents. Some authors refer to this discipline as mathematical ecology (Anderson and May, 1991), but since mathematics is an integral and significant feature of all epidemiology, the phrase does not describe the distinctive features of ecological epidemiology. Inevitably, the different branches of epidemiology have developed their own terminology, which is unfortunate because such developments foster the impression that the various subdisciplines have little in common.

Parasitic and Infectious Diseases

Ecological epidemiology is the principal focus of this book. One of the principal tools of the ecological epidemiologist is the mathematical model. The purpose of this chapter is illuminating the relationships that exist between standard epidemiological measures such as incidence, prevalence, and attack rate (cumulative incidence) and the mathematical models that recur throughout these chapters. The explanations require no more mathematics than already have been set out in the previous chapters. The interested reader is referred to the more rigorous treatment given by Preston (1987), Freeman and Hutchinson (1980), Bailey (1975), and Anderson and May (1991).

II. Incidence and the Force of Infection

Incidence and prevalence are the two principal measures of disease occurrence. Of the two, incidence is the more difficult to define. A clear explanation requires addressing first a rather unrealistic example. Imagine a cohort of uninfected individuals. A cohort is a group of individuals (human or nonhuman) whose membership is defined by the occurrence of a particular event during some period of time. For example, the defining event might be the diagnosis or contraction of a particular disease (Preston, 1987); also it might be birth. As a particular instance, consider a group of animals all born on the same day. These animals contract a foodborne infection from contaminated feed. Since the contaminated feed is present in only one of several troughs and since the animals feed at random from the troughs, the probability of infection is approximately constant over time. Figure 1 shows the fate of an imaginary cohort of initially uninfected animals. The temporal change in the number of animals at risk of infection (X) is modeled easily:

$$\frac{dX}{dt} = -iX, \tag{1}$$

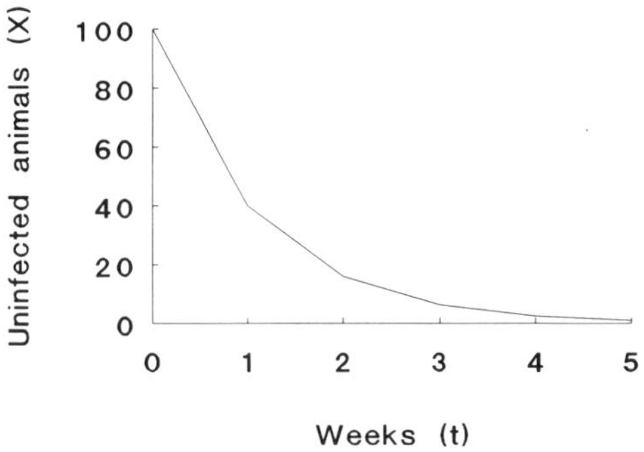

Figure 1 Change in the number of susceptible hosts during a foodborne disease episode (incidence, $i = 0.92$/wk).

where i is the instantaneous *per capita* rate of change with respect to time in the number of animals at risk of infection. As we have already seen, differential equations such as Eq. 1 have a standard analytical solution:

$$X_t = X_0 e^{-it}, \tag{2}$$

where X_0 is the number of uninfected animals at the start of the observation period. Remember, the exponential term e^{-it} is the proportion of animals not yet infected. The average time to infection (τ) can be calculated by integrating this exponential term:

$$\tau = \int_0^\infty e^{-it} = i^{-1}. \tag{3}$$

The answer is simply the reciprocal of i, a fact that will be useful later. The discussion of incidence can now proceed. Incidence (I) is the most important of all the standard epidemiological measures of disease occurrence, and frequently is defined as

$$I = \frac{\text{number of cases in time } t}{\text{sum for each individual of the time to disease onset}}.$$

We can see the meaning of this more clearly if we consider the example in Table 1. This table follows the fate of a subset of

Table 1 Fate of Several Susceptible Hosts Represented in Figure 1

Animal	Time to disease onset (months)
A	1
B	1
C	2
D	3
E	5
Total	12 animal months
Average time to disease onset	12/5 months
Incidence	5/12 months^{-1}

the cohort represented in Fig. 1, beginning with five uninfected animals. Two of the animals become infected at the end of the first month of observation, one at the end of the second month, one at the end of the third, and one at the end of the fifth. The sum of the time to infection for each individual is $1 + 1 + 2 + 3 + 5 = 12$. The number of cases is 5 because the study used five animals and all became infected by the end of the observation period. Thus,

$$I = \frac{5}{12} \tag{4}$$

or

$$\frac{1}{I} = \frac{12}{5}, \tag{5}$$

which states that the reciprocal of the incidence (I) is equal to the average time to infection. Look again at Eq. 3. This equation shows that the average time to infection in the model of the foodborne infection is simply the reciprocal of the parameter i. Thus, we can conclude that incidence (I) and the parameter i are identical. Incidence, which is sometimes called the "force of infection," is the instantaneous *per capita* rate of change with respect to time in the number of animals at risk of infection, as demonstrated by rearranging Eq. 1 and dividing both sides by X:

$$i = -\frac{dX}{dtX}. \tag{6}$$

III. Incidence in Models of Epidemic and Endemic Disease

In the example given earlier, incidence was constant. This simplification was useful but the reader should not assume that incidence is always constant. In contagious infections, for example, incidence is related to the density of infectious individuals, which may or may not be constant. Consider a simple epidemic. A single infected animal introduces the infection into a population consisting of a large number of susceptible individuals. As the infection spreads, we can divide the population into the susceptible (X), infectious (Y), and recovered/immune (Z) animals (Fig. 2). The infection is of short duration compared with the life-span of the host and causes no fatalities, that is, the total population size ($N = X + Y + Z$) remains constant. Following the conventions outlined in Chapter 2, this system can

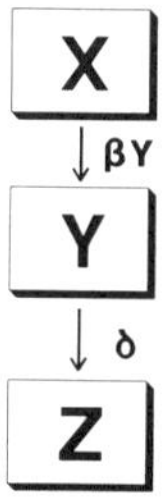

Figure 2 Flow diagram for a simple epidemic disease model X, Susceptible hosts; Y, infected/infectious hosts: Z, recovered/immune hosts; βY, force of infection; δ rate of recovery.

be represented as

$$\frac{dX}{dt} = -\beta YX \quad (7)$$

$$\frac{dY}{dt} = \beta YX - \delta Y \quad (8)$$

$$\frac{dZ}{dt} = \delta Y. \quad (9)$$

These equations have no analytical solutions. The solution shown in Fig. 3 was obtained with the help of a computer algorithm for numerical integration. The incidence (or "force of infection") is obtained by rearranging Eq. 7:

$$i = -\frac{dX}{dtX} = \beta Y. \quad (10)$$

Figure 3 shows that incidence tracks the number of infectious individuals, which follows the typical epidemic pattern. In the example given, incidence and the number of infectious individuals eventually decline to zero because the supply of susceptible individuals is never replenished, by birth or by loss of immunity.

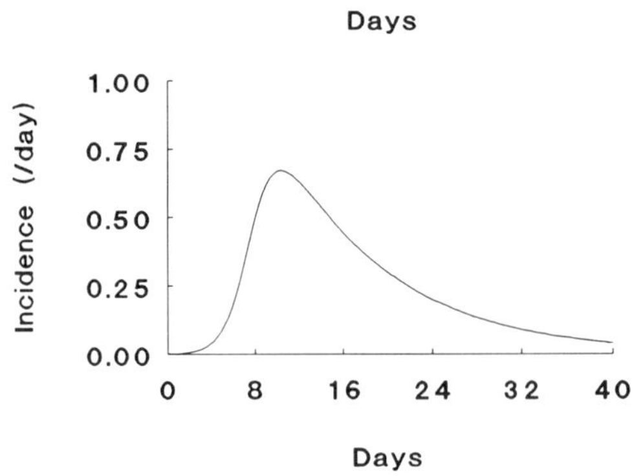

Figure 3 Numerical solutions of Eqs. 7 and 8 (*top*) compared with the calculated changes in incidence (βY) (*bottom*). Susceptible hosts, ——; infected hosts, - - -.

In cases in which the density of susceptible individuals is maintained at a sufficiently high level, the infection may not die out as it does in the simple epidemic. Instead, an endemic steady state is achieved. A simple but unrealistic model can be used to illustrate this point. Imagine an infection that has afflicted a closed population of size N. The duration of the infection is very much less than the life-span of the host and individual hosts can become infected many times because infection does not confer immunity (Fig. 4). The equations describing this system are

$$\frac{dX}{dt} = -\beta YX + \delta Y \quad (11)$$

$$\frac{dY}{dt} = \beta YX - \delta Y. \quad (12)$$

If population size ($N = X + Y$) and recovery rate (δ) remain constant, the population eventually will reach a steady state with respect to the infection (Fig. 5). When a steady state prevails, no change in the number of susceptible or infectious individuals occurs. Such a population is sometimes called stationary, but this description is misleading because movement between the two classes of host still occurs. Individuals are being recruited to the infectious class all the time, and an exactly equal number of individuals

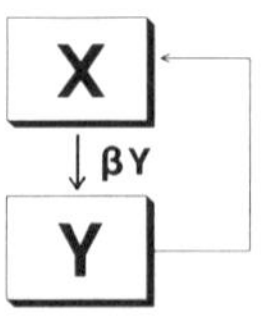

Figure 4 Flow diagram for a simple endemic disease model. X, Susceptible hosts; Y, infected/infectious hosts; βY, force of infection; δ, rate of recovery.

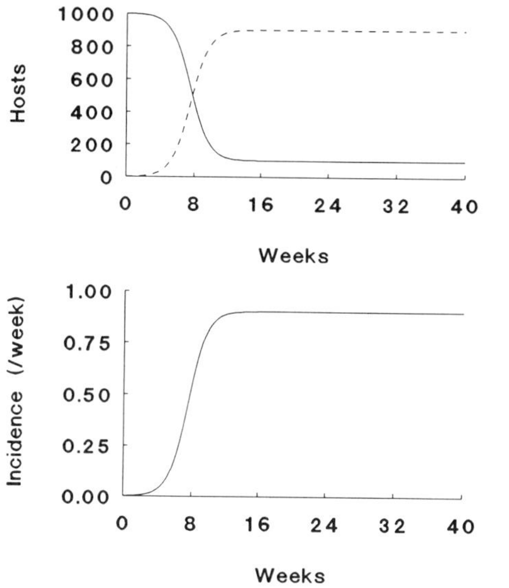

Figure 5 Numerical solutions of Eqs. 11 and 12 (*top*) compared with the calculated changes in incidence (βY) (*bottom*). Susceptible hosts, ——; infected hosts, - - -.

is recovering from infection. Thus, the net rate of change is zero:

$$\frac{dX}{dt} = 0 = -iX^* + \delta Y^*$$

$$\frac{dY}{dt} = 0 = iX^* - \delta Y^*. \qquad (13)$$

Here, the "force of infection", βY, has been replaced by i, the parameter representing incidence. In the period before the steady state is established, X and Y are changing all the time. When the system finally settles at a steady state X and Y are constant (X^*, Y^*). Thus, incidence (i) is also constant:

$$iX^* = \delta Y^*$$

$$i = \delta \frac{Y^*}{X^*}. \qquad (14)$$

By analogy with incidence (i), which is the reciprocal of the average time to disease onset, we can see that δ is the reciprocal of the average time (D) to recovery. Hence,

$$i = \frac{Y^*}{DX^*}, \qquad (15)$$

where Y^* is the number of infectious individuals (cases) when a steady state prevails, X^* is the number of susceptible individuals (at risk), and D is the duration of infectiousness. Does the relationship shown in Eq. 15 have any application outside the "simple but unrealistic model" from which it was derived? Freeman and Hutchinson's (1980) more rigorous derivation of the same expression shows that it does, provided only that incidence, population size, and the distribution of recovery times remain constant and that this constancy lasts at least as long as the longest recovery time included in the model.

IV. Prevalence

Prevalence (P) is defined straightforwardly as the proportion of the population that has the disease at one specific point in time. In the simple cohort model for a food borne infection, the possibility that individuals might recover from the infection was not considered. If they do not, and the infection becomes chronic, then prevalence depends only on incidence and the elapsed time since the infection was first introduced. Equation 2 yields

$$P \frac{(X_0 - X_t)}{X_0} = 1 - e^{-it}. \qquad (16)$$

On the other hand, if infected individuals recover, prevalence also will depend on the duration of infection, as seen most clearly in the steady-state model. Equilibrium prevalence (P^*) is

$$P^* = \frac{Y^*}{(X^* + Y^*)}, \tag{17}$$

and Eq. 15 states that

$$Y^* = iDX^*. \tag{18}$$

Therefore,

$$P = \frac{iD}{(1 + iD)}. \tag{19}$$

The elapsed time (t) is not included in this equation because, in the steady state, P is constant.

V. Attack Rate

The attack rate, sometimes known as the cumulative incidence, is simply the proportion of susceptible individuals that become infected over the period of interest, and defined as

$$\text{Attack rate} = \frac{\text{number of individuals that become infected}}{\text{number susceptible at the beginning}}.$$

In the simple cohort model, the attack rate is the same as the prevalence, but only because the infection is chronic. If the infection is not chronic, the right side of Eq. 16 would define the attack rate (A) and not the prevalence:

$$A = (1 - e^{-it}). \tag{20}$$

The attack rate is particularly important in many cohort studies because it is used as an estimate of incidence (given the assumption that incidence is both small and constant). This convention is odd because, despite its name, the attack rate is a proportion not a rate. However, suppose we were to measure the attack rate at the end of only one time unit (for example, 1 mo.). Then

$$A = (1 - e^{-1}). \tag{21}$$

Since

$$i \simeq (1 - e^{-i}), \tag{22}$$

when i is less than 0.1, A can be used as an approximate estimate of i. The reader should remember that attack rate and incidence are not synonymous terms, it is merely that they are numerically very similar when incidence is small. Unfortunately, this use of an approximate measure for incidence has led to much confusion. For example, in the parasitological literature, seeing incidence defined as the number of new cases of infection in a stated period of time is not unusual. Although similar in some ways to the attack rate, this parasitological measure does not serve as an estimate of true incidence in the sense used here.

A. Attack Rate in a Simple Epidemic

The attack rate for a simple epidemic is more difficult to determine. As noted earlier, Eqs. 7–9 have no analytical solution. However, we can use a mathematical trick to transform one of these equations into a form that can be solved: dividing Eq. 7 by Eq. 9. This manipulation eliminates Y from Eq. 7 and results in an equation that defines the rate of change in X *with respect to* Z:

$$\frac{dX}{dZ} = \frac{\beta X}{\delta}. \tag{23}$$

Since β and δ are both constants, Eq. 23 is identical in form to Eq. 1. As shown already, this kind of equation has a standard solution:

$$X_t = X_0 e^{-\frac{\beta}{\delta} Z_t}. \tag{24}$$

The attack rate (A) refers to the fraction of the original susceptible population ($X_0 = N$) that became infected. Thus, the final values

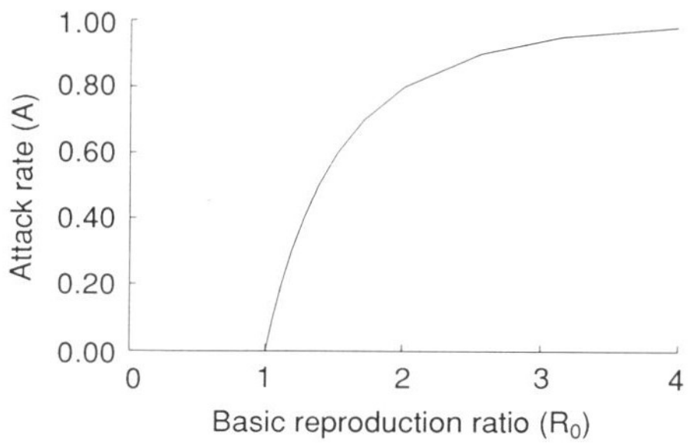

Figure 6 Relationship between the attack rate in a simple epidemic disease model and the basic reproduction ratio. Values from Bailey (1975).

of X and Z (X_∞ and Z_∞, respectively) can be defined in terms of N and A:

$$X_\infty = N(1 - A)$$

$$Z_\infty = NA. \qquad (25)$$

Substituting X_∞ for X_t, Z_∞ for Z_t, and N for X_0 in Eq. 24 yields

$$N(1 - A) = Ne^{-\frac{\beta N}{\delta}A}. \qquad (26)$$

This expression can be simplified by canceling the parameter N on each side and rearranging:

$$A = 1 - e^{-\frac{\beta N}{\delta}A}. \qquad (27)$$

The equation can be simplified more, remembering that $\beta N/\delta$ is the basic reproduction ratio (R_0) of the infection (see Chapter 3):

$$A = 1 - e^{-R_0 A}. \qquad (28)$$

Thus, the attack rate for the simple epidemic model depends only on the basic reproduction ratio, R_0. If R_0 is known, the attack rate can be calculated from Eq. 28 by iterative (trial and error) means. If R_0 is greater than 3, the attack rate is 95% or more. A table of attack rates for a simple epidemics is given by Bailey (1975; who called them "intensities"); Anderson and May (1991) provide a diagram of attack rates for values of R_0 between 1 and 10. Interestingly, Eq. 28 suggests that some individuals may survive an epidemic uninfected, even without prior immunity. Some examples of the relationship between R_0 and attack rates are given in Fig. 6.

VI. Summary

Each of the commonly used measures of disease occurrence can be represented within the usual mathematical framework of ecological epidemiology. Indeed, estimating incidence (the "force of infection") is a problem that preoccupies workers in most epidemiological subdisciplines. The derivations described here also set out the assumptions that are implicit in the cohort or case–control studies described in standard epidemiological texts, but which are documented only cursorily or infrequently for students beginning a course in analytical epidemiology.

References

Anderson, R. M., and May, R. M. (1991). "Infectious Diseases of Humans: Dynamics and Control." Oxford University Press, Oxford.

Bailey, N. T. J. (1975). "The Mathematical Theory of Infectious Diseases and Its Applications," Griffin, London.

Freeman, J., and Hutchinson, G. B. (1980). Prevalence, incidence, and duration. *Am. J. Epidemiol.* **112,** 707–720.

Preston, S. H. (1987). Relations among standard epidemiologic measures in a population. *Am. J. Epidemiol.* **126,** 336–345.

Rothman, K. J. (1986). "Modern Epidemiology," Little Brown, Boston.

Smith, R. D. (1991). "Veterinary Clinical Epidemiology: A Problem-Oriented Approach," Butterworth-Heinemann, Stoneham, Massachusetts.

Joan L. Aron & Barry A. Silverman

Models and Public Health Applications

7

I. Introduction

A model creates an analog of a system for the purposes of analysis, evaluation, and prediction. A model is not intended to replicate all components of a real system, but is designed to extract key features needed to answer specific questions. Models often are criticized for omitting complex interactions, but simply replicating the real system serves no purpose. A more complex model is not necessarily better (Aron, 1989). The disparity between the simplicity of models and the complexity of the real world is best summarized in the old aphorism that all models are wrong, but some are useful. Thus, the most important step in model construction is the examination of objectives.

II. Objectives of Using Models in Parasitic Disease Control

In the context of public health problems, infectious diseases lend themselves particularly well to the systems modeling approach. Although the basic interaction is always the transmission from an infectious host to a susceptible host, the consequence of this interaction on disease in the host population is highly variable. Experimental approaches in the laboratory and in the field are extremely important, but cannot capture the full range of host–parasite interactions. Thus, models are valuable in studying the population dynamics of infectious diseases.

Generally, the objectives of infectious disease models are twofold: (1) scientific understanding of transmission and (2) decision making for the management of disease control operations. Successful application of models of public health problems requires that both objectives be well developed. Disease control management should be built on a solid understanding of the basic biological and social aspects of the disease.

Models designed for scientific understanding examine the host–parasite interaction with emphasis on the distribution and determinants of infection. Major issues addressed by these models include the effect of population heterogeneity on transmission and the dependence of transmission on population density (Dietz, 1988; Anderson and May, 1991). The populations studied may include not only human hosts, but also parasites, vectors, reservoirs, and intermediate animal hosts. Scientific models then can be used to analyze the different possible interventions in the cycle of disease transmission and to compare their potential effectiveness on reducing disease. For example, program managers often must compare the relative effectiveness of environmental modification such as elimination of vector breeding sites with that of other interventions such as drugs, vaccines, personal protection, and health education.

The application of models to management of disease control problems, however, requires more than an analysis of models. The results of models must be relevant to the type of information received by a manager.

III. Models and Information Systems

Models have become important in developing the scientific basis for information systems and in helping define the interrelationships of the parameters of the disease control paradigm supported by these systems. Scientific understanding of transmission and decision making for the management of disease control operations converge in health and management information systems. For parasitic diseases, a Health Information System (HIS) contains information about disease epidemiology, entomology, ecology, weather, demography, service statistics, and other indicators of population health status. Management Information Systems (MIS) are used to analyze financial, economic, accounting, human resources, property, inventory, and purchasing data. Microcomputers have made the integration of these categories of information more accessible to health managers and have led to the possibility of management decision making based on data collected, processed, and analyzed in "near real time." Computer-based information systems have the ability to monitor health indicators and program resources rapidly, and to evaluate the effectiveness of disease control operations.

The example of malaria illustrates the changes occurring at the interface between information systems, models, and disease control operations. As the eradication strategy was replaced by one of disease control, the goal shifted from one of reducing transmission and prevalence to one of reducing mortality and morbidity. Malaria information systems, originally developed for eradication programs, only recently have begun to be altered to reflect this strategy change (Silverman, 1990).

Malaria control information systems must be sensitive not only to changes in the number and severity of disease cases but also to changes in relevant predictive or causal indicators. An extensive range of variables influences the malaria problem and any potential interventions for its control. The World Health Organization (WHO) Expert Committee on Malaria has suggested a classification of indicators that includes the following:

1. Indicators for defining the problem and selecting the control approach. These indicators might include broad estimates or qualitative information of a geographical, ecological, anthropological, social, and economic nature to complement specific demographic, health, and malaria information.
2. Indicators for the evaluation of progress in the control of malaria. These indicators might include information on general health status of the population or on measured progress in antimalarial operations. These indicators also might include information on specific changes in the malaria situation. Possible indicators are malariometric, entomological, social, and economic indicators; indicators of special technical problems; and indicators of changes in the malaria status of particular human subpopulations.

Members of the WHO Division of Control of Tropical Diseases and representatives of the National Academy of Sciences Institute of Medicine (IOM) Committee on Malaria Prevention and Control proposed a model for simplifying the use of the indicators recommended by the Expert Committee (IOM, 1991). This model, first discussed in 1989, classifies the disease into a limited number of major types or "paradigms." The stated purpose for this model is to "enable program managers and nonspecialists to better understand the variability of malaria" and to provide a method of matching specific situations with resources and desired outcomes of intervention. The model is supposed to allow for stratification; the efficacious allocation of human, financial, and material resources; and targeting of services with the goal of preventing epidemics rather than merely responding to their occurrence. Elements of the "paradigm" model include types of malaria as defined by ecotype (e.g., forest or urban malaria); level of endemicity; parasite species; population characteristics; social, behavioral, and economic characteristics; level of health infrastructure; type of development activities (e.g., dam construction); and interventions. The IOM Committee Report indicated that the paradigm approach is in an early stage of development and requires field testing to develop further the algorithms to be used in the model.

Models that incorporate these indicators and that examine the interdependence of the control of parasitic disease and socioeconomic development will be useful in the design and selection of effective disease control/development intervention strategies. Parker (1983) proposed a "working framework" for such a program selection–resource allocation model that is based on the integration of disease control efforts with socioeconomic development. This model was an attempt to identify and organize the basic components of this integrated approach to disease control. A multiobjective planning and resource allocation model was formulated to help identify the most effective program-based disease/development intervention strategy. Economic models also have been used to examine the economic impact of parasitic disease, to justify program selection and resource allocation (Shepard *et al.*, 1991).

New information systems and methods of analysis will strengthen decision making based on indicators of local conditions. Small-area analysis is a geographical analysis of health data that attempts to detect clustering of disease cases in small discrete areas. The combination of small-area analysis and the use of geographic information systems (GIS) potentially can serve as an early warning system of untoward health effects (Elliot, 1988). GIS are computerized mapping systems for the capture, storage, management, analysis, and display of spatial and descriptive data. GIS technology has been used in the study of vector-borne livestock diseases in Africa including trypanosomiasis and theileriosis, of malaria in Mexico, and of dracunculiasis in Africa.

The integration of modern information systems and models will create new opportunities for the use of models as teaching tools. As more information becomes readily available in real time, the use of models for teaching will overlap more and more with the use of models in disease control management, strengthening both activities. The training of managers will be improved because this training will be connected more closely to actual operations. The more intensive use of computers continues the evolution of the use of models in training, which has its academic origins at the interface between mathematics and biology (National Science Foundation, 1991). Textbooks for teaching models have provided the basic underpinnings for mathematical biology in general (Edelstein-Keshet, 1988) and in the specific context of infectious disease control (Anderson and May, 1991). However, as computers have become more widely available, computer-based models have been used to teach concepts in epidemiology and disease control. For example, the model developed for the Garki Project has been used extensively for training (Molineaux and Gramiccia, 1980). Interactive video technologies are beginning to influence the design of computer-based models. For example, the Schistosomiasis Unit in the WHO Division of Control of Tropical Diseases has plans to develop management training software using these technologies (S. Yoon, personal communication).

The only danger in these technological advances is that the technology of information systems becomes an end in itself. To counter this effect, it is instructive to harken back to the application of successful models.

IV. Case Studies of Successful Models

Successful disease control programs depend not only on an understanding of the biology of transmission but also on the integration of disease control into the larger social, economic, political, and technological context. Two case studies—the malaria control efforts of the Tennessee Valley Authority (TVA) in the United States and the Onchocerciasis Control Programme (OCP) in West Africa—have been selected to illustrate these features. Although the selection of these two cases can be criticized because they commanded levels of financial support that are unusual for disease control operations, important lessons can be learned from these situations that have broader applicability. These disease control programs depend on scientific models of disease transmission, exhibit superb technical coordination, and contribute to an overall scheme for economic development. In these projects, the successful use of models is characterized by an interplay between science and the larger context for decision making.

A. Tennessee Valley Authority

The TVA was created as a regional development agency in 1933 to generate economic growth in a poorly developed region of the country (Kitron, 1987). The TVA had several objectives: (1) development of a reliable navigable waterway; (2) reservoir storage and flood control; (3) power generation; (4) reforestation; and (5) agricultural development. The TVA program recognized from the beginning that endemic malaria posed an obstacle to the achievement of these objectives. Water impoundments from its projects could create excellent breeding habitats for the vector species *Anopheles quadrimaculatus*. The problem facing the TVA was determining how to incorporate malaria control into the very design of the water projects. The solution of this problem was found using models focused on environmental measures to eliminate mosquito breeding.

The TVA model for malaria control has a solid scientific foundation. All TVA projects began with a survey of epidemiological, biological, and engineering aspects of malaria distribution, vector production, and control. The TVA used experience, biological studies, and water requirements to establish a model that links water levels with the availability of habitat for the breeding of *A. quadrimaculatus*. On the basis of the model, water levels were manipulated for the purpose of eliminating the vector breeding sites. These environmental management measures are still being used in a comprehensive water level management schedule for vector control and malaria prevention.

The water management scheme of the TVA consists of four distinct phases. During the first phase, which occurs in early spring, the impoundments are refilled; additional water up to 30 cm above the full pool level is included to retard plant growth. No mosquito breeding occurs during this phase. The second phase is characterized by a constant pool level. This phase, which lasts until the beginning of the mosquito breeding season, effectively controls plant growth and inhibits semiaquatic vegetation at the shoreline. The third stage begins with the onset of the breeding season and involves the weekly removal of ~30 cm of water and refilling to the original pool level on the weekends. The fourth phase begins a few weeks after the initiation of the third stage. Increases in water demand and reduced rainfall cause a constant drop in the pool water level called "recession." Weekly fluctuations combined with seasonal recessions destroy mosquito eggs and larvae, reduce mosquito breeding areas, and help produce vegetation-free shorelines. The TVA has been able to use this water management model to eliminate malaria from its catchment area at a cost that is insignificant compared with the cost of alternative methods such as the use of larvicide.

The water control projects of the TVA are among the most successful examples of the use of a model for the control of malaria. In recent decades, the use of DDT and other insecticides for spraying has directed attention away from environmental management. However, interest has been renewed in incorporating an impact statement for malaria into agricultural development projects (Gwadz, 1991). Of course, simply replicating the past is not possible. The challenge is adapting the water management concepts of the TVA to the variety of ecological settings, socioeconomic constraints, and control options of today.

B. Onchocerciasis Control Programme

The OCP in West Africa is especially instructive in examining the role of models in disease control because, like the TVA, this program has been a centralized operation with excellent documentation of its control activities and supporting research. The creation of the OCP in 1974 was a result of the high priority given to the control of this disease by the United States Agency for International Development, the World Bank, and the United Nations Development Programme. Consider that a study by the World Bank estimated (in U.S. dollars) that the annual *per capita* cost of preventing onchocerciasis is $0.45 compared with a budget of $1.00 for all other diseases (Le Berre *et al.*, 1990). Before OCP, onchocerciasis control had received modest scientific attention in the form of vector control studies by the French and drug trials by the British. OCP built directly on the French approach of larviciding streams and rivers, but applied this technique on a much grander scale. Only a large program had a chance of interrupting transmission.

With this structure of vector control in place, one of the most important remaining questions was how long the program would need to be maintained to virtually rid the population of parasites. In this context, the application of epidemiological models was relevant. The following discussion focuses on the contribution of a modeling activity (led by statistician J. Remme) that was integrated into the activities of OCP. Since OCP was very successful at interrupting (or at least seriously reducing) transmission, the dynamics of transmission were not the key element in the model.

The most important feature of the epidemiological model for the reduction of parasites in the population is the longevity of the female parasites living in the human hosts. The starting estimate of 20 yr was based empirically on the experience gained during the eradication of onchocerciasis in Kenya. Remme *et al.* (1990) examined this longevity by considering the interval between inoculation of the parasite into the human host (by biting blackflies) and the disappearance of microfilariae, the immature stage of the parasite responsible for transmission. Based on OCP data, these researchers estimated this interval to be, on average, 10.4 yr, with 95% of the infections having a duration of less than 15 yr. The pattern of variability in the duration of life of the parasite is as critical as the mean duration of life, primarily because of the way this variability determines the maximum life-span. [For example, Dietz (1982) used a model with a shorter mean duration of life of 8.3 yr for the worms that, nevertheless, had a greater potential for a longer life-span, so 5% of the parasites lived longer than 25 yr.]

Now that the OCP is approaching the end

of the original plan for spraying, the risk and dynamics of recrudescence assume greater importance. The OCP model has been extended to include the complete cycle of transmission (Plaisier *et al.*, 1991). The model provides guidelines on how quickly infection would build up to a level of public health importance; the recrudescence time increases exponentially as the duration of the transmission control approaches 14 yr. Note that an estimate of the probability of recrudescence by itself is neither alarming nor reassuring, but provides a parameter for the political discussions about the fate of onchocerciasis control.

The introduction of ivermectin as a safe and effective drug to reduce microfilarial infection is an important new tool in onchocerciasis control. Merck Pharmaceuticals is donating the drug for use by affected countries in a program managed by the Mectizan Committee at Global 2000 at the Carter Center in Atlanta, Georgia. By reducing microfilarial infection, transmission can be reduced (Taylor and Greene, 1989). Now that ivermectin is available, its effect on transmission becomes a critical component for evaluation. Model estimates currently suggest that ivermectin cannot substitute for vector control, but may be helpful in controlling foci of infection (Plaisier *et al.*, 1991).

The OCP has employed models as an integral component of their planning and field operations. As the responsibility for disease control is assumed by its member countries, the challenge will be to adapt these models and information systems to the local conditions. (For further discussion, see Chapter 13.)

V. Limitations of Models

We have seen the successful application of models to the control of malaria and onchocerciasis, but it is important not to view a model as a static accomplishment that will last for all time. Since the function of models is extracting key features from a system, and since the scope of disease problems is changing constantly, the development of models must be a dynamic process. In this section, we describe current problems posed by biological phenomena and the sociopolitical context that are not addressed adequately by available models.

The ability of organisms to develop resistance to drugs or insecticides has confounded hopes of disease eradication. Mathematical models of transmission based on populations that do not adapt have been applied with the expectation that the eradication program would act so swiftly and completely that populations would not have time to adapt. For both malaria and onchocerciasis, resistance has become a major operational issue. Considerable modeling of resistance has taken place, drawing from the literature of population genetics (Anderson and May, 1991). For example, Curtis and Otoo (1986) emphasized the role of linkage between two resistance genes, resulting in a selective build-up of doubly resistant insects in response to simultaneous exposure to two different pesticides. Resistance of *Plasmodium* to drugs also has received a similar, if somewhat limited, theoretical treatment (Curtis and Otoo, 1986). However, such models have yet to be incorporated into international efforts to control these diseases.

Models of resistance to chemicals used in disease control can be enhanced by drawing on the vast literature on agricultural pest control. The use of models in agriculture has gone beyond mere theory and has had an impact on agricultural practices. The study of the resistance of crop pests to chemical controls integrates basic biology, experimental agricultural science, and agricultural crop management (Gould, 1991). One important area of study has been an examination of the strategy of reducing the development of resistance by subjecting a pest to two pesticides. Although this approach is intuitively appealing because the pest must overcome more barriers before becoming resistant, the success of this approach depends on characteristics of both the pests and the pesticides. For example, success of the two-pesticide strategy requires that the inheritance of resistance to each pesticide is recessive, that the doses of pesticide are sufficient to kill pests even if they are resistant to one of the pesticides, and that some untreated pest habitats exist that provide refuges. The two-pesticide strategy also requires that the pesticides used be different enough biochemically that a single genetic change will not confer resistance to both pesticides.

Another factor that has limited eradication programs is re-introduction of the vector or the parasite populations. Since vector control in the OCP began in 1975, *Simulium sirbanum* and *Simulium damnosum s. str.* have re-invaded the rivers in the area that were treated with larvicide. Re-invasion has necessitated continuous treatment of river breeding sites with larvicide and has increased the potential for development of resistance. The management of *Simulium* resistance to a whole series of larvicides has become critical to the successful implementation of OCP strategy (WHO, 1985). The most recent model of transmission from the OCP project has considered recrudescence from the remaining reservoir of parasites, but has not yet incorporated re-invasion of infective flies or infective individuals (Plaisier *et al.*, 1991).

Also, epidemics can occur because of the interplay of spatial or temporal variation in transmission and the level of immunity in a host population. For example, in one region, disease may be present and the population may be partially or wholly immune whereas, in another region, the disease may be absent and the population may be susceptible. When susceptible people from the uninfected region move to a focus of disease transmission, the number of cases can increase rapidly. The current malaria problem in Brazil is caused in large part by the movement of nonimmune urban residents to former jungle areas cleared for agriculture or mining (Pan American Health Organization, 1991). Reduction in the rate of transmission in an area also can allow the build-up of susceptible individuals to set the stage for an epidemic. For malaria, the build-up is the result of an increase in the number of unexposed individuals as well as an increase in the number of exposed individuals who have lost some of their immunity. Models of the dynamics of the loss of immunity have been explored (Aron, 1988; Struchiner *et al.*, 1990), but no programmatic focus on these activities has occurred to date.

When eradication cannot be achieved, the issue becomes one of controlling a disease to an acceptable level. Successful modeling of an acceptable level of control must take into account social, political, and economic as well as biological parameters. Numerous examples exist in which social and cultural factors limit the success of disease control programs. In Southern Africa, domiciliary spraying of DDT against malaria vectors occurs annually in October and November, but the insides of the homes often are repainted in December in preparation for the Christmas holiday return of the men who work elsewhere in the mines. In other settings, where the insecticide is still biologically effective against the mosquito vector, spraying may be unpopular because it is no longer effective against other nuisance pests such as bedbugs and roaches. Ecological priorities for protecting the natural environment may outweigh disease control arguments when considering the use of insecticide spraying or source reduction for vector control.

VI. Conclusion

This chapter reviewed the objectives of using models in disease control and presented two case studies of the successful use of models in the control of parasitic diseases. The case studies provide a background against which to examine limitations on current models. Making models more relevant to disease control in the future will require integration of models into computer-based management and health information systems. Models of biological processes must be understood as components of a larger social, economic, and political system. Disease control can be effective in isolation but its sustainability depends on the overall level of development.

Acknowledgments

Preparation of this chapter was sponsored in part by the Agency for International Development, Bureau of Africa under Cooperative Agreement No. AFR-0474A009041-00 (JLA) and in part by the Vector Biology and Control Project under Contract No. DPE-5948-C-00-5044-00 to Medical Service Corporation International, Arlington, Virginia, USA, for the Agency for International Development, Office of Health, Bureau of Research and Development (BAS). The opinions expressed here are those of the authors and do not necessarily reflect those of the Agency for International Development or its contractors.

References

Anderson, R. M., and May, R. M. (1991). "Infectious Diseases of Humans: Dynamics and Control." Oxford University Press, Oxford.

Aron, J. L. (1988). Acquired immunity dependent upon exposure in an SIRS epidemic model. *Math. Biosci.* **88,** 37–47.

Aron, J. L. (1989). Simple versus complex epidemiological models. *In* "Applied Mathematical Ecology" (S. A. Levin, T. G. Hallam, and L. J. Gross, eds.), pp. 176–192, Springer-Verlag, Berlin.

Curtis, C. F., and Otoo, L. N. (1986). A simple model of the build-up of resistance to mixtures of anti-malarial drugs. *Trans. R. Soc. Med. Hyg.* **80,** 889–892.

Dietz, K. (1982). The population dynamics of onchocerciasis. *In* "Population Dynamics of Infectious Diseases: Theory and Applications" (R. M. Anderson, ed.), pp. 209–241. Chapman and Hall, London.

Dietz, K. (1988). Mathematical models for transmission and control of malaria. *In* "Malaria: Principles and Practice of Malariology" (W. H. Wernsdorfer and I. McGregor, eds.), Vol. 2, pp. 1091–1133. Churchill Livingstone, Edinburgh.

Edelstein-Keshet, L. (1988). "Mathematical Models in Biology." Random House, New York.

Elliot, P. (1988). The small area health statistic unit. *In* "Methodology of Enquiries into Disease Clustering" (P. Elliot, ed.), pp. 53–56. London School of Hygiene and Tropical Medicine, London.

Gould, F. (1991). The evolutionary potential of crop pests. *Am. Scientist* 79, 496–507.

Gwadz, R. W. (1991). Malaria and development in Africa: A new strategy for research, training and control. *In* "Malaria and Development in Africa," pp. 99–103. American Association for the Advancement of Science, Washington, D.C.

Institute of Medicine (1991). "Malaria: Obstacles and Opportunities." National Academy Press, Washington, D.C.

Kitron, U. (1987). Malaria, agriculture, and development: Lessons from past campaigns. *Int. J. Health Services* **17,** 295–326.

Le Berre, R., Walsh, J. F., Philippon, B., Poudiougo, P., Hendrickx, J. E. E., Guillet, P., Seketeli, A., Quillevere, K, Grunewald, J., and Cheke, R. A. (1990). The WHO Onchocerciasis Control Programme: Retrospect and prospects. *Phil. Trans. R. Soc. London B* **328,** 721–730.

Molineaux, L., and Gramiccia, G. (1980). "The Garki Project: Research on Epidemiology and Control of Malaria in the Sudan Savanna of West Africa." World Health Organization, Geneva.

National Science Foundation (1991). "Workshop on Mathematical Biology: The Interface—Challenges and Opportunities." National Science Foundation Press, Washington, D.C.

Pan American Health Organization (1991). "Status of Malaria Programs in the Americas: XXXIX Report." Pan American Health Organization, Washington, D.C.

Parker, B. R. (1983). A program selection/resource allocation model for control of malaria and related parasite diseases. *Computers Operations Res.* **10,** 375–389.

Plaisier, A. P., van Oortmarssen, G. J., Remme, J., Alley, E. S., and Habbema, J. D. F. (1991). The risk and dynamics of onchocerciasis recrudescence after cessation of vector control. *Bull. WHO* **69,** 169–178.

Remme, J., De Sole, G., and van Oortmarssen, G. J. (1990). The predicted and observed decline in onchocerciasis infection during 14 years of successful control of *Simulium* spp. in West Africa. *Bull. WHO* **68,** 331–339.

Shepard, D. S., Ettling, M. B., Brinkmann, U., and Sauerborn, R. (1991). The economic cost of malaria in Africa. *Trop. Med. Parasitol.* **42,** 199–203.

Silverman, B. A. (1990). Review of the data collection

system of the Swaziland Malaria Control Unit. Vector Biology and Control Project Report No. 142, Washington, D.C.

Struchiner, C. J., Halloran, M. E., Robbins, J. M., and Speilman, A. (1990). The behavior of common measures of association used to assess a vaccination programme under complex disease transmission patterns—A computer simulation study of malaria vaccines. *Int. J. Epidemiol.* **19,** 187–196.

Taylor, H. R., and Greene, B. M. (1989). The status of ivermectin in the treatment of human onchocerciasis. *Am. J. Trop. Med. Hyg.* **41,** 460–466.

World Health Organization (1985). Ten years of onchocerciasis control in West Africa: Review of the work of the Onchocerciasis Control Programme in the Volta River Basin Area from 1974 to 1984. OCP/GVA/85.1B. World Health Organization, Geneva.

D. Wakelin

8

Host Populations: Genetics and Immunity

I. Introduction

The host–parasite relationship represents the interaction of two genetic lineages, those of the parasite and those of the host. The relationship itself exerts a selection pressure on both lineages, because survival in this interaction is in the interests of both partners, unless parasite transmission requires host incapacity or death (as is the case, for example, for certain larval helminths; see Chapter 18). Although the parasite is exploitative, the ability of the host to eliminate infection completely may be gained only at the expense of diverting too high a proportion of resources into the immune response, and may have severe immunopathological consequences, reducing host fitness. Although host immunity is antagonistic to the parasite, reduction of immunogenicity to a level at which no host response occurs may result in an uncontrolled growth of parasite population, thus increasing density-dependent effects on parasite reproduction and threatening host survival. Parasite population growth also may become excessive if host responsiveness is reduced by too great a degree. If parasite immunogenicity is increased, the response of the host may threaten the survival of the parasite and, if accompanied by severe pathology, may threaten the survival of the host as well.

Evolution acts on the genetic heterogeneity within each lineage involved in the host–parasite relationship, selecting those genes which allow each partner to approach optimal fitness under the circumstances existing at any given time. Variability in the parasite allows the selection of genotypes that maintain maximum reproductive potential as host influences change; variability in the immune response allows the host to combat the more rapid evolution of diversity in parasite populations, and maintain its own fitness (see Chapter 23). Adaptation favoring one partner at the expense of the other will be followed by counteradaptation in the other partner—the adaptation–counteradaptation "arms race" (Barnard, 1984). Although situations will arise in which short-term gains in parasite transmission might favor sacrificing the host, a coevolved equilibrium also can be selected, because optimal fitness for the parasite necessarily involves continuing survival of the host and optimal fitness for the host may require less than complete elimination of the parasite. The oscillations of the adaptation–counteradaptation race may

Parasitic and Infectious Diseases

become progressively damped, creating a "balanced" host-parasite relationship. The existence of "balanced" host-parasite relationships, in which hosts and parasites coexist successfully, has sometimes, and erroneously, been interpreted as the outcome of movement of host and/or parasite toward a desirable evolutionary goal. Some researchers have seen balanced relationships as a form of victory for the parasite, modification of immunogenicity allowing long-term survival at the expense of the host (Dineen, 1963; Sprent, 1969). The converse interpretation, that balanced relationships reflect an accommodation in the immune responsiveness of the host, allowing a harmless symbiosis, is equally misleading (Davies, 1986). Both views overlook the fact that balanced relationships reflect evolutionary change in both partners, arising from selection acting on each genome in the context of a continuing, and continually changing, confrontation. The "goal" of these changes is not a balance as such, but optimal fitness in each partner, allowing optimal replication of each gene lineage (Dawkins, 1990).

At any one time, host and parasite populations will possess a range of alleles that have been selected in the past to provide each organism with optimal fitness and which provide the plasticity for future evolutionary change. Since individual species of hosts commonly are infected with several species of parasite, as well as with other pathogens, the genes regulating host immune responses will be those selected to provide protection against all the infections that pose threats to survival. As requirements of protection against one pathogen may conflict with those needed for defense against another, the alleles possessed by any particular individual therefore represent a compromise, selected to allow optimal survival in the face of multiple challenges. Alleles possessed by parasite populations similarly represent the compromise necessary to allow successful survival and reproduction in a range of host species, the degree of compromise presumably reflecting the degree of host specificity.

This pool of genetic variability in both host and parasite populations is reflected in individual variation in host response. In any population, some individuals are more, and others less, resistant to any given infection. There may be a continuous gradation in response phenotype, or there may be discontinuous variation in which the population splits into responders and non-responders. In populations exposed to infection under natural conditions, low-response phenotypes can be interpreted as the cost of maintaining host fitness in the face of multiple infection. In populations of humans, and of domestic animals, in which normal transmission patterns have been disrupted, the range of infections altered, and selection pressures drastically modified, low-response phenotypes can constitute a major factor in determining the clinical and economic consequences of infection. Here, the host-parasite relationship easily is seen as a battle, with outright victors and losers, but it should be remembered, however, that this situation represents an unnatural and recent distortion of a co-adapted interspecific association. Ironically, awareness of this distortion is leading to a greater interest in the genetic effects that influence susceptibility and resistance to parasitic infections. Much of this interest is goal oriented. Genetic manipulation of domestic animals (by selective breeding or by construction of transgenic animals) now is seen as a possible alternative to the use of antiparasitic agents for the control of infection. Greater understanding of the genetic determinants of parasitic disease in humans is seen as necessary for the design and use of rational prophylactic strategies. Both approaches are contributing enormously to the overall understanding of parasite immunogenetics.

The development of interest in parasite immunogenetics that has taken place in the last 10-15 yrs has brought with it a rapid growth of knowledge about the range of variability affecting host immune responses that exists in host and parasite populations. However, far more still is known about host variation in responsiveness than about variation in parasite immunogenicity; knowledge of both is necessary for a full understanding of the impact of variation on the epidemiology of parasitic infections. Although this chapter concentrates exclusively on the host, the information given must be interpreted as a necessarily incomplete view of the factors influencing host–parasite relationships.

II. Immunity to Infection: Mechanisms and Genetic Control

A. Mechanisms

The ability of parasites to establish, develop, mature, and survive in their hosts is determined by many different factors. The behavior, ecology, structure, and physiology of potential hosts all may impose constraints on successful parasitism, and can be seen as aspects of innate resistance to infection. In evolution, however, parasites have become adapted to these characteristics of their natural hosts, which no longer act as major constraints. Hosts, therefore, must rely on their protective responses if they are to influence parasite survival. In vertebrates, host-protective responses include components of both the adaptive and innate immune systems, whereas invertebrates rely essentially on their non-adaptive defenses. This chapter, for brevity, focuses on the mammalian adaptive immune responses and the associated components of the innate immune system.

The mammalian immune system can mount an impressive armory of responses against parasites (see Warren, 1993). Many of these responses may have no, or only marginal, activity against the target parasite; some may be detrimental to the host; probably only a few constitute the major effectors of resistance. To a certain extent, the potential effectiveness of a given component is determined by the size and characteristics of the parasite concerned. Small plasma membrane-bound protozoans, for example, constitute a target very different from large cuticle-covered nematodes. Equally, the range of effectors available for effective antiparasite activity is constrained by the niche occupied by the parasite. Intracellular parasites are inaccessible to components capable of interacting with extracellular organisms; intestinal species are protected from direct interactions with cytotoxic cells. In many host–parasite relationships, including several of major economic and medical significance, little is known of the components of the protective response that actually do, rather than may, have a significant function *in vivo.* Certainly , in cases for which some knowledge is available, host protection does not rely on one effector alone; this situation probably will be found to exist in all cases. A summary of important effector mechanisms is given in Table 1.

1. Antibody

A variety of defense mechanisms is mediated or enhanced by antibody recognition of host antigens. Combination of immunoglobulin with target antigens on the parasite surface facilitates direct interaction with macrophages and granulocytes, resulting in phagocytosis, encapsulation, or membrane damage through antibody-dependent cellular cytotoxicity (ADCC). Some antibody isotypes also can "arm" effector cells directly, providing them with the ability to recognize target antigens. Activation of complement by bound antibody can produce similar effects, and can lead to direct cytotoxicity after insertion of the lytic complex into cell membranes. In many cases, investigators believe that binding of antibody to molecules that play critical roles in the biology of the parasite (e.g., enzymes, receptors, sense organs, transport systems) may be sufficient to interfere with normal development or metabolism, and thus may reduce parasite survival.

Table 1 Adaptive Immune Responses to Parasitic Infections

Effector	Function	Mechanism	Target
Antibody	Binding to target molecule or cell receptor	ADCC[a]	Helminths
		Binding	Protozoa/helminths
		Lysis and C′[b]	Protozoa/cestodes
		Opsonization	Trypanosomes
T cells	Release of cytokines	Activation of macrophages	*Leishmania*/helminths
		Inflammation	Helminths/arthropods
		Interferon	Malaria/*Eimeria*
Macrophages	Intracellular digestion	Phagocytosis	Protozoa
	Release of enzymes	ADCC	Helminths
Eosinophils	Release of mediators	ADCC	Helminths
Mast cells and basophils	Release of mediators	Inflammation	Helminths/arthropods

[a] Antibody-dependent cellular cytotoxicity.
[b] Complement.

2. T Cells

Direct T-cell-mediated cytotoxicity rarely plays a role in antiparasite immunity. T cells do, however, play a significant role in other ways, largely through release of cytokines that orchestrate other components of the immune and inflammatory systems. Not only do T cells initiate and regulate antibody responses, they activate macrophages and other host cells through release of interferon γ (IFNγ), enabling them to kill intracellular organisms. Cytokines such as IFNγ also can kill parasites directly. In addition, T cells control the development, differentiation, and activity of cells concerned in inflammatory responses (basophils, eosinophils, mast cells), the consequences of which can exert powerful antiparasite activity.

3. Macrophages

Phagocytosis, an ancestral means of defense, remains an important component of the antiparasite response, particularly against protozoans. Macrophages are not able to attack larger parasites directly, unless the target is opsonized or the cell itself is armed by antibody. Under these conditions, multicellular parasites can be killed by ADCC. Many protozoans live within macrophages, avoiding the normal defenses of the cell in a variety of ways. Exposure to cytokines such as IFNγ can reconvert the parasitized macrophage into an effective killing agent.

4. Eosinophils

Although eosinophils can phagocytose, their major role in antiparasite activity is linked with the ability to release a variety of potent mediators. Some of these (e.g., major basic protein, peroxidase) can damage parasites directly and are major effectors in eosinophil-mediated ADCC reactions. Other components initiate inflammatory responses that may be detrimental to parasite survival.

5. Mast Cells and Basophils

Immediate-type hypersensitivity responses involving these amine-containing cells are almost invariable accompaniments of infection with helminth and arthropod parasites. The release of preformed mediators such as histamine or serotonin (5HT) after the combination of surface-bound antibody with antigen, and the *de novo* generation of other mediators, initiates powerful local inflammatory changes that may reduce parasite survival. Some of the mediators also may bring about direct damage themselves.

B. Genetic Control

All aspects of immune and inflammatory responses, from initiation to ultimate expression, are known to be subject to genetic control (Fig. 1). This control is largely polygenic, reflected in continuous quantitative variation, but in some cases single gene control is involved and variation is discontinuous, reflected in qualitative differences between individuals. The degree of polymorphism in the genes concerned is variable, but some genes—notably those in the major histocompatibility complex (MHC), those responsible for diversity in T and B cell receptors (V genes), and those in the complement system—are highly polymorphic.

The visible outcome of the activity of genes controlling immune function depends entirely on the point in development at which they operate and the nature of the process regulated. At one extreme are the major developmental genetic defects, which result in extreme immunodeficiencies such as the absence of T cells or B cells. The severity of these deficiencies in humans is such that the individual concerned is unlikely to survive without major medical intervention; animals with such deficiencies do not survive in the wild. At the other extreme are the small quantitative variations which reflect genetic diversity in heterogeneous populations, but normally have only a slight influence on immune function. More significant variations in immune response also exist which, although fully compatible with survival, can reduce markedly the ability to develop and express resistance against particular parasites.

Genes controlling the immune and inflammatory responses occur throughout the genome. In relatively few cases have the genes concerned been identified and mapped (Blackwell, 1988a). Immunogeneticists (somewhat irrationally) tend to consider two groups of genes in the context of the immune response: those linked to the MHC and the rest, the non-MHC or background genes. The fact that the MHC occupies 0.5 centimorgans (CM) on one chromosome and that the "background" constitutes the remainder of the genome does not invalidate this distinction, because MHC-linked genes control the immune response at a very fundamental level, that of antigen recognition (Schwartz, 1985,1986). Cell-surface molecules encoded by MHC-linked genes (the class I and class II MHC molecules) determine in an absolute way the capacity of T cells to recognize epitopes. If an epitope cannot be seen, because the "right" MHC molecule is not available, this

Figure 1 Points at which genetically determined influences can act on the immune and inflammatory mechanisms of resistance to parasites. 1, Stem cell deficiencies; 2, processing and presentation of antigen; 3, production of cytokines; 4, response to cytokines; 5, antibody production (isotype, specificity, level, affinity); 6, complement components.

epitope cannot initiate or be the target of an immune response. Conversely, if epitopes are recognized that normally are not seen, inappropriate T cell responses may be generated and immunopathological changes may result. Although highly polymorphic within populations, MHC molecules are easily identifiable; individuals can be typed for the molecules they possess. If susceptibility to infection results from a failure of T cells to recognize a critical epitope, or if immunopathology results from abnormal recognition, these conditions will show MHC linkage. In terms of infection generally, and parasitic infection in particular, molecules relevant to protective immunity can be expected to carry many epitopes. Equally, individual hosts can be expected to carry a variety of MHC molecules capable of presenting epitopes. Therefore MHC linkage of susceptibility to parasitic infection, or of immunopathological responses, might be expected to be rare. In fact, although not common, such linkages are well known, having been identified in humans (Table 2) and demonstrated in a number of experimental infections (Wassom and Kelly, 1990).

Variations in resistance, susceptibility, and immunopathology more often reflect the activities of background genes, but very few of the genes concerned have been identified and few markers are available to enable linkage studies to be performed. These genes regulate all immunological and inflammatory cell populations, determining their development and frequency and controlling their production of receptors and mediators. These critical functions may determine in an absolute sense whether or not a protective response can be made; more often, the expression of these genes is seen in quantitative variations. Identification of these genes, their gene products, and their contribution to the control of resistance or susceptibility is an area of study in which substantial progress can be expected, using the techniques of molecular genetics.

Whereas small differences in immune response to defined antigens can be detected under controlled laboratory conditions (for example, by measuring levels of antibody production or T cell proliferation), differences relevant to the outcome of parasitic infections can be detected only by relatively crude criteria such as host survival, level of infection, resistance to re-infection, or degree of pathology. These criteria represent the end points of a complex series of events, initiated by and targeted at multiple complex antigens. It might, therefore, be expected that significant genetically determined differences in resistance and susceptibility seldom would be detected. In practice, this is not the case; such differences have been identified wherever they have been sought. Several examples of major, qualitative differences in ability to resist infection are known (for reviews, see Rosenstreich *et al.*, 1982; Wakelin and Blackwell, 1988). Consideration of these facts leads to several important conclusions.

Table 2 Parasitic Infections in Humans with Linkage to Major Histocompatibility Complex Genes

Parasite	Disease
Protozoa	Giardiasis
	Malaria
Helminths	Schistosomiasis
	Ascariasis
	Filariasis
	Trichuriasis
Arthropods	Scabies

1. Of the multiple antigens and multiple epitopes presented by parasites to their hosts, only a small number are of critical importance to the host–parasite relationship.
2. Of the multiple immune and inflammatory responses elicited by infection, only certain responses exert a significant influence on the parasite concerned.

3. Variation in critical responses to the critical antigens can be expected in normal, genetically heterogeneous populations of hosts.
4. The degree of variation in these responses is sufficient to result in substantial individual variation in susceptibility to infection.
5. Variation in level of parasitism between individuals will influence the epidemiology of infections in host populations, and will be subject to evolutionary selection pressures.

The first four of these conclusions can be validated by data obtained from experimental studies in the laboratory and from extensions of these studies into the field. The fifth is the focus of more theoretical approaches.

III. Genetic Influences on Immunity to Infection

Evidence that significant genetically determined variations exist in levels of parasitic infection has been obtained from many field studies of human and animal populations (see Chapters 14, 15, 23), as well as from detailed experimental analysis in laboratory models (see Chapter 20). Although in field studies correlations can be made between levels of infection and particular parameters of the host immune and inflammatory responses, only in well-controlled laboratory studies can causal links be established. Experimental studies in the mouse, the best-defined laboratory host in terms of both immunology and genetics, have provided the most detailed information. A full review of such work is given by Wakelin and Blackwell (1988,1993). For the purposes of this chapter, some key points from a limited number of studies are summarized to indicate the range of genetically determined processes known to influence host responsiveness to infection.

A. Protozoa

Protozoan parasites, with other microparasites, have the capacity to replicate within the host. Therefore, levels of infection within individuals are determined primarily by the capacity of the host to control this replication. The response components necessary to achieve this control vary among systems, depending on the type of organism and its location within the host.

1. Malaria

Genetic influences operate at all stages of infection in malaria. A great deal of information is available about the effects of genetic polymorphisms in target red cells that prevent invasion or inhibit development, and thus determine levels of infection independently of immunity (Weatherall, 1987). Variations in immune responses also influence the development and severity of infection. In humans MHC (HLA) linkages have been described (Hill *et al.,* 1991), but little is known of the underlying mechanisms. Experiences with malaria vaccines employing recombinant antigens have shown the importance of genetic restriction on epitope recognition at the T-cell level, which would reflect MHC gene expression (Good *et al.,* 1988). This restriction could exercise an effect on the response to infection with specific strains of malaria, given the limited epitopes available in the initial infective stage, the sporozoite. At the population level, however, restriction of recognition is unlikely to have a significant influence on resistance to infection, because substantial epitope polymorphism exists within parasite populations.

Mouse-model work has shown clear-cut genetic influences in resistance to malaria (Fig. 2a). Both polygenic and single gene control have been demonstrated. The effector mechanisms involved remain unclear. Some evidence suggests that quantitative variation in antibody production (titer) may determine levels of resistance, and that

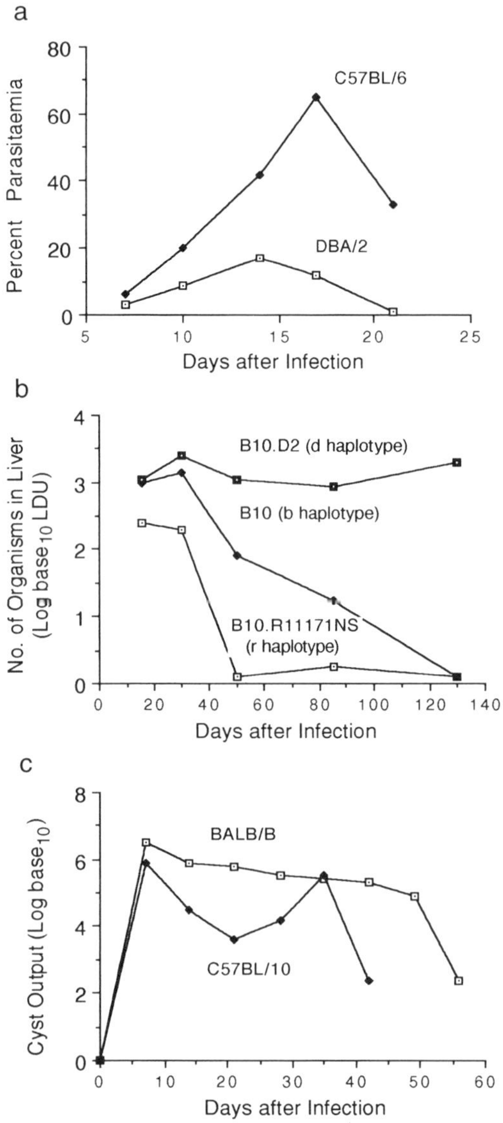

Figure 2 Examples of genetic influences on the course of infection with protozoan parasites in strains of mice. (a) *Plasmodium yoelii* (nonlethal strain 17X) in DBA/2 (H-2d) and C57BL/6 (H-2b), showing primary influence of non-MHC genes on level of parasitemia. Data from Sayles and Wassom (1988). (b) *Leishmania donovani* in three H-2 congenic strains on the B10 genetic background, showing influence of MHC (H-2)-linked genes on recovery from infection. Data from Blackwell (1988b). (c) *Giardia muris* in MHC (H-2b)-compatible BALB/b and C57BL/10, showing primary influence of non-MHC genes on duration and level of infection. Data from Venkatesan *et al.* 1993.

qualitative (isotype, antigen recognition) variation also may be important (Taylor *et al.,* 1988). The focus on T-cell-mediated responses against liver stages (Schofield, 1989) suggests that variation in cytotoxic potential, or production of cytokines such as IFNγ, could play a significant role in determining overall levels of resistance.

2. Leishmania

The spectra of diseases observed in individuals infected with the species that cause human leishmaniasis are suggestive of profound variation in response to infection. Analysis of infections in the mouse has confirmed that genetically determined variation can generate a range of clinical conditions, providing a detailed picture of the genes that regulate responsiveness and of the mechanisms through which their effects are expressed (reviewed by Blackwell, 1988b). Both MHC-linked and background genes are involved, the latter controlling the suitability of host macrophages for parasite growth and replication. In resistant strains of mice, parasite invasion occurs but subsequent development proceeds slowly. In contrast, in susceptible strains, parasite replication is rapid and infection spreads quickly between cells. In the case of the visceral leishmaniasis caused by *Leishmania donovani,* this fundamental difference between strains is determined by a single gene *Lsh,* mapped to chromosome 1, which has dominant resistant (*r*) and recessive susceptibility (*s*) alleles. *Lsh* now is known to be the same as two other genes identified in relation to the control of resistance and susceptibility to infection with *Salmonella typhimurium (Ity)* and *Mycobacterium bovis (Bcg)* (Mock and Seldin, 1989). *Lsh* clearly controls a central property of macrophages that determines their suitability as an environment for intracellular pathogens. This property remains undefined, but its ability to influence the development of phylogenetically distinct pathogens clearly places a premium on selection for the resistant allele in target populations. Indeed, a survey of 253 wild mice from *Leishmania*-endemic and *Leishmania*-free countries failed to detect the *Lsh*[s] allele, suggesting that strong selection had occurred for resistance to *S. typhimurium* and/or *M. bovis* which, in contrast to *Leishmania,* are natural pathogens of the mouse (Blackwell and Plant, 1986). Curiously, the *Lsh* gene is without effect on the development of *Leishmania major,* responsible for the cutaneous form of leishmaniasis. Instead, the extent of parasite growth within macrophages in this species is controlled by a distinct gene, designated *Scl-1* and mapped to chromosome 8.

In the case of *L. donovani,* mice carrying *Lsh*[s] subsequently may express resistance to infection as a result of MHC-influenced, T-cell (IFNγ)-mediated activation of infected macrophages, allowing normal killing mechanisms to occur. The level of immune-mediated resistance is determined by the MHC (H-2) haplotype of the mice; some haplotypes (e.g., *r* and *b*) allow cure to occur whereas others (such as *d*) fail to induce the necessary macrophage responses (Fig. 2b). These genetic influences probably act at the level of antigen presentation, determining the subset of T helper cells stimulated and, thus, the cytokines released. T-cell-mediated responses also are involved in the control of *L. major;* for both species, clear MHC-linked differences occur in the expression of this immunity. Detailed studies with *L. major* have shown that these differences also operate at the level of the T helper cell subsets and are mediated through patterns of cytokine secretion (Locksley and Scott, 1991), some of which allow macrophage activation and thus provide immunity whereas others lead to suppression and continuing infection.

3. Intestinal Protozoa

Intestinal parasites are among the most common of all infections and therefore can exercise strong selection pressures on host populations. Clear genetically determined differences in response to infection have been defined for several intestinal species, including *Giardia* in humans and mice (Belosevic *et al.,* 1984) and *Eimeria* in chickens and mice (Wakelin and Rose, 1990; Fig. 2c). For *Giardia* infections, differences between resistant and susceptible mice are such that the latter show chronic prolonged cyst output, whereas resistant mice clear infection rapidly. This difference resembles the variation seen in infected humans; most individuals are asymptomatic or, if symptomatic, recover rapidly, but others have long-lasting chronic infection with intermittent cyst release. Individuals particularly at risk are those deficient in antibodies, especially IgA, although other factors also are involved. In *Eimeria* infections, genetic influences on resistance appear to operate primarily through the speed at which T-cell (IFNγ)-mediated responses are expressed (Wakelin *et al.,* 1993). Resistant and susceptible strains of mice show similar responses, but the latter express these too late to influence the outcome of infection. On challenge, both strains express effective immunity.

B. Helminths

The range of immune responses capable of influencing the outcome of infections with parasitic worms is necessarily different from that appropriate to protozoan infections. Not only are the targets larger and often protected by surface layers that are difficult to breach, but their positions in the body may circumscribe the mechanisms that are otherwise able to exert deleterious effects.

1. Schistosomes

Epidemiological studies in human populations infected with *Schistosoma mansoni* and *Schistosoma haematobium* have shown that considerable variation in levels of parasitism can occur between individuals who have experienced comparable exposures to infection (Butterworth *et al.,* 1989; Chapters 10, 15). These detailed re-infection studies provide some of the strongest evidence for the existence of an effective immunity in humans, and have provided some indications of the mechanisms through which genetically determined variation may be expressed (Abel *et al.*, 1991). A significant correlation has been made between resistance and the ability to mount antischistosome IgE responses (Fig. 3). That an inverse relationship exists between production of IgE and production of IgG_4 is well established. The relative levels of each are determined by the balance of active T cell subsets and their associated cytokine profiles. One possibility is, therefore, that susceptible individuals are programmed to produce the "wrong" T cell subset response. The "wrong" antibody responses can, in fact, block protective immune mechanisms and maintain chronic infection in the host. Studies with mouse models, although they confirm the operation of genetically determined variation in resistance, have yet to provide a clear-cut picture of the factors involved. Resistance to infection in mice is brought about largely as the result of immunopathological changes in the liver; genetic differences therefore are likely to reflect differences in pathological responsiveness, perhaps because of differences in T-cell responsiveness to the antigens associated with anti-egg granuloma formation (Dean *et al.*, 1981). Such an association has been made in humans infected with *Schistosoma japonicum* and is under the control of MHC-linked genes (Ohta *et al.*, 1990). MHC linkage also has been claimed for the responses leading to hepatosplenomegaly in schistosomiasis mansoni (Mahmoud, 1980).

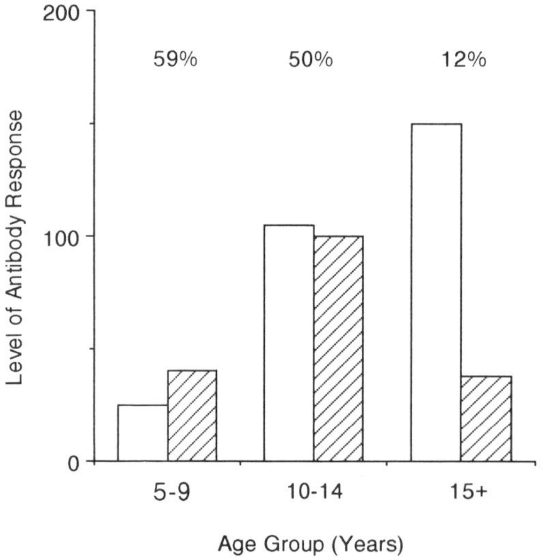

Figure 3 Levels of anti-schistosome IgE (open bars) and IgG_4 (hatched bars) antibodies in children exposed to *Schistosoma haematobium.* Level is expressed on an arbitrary scale based on measured optical density in an ELISA. The value over each column is the percentage of individuals who had become reinfected 15 mo after chemotherapy. Data from Hagan *et al.* (1991).

A less complicated approach to examining differences in resistance to schistosomes has been that of immunizing mice, using irradiated larvae, and then challenging with normal parasites. Irradiated vaccines elicit an immunity that is not dependent on pathology, but reflects conventional antiparasite immune responses. Use of this model has shown substantial differences in resistance between strains of mice, to which both MHC-linked and background genes contribute (Sher *et al.*, 1984). At present, no consensus exists on the responses that might be subject to variation in this situation, but as ADCC is likely to be an important effector, variation in antibody isotype and specificity and variation in effector cell populations are likely to be involved. Experimental evidence exists for variation in each of these components. Some authors consider that macrophages have an important antiparasite role in vaccinated mice. Variations affecting cell numbers or their state of activation [e.g., levels of the T cell cytokine IFNγ have been demonstrated in strains showing contrasting response phenotypes (James *et al.*, 1984)].

2. Tapeworms

It was shown many years ago that the establishment of larval *Taenia taeniaeformis* in strains of mice was highly variable. Some strains supported the growth and development of large viable populations of cysts in the liver, whereas others were almost totally refractory. The genetic control of this variation now is known to be expressed through background genes that regulate the speed at which antiparasite antibody is produced (Mitchell *et al.,* 1980). In rapid responders, complement-fixing antibodies become available before the larval worms develop an effective anticomplementary activity. The host is, therefore, able to destroy the parasites through a combination of complement-mediated lysis and ADCC. In slow responders, parasites have time to become resistant to host attack before the antibody response is developed fully. The association between speed of response and effective resistance holds for primary infections only. On challenge, both rapid and slow responders are capable of making an effective antibody response well before the parasites establish their defenses. Little is known about the genetic control of this variation in response, but it may be analogous to that controlling the antibody response phenotype in the Biozzi lines of high- and low-antibody responder mice (Biozzi *et al.,* 1972).

One of the first studies explicitly to relate observed genetic variation in resistance to the epidemiology of a parasitic infection (other than early papers concerned with veterinary parasites) was that of Wassom and colleagues (1973,1974), who examined the distribution of the adult tapeworm *Hymenolepis citelli* in deer mice *(Peromyscus maniculatus).* The distribution of the worm was highly aggregated (Fig. 4), less than 2% of the mice having mature tapeworms. Laboratory studies showed that aggregation reflected the inability of a few individuals to develop the effective resistance shown by the majority. Resistance was under the control of a single dominant allele. Although nothing is known of the mechanisms underlying resistance in this system, the demonstration of qualitative differences in resistance in a wild host population illustrates the strength of selection pressures that can operate on antiparasite responses and validates some of the conclusions drawn from laboratory-based studies (Wassom *et al.,* 1986).

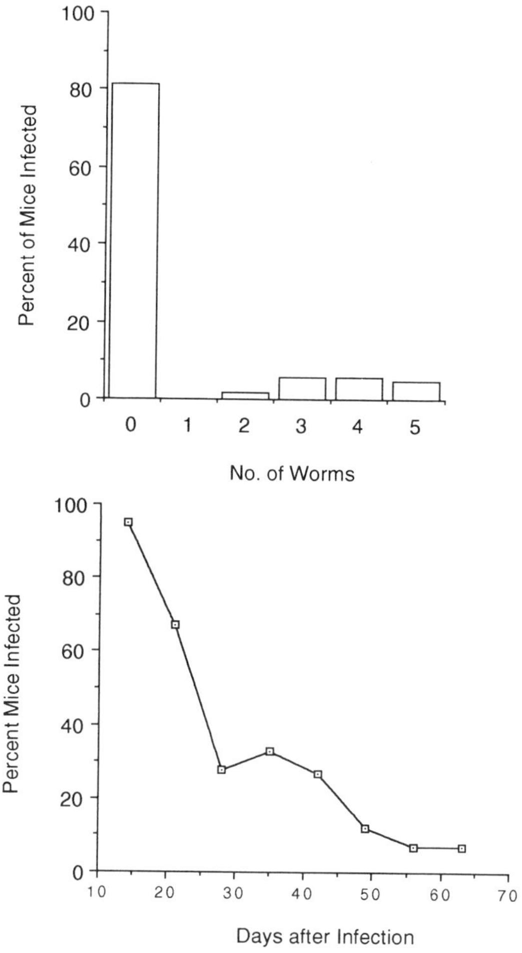

Figure 4 Distribution of mature *Hymenolepis citelli* in experimentally infected *Peromyscus maniculatus,* showing marked aggregation and time course of worm loss. Data from Wassom *et al.* (1986; *top*), (1973; *bottom*).

3. Gastrointestinal Nematodes

Infections with gastrointestinal nematodes are among the most common experienced by human and animal hosts (see Chapter 14). The ease of assessing levels of infection in field studies and of establishing experimental infections in the laboratory immunogenetic analysis. Data from field studies repeatedly have shown that infections are aggregated in the host population (Bundy, 1988; Chapter 14), and laboratory studies have confirmed that aggregated distributions can arise from variation in levels of resistance to infection (Fig. 5). Now a great deal is known from this experimental work about the mechanisms involved in immunity and of their genetic control.

Several gastrointestinal nematodes reach the intestine after parenteral migrations (e.g., *Ascaris,* hookworms, *Nippostrongylus, Strongyloides*). Antiworm responses can act against these migrating stages, particularly during reinfections. These responses involve direct antibody activity and various forms of ADCC, acting in the skin or in the lungs. Responses operative in the intestine are relevant to all these species, as well as species

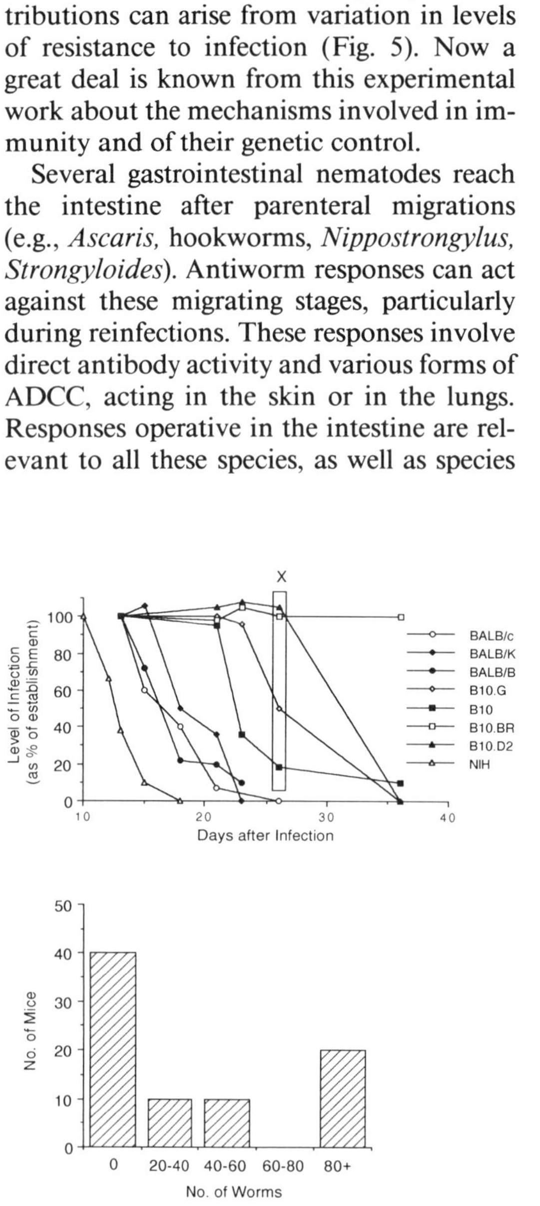

Figure 5 Time course of infections with *Trichuris muris* in 8 strains of inbred mice. Level of infection is expressed as percentage of worm establishment measured on day 10 (NIH) or 13 (other strains). The histogram shows the aggregated distribution of worms that would be obtained if 10 mice from each strain had been killed at time *X*. Data from Else and Wakelin (1988).

such as *Trichinella* and *Trichuris* which establish directly in the gut (Rothwell, 1989). In no case are the mechanisms of resistance fully understood. Although many components of the immune and inflammatory responses can act against the parasite, certain responses are likely to predominate and be the major effectors of immunity. However, other responses are likely to take over when the major components are inoperative. Antibodies may play an important role against a number of gastrointestinal nematodes. The binding of antibody to target molecules present on the cuticle and intestinal epithelium, or released from the worm, can interfere with a variety of essential activities, including behavior, nutrition, reproduction, and defense mechanisms. IgG isotypes appear to be important in this context; IgA also can mediate protective activity. Gastrointestinal nematodes elicit strong IgE responses, and this isotype long has been implicated in protective immunity through its involvement in immediate hypersensitivity reactions. T-cell-mediated inflammatory responses, in which mast cells and eosinophils predominate, also are thought to play a significant role, primarily in terms of changing the intestinal environment to the detriment of worm survival. Several changes downstream from these immunologically induced responses may be critical in this respect, including mucus secretion, increased fluid flux, mediator release, alteration in mucosal structure, and increased peristalsis (Miller, 1984). Evidence for genetically determined variation in these components of the protective response to gastrointestinal nematodes comes primarily from studies using laboratory systems (especially mice), but many detailed analyses on infections in domestic animals give comparable results (see Wakelin, 1988a).

The complexity of the responses with potential to influence these parasites suggests that clear-cut variation between individuals, strains, or breeds might be difficult to detect. In fact, variation appears to be the norm and can result in response phenotypes ranging from highly resistant, in which initial infections are short-lived and re-infection is virtually impossible, to highly susceptible, in which the host appears to have little effective immunity at any time. Much of this variation is quantitative and under polygenic control, but qualitative variation, implying control by a small number of genes, also has been described (e.g., Wakelin, 1975; Albers *et al.,* 1987). A clear pattern that emerges from immunogenetic studies with gastrointestinal nematodes is that, although background genes exercise the strongest influence on resistance and susceptibility, MHC-linked genes can exert a powerful moderating effect. The expression of these genetic influences can be through any of the components listed earlier. The kinetics of antibody production, the isotype produced, the specificity of recognition, and the ability to generate populations of inflammatory cells are among the most important variables. It is becoming apparent that selection and activation of T helper cell subsets, and the pattern of cytokines produced, can play a central role in determining the response phenotype.

4. Filarial Nematodes

As for the gastrointestinal nematodes, epidemiological data from field studies have shown extensive variation in human responsiveness to filarial infection (Piessens and Mackenzie, 1982). Even when immunological evidence confirms exposure to infected vectors, individuals may show no parasitological or pathological signs of infection. Other individuals in the same population may show heavy and chronic infections, whereas still others may suffer severe pathology. This variation can be explained most easily in terms of genetically determined variation in immune responsiveness, a view that is supported by laboratory studies in a variety of experimental hosts including cats, hamsters, rats, and mice (Philipp *et al.,* 1984). Infective larvae seem to be important targets of immune responses leading to resistance, whereas responses directed against adult and microfilarial stages may lead to pathology. Many studies have implicated variation in antibody production (kinetics, level, isotype, affinity, specificity) as an important determinant of variation in responsiveness, which may reflect genetic control of the pattern of cytokine release (King and Nutman, 1991). Clearly, given the probable involvement of ADCC mechanisms, variation in inflammatory cell populations also may be involved. MHC linkage has been identified in some of the human studies and confirmed in some laboratory work (Kwan-Lim and Maizels, 1990), but other studies have shown no MHC effect.

C. Arthropods

Individual and breed variation in resistance to ectoparasitic arthropods has been recognized empirically for many years in humans, domestic animals, and laboratory mice (reviewed by Wakelin, 1988b). Susceptible hosts can acquire very heavy infestations, whereas resistant hosts remain only lightly parasitized. Work with ticks has shown that immediate-type inflammatory responses at the site of feeding form a major component of host resistance. Studies in mice and guinea pigs suggest that genetically determined differences in ability to mount these responses, primarily through the production and localization of cells such as mast cells, basophils, and eosinophils and of IgE isotype antibody, may be important determinants of host response phenotype (Matsuda *et al.,* 1990). Genetic control of these parameters appears to be largely through the activity of background genes.

IV. Immunogenetics and Epidemiology

As shown in the preceding sections, evolutionary considerations predict that host populations should contain the genetic potential for a range of response phenotypes, from the highly resistant to the highly susceptible. Experimental studies confirm this hypothesis. When infected under controlled laboratory conditions, hosts express these phenotypes in terms of differential parasite burdens or differential parasite-induced pathology. Epidemiological studies repeatedly find a similar phenotypic range in natural host populations, whether human or domestic or wild animal. A crucial question is the extent to which such marked differences in levels and consequences of parasitism under field conditions are determined by genetic differences in immune response. Obviously many other factors may be involved (see Chapter 20). When transmission involves direct uptake of infective stages, behavioral and socioeconomic factors can influence exposure to infection. The distribution of infective stages, by virtue of the processes involved in their release and distribution, is unlikely to be uniform (Wong and Bundy, 1990). The fact that children are always the most heavily infected with gastrointestinal parasites (see Chapter 14) reflects the fact that their behavior results in greater contact with soilborne infective stages than is experienced by adults. Aggregated distributions of infection within children therefore may arise from unequal exposure to infection. Although data support this possibility (Wong *et al.,* 1991), how widely these data may be extrapolated, or whether similar factors influence aggregation of infection in adults, is not yet certain. Intuitively, the universal observation of aggregated distributions in human infections would seem to point more to factors affecting parasites after infection has taken place than to major behavioral influences on exposure. Support for this view can be taken from data on infections in domestic stock. The aggregated distribution of gastrointestinal infections in domestic animals maintained intensively under field conditions, where the density of infective stages in restricted areas should result in a more uniform transmission, can be considered good evidence for the operation of individually variable, postinfection influences on parasite burdens.

In those parasites transmitted by vectors, factors affecting exposure to the vector are important determinants of levels of infection. However, exposure to transmission, for example by an insect vector, may be relatively uniform within a given population, being determined less by the host than by vector behavior. Many studies, particularly with filarial nematodes, in which level of infection is related to the total number of infection events, have shown that the distribution of parasites in the population is not uniform even when, as discussed earlier (see also Ottesen, 1984), good immunological evidence exists for exposure to infection in the parasitologically negative individuals. In human schistosomiasis, where potential exposure to infection can be monitored accurately by measuring frequency of water contact, it has been shown unequivocally that variation in level of infection can be independent of degree of exposure (Tingley *et al.,* 1988).

Each of these observations can be supported by parallel studies in experimental models using protozoan, helminth, or arthropod parasites. Aggregated distributions of intestinal protozoa and nematodes are reproducible when infections are established in panels of inbred strains of mice, each with distinct response phenotypes, using standardized inocula (Figs. 4,5). Similar work, using groups of domestic animals, has provided equivalent data. Experimental studies, in which vector-borne parasites can be established directly by syringe passage or by implantation, reveal profound differences in individual or strain response, both immunological and pathological. Whenever examined in experimental animals, genetic variation in immune responsiveness to parasitic infection and variation in resistance or susceptibility has been identified (see Wakelin and Blackwell, 1988). Seroepidemiological surveys of human populations infected with parasites reveal extensive variations in level, isotype, and specificity of response (e.g., Haswell-Elkins *et al.,* 1989). Similar variation between individuals is apparent in the more limited studies carried out on T-cell recognition or on levels of inflammatory response. All these facts point irresistibly to the likelihood that genetic variability plays a key role in determining parasite distribution, and therefore parasite transmission, under field conditions—it is not simply an artifact of laboratory studies.

V. Conclusions

Epidemiologists are concerned with observing, describing, analyzing, and predicting the patterns of distribution and transmission of parasites in their host populations. Among the many insights required by these activities is an understanding of the factors that regulate the level of parasitism in individual members of the host species. Once infection has taken place, and parasites have become established, the immune response becomes one of the most important influences on the survival and reproductive potential of the parasite and, thus, an important determinant of distribution and transmission. The evidence that hosts are heterogeneous in their capacity to develop and express protective responses, that this variation is genetically determined and therefore subject to evolutionary change, has many important implications concerning epidemiology, both as a theoretical discipline and as a basis for the development of predictive models for practical parasite control.

The existence of genetic heterogeneity in response to infection represents both the result of, and the vehicle for, constant adjustment of host–parasite relationships to maintain optimal fitness. This heterogeneity does not represent a failure by the host population to come to terms with parasites. At this level, the epidemiology of a particular infection must be interpreted in the context of all the infections that exert selective influences on the host. In addition, the epidemiologist must understand the cost–benefit relationship of each interaction from the perspectives of both organisms concerned, and must see these relationships as dynamic, not static. The realization that relatively few genes exert biologically significant effects on host resistance makes it possible to see that evolutionary change in the balance of a relationship can occur as a result of small alterations in parasite or host activity. The patterns of parasite transmission, although apparently stable for periods of time, therefore have the potential to change dramatically.

At a more applied level, the realization that populations are heterogeneous with respect to infection levels, even in the face of uniform exposure, provides a predictive framework for approaches to prophylactic control, whether through environmental measures, chemotherapy, or vaccination (Anderson and May, 1985). It also alters perception of the factors responsible for immunopathological responses to infection as these, like resistance and susceptibility, are genetically determined.

Immunogenetic analysis is approaching the level at which aspects of the regulation of host–parasite relationships will be explicable in molecular terms. For example, recognition of parasite antigens, the decisive initial event, now can be interpreted in terms of the genes, gene products, and epitopes that determine antigen presentation to T cells. The consequences of molecular change in any of these components soon will be predictable on an informed basis. Progress in cellular immunology allows analysis of genetic variation of subsequent events in the immune and inflammatory responses to be undertaken at the level of cytokine–cytokine receptor interactions, as well as permitting assessment of availability of the precursor populations necessary to generate effectors of protective responses. Understanding all these events will clarify the balance between protective and pathological components of immune responses to infection, a balance that is crucial both in human terms, as a determinant of disease, and in theoretical terms, as a parameter of evolutionary selection.

A significant factor in the development of immunogenetic studies has been the application of the techniques and approaches of molecular genetics. These techniques are making possible rapid and precise genetic analysis of large populations, sampled under field conditions, allowing linkage studies to be undertaken more easily and more reliably. Genes responsible for specific effects on immune responses now can be identified, mapped, and manipulated by molecular methods. In the laboratory, short cuts to the development of congenic lines for examining the influence of particular alleles on particular genetic backgrounds are now available. Transgenic animals provide a direct route for examining gene function under precisely controlled conditions. Extrapolation of the data from such studies to humans becomes increasingly realistic as more and more of the human genome is mapped, and as more and more syntechnic relationships between animal and human genomes are identified. All these considerations lead to the view that parasite immunogenetics has a large and increasingly important role to play in all aspects of our growing understanding of the host-parasite relationship, not least in informing and shaping epidemiological interpretations of patterns of infection in human and animal populations.

Acknowledgments

I am grateful for the research support that has been provided by the British Medical Research Council and the Wellcome Trust for my own work in this field. I acknowledge also the invaluable contributions that have been made by my research collaborators at Nottingham and elsewhere.

References

Abel, L., Demenais, F., Prata, A., Souza, A. E., and Dessein, A. (1991). Evidence for the segregation of a major gene in human susceptibility/resistance to infection by *Schistosoma mansoni*. *Am. J. Hum. Genet.* **48**, 959-970.

Albers, G. A. A., Gray, G. D., Piper, L. R., Barker, J. S. F., Le Jambre, L. F., and Barger, I. A. (1987). The genetics of resistance and resilience to *Haemonchus contortus* infection in young Merino sheep. *Int. J. Parasitol.* **17**, 1355-1363.

Anderson, R. M., and May, R. M. (1985). Helminth infections of humans: Mathematical models, population dynamics, and control. *Adv. Parasitol.* **24**, 1-101.

Barnard, C. J. (1984). "Producers and Scroungers: Strategies of Exploitation and Parasitism." Chapman and Hall, London.

Belosevic, M., Faubert, G. M., Skamene, E., and MacLean, J. D. (1984). Susceptibility and resistance of inbred mice to *Giardia muris*. *Infect. Immun.* **44**, 282-286.

Biozzi, G., Stiffel, C., Mouton, D., Bouthillier, Y., and Decreusefond, C. (1972). Genetic regulation of the function of antibody-producing cells. *In* "Progress in Immunology" (B. Amos, ed.), pp. 529-545. Academic Press, New York.

Blackwell, J. M. (1988a). Conclusions. *In* "Genetics of Resistance to Bacterial and Parasitic Infection" (D. Wakelin and J. M. Blackwell, eds.), pp. 265-275. Taylor and Francis, London.

Blackwell, J. M. (1988b). Protozoan infections. *In* "Genetics of Resistance to Bacterial and Parasitic Infection" (D. Wakelin and J. M. Blackwell, eds.), pp. 103-151. Taylor and Francis, London.

Blackwell, J. M., and Plant, J. E. (1986). Expression of the natural resistance gene *(Lsh)* in wild mice infected experimentally with *Leishmania donovani* or *Salmonella typhimurium*. *Curr. Top. Microbiol. Immunol.* **127**, 323-330.

Bundy, D. A. P. (1988). Population ecology of intestinal helminth infections in human communities. *Phil. Trans. R. Soc. London B* **321**, 405-420.

Butterworth, A. E., Corbett, E. L., Dunne, D. W., Fulford, A. J. C., Kimani, G., Gachuhi, R. K., Klumpp, R., Mbugua, G., Ouma, J. H., arap Siongok, T. K., and Sturrock, R. F. (1989). Immunity and morbidity in human schistosomiasis. *In* "New Strategies in Parasitology" (K. P. W. J. McAdam ed.), pp. 193-210. Churchill Livingstone, Edinburgh.

Davies, A. J. S. (1986). Evolution and functions of the immune system. *In* "The Ruminant Immune System in Health and Disease" (W. I. Morrison, ed.), pp. 1-31. Cambridge University Press, Cambridge.

Dawkins, R. (1990). Parasites, desiderata lists, and the paradox of the organism. *Parasitology* **100**, S63-S73.

Dean, D. A., Bukowski, M. A., and Cheever, A. W. (1981). Relationship between acquired resistance, portal hypertension, and lung granulomas in ten strains of mice infected with *Schistosoma mansoni. Am. J Trop. Med. Hyg.* **30,** 806–814.

Dineen, J. K. (1963). Antigenic relationship between host and parasite. *Nature (London)* **197,** 471–472.

Else, K. J., and Wakelin, D. (1988). The effects of H-2 and non-H-2 genes on the expulsion of the nematode *Trichuris muris* from inbred and congenic mice. *Parasitology* **96,** 543–550.

Good, M. F., Kumar, S., and Miller, L. H. (1988). The real difficulties for malaria sporozoite vaccine development: Non-responsiveness and antigenic variation. *Immunol. Today* **9,** 351–355.

Hagan, P., Blumenthal, U. J., Dunn, D., Simpson, A. J. G., and Wilkins, H. A. (1991). Human IgE, IgG4 and resistance to reinfection with *Schistosoma haematobium. Nature (London)* **349,** 243–245.

Haswell-Elkins, M. R., Kennedy, M. W., Maizels, R. M., Elkins, D. B., and Anderson, R. M. (1989). The antibody recognition profiles of humans naturally infected with *Ascaris lumbricoides. Parasite Immunol.* **11,** 615–627.

Hill, A. V. S., Allsopp, C. E. M., Kwiatkowski, D., Anstey, N. M., Twumasi, P., Rowe, P. A., Bennett, S., Brewster, D., McMichael, A. J., and Greenwood, B. M. (1991). Common West African HLA antigens are associated with protection from severe malaria. *Nature (London)* **352,** 595–600.

James, S. L., Correa-Oliveira, R., and Leonard, E. J. (1984). Defective vaccine-induced immunity to *Schistosoma mansoni* in P strain mice. *J. Immunol.* **133,** 1587–1593.

King, C. L., and Nutman, T. B. (1991). Regulation of the immune response in lymphatic filariasis and onchocerciasis. *Immunoparasitol. Today* A54–A58.

Kwan-Lim, G. E., and Maizels, R. M. (1990). MHC and non-MHC restricted recognition of filarial surface antigens in mice transplanted with adult *Brugia malayi* parasites. *J. Immunol.* **145,** 1912–1920.

Locksley, R. M., and Scott, P. (1991). Helper T-cell subsets in mouse leishmaniasis: Induction, expansion and effector function. *Immunoparasitol. Today* A58–A61.

Mahmoud, A. A. F. (1980). Genetics of schistosomiasis. *In* "Modern Genetic Concepts and Techniques in the Study of Parasites" (F. Michal, ed.), pp. 303–322. Schwabe, Basel.

Matsuda, H., Watanabe, N., Kiso, Y., Hirota, S., Ushio, H., Kannan, Y., Azuma, M., Koyama, H., and Kitamura, Y. (1990). Necessity of IgE antibodies and mast cells for manifestation of resistance against larval *Haemaphysalis longicornis* ticks in mice. *J. Immunol.* **144,** 259–262.

Miller, H. R. P. (1984). The protective mucosal response against gastrointestinal nematodes in ruminants and laboratory animals. *Vet. Immunol. Immunopathol.* **6,** 167–259.

Mitchell, G. F., Rajasekariah, G. R., and Rickard, M. D. (1980). A mechanism to account for mouse strain variation in resistance to the larval cestode, *Taenia taeniaeformis. Immunology* **39,** 481–489.

Mock, B., and Seldin, M. (1989). A comparison of genetic linkage maps surrounding the *Lsh-Ity-Bcg* disease resistance locus. *Resist. Immunol.* **140,** 769–774.

Ohta, N., Edahiro, T., Ishii, A., Yasukawa, M., and Hosaka, Y. (1990). HLA-DQ-controlled T cell response to soluble egg antigen of *Schistosoma japonicum* in humans. *Clin Exp. Immunol.* **79,** 403–408.

Ottesen, E. A. (1984). Immunological aspects of lymphatic filariasis and onchocerciasis in man. *Trans. R. Soc. Trop. Med. Hyg.* **78** *(Suppl.),* 9–18.

Philipp, M., Worms, M. J., Maizels, R. M., and Ogilvie, B. M. (1984). Rodent models of filariasis. *Contemp. Top. Immunobiol.* **12,** 275–321.

Piessens, W. F., and Mackenzie, C. D. (1982). Immunology of lymphatic filariasis and onchocerciasis. *In* "Immunology of Parasitic Infections" (S. Cohen and K. S. Warren, eds.), pp. 622–653. Blackwell Scientific Publications, Oxford.

Rosenstreich, D. L., Weinblatt, A. C., and O'Brien, A. D. (1982). Genetic control of resistance to infection in mice. *Crit. Rev. Immunol.* **3,** 263–330.

Rothwell, T. L. W. (1989). Immune expulsion of parasitic nematodes from the alimentary tract. *Int. J. Parasitol.* **19,** 139–168.

Sayles, P. C., and Wassom, D. L. (1988). Immunoregulation in murine malaria. Susceptibility of inbred mice to infection with *Plasmodium yoelii* depends on the dynamic interplay of host and parasite genes. *J. Immunol.* **141,** 241–248.

Schofield, L. (1989). T cell immunity to malaria sporozoites. *Exp. Parasitol.* **68,** 357–364.

Schwartz, R. H. (1985). T-lymphocyte recognition of antigen in association with gene products of the major histocompatibility complex. *Annu. Rev. Immunol.* **3,** 237–261.

Schwartz, R. H. (1986). Immune response *(Ir)* genes of the murine major histocompatibility complex. *Adv. Immunol.* **38,** 31–201.

Sher, A., Hieny, S., and James, S. L. (1984). Mechanisms of protective immunity against *S. mansoni* infection in mice vaccinated with irradiated cercariae. VI. Influence of the major histocompatibility complex. *Parasite Immunol.* **6,** 319–328.

Sprent, J. F. A. (1969). Evolutionary aspects of immunity in zooparasitic infections. *In* "Immunity to Parasitic Animals" (G. J. Jackson, R. Herman, and I. Singer, eds.), Vol. 1, pp. 3–62. North-Holland, Amsterdam.

Taylor, D. W., Pacheco, E., Evans, C. B., and Asofsky, R. (1988). Inbred mice infected with *Plasmodium yoelii* differ in their antimalarial immunoglobulin isotype response. *Parasite Immunol.* **10,** 33–46.

Tingley, G. A., Butterworth, A. E., Anderson, R. M., Kariuki, H. C., Keoch, D., Mugambi, M., Ouma, J. H., arap Siongok, T. K., and Sturrock, R. F. (1988). Predisposition of humans to infection with *Schistosoma mansoni:* Evidence from the reinfection of individuals following chemotherapy. *Trans. R. Soc. Trop. Med. Hyg.* **82,** 448–452.

Venkatesan, P., Finch, R. G., and Wakelin, D. (1993). MHC haplotype influences primary *Giardia muris* infections in H-2 congenic strains of mice. *Int. J. Parasitol.* **23,** 661–664.

Wakelin, D. (1975). Genetic control of immune responses to parasites: Selection for responsiveness and non-responsiveness to *Trichuris muris* in random-bred mice. *Parasitology* **71,** 377–384.

Wakelin, D. (1988a). Helminth infections. *In* "Genetics of Resistance to Bacterial and Parasitic Infection" (D. Wakelin and J. M. Blackwell, eds.), pp. 153–224. Taylor and Francis, London.

Wakelin, D. (1988b). Arthropod parasites. *In* "Genetics of Resistance to Bacterial and Parasitic Infection" (D. Wakelin and J. M. Blackwell, eds.), pp. 225–232. Taylor and Francis, London.

Wakelin, D., and Blackwell, J. M. (1988). "Genetics of Resistance to Bacterial and Parasitic Infections." Taylor and Francis, London.

Wakelin, D., and Blackwell, J. M. (1993). Genetic variation in parasitic infection. *In* "Immunology and Molecular Biology of Parasitic Infections" (K. S. Warren, ed.). pp. 3–22. Blackwell Scientific Publications, Oxford.

Wakelin, D., and Rose, M. E. (1990). Immunity of coccidiosis. *In* "Coccidiosis" (P. L. Long, ed.), pp. 281–306. CRC Press, Boca Raton, Florida.

Wakelin, D., Rose, M. E., Hesketh, P., Else, K. J., and Grencis, R. K. (1993). Immunity to coccidiosis: Genetic influences on lymphocyte and cytokine responses to infection with *Eimeria vermiformis* in inbred mice. *Parasite Immunol.* **15,** 11–19.

Warren, K. S. (ed.) (1933). "Immunology and Molecular Biology of Parasitic Infections." Blackwell Scientific Publications, Oxford.

Wassom, D. L., and Kelly, E. A. B. (1990). The role of the major histocompatibility complex in resistance to parasite infections. *Crit. Rev. Immunol.* **10,** 31–52.

Wassom, D. L., Guss, V. M., and Grundmann, A. W. (1973). Host resistance in a natural host-parasite system. Resistance to *Hymenolepis citelli* by *Peromyscus maniculatus. J. Parasitol.* **59,** 117–121.

Wassom, D. L., DeWitt, C. W., and Grundmann, A. W. (1974). Immunity to *Hymenolepis citelli* by *Peromyscus maniculatus:* Genetic control and ecological implications. *J. Parasitol.* **60,** 47–52.

Wassom, D. L., Dick, T. A., Arnason, N., Strickland, D., and Grundmann, A. W. (1986). Host genetics: A key factor in regulating the distribution of parasites in natural host populations. *J. Parasitol.* **72,** 334–337.

Weatherall, D. J. (1987). Common genetic disorders of the red cell and the "malaria hypothesis." *Ann. Trop. Med. Parasitol.* **81,** 539–548.

Wong, M. S., and Bundy, D. A. P. (1990). Quantitative assessment of contamination of soil by the eggs of *Ascaris lumbricoides* and *Trichuris trichiura. Trans. R. Soc. Trop. Med. Hyg.* **84,** 567–570.

Wong, M. S., Bundy, D. A. P., and Golden, M. H. N. (1991). The rate of ingestion of *Ascaris lumbricoides* and *Trichuris trichiura* eggs in soil and its relationship to infection in two children's homes in Jamaica. *Trans. R. Soc. Trop. Med. Hyg.* **85,** 89–91.

Noel W. Solomons & Marilyn E. Scott

9

Nutritional Status of Host Populations Influences Parasitic Infections

I. Introduction

Much has been, and is being, written about the interaction between nutrition and infection. By definition, the parasite depends on the host for nutritional sustenance. Therefore, it should not be a surprise to find that host nutritional status plays a very important role in the dynamic interaction between host and parasite populations.

From a historical perspective, an interaction between parasitism and host nutrition long has been suspected. Scrimshaw, Taylor, and Gordon, in their classic reviews (Scrimshaw *et al.,* 1959, 1968), set out a conceptual framework for the interaction between infection and host nutritional status, based on the assumption that interactions are more than the additive contribution or coexistence of nutritional factors and infection. These investigators classified the majority of interactions of malnutrition and infection as synergistic, that is, infection aggravated nutritional deficiencies and malnutrition aggravated infections. In contrast, these researchers also identified antagonistic interactions, that is, increasing degrees of malnutrition led to reduced levels or severity of infection. (Imagine also a second form of antagonistic interaction, in which increasing levels of infection may lead to reduced degrees of malnutrition. This situation might occur when parasites synthesize nutrients that are required by the host, such as folate, biotin, or vitamin K.)

Over the past 30 yrs, we have learned a great deal about how parasitic infection affects host nutritional status (see review by Stephenson, 1987). In contrast, we understand much less about how host undernutrition influences parasitic infections. Host malnutrition leads to impaired immunity, and a functional immune system is required to control infections. Therefore, host nutritional status must be important for the parasite. This premise has formed the basis for the design and interpretation of most of the limited research in this area. Hypothetically, host nutritional status affects the parasite in many other ways as well. In this chapter, we will explore a spectrum of ideas, many of which have not been tested but are logical extensions of current knowledge.

II. Quantifying Infection and Nutritional Status

One of the most challenging aspects of research in nutrition–infection interactions

Parasitic and Infectious Diseases

concerns methodologies. In the field and in the laboratory, investigations rely on qualitative or quantitative measures of infection and nutritional status, but the reliability with which chosen indicators actually measure infection or nutritional status is often low. Many of the indicators most valuable for assessing nutritional status are invasive, so their use is not warranted in the majority of human studies. In the following sections, we discuss a few of the issues related to choosing appropriate indicators of infection and nutritional status, and to interpreting the resulting data. For a discussion of issues related to study design, refer to the review by Stephenson (1987).

A. Infection Status

For macroparasites, intensity of infection is measured most often by quantifying the numbers of eggs or larvae produced by the adult parasites. Only rarely can actual numbers of worms be counted (collection of worms following anthelmintic expulsion). Considerable discussion has taken place over the years about the accuracy of egg counts as indicators of numbers of adult worms. Reasons for concern include the effects of parasite density on *per capita* egg or larval release from adult worms (see reviews by Keymer and Slater, 1987; Shostak and Scott, 1993) and day-to-day variability when the same technique is used repeatedly on an individual (Hall, 1981). Accuracy also is affected to some extent by the uneven distribution of eggs within a stool sample. Despite these considerations, clearly such data do provide valid indicators for infection intensity. If we use *Ascaris* as an example, we would feel confident in stating that an individual passing 1000 eggs per gram of stool has fewer adult worms than an individual passing 100,000 eggs per gram of stool. In contrast, we would not attempt to distinguish between individuals passing 50,000 and 55,000 eggs per gram. In a statistical sense, however, with respectable sample sizes, the generalization that individuals passing more eggs have more worms has been shown repeatedly to be reasonable (see Chapter 14).

In microparasite infections, parasitemia often is measured. Although this measure is an important indicator of current status, care must be taken in extrapolating this infection level into the past or the future. Although infection levels for macroparasites (in the absence of intervention) tend to remain relatively stable, parasitemia within an individual characteristically rises and falls over a period of a few days or weeks. Correlating various nutritional indicators with parasitemia at a single point in time is appropriate, but parasitemia should be used with caution in attempting to make associations or inferences about more chronic conditions. The exception to this generalization would be when parasitemia in individuals is monitored over time.

Other complications are of relevance to this discussion. The first concerns the possible effect of host nutritional status on rates of egg or larval production. Under some circumstances, the undernutrition of the host leads to inadequate nutrition of the parasite. Under such conditions, parasite reproduction is likely to be affected (see Section III,E). Thus, researchers should be aware that egg production in severely malnourished individuals may not be a reliable indicator of worm numbers.

A second complication applies to microparasite infections. With *Giardia,* for example, detection of infection depends on identifying cysts or trophozoites in the stool. Cyst release in humans is characteristically intermittent, meaning repeated stool samples must be examined to identify individuals who are uninfected.

Finally, in some instances a parasite may cause nutritional problems even when not present in the host. For example, during a *Giardia* infection of laboratory rodents, activity of mucosal disaccharidases is reduced. This effect occurs even when the rodent is given a parasite extract rather than the actual parasite (Belosevic *et al.,* 1986) and presumably is caused by some intestinal immune response to exposure to parasite antigen. This response has important implications when researchers are attempting to study lactose intolerance, for example. Knowing that *Giardia* infection leads to reduced lactase activity, researchers generally exclude *Giardia*-infected individuals from the study. However, they may not exclude individuals who are ingesting *Giardia* cysts, but remain uninfected, and are experiencing a drop in lactase activity as a consequence. As such, the prevalence of a genetic lactose intolerance in a given population may be overestimated.

When assessing infection level for populations, note that macroparasites are almost always clumped in the host population so only a few individuals are infected with large numbers of worms (see Chapter 4). By necessity, this distribution affects the sample size. Note also that many infections (including *Ascaris*) have been shown to be clustered in families (Forrester *et al.,* 1988), which also significantly complicates sampling of the population and study design.

B. Diet and Nutritional Status

To the uninitiated, the terms "diet" and "nutritional status" may be used somewhat interchangably. We prefer, however, to make a distinction between the two. Measures of diet describe the pattern of food consumption and the type and quantity of nutrients ingested. Measures of nutritional status quantify the total body reserves of a nutrient.

1. Diet

Both qualitative and quantitative estimates of dietary intake are fraught with measurement errors. Precision (reproducibility of repeated measurements) is generally more difficult to achieve than accuracy (ability to obtain the correct value) (Tangney *et al.,* 1987). People do not eat the same foods or the same spectrum of nutrients on a day-to-day basis. Thus, information must be collected over several days to obtain accurate estimates of intake for a given individual (Beaton *et al.,* 1979). The extent to which this method poses problems varies depending on the nutrient of interest. For example, intake of protein and energy is more constant from day to day than intake of vitamin A (Sempos *et al.,* 1985). Precision becomes particularly important for population-based studies in which, for example, correlations between dietary intake and disease state are being attempted. People living in the same population have very similar average intakes of nutrients; therefore, precise individual estimates are required to differentiate individuals and rank them in a stable sequence according to intake of a given nutrient or dietary substance.

Since many parasites are foodborne, dietary intake is a useful measure in studies of nutrition–infection studies. Food preference and loss of appetite have direct implications for exposure to infection. Dietary constituents also determine the physicochemical environment in which many parasites live. On the other hand, dietary intake is not a reliable indicator or predictor of nutritional status (Tangney *et al.,* 1987). Because nutritional status is a function of retention as well as intake, estimates of intake indicate little about storage and availability for utilization in the organism.

2. Nutritional Status

Assessment of nutritional status is complicated because of problems of validity and reliability. Height-for-age commonly is compared with normative curves of the National Center for Health Statistics of the United States as an indicator of "chronic malnutrition." Although height and age both can be determined very precisely and reproducibly, the relationship between height and age is, in part, determined by genetic factors. Therefore, height-for-age is more than simply an indicator of nutritional status. Biochemical indices of nutritional status are even more problematic. For instance, analysis of zinc levels in hair is used as an index of total body zinc reserves. However, hair growth is arrested in zinc deficiency, which can confound the nutritional interpretation of the content of zinc in human hair (Solomons, 1979). In addition, contamination with zinc from the external environment is difficult to control, again complicating the interpretation of hair zinc levels. Plasma zinc concentration is the most widely used index of human zinc status, but also is affected by a wide number of other factors. For example, hemolysis of red cells in the sample leads to a false increase in zinc concentration in plasma, whereas many inflammatory responses lead to a decrease in plasma concentration that gives a false impression of a decrease in body stores (Solomons, 1983).

Functional indices are projected as sensitive for the detection of nutrient deficiencies since they occur early, before biochemical or clinical manifestations of malnutrition have appeared (Solomons and Allen, 1983). However, a low value for such indices is not always indicative of a nutritional problem, but may result from physical or physiological disorders unrelated to nutritional status. For example, dark adaptation is cited as a functional index of vitamin A deficiency, but abnormal adaptation may result from nerve damage that is unrelated to a deficiency of vitamin A. The precision of measurement varies among functional tests. Behavioral tests often have a "learning effect" in which the subject shows improved performance merely from repeated testing. Most importantly, functional indices vary markedly from subject to subject. If individuals in a population are to be ranked with respect to some functional index, the percentage of adequacy in relation to the maximal value in the population is a far better score than the simple performance score.

Choice of functional, biochemical, or clinical signs to identify and gauge the severity of a nutrient deficiency depends on how the body responds to a deficiency of that nutrient. As an individual becomes more and more malnourished, a sequence of physiological and biochemical adaptations occurs in the body. Golden (1991) classified nutrient deficiency states into two broad categories, Type I and Type II, on the basis of the pattern of adaptations. Deficiencies classified as Type I are characterized by a progressive desaturation of the nutrient in tissue, whereas those classified as Type II are highly conserved and show stable tissue concentrations until the deficiency becomes very severe.

The first indication of deficiency of a Type I nutrient is impairment of functions that require that nutrient. Functional tests of nutritional status are the most sensitive, and the earliest, of the indices to reveal nutrient scarcity. Using iron as an example, some of the earliest changes are alterations in cognitive function, muscle metabolism, and immune defenses. These changes occur long before the red cell mass is compromised. With a greater depletion of reserves, biochemical indices of tissue or transport desaturation become measureable. With iron deficiency, a reduced concentration of iron in circulation, an increase in the amount of incomplete hemoglobin components in red cells, and an increased concentration of tissue receptors that capture the increasingly scarce iron from circulation can be detected. Only when the reserves are depleted totally and the body is trying to overextend its resources do major clinical manifestations of deficiency appear. For iron, these symptoms involves clinically evident anemia with pallor, exertional shortness of breath, and tachycardia. When depleted, deficiencies of nutrients such as copper, selenium, and virtually all vitamins show this pattern of progression from functional changes, to biochemical alterations to overt deficiency signs and symptoms; therefore, these nutrients show a Type I response during deficiency.

However, not all nutrient deficiencies conform to the model of progressive desaturation of tissues. Sodium, potassium, chloride, magnesium, zinc, and nitrogen are among the nutrients for which the body shows a Type II response to deficiency (Golden, 1991). At almost the first sign of a stress on body stores, physiological mechanisms intercede that enhance the conservation of the nutrient. For example, absorption studies often show enhanced uptake by the gut, and studies of nutrient turnover using radiolabeled isotopes show decreased rates of tissue exchange and reduced excretion compared with well-nourished individuals. Functional tests that indicate deficiency of Type II nutrients usually measure the process of conservation itself. As a consequence of these conservation processes, nutrient concentrations in cells and tissues are not reduced and total body levels are maintained at the expense of growth and development. In children, growth failure appears at the early stages of deficiency of all Type II nutrients and is, therefore, a nonspecific clinical indicator of Type II deficiencies. In adults, other early clinical manifestations are seen, for example, muscle wasting. The progressive impairment of nutrient-dependent biochemical pathways observed in Type I nutrient deficiencies is not seen in Type II deficiencies, and therefore cannot be used as a reliable indicator of the severity of these deficiencies.

III. Host Nutrition and the Infection Cycle

As has been noted in virtually every other chapter in this volume, the dynamic interaction between host and parasite populations depends on the balance between factors that increase the numbers of parasites (reproduction, immigration) and factors that decrease the numbers of parasites (death, emigration). How is host nutritional status likely to affect each of the biological steps involved in determining the outcome of a parasitic infection?

A. Contact between Host and Parasite

Successful transmission of parasites to a new host requires, as a first step, physical contact between the infective stage of the parasite and the host. Contact leading to infection occurs through five different processes, depending on the type of infectious agent: (1) direct contact between infected and uninfected hosts (e.g., measles); (2) ingestion of free-living infective stages (e.g., *Ascaris*); (3) ingestion of infected intermediate hosts (e.g., *Taenia*); (4) penetration of free-living stages present in the environment (e.g., hookworm or schistosomes); and (5) vector-borne transmission (e.g., malaria). A range of host behaviors is influenced by nutritional deficiencies (Husaini *et al.,* 1991), and we therefore predict that contact between potential hosts and infective stages of parasites also will be influenced by nutritional deficiencies. The effect of deficiencies on eating habits and levels of physical activity are of particular interest.

Host anorexia induced by zinc deficiency (Humphries *et al.,* 1989) will reduce exposure to infective stages in food. This situation is an example of an antagonistic interaction in which malnutrition reduces exposure to infection. On the other hand, a synergistic interaction occurs when the voracious appetite induced by marasmus (Humphries *et al.,* 1989) increases exposure to infective stages in food. Ingestion of soil (geophagia) induced by mineral deficiencies such as iron and zinc (Reid, 1992) is predicted to increase exposure to infective stages of soilborne pathogens. This hypothesis was tested by Wong *et al.* (1988), who measured soil ingestion in Jamaican children by quantifying silica in stool. These investigators reported that some children ingested more than 10 g soil per day. In a follow-up study (Wong *et al.,* 1991), these researchers made crude estimates of the rates of ingestion of nematode eggs, based on rates of soil ingestion per child and average number of eggs per g soil, and correlated these estimates with the intensity of *Ascaris* and *Trichuris* infections in the same children. These investigators detected a significant positive correlation in one of their two study populations. The authors suggested that the lack of significance in the second population may have been the result of poor sensitivity of their measure of exposure.

Nutritional deficiencies influence food preference because of a physiological aversion or attraction to certain foods (Pope *et al.,* 1992) or because of a culturally based concept of "good" and "bad" foods during illness (Cosminsky, 1978). If the diet of a malnourished person is altered so intake of foods containing parasite infective stages is reduced (or increased), then exposure to foodborne infections will be reduced (or increased).

Physical activity of the malnourished individual often is reduced, leading to an altered pattern of interaction with the environment or with other people in the community. This change affects the probability of contact with infective stages of a variety of types of parasites. Children with protein-energy malnutrition, for example, tend to play outdoors less (Viteri and Torun, 1981) and therefore will be less exposed to penetrating stages of hookworms and schistosomes present in the soil or in rivers or lakes. If malnutrition causes children to socialize less, or to remain home from school, the risk of exposure to directly transmitted infections such as measles will be reduced. Transmission of vector-borne diseases requires that people be in the same location as the vectors. For example, reduced outdoor activity will reduce contact with blackflies and, thus, reduce the risk of exposure to onchocerciasis that is transmitted by the blackflies (Remme and Zongo, 1989). At the same time, reduced outdoor activity will increase contact with triatomid bugs that live in thatched homes (Marsden, 1983) and thus increase the risk of exposure to Chagas' disease transmitted by these vectors. If children are so sick that they do not bother to brush away biting insects, the probability of exposure to vector-borne infections will increase.

Surprisingly few studies have attempted to link nutritional deficiencies directly to risk of exposure to infection. It is important to study these links more thoroughly, because, in the absence of exposure, infection and disease cannot occur.

B. Establishment of Infection

Once contact has occurred, the actual establishment of the parasite depends on the host providing a hospitable environment for the parasite. The environment provided by the host for the incoming parasite is influenced by host physiology and by specific and nonspecific defense mechanisms that normally protect the host against invading organisms.

Altered gastrointestinal tract physiology frequently occurs during nutritional deficiency (Rosenberg and Solomons, 1984). This change will influence the establishment of the parasite, because the gastrointestinal environment provides chemical triggers that inform many parasites of their new environment. Excystment of digeneans or tapeworms, or exsheathment of nematodes, is triggered by specific cues such as a particular pH or partial pressure of carbon dioxide in the intestine, or presence of certain enzymes or certain bile salts (Hawdon and Schad, 1991). If the appropriate cue is missing, the larval parasite will pass directly through the intestine without establishing an infection.

Altered gut physiology can, on the other hand, facilitate parasite establishment when nonspecific host defenses that normally protect the host from parasites are disrupted. Some nutritional deficiencies (e.g., vitamin A) interrupt epithelial tissue integrity and, thus, facilitate access of the parasite to host tissues (Darip *et al.,* 1979). In other cases, nutritional deficiencies reduce the production of the nonspecific defense substances such as mucin, allowing parasites to establish when they otherwise may not. This change is particularly important in the context of parasitic disease, since organisms that are not usually highly pathogenic can do much more damage in a malnourished host than in a well-nourished host.

C. Parasite Survival

Having established within the host, parasite survival is dependent on obtaining nutrients from the host environment and on evading host defense mechanisms. The requirement of parasites for specific nutrients (e.g., starch, casein, and *para*-aminobenzoic acid) has been demonstrated in several cases, some of which were reviewed by Crompton (1991). When these nutrients are not present in adequate amounts in the host, either because of deficient dietary intake or because their synthesis by the host is impaired, parasite survival is affected adversely.

During the coevolution of host and parasite, the parasite life cycle evolved to depend on "normal" physiological processes of the host. When nutritional deficiencies alter such processes, the parasite finds itself in an abnormal environment and often is unable to survive. A particularly interesting example concerns the effect of vitamin E deficiency on malaria infection in rodents. Levander and colleagues (1990) showed that mice maintained on vitamin E-deficient diets had reduced parasitemia and reduced mortality induced by *Plasmodium berghei.* These researchers postulated that vitamin E deficiency leads to premature rupture of infected erythrocytes. The parasites thus released into the circulation are not able to re-infect other red cells, and progression of the infection is interrupted. This situation is an example of an antagonistic association in which poor nutritional status is apparently harmful to the parasite and, thus, has some beneficial outcomes for the host.

In mild to moderate nutritional deficiencies, synergistic interactions also have been reported in which the harmful effects of the nutritional deficiency are exacerbated by the harmful effects of the parasite. Two examples are enhanced survival of the parasite in the deficient host, compared with well-nourished hosts, for mouse trichostrongylids in protein-deficient animals (see Chapter 20), and delayed expulsion of the intestinal nematode *Trichinella spiralis* in zinc-deficient rats (Fenwick *et al.,* 1990).

Under conditions of more severe nutritional deficiency, the advantage to the parasite of impaired immunocompetence of the host may be offset because the parasite itself may become malnourished. Remarkably little is known about the nutritional requirements of parasites. Fortunately, with increasing success in growing and maintaining parasites under artificial *in vitro* conditions, more and more will gradually be understood about their nutritional requirements (Wang, 1991).

Medicinal herbs are used to combat parasitic infection in many cultures. Their presence in the diet affects the survival of the parasite, but also has implications for the nutritional status of the host. Many of these herbs have nutritive value (Etkin and Ross, 1982), but many interfere with nutrient uptake or utilization because they contain compounds, such as thiaminases, biotin binders, or alkaloids, that interfere with digestion or absorption (Johns, 1990). This interaction applies equally to animal populations, in which dietary preferences may have evolved to include plants with medicinal properties (Johns, 1990).

D. Nutrition and Host Immunity to Infection

A vast literature exists on the effects of nutritional deficiency on specific host immunological responses (see Isliker and Schürch, 1980; Chandra, 1988). For example, zinc-deficient animals have impaired cell-mediated cytotoxicity and T helper cell function (Frost *et al.,* 1981; Fraker *et al.,* 1982). The intestinal immune system is disrupted in vitamin A-deficient animals (West *et al.,* 1989), presumably weakening the host defense against enteric parasites. The profile of cytokine production by cells from vitamin A-deficient animals also is altered substantially (Carman *et al.,* 1992), leading to profound changes in the regulation of immune cell function. Protein-energy malnutrition in humans and severe protein deficiency in experimental animals are associated most notably with depressed T-cell-mediated immune responses, atrophy of lymphoid organs, decreased numbers of T cells, decreased thymic hormone activity, and impaired delayed hypersensitivity (Gross and Newberne, 1982; Slobodianik *et al.*, 1984). Although immunodepression frequently is observed in patients with severe protein-energy malnutrition (Chandra, 1988), more moderate protein deficiency in experimental animals has been shown to enhance cell-mediated immunity in comparison with well-nourished animals (Petro, 1985). Whether this result can be explained by as yet poorly understood regulatory mechanisms, or whether the immunodepression observed in human patients is linked not only to the protein-energy deficit but to other concomitant nutritional deficiencies, is not known.

Host immunity is a key mechanism that controls parasitic infections in the host (see Chapter 8). The impairments in immunity that occur during nutritional deficiencies will facilitate the establishment and survival of parasites. Therefore, increased levels of infection in malnourished hosts usually are attributed to a malfunctioning immune system, even when no measures of immune function have been made. We have shown, however, that other physiological, biochemical, and behavioral changes occur in the malnourished host, and that these also can lead to elevated levels of infection. Therefore, assuming that impaired host immunity explains all host–parasite interactions is an oversimplification.

E. Parasite Reproduction

1. Macroparasites

Reproduction of the parasite also has been shown to be influenced by host nutritional status. Macroparasite reproduction is measured by the rate of egg or larval release from the host. For example, delayed onset of reproduction and prolonged embryogenesis by a filarial nematode were observed in protein-deficient cotton rats (Storey, 1982). Egg production by the acanthocephalan parasite *Moniliformis moniliformis* was reduced in rats fed a low carbohydrate diet (Crompton *et al.,* 1988). In both studies, worms in the deficient host were smaller, which may explain the reduced fecundity. Alternatively, the disrupted gastrointestinal environment had an effect on the reproductive physiology of the parasite.

Negative effects of host malnutrition on parasite reproduction are not always found. In fact, a study by El-Hag *et al.* (1989) revealed that egg production of the intestinal nematode *Nippostrongylus brasiliensis* was significantly enhanced in zinc-deficient rats relative to zinc-sufficient rats or pair-fed controls. Interestingly, these authors reported no effects of the zinc deficiency on parasite survival.

Determining the extent to which host nutritional status affects parasite reproduction in human populations is more difficult. Bundy and Golden (1987) monitored egg output in severely malnourished infants undergoing hospitalized refeeding regimes in Jamaica. These investigators noted that the numbers of *Ascaris* eggs passed by some children decreased during the period of refeeding. Their previous work (Golden and Golden, 1981) showed that, during such refeeding periods, children become increasingly deficient in zinc because of elevated demands for zinc necessary for metabolism. These authors hypothesized that the zinc deficiency state induced in these children led to the decreased rate of egg production. Therefore, we might, for example, postulate that the worms themselves become deficient in zinc, and that this reduces their reproduction. Will a deficit of zinc in host tissues be detected by a parasite in the lumen of the intestine? The efficiency of zinc absorption through the intestinal mucosa is elevated during zinc deficiency (Linder, 1991), so the host may "outcompete" the parasite for dietary zinc. It seems unlikely, however, that less zinc would be available for the parasite during the refeeding period than prior to refeeding, when the children were severely malnourished.

We understand little about the adaptive physiological responses of parasites to environmental stress. In the severely malnourished child, *Ascaris* may respond by investing limited resources into reproduction, as occurs in many other animals. The physiological response of the parasite when nutrients then become available (during the refeeding period) may be to reallocate resources to growth or tissue maintenance rather than to reproduction, leading to a reduced rate of egg production during the refeeding period. The observation made by Bundy and Golden (1987) was simple. We do not understand why *Ascaris* egg output declined, and many possible explanations have been put forward that require further study. Nevertheless, the underlying concept is important. If the declining egg output resulted from a temporary zinc deficit in the host, then improving the zinc status of the individual could lead to an increase in parasite fecundity and, in turn, an increase in rate of parasite transmission. If the declining egg output resulted from a shift in the parasite from reproduction to growth, the improvement in nutritional status of the host would, in the long term, improve the "health" of the parasite and again increase the rate of transmission. These considerations are critically important when designing nutritional intervention programs that are expected to reduce infection.

2. Microparasites

In microparasitic infections, parasitemia reflects the rate of division of the parasitic organism within the host. Data for *Trypanosoma musculi* in zinc-deficient mice (Lee *et al.*, 1983) and for *P. berghei* in vitamin A-deficient rats (Krishnan *et al.*, 1976) show that the parasitemia rises to a much higher level in deficient animals than in well-nourished and pair-fed animals. These studies both suggest, therefore, that the nutritional deficiency enhances the ability of the parasite to replicate, presumably because of an impaired immune response by the host.

In contrast, Kaikai and Thurnham (1983) found that *P. berghei* parasitemia was lower in riboflavin-deficient rats than in control animals. Riboflavin is important for reticulocytosis, and reticulocytes are the preferred host cell for *P. berghei.* Impaired reticulocytosis, therefore, may have been responsible for the reduced parasitemia. Alternatively, the effect associated with riboflavin deficiency may have been caused by effects on the enzyme glutathione reductase. This enzyme is present in unusually high levels in infected red cells, suggesting that it has some important function for the parasite. If the riboflavin deficiency leads to reduced levels of this enzyme, the parasite may be affected. Somewhat similar results have been observed in a small study in Papua New Guinea, where erythrocyte glutathione reductase activity (used as an indicator of riboflavin status) was lower in infants infected with malaria than in those who were uninfected (Thurnham *et al.*, 1983).

F. Parasite Survival in the External Environment

To complete the cycle, the survival of the parasite outside the host often depends on nutrient stores that are obtained from the previous host. This rule applies, for example, to encysted stages of protozoa, to free-swimming larval stages of digeneans, or to *Ascaris* eggs. If the host was malnourished, we suspect that the nutritional stores present in these larval stages are likely to be reduced. To our knowledge this has not been investigated.

G. Nutritional Effects and Pathogenicity of Infectious Diseases

For parasite transmission, a host population must exist. The more hosts present in the population, the higher the net rate of transmission. Therefore, any change that influences the survival or reproduction of the host population will have an impact on the transmission dynamics of the parasite.

Nutritional status of the host population has a direct effect on host population size. For example, malnourished women give birth to low-weight babies; these babies are less likely to survive the first few years of life. Nutritional status also has an indirect effect on host population size, because the mortality resulting from infectious diseases is even higher in malnourished populations than in well-nourished populations. Examples of this effect are abundant, including the lethality of measles in many countries with high rates of malnutrition (McLean and Anderson, 1988) and the outbreaks of epidemics under famine conditions (Shears, 1991).

However, this synergistic association between worsening nutritional status and worsening disease is not applicable to all situations. From time to time, conditions arise in which nutritional deficiencies appear to reduce the severity of parasitic diseases. Several of the more commonly cited examples are reviewed by Crompton (1991) and include the possible protective effects of iron deficiency with respect to malaria (a topic under current critical discussion). As scientists, we seek to understand why a nutritional deficiency may protect the host against infection or disease, but also recognize the important public health implications. Will improving the nutritional status actually worsen disease in some instances? This issue is discussed in Section V.

In conclusion, every major biological process involved in the transmission, survival, and reproduction of parasites is affected by host nutritional deficiencies. Most of the effects are beneficial to the parasite, but some appear to be detrimental. In fact, the same nutritional deficiency may be beneficial at one stage of the life cycle and detrimental at another, or may be beneficial to one parasite species and detrimental to another.

IV. Nutritional Status and the Effectiveness of Control Programs

That parasitic infection and malnutrition co-occur in the same populations and the same individuals is well known. Therefore, most antiparasite control programs are implemented in populations suffering from varying degrees of nutritional deficiencies. The most commonly used control measures for parasitic diseases involve chemotherapy. The dose of drug recommended by the manufacturer is assumed to be the appropriate dose for use in these populations. However, the efficacy many of drugs, including those used against parasites, can be altered (either enhanced or reduced) in the malnourished individual. An entire body of literature is devoted to the interaction between nutrients and drugs (see Roe, 1985). The rate of uptake of drugs, their distribution through the body, their rate of metabolism into active and inactive compounds, their mode of action, and their rate of excretion all depend on physiological and biochemical activities of the body. Drug efficacy also depends on parasite physiology, which is affected by host nutritional status. Prichard *et al.* (1981) reported reduced efficacy of benzimidazole anthelmintics in protein- and iron-deficient rats, due to reduced absorption of the drug by the parasite *N. brasiliensis.* These researchers speculated that the metabolic rate of the parasite in the deficient host may be reduced, leading to reduced drug uptake.

Surprisingly, the potential reduced efficacy of antiparasite drugs in malnourished individuals has received virtually no attention from scientific or health professionals. This area of investigation is very important, not only because of the need to use drugs appropriately for successful control of infection but also because use of drugs with reduced efficacy leads rapidly to selection for drug-resistant strains of parasite (Prichard, 1990). Selection for resistance occurs when some of the worms present in the host are not killed by the drug given to the host. If the efficacy of the drug is reduced in malnourished hosts, then the "recommended" dose may kill fewer parasites than in a well-nourished host; use of the drug in a malnourished population could lead inadvertently to selection of a drug-resistant strain of parasite.

Malnutrition also is known to reduce the success of vaccination programs. Vaccination is used to stimulate host protective immunity before natural exposure to a specific pathogen. Success of such a program depends on the ability of the vaccinated individual to mount an immune response. However, as described earlier, many nutritional deficiencies have direct effects on host immunological systems. Therefore, vaccination in such individuals is ineffective (Katz, 1977; Keusch, 1990). For example, Keusch (1991) estimated that 50% of individuals vaccinated against measles were not protected, simply because of a nutritionally based inadequacy to respond to the vaccine.

V. Will Nutritional Intervention Reduce Parasitic Infection?

We have shown above that malnutrition reduced host immunity, and improvement in host nutritional status improves the ability to combat infectious agents. Therefore, a nutritional intervention that improves nutritional status should reduce parasitic infection. However, host nutritional status interfaces with the infection process in many ways. Depending on the severity of the deficiency, and the specific infection of concern, under some circumstances an improvement in nutritional status may have no detectable effect on infection or disease, or may increase the risk of infection and disease.

Perhaps this feature accounts, in part, for the inconsistent results reported following nutritional interventions in human populations. For example, supplementation with vitamin A has occurred in many populations with endemic hypovitaminosis A, but the effects on infectious mortality have not been consistent (see review by Beaton *et al.*, 1993). Some studies have shown reduced mortality following supplementation, including studies by Sommer *et al.* (1986) in rural Indonesia, Rahmathullah *et al.* (1990) near Tamul Nadu in India, and West *et al.* (1991) in rural Nepal. In contrast, in Hyderabad, India (Vijayaraghavan *et al.*, 1990), and in rural Sudan (Herrera *et al.*, 1992), vitamin A supplementation had no effect on mortality. Beaton *et al.* (1993) considered a wide range of possible explanations for the lack of effect in the latter two studies. The specific sample size and the formulation and dose of vitamin A varied among studies. Also, differences existed in the background vitamin A status of the populations, the distribution of underlying protein-energy malnutrition, the spectrum and virulence of infectious agents in the populations, and genetic factors of the study populations. The conclusion of the investigators, however, was that in both of the "no-effect" studies, the expected difference in vitamin A status between the treated and untreated children at the end of the studies was not present. The researchers suggested that this measure likely accounted for the lack of significant effect of supplementation on infectious mortality.

Most nutrition intervention programs have not evaluated the impact of infectious diseases. In the future, monitoring infection status during the implementation of nutritional interventions will be extremely important. The health team should be aware that unexpected increases in certain infections may occur, and should be able to respond with appropriate preventative or treatment measures. Both positive and negative effects should be reported, and considered in any cost–benefit analysis.

VI. Conclusion

Just as the nutritional status of an individual results from a series of biochemical and physiological feedbacks that regulate, to a greater or lesser extent, the quantity and distribution of a given nutrient, so the magnitude of parasitic infection in a host population is regulated by a variety of factors, most of which are affected by host nutritional status. Over time, the nutritional status of the host population may improve or deteriorate. This will lead to changes in the rates of parasite establishment, survival, and reproduction. We believe that the magnitude of these effects is rarely appreciated, yet the effects have very important implications for the choice of intervention strategies designed to improve the health and nutrition of populations in the developing world. We urge nutritionists and infectious disease specialists to work together, and consider the more neglected aspects of nutrition-infection interactions that have been outlined in this chapter.

Acknowledgments

We are grateful for many useful discussions with our colleagues and students. We also thank R. W. Harvey for his critical review of an earlier draft. M. E. Scott acknowledges research support from the Natural Sciences and Engineering Research Council of Canada and from the Fonds FCAR pour l'Aide à la Recherche.

References

Beaton, G. H., Martorell, R., L'Abbé, K. A., Edmonston, B., McCabe, G., Ross, A. C., and Harvey, B. (1993). "Effectiveness of Vitamin A Supplementation in the Control of Young Child Morbidity and Mortality in Developing Countries." Final Report to Canadian International Development Agency. University of Toronto.

Beaton, G. H., Milner, J., Corey, P., McGuire, V., Cousins, M., Stewart, E., de Ramos, M., Hewitt, D., Grambsch, P. V., Kassim, N., and Little, J. A. (1979). Sources of variance in 24-hour dietary recall data: Implications for nutrition study design and interpretation. *Am. J. Clin. Nutr.* **32,** 2546-2559.

Belosevic, M., Faubert, G. M., and MacLean, J. D. (1986). Disaccharidase activity in the small intestine of gerbils *(Meriones unguiculatus)* during primary and challenge infections with *Giardia lamblia. Gut* **30,** 1213-1219.

Bundy, D. A. P., and Golden, M. H. N. (1987). The impact of host nutrition on gastrointestinal helminth populations. *Parasitology* **95,** 623-635.

Carman, J. A., Pond, L., Nashold, F., Wassom, D. L., and Hayes, C. E. (1992). Immunity to *Trichinella spiralis* infection in vitamin A-deficient mice. *J. Exp. Med.* **175,** 111-120.

Chandra, R. K. (1988). "Nutrition and Immunology." Liss, New York.

Cosminsky, S. (1978). The impact of methods on the analysis of illness concepts in a Guatemalan community. *Social Sci. Med.* **11,** 325-332.

Crompton, D. W. T. (1991). Nutritional interactions between hosts and parasites. *In* "Parasite-Host Associations. Coexistence or Conflict?" (C. A. Toft, A. Aeschlimann, and L. Bolis, eds.), pp. 228-257. Oxford University Press, Oxford.

Crompton, D. W. T., Keymer, A. E., Walters, D. E., Arnold, S. E., and Marrs, R. W. (1988). Factors influencing the fecundity of *Moniliformis moniliformis* (Acanthocephala): Constant dose and varied diet. *J. Zool. London* **214,** 221-234.

Darip, M. D., Sirisinha, S., and Lamb, A. J. (1979). Effect of vitamin A deficiency on susceptibility of rats to *Angiostrongylus cantonensis. Proc. Soc. Exp. Biol. Med.* **161,** 600-604.

El-Hag, H. M. A., Macdonald, D. C., Fenwick, P., Aggett, P. J., and Wakelin, D. (1989). Kinetics of *Nippostrongylus brasiliensis* infection in the zinc-deficient rat. *J. Nutr.* **119,** 1506-1512.

Etkin, N. L., and Ross, P. L. (1982). Food as medicine and medicine as food. *Social Sci. Med.* **16,** 1559-1573.

Fenwick, P. K., Aggett, P. J., Macdonald, D., Huber, C., and Wakelin, D. (1990). Zinc deficiency and zinc repletion: Effect on the response of rats to infection with *Trichinella spiralis. Am. J. Clin. Nutr.* **52,** 166-172.

Forrester, J. E., Scott, M. E., Bundy, D. A. P., and Golden, M. H. N. (1988). Clustering of *Ascaris lumbricoides* and *Trichuris trichiura* infections within households. *Trans. R. Soc. Trop. Med. Hyg.* **82,** 282-288.

Fraker, P. J., Caruso, R., and Kierszenbaum, F. (1982). Alteration of the immune and nutritional status of mice by synergy between zinc deficiency and infection with *Trypanosoma cruzi. J. Nutr.* **112,** 1224-1229.

Frost, P., Rabbani, P., Smith, J., and Prasad, A. (1981). Cell-mediated cytotoxicity and tumor growth in zinc-deficient mice. *Proc. Sci. Exp. Biol. Med.* **167,** 333-337.

Golden, B. E., and Golden, M. H. N. (1981). Plasma zinc, rate of weight gain, and the energy cost of tissue deposition in children recovering from severe malnutrition on a cow's milk or soya protein based diet. *Am. J. Clin. Nutr.* **34,** 892-899.

Golden, M. H. N. (1991). The nature of nutritional deficiency in relation to growth failure in poverty. *Acta Paediatr. Scand. (Suppl.)* **374,** 95-110.

Gross, R. L., and Newberne, P. M. (1982). Role of nutrition in immunoregulation function. *Physiol. Rev.* **60,** 188–302.

Hall, A. (1981). Quantitative variability of nematode egg counts in faeces: A study among rural Kenyans. *Trans. R. Soc. Trop. Med. Hyg.* **75,** 682–687.

Hawden, J. M., and Schad, G. A. (1991). Developmental adaptations in nematodes. *In* "Parasite–Host Associations. Coexistence or Conflict?" (C. A. Toft, A. Aeschlimann, and L. Bolis, eds.), pp. 274–298. Oxford University Press, Oxford.

Herrera, M. G., Nestel, P., el Amin, A., Fawzi, W. W., Mohamed, K. A., and Weld, L. (1992). Vitamin A supplementation and child survival. *Lancet* **340,** 267–271.

Humphries, L., Vivian, B., Stuart, M., and McClain, C. J. (1989). Zinc deficiency and eating disorders. *J. Clin. Psychiatry* **50,** 456–459.

Husaini, M. A., Karyadi, L., Husaini, Y. K., Sandjaja, Karyadi, D., and Pollitt, E. (1991). Development of short-term supplementary feeding in nutritionally-at-risk Indonesian infants. *Am. J. Clin. Nutr.* **54,** 799–804.

Isliker, H., and Schürch, B. (eds.) (1980). "The Impact of Malnutrition on Immune Defense in Parasitic Infestation." Nestlé Foundation Publication Series Vol. 2. Hans Huber Publishers, Bern.

Johns, T. (1990). "With Bitter Herbs They Shall Eat It. Chemical Ecology and the Origins of Human Diet and Medicine." University of Arizona Press, Tucson.

Kaikai, P., and Thurnham, D. I. (1983). The influence of riboflavin deficiency on *Plasmodium berghei* infection in rats. *Trans. R. Soc. Trop. Med. Hyg.* **77,** 680–686.

Katz, M. (1977). Immunization in malnourished children-a review. *In* "Malnutrition and the Immune Response" (R. M. Suskind, ed.), pp. 421–427. Raven Press, New York.

Keusch, G. T. (1990). Micronutrients and susceptibility to infection. *Ann. N.Y. Acad. Sci.* **587,** 181–188.

Keusch, G. T. (1991). Nutritional effects on response of children to respiratory tract pathogens: Implications for vaccine development. *Rev. Infect. Dis.* **13,** S486–S491.

Keymer, A. E., and Slater, A. F. G. (1987). Helminth fecundity: Density dependence or statistical illusion? *Parasitol. Today* **3,** 56–58.

Krishnan, S., Krishnan, A. D., Mustafa, A. S., Talwar, G. P., and Ramalingaswami, V. (1976). Effect of vitamin A and undernutrition on the susceptibility of rodents to a malarial parasite *Plasmodium berghei. J. Nutr.* **106,** 784–791.

Lee, C. M., Humphrey, P. A., and Aboko-Cole, G. F. (1983). Interaction of nutrition and infection: Effect of zinc deficiency on resistance to *Trypanosoma musculi. Int. J. Biochem.* **15,** 841–847.

Levander, O. A., Ager, A. L., Jr., Morris, V. C., and May, R. G. (1990). *Plasmodium yoelii:* Comparative antimalarial activities of dietary fish oils and fish oil concentrates in vitamin E-deficient mice. *Exp. Parasitol.* **70,** 232–329.

Linder, M. C. (1991). Nutrition and metabolism of trace elements. *In* "Nutritional Biochemistry and Metabolism with Clinical Applications" (M. A. Linder, ed.), pp. 215–276. Elsevier, New York.

Marsden, P. D. (1983). The transmission of *Trypanosoma cruzi* infection to man and its control. *In* "Human Ecology and Infectious Diseases" (N. A. Croll and J. H. Cross, eds.), pp. 253–289. Academic Press, New York.

McLean, A. R., and Anderson, R. M. (1988). Measles in developing countries. I. Epidemiological parameters and patterns. *Epidemiol. Infect.* **100,** 111–133.

Petro, T. M. (1985). Effect of reduced dietary protein intake on regulation of murine *in vitro* polyclonal T lymphocyte mitogenesis. *Nutr. Res.* **5,** 263–276.

Pope, J. F., Skinner, J. D., and Carruth, B. R. (1992). Cravings and aversions of pregnant adolescents. *J. Am. Diet. Assoc.* **92,** 1479–1482.

Prichard, R. K. (1990). Anthelmintic resistance in nematodes: Extent, recent understanding, and future directions for control and research. *Int. J. Parasitol.* **20,** 515–523.

Prichard, R. K., Kelly, J. D., Bolin, T. D., Duncombe, V. M., and Fagan, M. R. (1981). The effect of iron and protein deficiency on plasma levels and parasite uptake of [^{14}C]fenbendazole in rats infected with *Nippostrongylus brasiliensis. Am. J. Epidemiol. Bacteriol. Kinet.* **59,** 567–573.

Rahmathullah, L., Underwood, B. A., Thulasiraj, R. D., Milton, R. C., Ramaswamy, K., Rahmathullah, R., and Babu, G. (1990). Reduced mortality among children in southern India receiving small weekly dose of Vitamin A. *N. Engl. J. Med.* **323,** 929–935.

Reid, R. M. (1992). Cultural and medical perspectives on geophagia. *Med. Anthropol.* **13,** 337–351.

Remme, J., and Zongo, J. B. (1989). Demographic aspects of the epidemiology and control of onchocerciasis in West Africa. *In* "Demography and Vector-Borne Diseases" (M. W. Service, ed.), pp. 367–386. CRC Press, Boca Raton, Florida.

Roe, D. A. (1985). Drug-food and drug-nutrient interactions. *J. Environ. Pathol. Toxicol. Oncol.* **5,** 115–135.

Rosenberg, I. H., and Solomons, N. W. (1984). Physiological and pathophysiological mechanisms in mineral absorption. *In* "Absorption and Malabsorption of Mineral Nutrients" (N. W. Solomons and I. H. Rosenberg, eds.), pp. 1–13. Liss, New York.

Scrimshaw, N. S., Taylor, C. E., and Gordon, J. E. (1959). Interactions of nutrition and infection. *Am. J. Med. Sci.* **237,** 367–403.

Scrimshaw, N. S., Taylor, C. E., and Gordon, J. E. (1968). "Interactions of Nutrition and Infection." World Health Organization, Geneva.

Sempos, C. T., Johnson, N. E., Smith, E. L., and Gilligan, C. (1985). Effects of intraindividual and interindividual variation in repeated dietary records. *Am. J. Epidemiol.* **121,** 120–130.

Shears, P. (1991). Epidemiology and infection in famine and disasters. *Epidemiol. Infect.* **107,** 241–251.

Shostak, A. W., and Scott, M. E. (1993). Detection of density-dependent growth and fecundity of helminths in natural infections. *Parasitology* **106,** 527–539.

Slobodianik, N. H., Cosarinsky, R. C., Langini, S. H., Roux, M. E., Río, M. E., and Sanahuja, J. C. (1984). Effect of severe protein deficiency on surface and intracellular markers of growing rats' lymphoid organs. *Nutr. Rep. Int.* **29,** 957–964.

Solomons, N. W. (1979). On the assessment of zinc and copper nutriture in man. *Am. J. Clin. Nutr.* **32,** 856–871.

Solomons, N. W. (1983). Recent progress in zinc nutrition research. *Nutr. Update* **1,** 123–145.

Solomons, N. W., and Allen, L. H. (1983). The functional assessment of nutritional status: Principles, practice and potential. *Nutr. Rev.* **41,** 33–50.

Sommer, A., Tarwotjo, I., Djunaedi, E., West, K. P., Jr., Loeden, A. A., Tilden, R., and Mele, L. (1986). Impact of vitamin A supplementation on childhood mortality. A randomized controlled community trial. *Lancet* **1,** 1169–1173.

Stephenson, L. S. (1987). "The Impact of Helminth Infections on Human Nutrition. Schistosomes and Soil-Transmitted Helminths." Taylor and Francis, New York.

Storey, D. M. (1982). The host–parasite relationships in normal and protein-malnourished cotton rats infected with *Litomosoides carinii* (Nematoda: Filarioidea). *Parasitology* **85,** 543–558.

Tangney, C. C., Shekelle, R. B., Raynor, W., Gale, M., and Betz, E. P. (1987). Intra- and interindividual variation in measurements of β-carotene, retinol, and tocopherols in diet and plasma. *Am. J. Clin. Nutr.* **45,** 764–769.

Thurnham, D. I., Oppenheimer, S. J., and Bull, R. (1983). Riboflavin status and malaria in infants in Papua New Guinea. *Trans. R. Soc. Trop. Med. Hyg.* **77,** 423–424.

Vijayaraghavan, K., Radhaiah, G., Prakasam, B. S., Sarma, K. V., and Reddy, V. (1990). Effect of massive dose vitamin A on morbidity and mortality in Indian children. *Lancet* **336,** 1342–1345.

Viteri, F. E., and Torun, B. (1981). Nutrition, physical activity and growth. *In* "The Biology of Normal Human Growth" (M. Ritzen, A. Aperia, and K. Hall, eds.), pp. 265–273. Raven Press, New York.

Wang, C. C. (1991). Metabolic deficiencies in parasites and their relation to host metabolism. *In* "Parasite–Host Associations. Coexistence or Conflict?" (C. A. Toft, A. Aeschlimann, and L. Bolis, eds.), pp. 258–273. Oxford University Press, Oxford.

West, K. P., Jr., Howard, G. R., and Sommer, A. (1989). Vitamin A and infection: public health implications. *Annu. Rev. Nutr.* **9,** 63–86.

West, K. P., Jr., Pokhrel, R. P., Katz, J., LeClerq, S. C., Khatry, S. K., Shrestha, S. R., Pradhan, E. K., Tielsch, J. M., Pandey, M. R., and Sommer, A. (1991). Efficacy of vitamin A in reducing preschool child mortality in Nepal. *Lancet* **338,** 67–71.

Wong, M. S., Bundy, D. A. P., and Golden, M. H. N. (1988). Quantitative assessment of geophagous behaviour as a potential source of exposure to geohelminth infection. *Trans. R. Soc. Trop. Med. Hyg.* **82,** 621–625.

Wong, M. S., Bundy, D. A. P., and Golden, M. H. N. (1991). The rate of ingestion of *Ascaris lumbricoides* and *Trichuris trichiura* eggs in soil and its relationship to infection in two children's homes in Jamaica. *Trans. R. Soc. Trop. Med. Hyg.* **85,** 89–91.

Victor Apanius & Gerhard A. Schad

Host Behavior and the Flow of Parasites through Host Populations

10

I. Introduction

If behavior is defined broadly as the complete repertoire of organismal responses to environmental, biotic, and social stimuli, then understanding host behavior provides a foundation for all considerations of parasite epidemiology. Several books using this expansive definition of behavior have been published (Barnard and Behnke, 1990; Loye and Zuk, 1991; Toft *et al.*, 1991) and provide a sweeping overview of the subject. This chapter focuses on those aspects of host behavior that are requisite for understanding patterns of parasitic infections in host populations. We view the behavioral decisions made by hosts as an optimization problem (Parker and Maynard Smith, 1990); the fitness cost of parasitism is but one of many ecological factors balanced by the host.

This chapter is based on the premise that selective forces acting on individual hosts will be reflected in populational responses in ecological as well as evolutionary time frames (Lomnicki, 1988; Anderson, 1991). The behavior of individuals is guided by a set of decision rules designed to maximize individual evolutionary fitness (Krebs and Davies, 1991). The composite of these individual-based outcomes determines the infection patterns that are observed at the population level by epidemiologists.

The successful control of parasitic disease in human populations ultimately depends on acceptance and compliance by individuals, yet individuals, or groups of individuals, may have motivations or goals that are in conflict with the goals of society as a whole. Some individuals may view compliance as an imposition on their daily routine or an unfair restriction to their personal freedom. Theoretically, not all individuals benefit from a vaccination program (Anderson and May, 1983; see also Chapter 11). This conflict between maximizing individual fitness in contrast to mean population fitness has been explored actively by evolutionary ecologists using game theory (Maynard Smith, 1982). This approach can be applied profitably to problems in parasite epidemiology, in which individual host behavior can frustrate intervention or programs and might appear to be counterproductive. The costs and benefits of preventative measures must be considered from the perspective of individuals embedded in social and economic networks that may favor behaviors that significantly affect parasite transmission. The primacy of individual-level selection will be

Parasitic and Infectious Diseases

apparent as the interlocking behavioral tactics that unite parasites and their hosts are considered. Each stage of this logical progression results from adaptations of parasites to increase transmission and ensuing counteradaptations by the host to resist infection. As with Combes (1991) use of the word "favorization," we will consider ethological processes that increase the probability of parasite transmission but care will be taken to distinguish between casual processes, such as adaptation, and simple consequences of disease, that is, physiological alteration of normal behavior. The distinction is important for epidemiologists because a mechanistic understanding of the processes that unite parasite with host should help clarify the patterns observed in nature. By focusing on the constraints as well as the opportunities for modifying host behavior, epidemiologists can better evaluate proposed intervention strategies.

II. Parasites Take Advantage of Inevitable Host Behavior to Maximize Their Transmission

A. Foraging Behavior

Numerous studies demonstrate the dependence of parasite transmission on host behavior (primary literature reviewed by Tinsley, 1990; Keymer and Read, 1991; Hart, 1992). A key generalization to emerge from these studies is that host behaviors that are necessary for survival, such as feeding or mating, are precisely the ones most likely to be used by parasites to effect their transmission. The reason is obvious: hosts that maximize their evolutionary fitness by outconsuming, or "outmating," their competitors will encounter a greater number of infective stages as a simple matter of probability.

If a host seeks alternative foods to avoid parasite exposure, it may have to accept food choices with lower nutritional value (Lozano, 1991). Because foraging decisions are thought to represent optimizations for maximizing survival and reproductive success (Stephens and Krebs, 1986), the pathophysiological cost of ingesting parasite-bearing food items may be balanced by a greater energetic profitability of consuming prey that contain infective stages (Drobney *et al.*, 1983; Moore and Bell, 1983). Female wood ducks *(Aix Sponsa)* increase their intake of aquatic invertebrates during the breeding season to supply the protein for their eggs; increased dietary protein leads to larger clutches of eggs (Drobney and Fredrickson, 1985). The increased intake of invertebrates is linked to an increase in the prevalence and intensity of gut helminths in females relative to males (Fig. 1). Conversely, the host may avoid foods that have a high probability of containing infective stages, as discussed in Section III.

Optimal foraging theory has been invoked to explain the prey choices expressed by three-spined sticklebacks *(Gasterosteus aculeatus)* parasitized by microsporidean or larval cestode parasites (Milinski, 1984). The fish consume smaller prey relative to uninfected controls because they presumably are inferior competitors.

Rationalizing the consumer decisions that affect dietary acquisition of parasites may require a complete accounting of costs and benefits inherent in that food. Dynamic programming methodology permits prediction of the optimal blend of dietary items, given adequate information about their relative costs and benefits (Mangel and Clark, 1988). In the future, the modeling of parasite transmission should consider trade-offs in consumer choice and will require integrating nutritional and ecological information with parasitological data.

B. Reproductive Behavior

Depending on the breeding system, hosts might maximize their fitness through multiple matings and thereby increase their risk of exposure to venereally transmitted parasites (Smith and Dobson, 1992). Models of

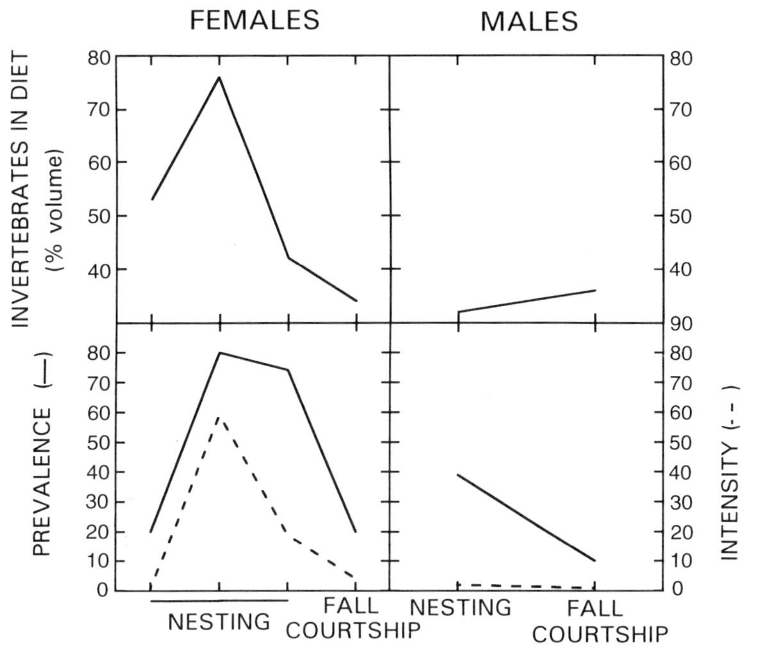

Figure 1 Relationship between foraging behavior and gut helminthiasis in wood ducks *(Aix sponsa)*. During the nesting period, show females ingest a greater proportion of invertebrates and display a concomitant increase in the prevalence (______) and intensity (– – –) of gut-dwelling cestodes and trematodes, compared with males. The dietary invertebrates are presumably the intermediate hosts for the helminths. Redrawn with permission from Drobney *et al.* (1983).

the epidemiology of the human immunodeficiency virus powerfully demonstrate the sensitivity of transmission dynamics to host sexual behavior (Anderson *et al.*, 1991). Attempts to increase their reproductive fitness through multiple matings increase the risk of parasite transmission in wildlife populations. Male spadefoot toads *(Scaphiophus couchii)* re-enter temporary pools of water in an attempt to fertilize multiple females. Males that are unsuccessful at mating spend even more time in the water than successful males. These differences in reproductive effort are reflected in the intensity of infection by the water-transmitted monogenean trematode *Pseudodiplorchis americanus.* Compared with females, infection intensity is greater in males and greatest in unsuccessful males (Tinsley and Jackson, 1988). Thus, parasitic intensity differs between the sexes because of divergent reproductive strategies and varies among males based on their reproductive effort (Zuk, 1990).

C. Social Behavior

Social aggregations of hosts provide exceptional transmission opportunities for many of their parasites, especially those with direct life cycles. The observed increase in ectoparasite load experienced by colonial animals appears to be offset by benefits derived from other ecological factors, such as information sharing and predator avoidance (Freeland, 1976; Hoogland and Sherman, 1976; Hoogland, 1979; Brown, 1986; Brown and Brown, 1986; Møller, 1987; Shields and Crook, 1987). On the other hand, social aggregations can benefit hosts by reducing the impact of biting flies through a dilution effect (fewer bites per host in large groups) as well as through the "selfish herd" effect

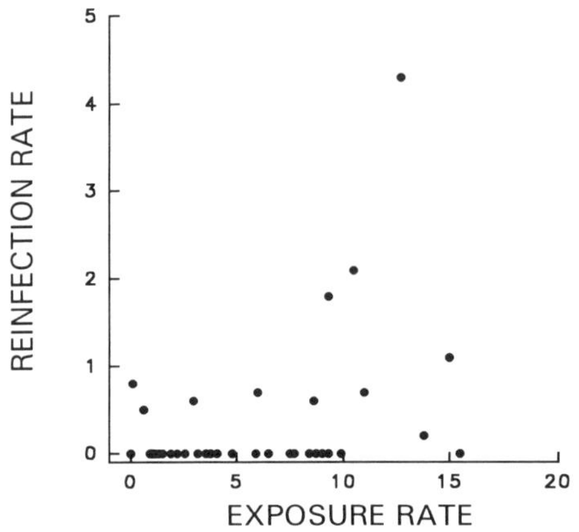

Figure 2 Water-contact activities of children increase the rate of re-infection by *Schistosoma haematobium* following praziquantel treatment. Exposure rate is a function of water contact duration, type of activity, time of day, and the density of infected snails. Re-infection rate refers to parasite egg output. Redrawn with permission from Chandiwana *et al.* (1991).

(hosts in center of herd receive fewer bites; reviewed by Hart, 1992).

D. Behavioral Specialization

Any division of behavioral roles within a population may require separate parameters in a model to account for heterogeneity in exposure for parasitic infection. For humans, the amount of time spent in contact with water varies considerably as a function of sex and age; adult males fish, adult females wash clothes, pre-adolescent children play in water, and so on (see Chapter 15). This variation in water contact has been correlated with the intensity of re-infection by *Schistosoma haematobium* following chemotherapeutic cure (Fig. 2, Chandiwana *et al.,* 1991).

In this case, this type of analysis depends on accurate estimates of infected snail density, and diurnal timing of cercarial release and human water-contact activities. When these sources of variation were measured, the resulting index of exposure accounted for 16% of the variation in egg output on re-infection (Chandiwana *et al.,* 1991). In other studies, sociological factors such as occupational exposure have accounted for up to 60% of the variance in hookworm egg output (Schad *et al.,* 1983). The unexplained variance may reflect genetic or immunological predisposition to infection, a topic considered in Section V. Future advances in this area of study will depend on experiments as well as on more sophisticated statistical models that examine the co-variance structure of longitudinal sampling designs (Hsiao, 1986; von Eye, 1990).

This chapter has addressed aspects of host behavior that parasites exploit for their transmission. In general, the host will attempt to modify behaviors to minimize the risk of parasite transmission, as discussed in the next section, but parasite avoidance may conflict with the host goal of maximizing its own lifetime reproductive success. When parasite transmission is linked to foraging, reproduction, or essential maintenance functions, the cost of parasitism may have to be tolerated in exchange for the fitness gained by these unavoidable activities.

III. Hosts Show Innate and Acquired Behavioral Resistance to Parasites

A. Parasite Avoidance

Some animals display the ability to discriminate food containing infective stages of parasites. Cattle ingest fewer nematode larvae than expected, compared with random samples of forage (Michel, 1955), and they avoid grazing near localized concentrations of *Boophilus microplus* tick larvae (Sutherst *et al.,* 1986). Aversion to food containing infective *Nippostrongylus brasiliensis* larvae can be learned through classical conditioning (Fig. 3; Keymer *et al.,* 1983), demonstrating that behavioral resistance to parasitism can be acquired phenotypically through learning. The mode of inheritance of these parasite-avoidance foraging behaviors deserves further investigation.

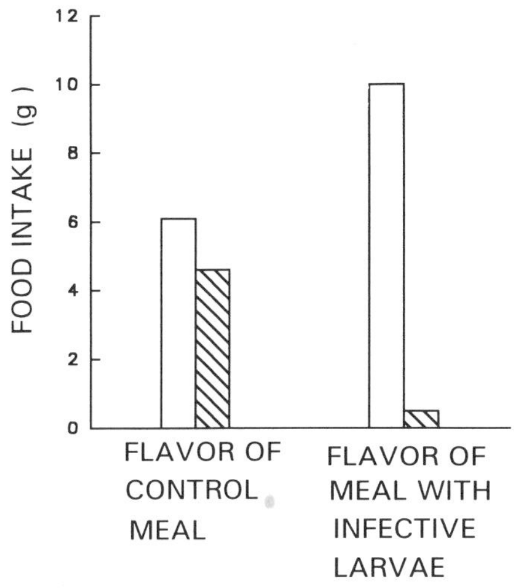

Figure 3 Learned taste-aversion of rats *(Rattus norvegicus)* in response to a flavorant associated with a meal containing infective *Nippostrongylus brasiliensis* larvae. Infected rats (hatched bars) reduced their food intake of subsequent meals containing the same flavorant but did not alter their intake of meals flavored with the control flavorant. Control rats are represented by open bars. Redrawn with permission from Keymer *et al.* (1983).

B. Parasite Removal

Self-maintenance behaviors inform the observer that the host is responding to ectoparasitism. Interference with the grooming ability of house mice *(Mus domesticus)* leads to a wider distribution of anopluran lice *(Polyplax serrata)* on the host body (Murray, 1961). The amount of time spent preening or grooming correlates with intensity of infection. One consequence of poor foraging performance is diminution in the amount of time available for grooming (Murray, 1990; Clayton, 1991a), resulting in further damage to the plumage or pelage and the loss of insulation, which may compromise the thermoregulatory ability of the host. Further, the unthrifty appearance characteristic of heavy ectoparasite burdens provides information on the health status of an individual to potential mates.

Allopreening or mutual grooming appears to be effective at reducing ectoparasite populations and may constitute a benefit of sociality, compensating for the increased ectoparasite transmission facilitated by group living. Social strife may upset the normal pattern of allogrooming and permit ectoparasite populations to increase on parts of the body that are inaccessible to self-grooming. As an extension of self-grooming, birds practice dust-bathing and anting, the application of ant body fluids to the plumage. These behaviors appear to have anti-ectoparasitic functions, although further study is needed (Murray, 1990).

C. Chemotherapy

Infected hosts can resist parasitism by utilizing plants that have parasitocidal activity. Some birds, especially raptors, bring plant foliage to their nests. European starlings introduced into North America preferentially obtain plants that contain secondary metabolites that have biological activity against ectoparasites (Fig. 4; Clark and Mason, 1988; Clark, 1991). Selective ingestion of plants containing secondary metabolites with anthelmintic activity has been noted for chimpanzees (*Pan troglodytes;* Wrangham and Nishida, 1983; Huffman and Seifu, 1989; Newton and Nishida, 1990) and Ethiopian baboons (*Papio* spp.; Phillips-Conroy, 1986). These investigations suggest that animals, like humans, are capable of self-administering pharmacologically active plants in response to parasitism (reviewed by Newton, 1991). Whether medicinal plant usage is more prevalent in social species, which are presumably at greater risk of infection with directly transmitted parasites, is not known.

D. Mate Choice

In addition to being wary of parasitism when making dietary choices, animals appear to choose mates based on their parasite load (Howard and Minchella, 1990; reviewed by Clayton, 1991b). Hamilton and Zuk (1982) hypothesized that the elaborate

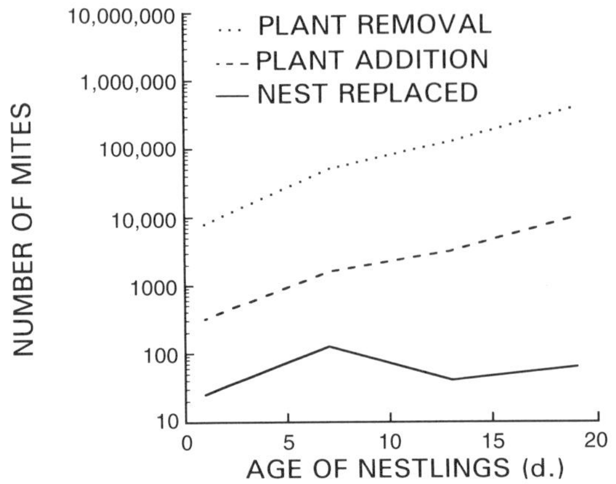

Figure 4 Mean number of *Ornithonysus sylvarium* mites in nests of European starlings *(Sturnus vulgaris)* in which the plant material brought to the nest by the birds was removed or supplemented. For the nest-replaced treatment, the entire nest was replaced with noninfested dry plant matter. Wild carrot *(Daucus carota)* was brought to the nests selectively by the birds and was added to the nests in the plant-addition treatment. Release of volatile compounds from this plant appears to control the populations of this hematophagous ectoparasite. Redrawn with permission from Clark and Mason (1988).

displays and ornamentations produced by sexual selection advertise parasite resistance genes to potential mates. The ciliate *Ichthyophthirius multifiliis* infects sticklebacks and causes white-spot disease. Infected males are colored less intensely and are less likely to be chosen by receptive females (Milinski and Bakker, 1990). Parasitism does not affect mate choice under green lighting conditions, which render indistinguishable the difference in red coloration between infected and uninfected fish (Fig. 5).

Compelling correlational and experimental evidence links mate choice and parasite burdens, especially for birds (Clayton, 1991b, and references therein). However, the interpretation hinges on a causal understanding of parasite resistance mechanisms. Parasitism may degrade the physiological performance of individuals and parasite-induced variation in mate quality may be advertised through the sexually selected traits (Howard and Minchella, 1990). Many ecological factors in addition to parasitism can affect the expression of the signal traits (Hill, 1991; Kirkpatrick and Ryan, 1991). Thus, whether these traits serve as an all-purpose grading scheme of phenotypic mate quality or whether the traits specifically advertise genetically based parasite resistance is not yet clear. Unfortunately, animal models that combine well-developed sexually selected traits with an established genetic basis for resistance to parasitism, such as avian coccidiosis (Lillehoj *et al.,* 1991), have been underutilized and leave the essential question unaddressed.

Rodents can discriminate between potential mates by olfactory clues that are linked genetically to the major histocompatibility complex (Yamazaki *et al.,* 1976; Egid and Brown, 1989; Potts *et al.,* 1991). This highly polymorphic gene complex influences immune-mediated disease resistance (Klein,

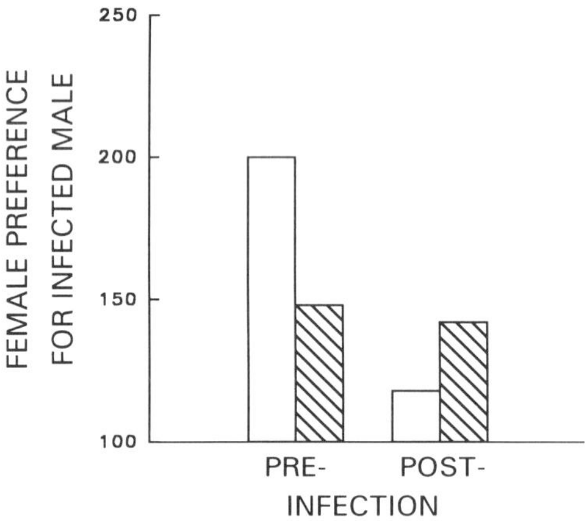

Figure 5 Female sticklebacks *(Gasterosteus aculeatus)* avoid mating with parasitized males under white light (open bars) but show no preference under green lighting (hatched bars). Parasitization by the ciliate *Ichthyophthirius multifiliis* reduces the red coloration of breeding males. The variation in coloration presumably is detected by females that are ready to spawn, as indicated by the avoidance of parasitized males as a function of illumination color. Redrawn with permission from Milinski and Bakker (1990).

1986; Wakelin and Blackwell, 1988; Potts and Wakeland, 1990; see Chapter 8). Whether this observation represents another independent verification of the linkage between mate choice and parasite resistance genes remains to be seen.

E. Suicidal Removal

Suicide is the ultimate behavioral response of an individual host to counter parasitism. Suicide can be adaptive to the host from the perspective of inclusive fitness (Smith Trail, 1980) or by group selection. If an infected host disperses or perishes before maturation of infective stages, it will spare kin or group members a presumably similar outcome. An experimental test of this hypothesis (McAllister and Roitberg, 1987) has been criticized on the grounds that the data are not sufficient to exclude alternative hypotheses (Latta, 1987; Tomlinson, 1987; but see McAllister and Roitberg, 1988). In addition to a better understanding of the host population genetic structure and the heritability of this trait, cladistic tests of adaptational hypotheses (Coddington, 1988) also might be fruitful approaches for verification.

IV. Parasites Alter Host Behavior to Enhance Their Transmission

A. Enhanced Transmission to Predatory Hosts

Compelling evidence suggests that some parasites modify host behavior to increase their transmission. Well-documented examples include protozoan (Hoogenboom and Dijkstra, 1987; Quinn *et al.*, 1987) and metazoan parasites, such as digenean trematodes (Brassard *et al.*, 1982; Curtis, 1987), nematodes (Rau, 1983a; Rau and Putter, 1984; Moore and Lasswell, 1986), and especially acanthocephalans (Bethel and Holmes, 1977; Moore, 1983a,b). In general, parasitic infection renders the intermediate host more susceptible to predation by the next host in the parasite life cycle. The prey becomes more visually conspicuous or normal avoidance behaviors are impeded. For hematophagous arthropod vectors, infection by parasites induces an increased probing rate, which increases the likelihood of transmission to the vertebrate host (reviewed by Molyneux and Jefferies, 1986; Dye, 1990).

B. Enhanced Transmission to Herbivorous Animals

In a few cases, the induced behavioral change is so complex that the claim of it being a transmission-enhancing adaptation by the parasite has not been contested. Cercariae of the digenean trematode *Dicrocoelium dendriticum* migrate into the nervous system of the intermediate host, various *Formica* spp., where typically only one of a cohort of worms localizes in the subesophageal ganglion (Schneider and Hohorst, 1971). In concert with invasion of the host nervous system by an individual worm, the behavior of the ant changes dramatically—as temperature decrease during the evening, infected ants uncharacteristically ascend and affix themselves to vegetation with clenched mandibles (Hohorst and Graefe, 1961; Hohorst, 1962; Hohorst and Lämmler, 1962). The obvious role of this behavioral change for positioning the intermediate host at an opportune site and time for consumption by the herbivorous definitive host, grazing sheep (*Ovis* spp.), is claimed as evidence for manipulation of intermediate host behavior by the parasite. Further, ants of the genus *Camponotus* appear to be lethargic after infection by another dicrocoelid trematode *Brachylecithum mosquensis,* thus making the ants more susceptible to the avian predators that are the definitive host for the fluke (Carney, 1969). This information suggests that closely related parasite species induce specific behaviors in a host species to match the transmission mode to the next host.

C. Mechanisms and Causality

Other examples of parasite-induced behavioral modification are less complex. Whether the behavioral alterations represent adaptive manipulation of host behavior by the parasite or a simple pathological consequence of infection is not always clear (Moore and Gotelli, 1990; Combes, 1991). Currently, the complexity or degree of behavior modification is considered a criterion for the adaptive nature of modifying intermediate host behavior. Cladistic tests of adaptational hypotheses (Coddington, 1988) could be applied profitably here, as suggested by Moore and Gotelli (1990).

The mechanisms by which parasites alter host behavior have received little attention (but see Romig *et al.,* 1980; Zohar and Rau, 1986; Helluy and Holmes, 1990). That foliage ascension and mandibular grasping behavior can be induced in ants by a trematode as well as by the fungus *Entomophthora* sp. (Loos-Frank and Zimmerman, 1976) suggests that a common neural mechanism may be manipulated by unrelated parasitic taxa.

Regardless of the mechanism, theoretical work has demonstrated the powerful effect that transmission-enhancing behavioral changes will have on transmission dynamics (Dobson, 1988). An important outcome of this modeling is the prediction that behavioral modification will sustain populations of indirect life cycle parasites at very low definitive host densities. This result coincides with the observation that most examples of behavioral modification involve large mobile predators with large home ranges and low population densities.

This chapter has addressed transmission-enhancing adaptations of parasites that utilize intrinsic and unavoidable host behaviors. Next, we examined the counter adaptations used by hosts to resist parasitic infection behaviorally. In this section, we considered the escalating tactic of modifying host behavior to maximize parasite transmission. (Note that this sequence does not necessarily represent the actual course of events in nature.) The underlying question is, "Who is manipulating whom?" This question is not only extremely difficult to answer, but may be overly simplistic if genetic adaptations are coevolutionarily coupled. In the next section, the concept that the relationship between host behavior and parasite transmission is evolutionarily dynamic and, at least for the host, may be decoupled from simple genetic determination is examined. For species exhibiting reciprocal adaptations for mediating ecological interactions, the application of game theory and the concept of evolutionary stable states provide the conceptual framework for understanding their long-term dynamics.

V. Host Behavioral Plasticity as an Evolved Response to Counter Parasitism

A. Escalation and Diversification

Except for learned responses, genetic determination for many of the behaviors discussed in previous sections was implicit. For example, the grooming response of mice is determined genetically, since mice born without forelimbs show characteristic neck motions associated with face-grooming movements (Fentress, 1973). A general feature of host–parasite relationships is the escalating nature of coevolved adaptations and counter-adaptations. These antagonistic ecologic interactions have been proposed to entrain mutual genetic changes, or co-adaptational cycles, through negative frequency-dependent selection (Hamilton, 1980,1982). This mode of natural selection not only is capable of preserving genetic diversity in populations, but may be a driving force that favors new variants that initially appear at very low frequencies. The selective advantage for rate variants under frequency-dependent selection may be equally applicable for genetically or culturally inherited traits.

Phenotypic plasticity is an evolved response of organisms for mediating ecological interactions on time scales shorter than those required for genetically fixed traits (Thompson, 1991). This characteristic may play a significant role in mediating behavioral resistance to parasitism. Diamond (1987) suggested that epigenetic behavioral specializations for foraging could be driven by the diversity of toxic metabolites in plant and animal diets. The selective advantages of behavioral plasticity, especially learning, are equally applicable to the escalating tactics that characterize host–parasite as well as plant–animal co-evolution. As an example, the phenotypic regulation of host growth has been proposed as a counter–adaptation against parasitic castration of molluscs by trematodes (Minchella *et al.*, 1985).

Game-theoretic analysis of a one parasite/one host system shows that natural selection acting on individuals is sufficient to explain the diversification of adaptive strategies between antagonistic species (Cohen and Newman, 1989). These results are generalizable to behavioral strategies as well as to genetic, physiological, and immunological ones (Cohen and Newman, 1989). Although additional work is required to test the importance of the simplifying assumptions, the promise of this approach is evident. Behavioral resistance to parasitism can be especially effective if it becomes de-coupled from genetic determination. Phenotypic plasticity provides a mode of behavioral resistance without requiring direct genetic determination for each trait. This feature avoids the predictability of genetically based behaviors that could be exploited by parasites. Learned responses, especially those that are transmitted culturally, permit an extremely rapid response for countering parasite transmission tactics, which are assumed to be genetically fixed. Speculating that behavioral flexibility of the host reduces the evolutionary importance of generation time for host responses to parasites is tempting. Just as phenotypic plasticity of the vertebrae immune system permits rapid somatic evolution of the lymphocyte-receptor repertoire, host behavioral plasticity may represent accelerated evolution of behavior-mediated resistance against parasites.

B. Learning and Cultural Transmission

The ability to prevent parasite transmission through culturally acquired behaviors is well-known for humans, but less so for wild animals, and may reflect the difficulty of obtaining adequate documentation for most animals. However, connections are almost made. Cultural transmission of feeding method has been observed for many animals, including the oystercatcher *Haematopus ostralegus* (see Sutherland, 1987). Oystercatchers have the capacity to recognize and discriminate against bivalves harboring infectious trematode sporocysts (Hulscher, 1982). Whether oystercatchers teach their offspring to avoid prey containing infective stages is unknown currently. In human populations, the relationship between cultural habits and parasite transmission has been documented throughout the world (reviewed by Croll and Cross, 1983; Bundy and Blumenthal, 1990; Nelson, 1990). The aversion to pork consumption in some cultures is an overworked example of the ancient association between cultural traditions and parasite avoidance behavior (Nelson, 1990). Variation in the age-dependent acquisition of hookworm larvae by children has been correlated with religious affiliation (Fig. 6) and corresponding defecation habits (Schad *et al.*, 1983; Hominick *et al.*, 1987). Variation in the intensity of roundworm *Ascaris lumbricoides* infections was associated with caste membership in an Indian fishing village (Haswell-Elkins *et al.*, 1989). To the extent that group members uniformly practice certain behaviors or reside in a specific locality, social group membership can yield

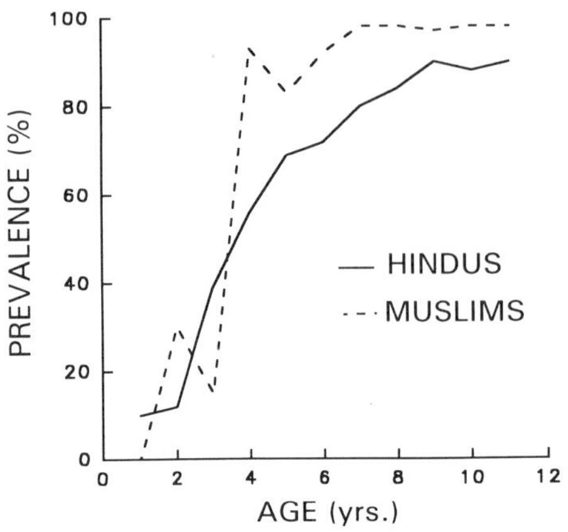

Figure 6 Difference in age-prevalence curves for human hookworm infections as a function of cultural practices. Infections of *Ancyclostoma duodenale* and *Necator americanus* are more prevalent in Muslims (– – –) who are thought to be less fastidious in their defecation habits than Hindus (——). Redrawn with permission from Schad *et al.* (1983).

predictive information about the risk of parasitic infection.

C. Social Status

Within a school, flock, tribe, sect, or any other grouping of animals or humans, the differences among individuals within a group can be as epidemiologically relevant as the differences among groups. The most prominent within-group behavioral differences often relate to dominance and subordinance relationships. Social structure based on a hierarchical ranking of individuals is a form of behavioral plasticity, since individuals modify foraging, reproductive, and social activities conditional on their social status.

The relationship between social relationships and parasitism is rather complex. Dominant individuals may enjoy better access to food resources but also may incur a greater risk of acquiring infectious stages (Schad, 1962; Hausfater and Watson, 1976; Halvorsen, 1986). On the other hand, mice experimentally infected with nematodes were unable to become behaviorally dominant (Freeland, 1981; Rau, 1983b) or lost their dominance ranking (Rau, 1984), indicating that parasitism can affect hierarchical social relationships causally.

Perturbation of social relationships can induce a physiological stress response leading to increased circulating levels of corticosteroids (Dantzer and Mormède, 1983; Mormède, 1990). Since corticosteroids have immunosuppressive activity (Khansari *et al.,* 1990), social stressors can compromise immune-mediated resistance against parasites. Empirical evidence of the linkage between social factors and antiparasite immunity is equivocal; however, this may be a result of different neurohormonal responses to acute and chronic stressors (Klein *et al.,* 1992). Despite the methodological difficulties, investigating the interaction between social relationships, endocrinology, and immunocompetence should be a provocative area for future research. At this time, concluding that a greater intensity of parasitic infection can be both a consequence as well as a cause of social status appears safe.

The socioeconomic structuring of human populations adds an additional layer of complexity to epidemiological analysis. That socioeconomic level is associated with genetic variation in blood group antigens (Beardmore and Karimi-Booshehri, 1983), and perhaps other genetic loci, complicates attempts to partition social and genetic effects on the susceptibility to helminth infections. The roles of genetic and social factors in determining the prevalence and intensity of helminth infections (Schad and Anderson, 1985; Keymer and Pagel, 1990; McCallum, 1990; Bundy and Medley, 1992) may prove to be indistinguishable since these two factors can be counfounded statistically. Further, the experimental verification that trichuriasis negatively affects some cognitive functions of school children (Fig. 7; Nokes *et al.,* 1992) underscores the interrelatedness of parasitism, cultural practices, and social advancement. Hopefully, future analyses will recognize multiple causality and will be

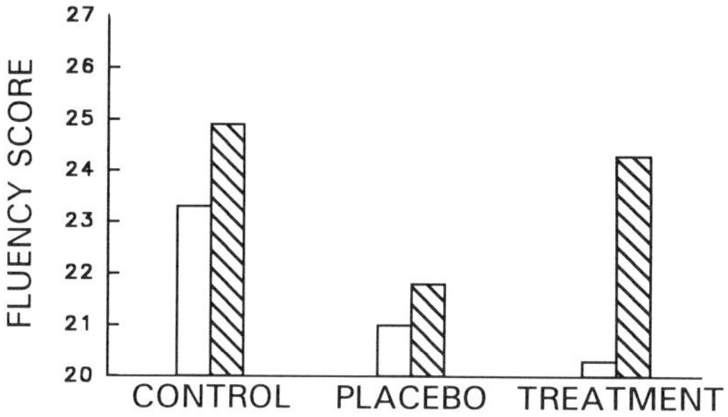

Figure 7 Effect of *Trichuris trichuria* infection on cognitive function of school children. Controls initially were noninfected and showed a greater fluency score than infected children (placebo and treatment) (open bars). After anthelmintic treatment (hatched bars) fluency ability significantly increased in the treatment but not the placebo groups. Redrawn with permission from Nokes *et al.* (1992).

fashioned in the style of variance components (McCallum, 1990) or path analysis (Loehlin, 1987), so the contribution of various causal factors can be compared. In addition to quantifying the risk of parasitic infection based on socioeconomic status, information about the genetic structure of the study population will be equally essential. Epidemiological analysis that can combine genetic, immunological, and behavioral information will greatly improve current understanding of the factors that control the flow of parasites through host populations.

References

Anderson, R. M. (1991). Populations and infectious diseases: Ecology or epidemiology? *J. Anim. Ecol.* **60,** 1–50.

Anderson, R. M., and May, R. M. (1983). Vaccination against rubella and measles: Quantitative investigations of different policies. *J. Hyg.* **90,** 259–325.

Anderson, R. M., May, R. M., Boily, M. C., Garnett, G. P., and Rowley, J. T. (1991). The spread of HIV-1 in Africa: Sexual contact patterns and the predicted demographic impact of AIDS. *Nature (London)* **352,** 581–589.

Barnard, C. J., and Behnke, J. M. (eds.) (1990). "Parasitism and Host Behaviour." Taylor and Francis, London.

Beardmore, J. A., and Karimi-Boosherhi, F. (1983). ABO genes are differentially distributed in socioeconomic groups in England. *Nature (London)* **303,** 522–524.

Bethel, W. M., and Holmes, J. C. (1977). Increased vulnerability of amphipods to predation owing to altered behavior induced by larval acanthocephalans. *Can. J. Zool.* **55,** 110–115.

Brassard, P., Rau, M. E., and Curtis, M. A. (1982). Parasite-induced susceptibility to predation in diplostomiasis. *Parasitology* **85,** 495–501.

Brown, C. R. (1986). Cliff swallow colonies as information centers. *Science* **234,** 83–85.

Brown, C. R., and Brown, M. B. (1986). Ectoparasitism as a cost of coloniality in cliff swallows *Hirundo pyrrhonota. Ecology* **67,** 1206–1218.

Bundy, D. A. P., and Blumenthal, U. J. (1990). Human behaviour and the epidemiology of helminth infections: The role of behaviour in exposure to infection. *In* "Parasitism and Host Behaviour" (C. J. Barnard and J. M. Behnke, eds.), pp. 264–289. Taylor and Francis, London.

Bundy, D. A. P., and Medley, G. F. (1992). Immunoepidemiology of human geoheminthiasis: Ecological and immunological determinants of worm burden. *Parasitology* **104,** S105–119.

Carney, W. P. (1969). Behavioral and morphological changes in carpenter ants harboring dicrocoeliid metacercariae. *Am. Midl. Nat.* **82,** 605–611.

Chandiwana, S. K., Woolhouse, M. E. J., and Bradley, M. (1991). Factors affecting the intensity of re-infection with *Schistosoma haemotobium* following treatment with praziquantel. *Parasitology* **102,** 73–83.

Clark, L. (1991). The nest protection hypothesis: The adaptive use of plant secondary compounds by European starlings. *In* "Bird–Parasite Interactions: Ecology, Evolution and Behavior" (J. E. Loye and M. Zuk, eds.), pp. 205–221. Oxford University Press, Oxford.

Clark, L., and Mason, J. R. (1988). Effect of biologically active plants used as nest material and the derived benefit to starling nestlings. *Oecologica* **77,** 174–180.

Clayton, D. (1991a). Coevolution of avian grooming and ectoparasite avoidance. *In* "Bird–Parasite Interactions: Ecology, Evolution and Behavior" (J. E. Loye and M. Zuk, eds.), pp. 258–289. Oxford University Press, Oxford.

Clayton, D. (1991b). The influence of parasites on host sexual selection. *Parasitol. Today* **7,** 329–334.

Coddington, J. A. (1988). Cladistic tests of adaptational hypotheses. *Cladistics* **4,** 3–22.

Cohen, J. E., and Newman, C. M. (1989). Host–parasite relations and random zero-sum games: The stabilizing effect of strategy diversification. *Am. Nat.* **133,** 533–552.

Combes, C. (1991). Ethological aspects of parasite transmission. *Am. Nat.* **138,** 866–880.

Croll, N. A., and Cross, J. (1983). "Human Ecology

and Infectious Diseases." Academic Press, New York.
Curtis, L. A. (1987). Vertical distribution of an estuarine snail altered by a parasite. *Science,* **235,** 1509–1511.
Dantzer, R., and Mormède, P. (1983). Stress in farm animals: A need for reevaluation. *J. Anim. Sci.* **57,** 6–18.
Diamond, J. M. (1987). Learned specializations of birds. *Nature (London)* **330,** 16–17.
Dobson, A. P. (1988). The population biology of parasite-induced changes in host behavior. *Q. Rev. Biol.* **63,** 139–165.
Drobney, R. D., and Fredrickson, L. H. (1985). Protein acquisition: A possible proximate factor limiting clutch size in wood ducks. *Wildfowl,* **36,** 122–128.
Drobney, R. D., Train, C. T., and Fredrickson, L. H. (1983). Dynamics of the platyhelminth fauna of wood ducks in relation to food habits and reproductive state. *J. Parasitol.* **69,** 375–380.
Dye, C. (1990). Epidemiological significance of vector–parasite interactions. *Parasitology* **101,** 409–415.
Egid, K., and Brown, J. L. (1989). The major histocompatibility complex and female mating preference in mice. *Anim. Behav.* **38,** 548–549.
Fentress, J. C. (1973). Development of grooming in mice with amputated forelimbs. *Science* **179,** 704–705.
Freeland, W. J. (1976). Pathogens and the evolution of primate sociality. *Biotropica* **8,** 12–24.
Freeland, W. J. (1981). Parasitism and behavioral dominance among male mice. *Science* **213,** 461–462.
Halvorsen, O. (1986). On the relationship between social status of host and risk of parasitic infection. *Oikos* **47,** 71–74.
Hamilton, W. D. (1980). Sex versus non-sex versus parasite. *Oikos* **35,** 282–290.
Hamilton, W. D. (1982). Pathogens as causes of genetic diversity in their host populations. *In.* "Population Biology of Infectious Diseases" (R. M. Anderson and R. M. May, eds.), pp. 269–296. Springer-Verlag, Berlin.
Hamilton, W. D., and Zuk, M. (1982). Heritable true fitness and bright birds: a role for parasites? *Science* **218,** 384–387.
Hart, B. L. (1992). Behavioral adaptations to parasites: An ethological approach. *J. Parasitol.* **78,** 256–265.
Haswell-Elkins, M., Elkins, D., and Anderson, R. M. (1989). The influence of individual, social group and household factors on the distribution of *Ascaris lumbricoides* within a community and implications for control strategies. *Parasitology* **98,** 125–134.
Hausfater, G., and Watson, D. F. (1976). Social and reproductive correlates of parasite ova emissions by baboons. *Nature (London)* **262,** 688–689.
Helluy, S., and Holmes, J. C. (1990). Serotonin, octopamine, and the clinging behavior induced by the parasite *Polymorphus paradoxus* (Acanthocephala) in *Gammarus lacustris* (Crustacea). *Can. J. Zool.* **68,** 1214–1220.
Hill, G. E. (1991). Plumage coloration is a sexually selected indicator of male quality. *Nature (London)* **350,** 337–339.
Hohorst, W. (1962). Die Rolle der Ameisen im Entwicklungsgang des Lanzettegels *(Dicrocoelium dendriticum). Z. Parasitenkd.* **22,** 105–106.
Hohorst, W., and Graefe, G. (1961). Ameisein–Obligatorische Zwischenwirte des Lanzettegels *(Dicrocoelium dendriticum). Naturwiss.* **48,** 229–230.
Hohorst, W., and Lämmler, G. (1962). Experimentelle Dicrocoeliose-Studien. *Tropenmed. Parasitol.* **13,** 377–397.
Hominick, W. M., Dean, C. G., and Schad, G. A. (1987). Population biology of hookworms in West Bengal: Analysis of numbers of infective larvae recovered from damp pads applied to the soil surface at defaecation sites. *Trans. R. Soc. Trop. Med. Hyg.* **81,** 978–986.
Hoogenboom, I., and Dijkstra, C. (1987). *Sarcocystis cernae:* A parasite increasing risk of predation of its intermediate host, *Microtus arvalis. Oecologia* **74,** 86–92.
Hoogland, J. L. (1979). Aggression, ectoparasitism and other possible costs of prairie dog (Sciuridae, *Cynomys* spp.) coloniality. *Behaviour* **69,** 1–35.
Hoogland, J. L., and Sherman, P. W. (1976). Advantages and disadvantages of bank swallow *(Riparia riparia)* coloniality. *Ecol. Monogr.* **46,** 33–58.
Howard, R. D., and Minchella, D. J. (1990). Parasitism and mate competition. *Oikos,* **58,** 120–122.
Hsiao, C. (1986). "Analysis of Panel Data." Cambridge University Press, Cambridge.
Huffman, M. A., and Seifu, M. (1989). Observations on the illness and consumption of a possibly medicinal plant *Vernonia amygdalina* (Del.) by a wild chimpanzee in the Mahale Mountains National Park, Tanzania. *Primates* **30,** 51–63.
Hulscher, J. B. (1982). The oystercatcher *Haematopus ostralegus* as a predator of the bivalve *Macoma balthica* in the Dutch Wadden Sea. *Ardea* **70,** 89–152.
Keymer, A. E., and Pagel, M. (1990). Predisposition to helminth infection. *In* "Hookworm Disease: Current Status and New Directions" (G. A. Schad and K. S. Warren, eds.), pp. 177–209. Taylor and Francis, London.
Keymer, A. E., and Read, A. F. (1991). Behavioural ecology: The impact of parasitism. *In* "Parasite–Host Associations: Coexistence or Conflict?" (C. A. Toft, A. Aeschlimann, and L. Bolis, eds.), pp. 37–61. Oxford University Press, Oxford.
Keymer, A., Crompton, D. W. T., and Sahakian, B. J. (1983). Parasite-induced learned taste aversion involving *Nippostrongylus* in rats. *Parasitology* **86,** 455–460.
Khansari, D. N., Murgo, A. J., and Faith, R. E. (1990).

Effects of stress on the immune system. *Immunol. Today* **11,** 170–175.
Kirkpatrick, M., and Ryan, M. J. (1991). The evolution of mating preference and the paradox of the lek. *Nature (London)* **350,** 33–38.
Klein, J. (1986). "The Natural History of the Major Histocompatibility Complex." Wiley, New York.
Klein, F., Lemaire, V., Sandi, C., Vitiello, S., Van der Logt, J., Laurent, P. E., Neveu, P., Le Moal, M., and Mormède, P. (1992). Prolonged increase of corticosterone secretion by chronic social stress does not necessarily impair immune function. *Life Sci.* **50,** 723–731.
Krebs, J. C., and Davies, N. B. (eds.) (1991). "Behavioral Ecology: An Evolutionary Approach." Sinauer Associates, Sunderland, Massachusetts.
Latta, B. (1987). Adaptive and non-adaptive suicide in aphids. *Nature (London)* **330,** 701.
Lillehoj, H. S., Ruff, M. D., Bacon, L. D., Lamont, S. J., and Jeffers, T. K. (1989). Genetic control of immunity to *Eimeria tenella.* Interaction of MHC genes and non-MHC linked genes influences levels of disease susceptibility in chickens. *Vet. Immunol. Immunopathol.* **20,** 135–148.
Loehlin, J. C. (1987). "Latent Variable Models: An Introduction to Factor, Path, and Structural Analysis." Erlbaum Associates, Hillsdale, New Jersey.
Lomnicki, A. (1988). "Population Ecology of Individuals." Princeton University Press, Princeton, New Jersey.
Loos-Frank, B., and Zimmerman, G. (1976). Über eine dem Dicrocoelium-Befall analoge Verhaltensänderung bei Ameisen der Gattung *Formica* durch einen Pilz der Gattung *Entomophthora. Z. Parasitenkd.* **49,** 281–298.
Loye, J. E., and Zuk, M. (eds.) (1991). "Bird-Parasite Interactions: Ecology, Evolution and Behavior." Oxford University Press, Oxford.
Lozano, G. A. (1991). Optimal foraging theory: A possible role for parasites. *Oikos* **60,** 391–395.
Mangel, M., and Clark, C. W. (1988). "Dynamic Modelling in Behavioral Ecology." Princeton University Press, Princeton, New Jersey.
Maynard Smith, J. (1982). "Evolution and the Theory of Games." Cambridge University Press, Cambridge.
McAllister, M. K., and Roitberg, B. D. (1987). Adaptive suicidal behavior in pea aphids. *Nature (London)* **328,** 797–799.
McAllister, M. K., and Roitberg, B. D. (1988). Assumptions about suicidal behavior of aphids. *Nature (London)* **332,** 494–495.
McCallum, H. I. (1990). Covariance in parasite burdens: The effect of predisposition to infection. *Parasitology* **100,** 153–159.
Michel, J. F. (1955). Parasitological significance of bovine grazing behavior. *Nature (London)* **175,** 1088–1089.
Milinski, M. (1984). Parasites determine a predator's optimal feeding strategy. *Behav. Ecol. Sociobiol.* **15,** 35–37.
Milinski, M., and Bakker, T. C. M. (1990). Female sticklebacks use male coloration in mate choice and hence avoid parasitized males. *Nature (London)* **344,** 330–333.
Minchella, D. J., Leathers, B. K., Brown, K. M., and McNair, J. N. (1985). Host and parasite counteradaptations: An example from a freshwater snail. *Am. Nat.* **126,** 843–854.
Møller, A. P. (1987). Advantages and disadvantages of coloniality in the swallow *Hirundo rustica. Anim. Behav.* **35,** 819–832.
Molyneux, D. H., and Jefferies, D. (1986). Feeding behavior of pathogen-infected vectors. *Parasitology* **92,** 721–736.
Moore, J. (1983a). Responses of an avian predator and its isopod prey to an acanthocephalan parasite. *Ecology* **64,** 1000–1015.
Moore, J. (1983b). Altered behavior in cockroaches *(Periplaneta americana)* infected with an archiacanthocephalan, *Moniliformis moniliformis. J. Parasitol.* **69,** 1174–1176.
Moore, J., and Bell, D. H. (1983). Pathology (?) of *Plagiorhynchus cylindraceus* in the starling *Sturnus vulgaris. J. Parasitol.* **69,** 387–390.
Moore, J., and Gotelli, N. J. (1990). A phylogenetic perspective on the evolution of altered host behaviors: A critical look at the manipulation hypothesis. *In* "Parasitism and Host Behavior" (C. J. Barnard and J. M. Behnke, eds), pp. 193–233. Taylor and Francis, London.
Moore, J., and Lasswell, J. (1986). Altered behavior in isopods *(Armadillidium vulgare)* infected with the nematode *Dispharynx nasuta. J. Parasitol.* **72,** 186–189.
Mormède, P. (1990). Neuroendocrine responses to social stress. *In* "Social Stress in Domestic Animals" (R. Zayan and R. Dantzer, eds.), pp. 203–211. Kluwer, Dordrecht, The Netherlands.
Murray, M. D. (1961). The ecology of the louse *Polyplax serrata* (Burm.) on the mouse *Mus musculus* L. *Austral. J. Zool.* **9,** 1–13.
Murray, M. D. (1990). Influence of host behaviour on some extoparasites of birds and mammals. *In* "Parasitism and Host Behaviour" (C. J. Barnard and J. M. Behnke, eds.), pp. 290–315. Taylor and Francis, London.
Nelson, G. S. (1990). Human behaviour and the epidemiology of helminth infections: Cultural practices and microepidemiology. *In* "Parasitism and Host Behavior" (C. J. Bernard and J. M. Behnke, eds.), pp. 234–263. Taylor and Francis, London.
Newton, P. (1991). The use of medicinal plants by primates: A missing link? *Trends Ecol. Evol.* **6,** 297–299.
Newton, P. N., and Nishida, T. (1990). Possible buccal administration of herbal drugs by wild chimpanzees, *Pan troglodytes. Anim. Behav.* **39,** 798–801.

Nokes, C., Grantham-McGregor, S. M., Sawyer, A. W., Cooper, E. S., and Bundy, D. A. P. (1992). Parasitic helminth infection and cognitive function in school children. *Proc. R. Soc. London B* **247,** 77–81.

Parker, G. A., and Maynard Smith, J. (1990). Optimality theory in evolutionary biology. *Nature (London)* **348,** 27–33.

Phillips-Conroy, J. E. (1986). Baboons, diet and disease: Food plant selection and schistosomiasis. *In* "Current Perspectives in Primate Social Dynamics" (D. M. Taub and F. A. King, eds.), pp. 287–304. Van Nostrand Reinhold, New York.

Potts, W. K., and Wakeland, E. K. (1990). Evolution of diversity at the major histocompatibility complex. *Trends Ecol. Evol.* **5,** 181–187.

Potts, W. K., Manning, C. J., and Wakeland, E. K. (1991). Mating patterns in seminatural populations of mice influenced by MHC genotype. *Nature (London)* **352,** 619–621.

Quinn, S. C., Brooks, R. J., and Cawthorn, R. J. (1987). Effects of the protozoan parasite *Sarcocystis rauschorum* on open-field behaviour of its intermediate vertebrate host, *Dicrostonyx richardsoni. J. Parasitol.* **73,** 265–271.

Rau, M. E. (1983a). The open-field behaviour of mice infected with *Trichinella spiralis. Parasitology* **86,** 311–318.

Rau, M. E. (1983b). Establishment and maintenance of behavioural dominance in male mice infected with *Trichinella spiralis. Parasitology* **86,** 319–322.

Rau, M. E. (1984). Loss of behavioural dominance in male mice infected with *Trichinella spiralis. Parasitology* **88,** 371–373.

Rau, M. E., and Putter, L. (1984). Running responses of *Trichinella spiralis*-infected CD-1 mice. *Parasitology* **89,** 579–583.

Romig, T., Lucius, R., and Frank, W. (1980). Cerebral larvae in the second intermediate host of *Dicrocoelium dendriticum* (Rudolphi, 1819) and *Dicrocoelium hospes* (Looss, 1907) (Trematodes, Dicrocoeliidae). *Z. Parasitenkd.* **63,** 277–286.

Schad, G. A. (1962). Helminth parasitism in a flock of domestic geese introduced to arctic summer conditions in Canada. *Can. J. Zool.* **40,** 1–4.

Schad, G. A., and Anderson, R. M. (1985). Predisposition to hookworm infection in humans. *Science* **228,** 1537–1539.

Schad, G. A., Nawalinski, T. A., and Rochar, V. K. (1983). Human ecology and the distribution and abundance of hookworm populations. *In* "Human Ecology and Infectious Diseases" (N. A. Croll and J. Cross, eds.), pp. 188–273. Academic Press, New York.

Schneider, G., and Hohorst, W. (1971). Wanderung der Metacercarien des Lanzett-egels in Ameisen. *Naturwiss.* **58,** 327–328.

Shields, W. M., and Crook, J. R. (1987). Barn swallow coloniality: A net cost for group breeding in the Adirondacks? *Ecology* **68,** 1373–1386.

Smith, G., and Dobson, A. P. (1992). Sexually transmitted diseases in animals. *Parasitol. Today* **8,** 159–166.

Smith Trail, D. R. (1980). Behavioral interactions between parasites and hosts: Host suicide and the evolution of complex life cycles. *Am. Nat.* **116,** 77–91.

Stephens, D. W., and Krebs, J. R. (1986). "Foraging Theory." Princeton University Press, Princeton, New Jersey.

Sutherland, W. J. (1987). Why do animals specialize? *Nature (London)* **325,** 483–484.

Sutherst, R. W., Floyd, R. B., Bourne, A. S., and Dallwitz, M. J. (1986). Cattle grazing behavior regulates tick populations. *Experientia* **42,** 194–196.

Thompson, J. D. (1991). Phenotypic plasticity as a component of evolutionary change. *Trends Ecol. Evol.* **6,** 246–249.

Tinsley, R. C. (1990). Host behaviour and opportunism in parasite life cycles. *In* "Parasitism and Host Behavior" (C. J. Barnard and J. M. Behnke, eds.), pp. 158–192. Taylor and Francis, London.

Tinsley, R. C., and Jackson, H. C. (1988). Pulsed transmission of *Pseudodiplorchis americanus* (Monogenea) between hosts *(Scaphiopus couchii). Parasitology* **97,** 437–452.

Toft, C. A., Aeschlimann, A., and Bolis, L. (eds.) (1991). "Parasite–Host Associations: Coexistence or Conflict?" Oxford University Press, Oxford.

Tomlinson, I. (1987). Adaptive and non-adaptive suicide in aphids. *Nature (London)* **330,** 701.

von Eye, A. (ed.) (1990). "Statistical Methods in Longitudinal Research." Academic Press, Boston.

Wakelin, D., and Blackwell, J. M. (eds.) (1988). "Genetics of Resistance to Bacterial and Parasitic Infection." Taylor and Francis, London.

Wrangham, R. W., and Nishida, T. (1983). *Aspilia* spp. leaves: A puzzle in the feeding behavior of wild chimpanzees. *Primates* **24,** 276–282.

Yamazaki, K., Boyse, E. A., Mike, V., Thaler, H. T., Mathieson, B. J., Abbott, J., Boyse, J., Zayas, Z. A., and Thomas, L. (1976). Control of mating preferences in mice by genes in the major histocompatibility complex. *J. Exp. Med.* **144,** 1324–1335.

Zohar, A. S., and Rau, M. E. (1986). The role of muscle larvae of *Trichinella spiralis* in the behavioral alterations of the mouse host. *J. Parasitol.* **72,** 464–466.

Zuk, M. (1990). Reproductive strategies and disease susceptibility: An evolutionary viewpoint. *Parasitol. Today* **6,** 231–233.

Angela R. McLean

Control of Microparasites through Vaccination

11

The question "How do vaccines work?" generally elicits a conventional answer that describes the immunological principles behind the protection of an individual by immunization. This chapter presents a different answer to the question, and describes how vaccines work to protect populations of individuals. The chapter starts with an overview of the impact of vaccination on host–parasite interactions at the population level. This overview includes a description of a simple mathematical model of the host–parasite population dynamics of a generalized microparasitic disease. Throughout the chapter, this model is used to support discussions of the impact of vaccination. The second section of the chapter explores the possible effects of vaccination, drawing on examples from the control of childhood infectious diseases. The third section discusses how theoretical insights can help in the planning of control programs. The last section addresses the future of this field, asking how theory might aid in efforts to control infectious diseases, and closes with a discussion of additional theoretical developments that will be needed.

I. Influence of Vaccines on the Population Dynamics of Microparasite–Host Interactions

The most successful vaccines have been against infections that themselves confer long-lasting immunity: smallpox, polio, measles, and other infections of childhood. For such diseases, considering the population to consist of three groups—susceptible, infectious, and immune—is useful. During the course of a natural infection, an individual progresses from the susceptible state through the infectious state, eventually becoming immune. From the perspective of an individual, vaccination moves a person directly from the susceptible state to the immune state, thus eliminating the unpleasantness and danger associated with full-blown disease. From the population perspective, the artificial removal of susceptible individuals to the immune state has effects beyond the prevention of disease in those vaccinated individuals. Chief among these effects is a reduction in the number of infectious people and, hence, in the amount of circulating virus. This effect has the im-

Parasitic and Infectious Diseases

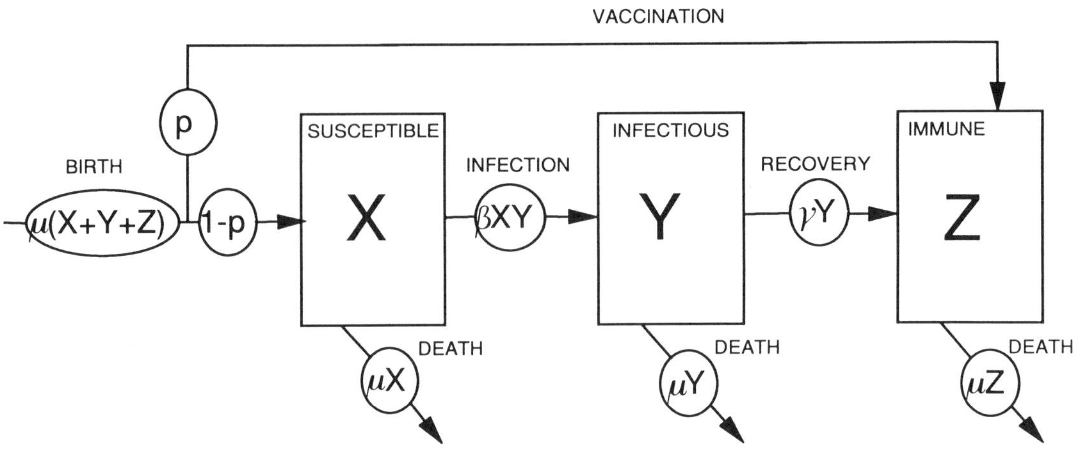

Figure 1 Structure of the simple model. The total population is divided into three groups: susceptible (X), infectious (Y), and immune (Z). The model describes the rates at which people move between groups. The number of newborns is equal to the total population ($X + Y + Z$) multiplied by a birth rate μ; the death rate in each group is also μ. The total population, therefore, remains fixed in size. Vaccination at birth splits newborns into a proportion, p, that enters the immune class, and the remaining proportion, $(1 - p)$, that is susceptible at birth. Susceptibles become infected at a *per capita* rate $\lambda_t = \beta Y_t$. Infectious individuals recover at rate γ to become immune; immunity is assumed to be lifelong.

portant consequence that vaccination has an indirect protective effect for individuals who have *not* been vaccinated. This important idea is referred to as herd immunity (Fox *et al.*, 1971). These population level effects are of interest in this discussion. Investigation of these effects is helped by the use of a simple mathematical model.

Suppose the three groups of the population (susceptible, infectious, and immune) are denoted by the letters X, Y, and Z, respectively. The rates of change of numbers of individuals in each group then can be written as three differential equations. The model assumes that all individuals are born susceptible and are subject to a constant *per capita* death rate (μ). The population is assumed to be of constant size; the *per capita* birth rate (μ) equals the death rate. The *per capita* rate at which susceptibles become infected (βY) is proportional to the number of infectious individuals present (Y). The equation for susceptibles is, therefore,

$$\frac{dX}{dt} = \mu(X + Y + Z) - \mu X - \beta XY. \qquad (1)$$

On leaving the susceptible group, newly infected individuals enter the infectious group. They recover to join the immune group at a constant *per capita* rate (γ) and are still subject to the constant death rate (μ). The equation for the infectious group is, therefore,

$$\frac{dY}{dt} = \beta XY - \mu Y - \gamma Y. \qquad (2)$$

Once recovered, an individual is assumed to be immune for life, giving the following equation for the immune group:

$$\frac{dZ}{dt} = \gamma Y - \mu Z. \qquad (3)$$

Vaccination is included easily in the model by splitting newborn individuals into a proportion, p, that is vaccinated at birth and enters the immune class and a proportion (1-p) that is not vaccinated and is therefore susceptible.

The structure of the model is summarized in Fig. 1. This simple model has three parameters: death rate (μ), recovery rate (γ), and transmission coefficient (β). The esti-

Table 1 Estimation of Parameters Used in Simple Microparasite Model

Parameter	Definition	Data source for estimation	Numerical value used in simulations
μ	*per capita* annual death rate = birth rate	life tables	0.02, i.e., expectation of life at birth is 50 yr
γ	*per capita* annual recovery rate	clinical data on duration of infectiousness	26, i.e., duration of infectiousness is 2 wk
β	transmission coefficient	age–incidence data	0.0029, i.e., average age at infection is 2 yr

mation of such parameters from epidemiological data is an important step that connects the theory of models with the reality of infectious disease as observed in populations. Methods for estimating these and other model parameters from epidemiological data are now well developed (see Chapter 4) and are surveyed in the book by Anderson and May (1991). Table 1 summarizes parameter definitions and the types of data from which parameter values might be estimated.

Figure 2 shows numerical solutions of this model over 12 yr, 6 without vaccination and 6 with mass vaccination programs. Programs that successfully vaccinate 75 and 95% of the population are modeled; the 95% program eradicates the infection, whereas the 75% program merely controls it. Because of herd immunity, vaccinating every individual is not necessary to achieve eradication. One obvious question is, "What fraction of the population must be vaccinated to eradicate an infectious disease?" Programs that control a pathogen without eradicating also are of interest. What will postvaccination dynamics be like? Do circumstances exist under which vaccination

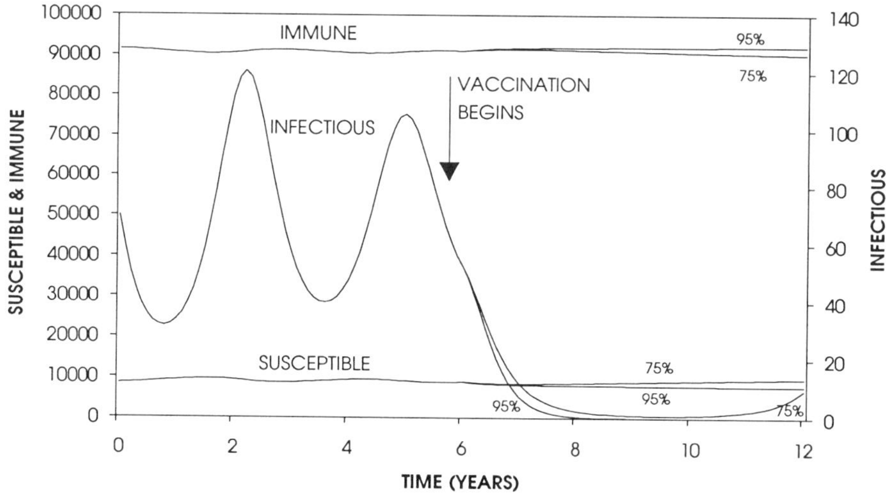

Figure 2 Numerical simulation of the differential equation model described in the text. Vaccination begins at year 6, and two vaccination strategies are modeled successfully vaccinating 75% and 95% of newborns. Notice that (1) the 95% strategy eradicates the disease but the 75% strategy does not; (2) the 75% strategy leads to a long period of low incidence, followed by a rebound in the number of cases; and (3) the number of susceptibles and immunes is barely changed by the introduction of immunization.

can lead to greater disease-related morbidity? An important point illustrated in Fig. 2 is that the total number of susceptibles remains roughly the same after mass vaccination as before. This theoretical prediction is matched by empirical studies (Fine and Clarkson, 1982) and is another reflection of the effects of herd immunity. Although the number of susceptibles remains the same, their age distribution is dramatically different after vaccination, a point that is addressed again later in this chapter.

The theoretical ideas that will be explored in this chapter are well illustrated by this simple model. Leaving out complexities introduced by, for example, age structure, seasonality, or spatial heterogeneity, the useful results of the model can be derived with minimal algebraic complexity and maximal intuitive insight. When models are used to address specific public health problems, such complications must be included. A substantial body of work exploring extensions of the model just described has been published and is summarized in the book by Anderson and May (1991). In this chapter, some results from age-structured models will be quoted. In age-structured models, the population still is divided into groups according to immunological status with respect to the pathogen under study, but the age distribution of individuals in each group also is considered. The models, therefore, describe changes over time of numbers *and* age distributions of individuals in each group.

II. Results of Vaccinating a Community

This section reviews the impact that vaccination can have on host–parasite interactions at the population level. The section is subdivided into three parts describing eradication, control without eradication, and the possibility that control of an infection by mass vaccination can worsen the impact of disease in a community. These three possibilities are exemplified by smallpox, measles, and rubella.

A. Smallpox

On December 9, 1979, the World Health Organization (WHO) announced the eradication of smallpox. Smallpox had been a greatly feared infection throughout the world, with a high case fatality rate (up to 20%; Fenner *et al.*, 1988) and survivors that often were disfigured. During 1936, smallpox transmission had been interrupted throughout the former Soviet Union; this precedent led a Soviet delegate (V. M. Zhdanov) to propose in 1958 that WHO undertake the global eradication of smallpox. This proposal was taken up and, in 1959, the 12th World Health Assembly passed a resolution undertaking the eradication of smallpox. Previous experience suggested that national mass vaccination programs that reached 80% of the population could eradicate smallpox in 4–5 yr, which was the target suggested in the resolution. In 1965, the United States decided to support the smallpox eradication effort in West and Central Africa in a joint program to eradicate smallpox and control measles. The program was remarkably successful, interrupting smallpox transmission in every country in the region (except Nigeria) within 3.5 yr of the start of field operation. In 1967, efforts were intensified globally. The strategy of the intensified program was to continue with a target of 80% vaccination rates in general, with higher targets in densely populated areas. By 1971, smallpox was endemic in only 12 countries and routine vaccination ceased in the United Kingdom and the United States. Major epidemics occurred in India, Pakistan, and Bangladesh in 1972, when it became clear that mass vaccination of 80% of the population was not going to eradicate smallpox in these parts of Asia—although the strategy had been successful in West and Central Africa. An explanation of this differential success, based on different population densities in the two regions, became apparent with hindsight and is discussed in the next section. The practical response to the failure of mass vaccination was the introduction of surveillance–containment measures in the 6 countries in which smallpox was still endemic; these measures began in 1973. Surveillance consisted of periodic searches to detect cases or to confirm their absence. These searches were carried out every 4–6 wk in endemic areas and every 2–3 mo in nonendemic areas. If a case of smallpox was found, containment measures were instigated that consisted of isolation of the case and complete vaccination coverage in the area.

Smallpox was particularly suited to this method of control because:

1. smallpox was very distinctive and easily recognizable.
2. patients were only infectious during the time when their rash was apparent; thus, no "silent spreaders" existed.
3. patients usually transmitted their infection to only a few others.
4. strong seasonality gave a "low-incidence" period during which the small number of cases could be dealt with easily.

Another 4 yr passed before smallpox finally was eradicated globally. The full story of the eradication of smallpox is recorded in the book by Fenner *et al.* (1988).

The eradication of smallpox is undoubtedly the greatest success of vaccination and must rank high among all public health measures in its impact. The success of the smallpox eradication program created high hopes for the eradication of other childhood infections, none of which have been fulfilled to date.

B. Measles

Measles is an example of a childhood infectious disease that has been controlled by mass vaccination, but not yet eradicated. Measles shares some of the properties that made the eradication of smallpox possible: no animal reservoir exists; a safe, cheap, and effective vaccine is available; and high disease-related morbidity and mortality provide a strong rationale for compliance with vaccination programs. However, other aspects of measles epidemiology make its eradication an even more formidable task than the eradication of smallpox. Chief among these features is the fact that measles is transmitted far more readily than smallpox; other characteristics being equal, a single case of measles gives rise to many more secondary cases than would a single case of smallpox. In the next section, a detailed discussion is presented describing the relationship between transmissibility of an infection and vaccine coverage required to eradicate it from a population.

Another barrier to eradication of measles is the so-called "window problem." This problem arises because very young children of immune mothers are protected from infection by maternally derived antibody. This antibody also prevents their successful vaccination with currently available vaccines, ruling out the possibility of vaccination against measles at birth. In many less developed countries, the average age at infection with measles is very low—typically 2–3 yr (reviewed by McLean and Anderson, 1988a). Prevention of early vaccination by maternally derived antibody, coupled with a low average age at infection, leaves a very small age range ("window") during which children can be vaccinated successfully, generating the window problem. A new generation of vaccines that are effective at younger ages is now available (reviewed by McLean *et al.*, 1991); a model of the same general form described in the first section has been used to study the predicted impact of such vaccines. This model and its results are described in the next section. The new vaccines will help alleviate the window problem, but experience in the United States shows that the goal of global eradication of measles will be very hard to achieve.

Immunization against measles began in 1963 in the United States. Despite intermediate coverage rates (60–70% by the late 1970s), by 1976 the age distribution of cases had changed dramatically: before mass immunization, 90% of cases occurred in children under 10 yr of age, but in 1976 this figure had dropped to 40%. In 1978, the United States announced a program to eliminate measles, the major thrust of which was mandatory immunization before school entry. This program has achieved coverage rates as high as 98%, yet sporadic outbreaks still occur, primarily among school-age children. In developing countries, the situation has been even more depressing. The combined smallpox–measles campaigns in West and Central Africa in the late 1960s had early successes in reducing incidence, but rebounds in the number of cases were observed often. A model-based explanation of these observations is presented in the next section. A report from Kinshasa, Zaire, exemplifies the difficulties of controlling measles in an urban setting in a developing country (Taylor *et al.*, 1988). Despite vaccination coverage rates of 50–60% for 5 yr, no change in age distribution of cases has been observed. Note that, although 60–70% measles vaccination coverage had a dramatic impact on the age distribution of cases in the United States, 50–60% coverage has had little impact in urban Africa. The reasons for this apparent incompatibility are presented in the subsequent discussion. Such a change in the age distribution of measles cases is desirable, since disease-related morbidity and mortality are greatest at younger ages.

Suggestions have been made that "ring immunization" could be successful in the control of measles, mirroring the success of containment programs in the eradication of smallpox. Inspection of the factors that allowed the success of such strategies for smallpox would indicate that, for measles, this is a poor idea. Measles has a much less distinctive presentation than smallpox; people infected with measles are infectious before they are sick and, most importantly, one infectious measles case can infect large numbers of people. In countries in which vaccination rates are already high, aggressive outbreak control can be seen as a strategy that augments mass vaccination, but ring immunization against measles should not be viewed as an alternative to mass vaccination.

C. Rubella

For infections in which the risk of serious disease rises with age, mass vaccination can increase the incidence of severe disease. The best understood example of such an infection is rubella (German measles) and congenital rubella syndrome (CRS). As an infection of childhood, rubella is comparatively benign, but if infection occurs during the first trimester of pregnancy the developing fetus risks severe damage. As discussed earlier, one effect of mass vaccination is raising the average age at infection among those not vaccinated successfully. For infections such as rubella, this possibility has important implications for the design of mass vaccination programs because certain ranges of vaccination coverage can increase the number of cases of CRS. Such coverage rates are high enough to raise the average age at infection among those not vaccinated into the high risk years of child-bearing, but not so high that the number of people put at risk in this way is reduced to counterbalance the increased risk of each unvaccinated individual. Although intuitive arguments can show that a mass vaccination program against rubella must reach a certain target coverage if the number of cases of CRS is to be reduced, more formal calculations are required to find that level of vaccine coverage. Such calculations have been performed for rubella using detailed information about age-specific patterns of transmission. Insights gained from such modeling work provide a sound scientific basis for the design of vaccination policies (Anderson and May, 1983a,b; Nokes and Anderson, 1987).

III. Advantages of Theoretical Insights

The previous section gave examples of mass vaccination programs and the resulting changes in the epidemiology of infectious diseases. This section presents theoretical results that can help predict those changes before they happen, or understand them afterward. Because the host–parasite interaction is complex, the effect of its disruption by vaccination is not always intuitively obvious. Fortunately, mathematical modeling can play a constructive role in such determinations.

A. Setting Targets

Why was smallpox eradicated with mass vaccination reaching 80% of the population in West and Central Africa, when these same levels of mass vaccination were not sufficient in India and other Asian countries? Why has smallpox been eradicated, but measles is still a major problem in many parts of the world? These questions can be answered by saying that population densities were higher in the relevant Asian countries than in West and Central Africa and that measles is more infectious than smallpox. How can the relationship between "infectiousness" and ease of eradication be quantified? The model defined at the beginning of the chapter can help.

Two new parameters will be useful in subsequent discussion. First, a way of encapsulating the different dimensions of infectiousness as a single number is required. This single number is the basic reproduction ratio, R_0, defined as "the number of secondary cases that would arise as a result of introducing one infectious individual into a wholly susceptible population." (For further discussion, see Chapter 3.) An individual is infectious for time $1/(\gamma+\mu)$ and, during that time, infects βN others (where N is the size of the total population, assuming everyone is susceptible). For the simple model of Eqs. 1–3, $R_0=\beta N/(\gamma+\mu)$. Notice how R_0 encompasses details of the physiology of the host–parasite interaction [duration of infectiousness, $1/(\gamma+\mu)$], host demography (total population size N), and parasite biology (coefficient of infectiousness, β). The second useful parameter is the force of infection, λ, defined as the per susceptible rate of infection. The force of infection is related to other model parameters by $\lambda_t=\beta Y_t$. Further, at equilibrium, $\lambda+\mu=\beta\mu N/(\mu+\gamma)$; hence, $\lambda=\mu(R_0-1)$. Thus,

$$R_0=\frac{\lambda}{\mu}+1. \qquad (4)$$

Finally, the numerical value of the force of infection is related to age–incidence data as the reciprocal of the average age at infection, A:

$$\lambda=\frac{1}{A}. \qquad (5)$$

Any control strategy that changes some aspect of the host–parasite interaction so that the actual reproduction ratio (R; number of secondary cases that arise as a result of one infectious individual) is less than 1 eventually will lead to eradication. Critical levels of control therefore can be calculated by finding the level of intervention at which $R=1$. Vaccination acts by reducing the size of the susceptible pool. If a fraction, p, of all newborns is vaccinated, the total number of susceptibles becomes $(1\text{-}p)N$. The critical vaccination level for eradication, p_c, thus is found by solving $(1-p_c)R_0=1$, to yield:

$$p_c=1-\frac{1}{R_0}. \qquad (6)$$

This desired result links degree of infectiousness, R_0, with vaccination coverage required for eradication, p_c. Estimates of R_0 based on epidemiological and demographic data can be used to set target levels of vaccination for eradication.

In an insightful article, Arita *et al.* (1986) point out that, because of different population densities in Asia and Africa, 80% vacci-

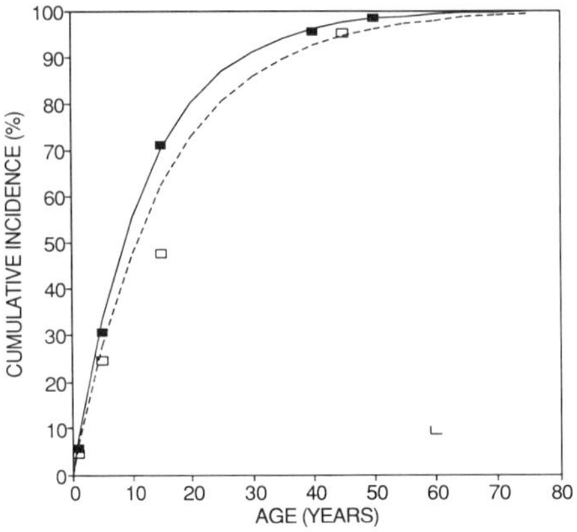

Figure 3 Smallpox age–incidence data from India (■) and Africa (□). The data and fitted model (India, ——; Africa, ---) reflect the different age–incidence patterns of smallpox in Africa and Asia. The lower average age at infection in Asia reflects higher population densities. Indian data are from Basu *et al.* (1979). African data are from Foege *et al.* (1975).

nation coverage would have reduced the density of susceptibles in West and Central Africa to 1.9/km^2 whereas the same mass vaccination coverage in Bangladesh would have resulted in 89 susceptibles per km^2. These researchers propose this fact as the reason that smallpox was eradicated from West and Central Africa at these levels of vaccination, but not from Asia. This same idea can be explored using the expressions just derived because the different population densities are reflected in different age distributions of cases in the two regions. Figure 3 shows cumulative age distributions of smallpox cases in India and Africa before mass vaccination. The African data yield an average age at infection of about 17 yr, whereas the Indian data yield a lower average age at infection of 12.5 yr. Average age at infection is calculated by fitting an exponential decay to the proportion susceptible. The reciprocal of the rate of decay yields the average age at infection. (Note that this value is a mean, not a median, so it is larger than the age at which cumulative incidence is 50%.) The lower average age at infection in India translates into a higher basic reproduction ratio (Eqs. 4, 5) and, thus, a greater vaccination coverage required for eradication (Eq. 6). By the same arguments, vaccination coverage required to eradicate measles from Africa will be much greater than the 80% that sufficed for smallpox because the average age at infection is so much lower; 2.5 yr is a typical value in an unvaccinated community. Table 2 summarizes these arguments with data on epidemiology and demography that are used to obtain estimates of the force of infection (λ) and the death rate (μ), which in turn yield estimates of the basic reproduction ratio (R_0) and, hence, of the critical vaccination coverage required for eradication, p_c.

B. Understanding Complex Dynamics

Measles control programs with high rates of coverage often experience great successes in the early stages, with a rebound in the numbers of cases after a few years (Foster and Pifer, 1971; Guyer, 1976). Numerical solutions of the simple model presented here predict the same pattern, with low incidence in the years immediately after the introduction of mass vaccination followed by

Table 2 Data on Epidemiology and Demography of Measles and Smallpox Infections in Africa and India

Infection	Region	Mean age at infection (A; yr)	Force of infection (λ)	Expectation of life at birth (yr)	μ (1/yr)	R_0	p_c (%)
Measles	Africa	2.5	0.4	45	0.022	18	95
Smallpox	Africa	17	0.06	45	0.022	3.73	73
Smallpox	India	12.5	0.08	60	0.017	5.71	83

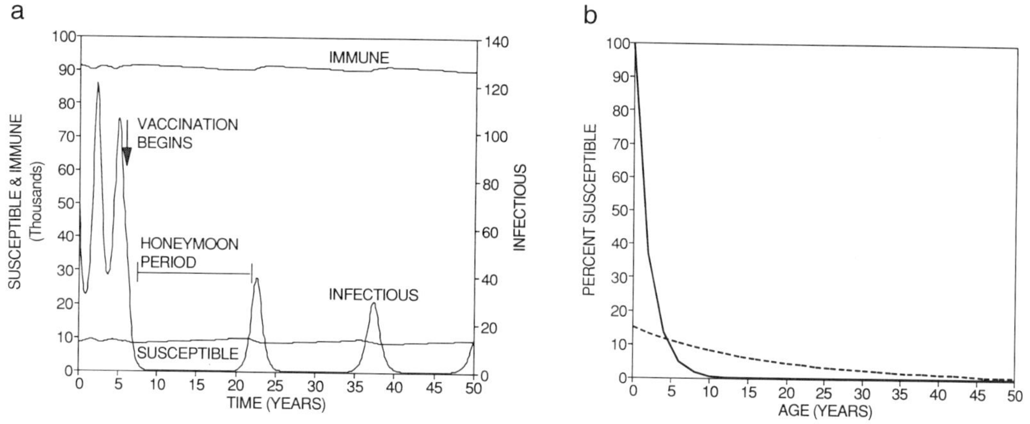

Figure 4 Numerical solutions of the vaccination model. (a) Mass vaccination and the honeymoon period. Programs of mass vaccination that are started rapidly and attain high levels of coverage can expect to result in a very low incidence for a few years, followed by rebounds in the number of cases. The initial period of low incidence has been dubbed the "honeymoon period" and is a result of the shift from the prevaccination to the postvaccination age distribution of susceptibles. This example uses parameter values shown in Table 1 with vaccination of 85% of newborns from year 6 onward. (b) The equilibrium age distributions of the numbers of susceptibles before (——) and after (---) vaccination. Notice that the same overall number of people are susceptible, but the distributions are different.

a rebound in the number of cases. This pattern is illustrated in Fig. 4a. The generation of this pattern can be explained in the following way. Theoretical and empirical studies (Fine and Clarkson, 1982; Anderson and May, 1982) suggest that a threshold density of susceptible people in a defined population exists, below which viral transmission cannot be maintained. In the absence of vaccination, the susceptible population oscillates about this threshold (as a consequence of epidemics of measles) with an age distribution that is determined by the rate of loss of protection by maternal antibody and the force of infection. After the introduction of mass vaccination, the susceptible population oscillates about a new equilibrium, with age distribution determined by the rate of loss of protection by maternal antibody, the force of infection, and the vaccination regimen. The new age distribution of susceptibles will be different (see Fig. 4b). The period of very low incidence that follows the introduction of a program of mass vaccination (the "honeymoon period") is generated during the shift from the prevaccination equilibrium to the postvaccination equilibrium (McLean and Anderson, 1988b). The important practical point is that a dramatic shortterm reduction in measles incidence after the start of a mass vaccination program is not an accurate reflection of the longer term impact of the same programs. After the initial period of very low incidence, the system will oscillate around a somewhat higher level of incidence with an interepidemic period slightly longer than that pertaining prior to control (see Fig. 4a).

This example illustrates how theoretical insights can help in understanding complex dynamics. With such understanding, events such as a sudden epidemic after apparent success with an immunization program can be predicted in advance.

Another seemingly puzzling observation that arose in measles control programs is that 60–70% mass vaccination had a big impact on the age distribution of measles in the United States, but 50–60% mass vaccination has had rather little impact on the age distribution of measles cases in urban Africa. This result is less perplexing after a study of the equilibrium average age at infection as a function of the proportion of children vaccinated at birth. The following relationship can be derived between average age at infection, A, and the proportion vaccinated at birth, p, from Eqs. 1–3:

$$A = \frac{(\mu + y)}{\mu[\beta(1 - p)N - (\mu + y)]}. \quad (7)$$

Notice that if $p = p_c$, the divisor of the right side of Eq. 7 is equal to zero and the average age at infection tends to infinity; thus, the disease is eradicated. The important point about Eq. 7 is that the relationship between average age at infection and vaccination coverage is highly nonlinear, and depends on all the parameters of the model. The relationship is illustrated in Fig. 5 with two different sets of parameter values representing prevaccination average ages at infection of 5 yr and 2 yr. Both curves start fairly flat, but the postvaccination average age at infection increases rapidly as vaccination coverage levels approach eradication levels. In the United States, where the average age at infection before immunization was 5–6 yr, vaccination of 60–70% is close to eradication levels and therefore has a large impact on the age distribution of cases. Vaccination of 50% of children in urban Africa is still vary far from the level that will be required for eradication; the changes in age distribution of cases are, therefore, negligible.

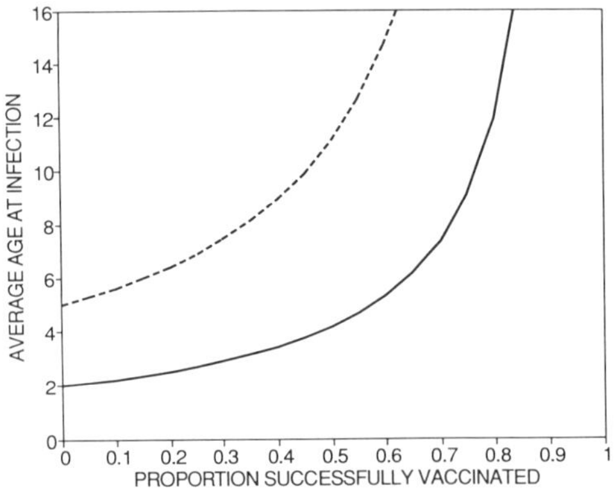

Figure 5 The average age at infection and vaccination coverage. The average age at infection rises with increasing vaccination coverage, but the relationship is nonlinear. Greater and greater effects are seen as the vaccination coverage level approaches that required for eradication. The two examples shown are for cases in which the prevaccination average age at infection is 2 yr (typifying the urban African situation; ——) and 5 yr (as in the United States; ---). The equation for the curves is Eq. 7 in the text.

C. Planning Strategies

When the targeting of vaccination is particularly complex (as with measles, because of the window problem), mathematical modeling provides a tool that can be used in planning vaccination strategies. This section discusses an example of such modeling. WHO has made a new recommendation for measles immunization in areas with very high rates of transmission. The new policy recommends a switch to the newer type of high potency Edmonston Zagreb (E–Z) vaccine (a strain with improved efficacy in the presence of maternally derived specific antibody), targeting immunization at 6 mo-old infants. A mathematical model of measles transmission has been developed that can be used to assess the likely impact of cohort immunization using the new strain of vaccine on the incidence of infection. The model is fully age structured and includes detailed descriptions of the immunological status of very young children with respect to both measles immunization and infection. The structure of the model is outlined in Fig. 6. [For a mathematical description of the model, see McLean *et al.* (1991).] The age-structured model allows the comparison of strategies that target different types of vaccines at different age groups. The target age at vaccination and age-specific vaccine efficacy can be varied quite specifically. As shown in Fig. 7, the modeling work can be used to lend support to the new recommendation. The mathematical model used to derive these recommendations has many epidemiological and demographic complexities and, therefore, many parameters that must be estimated from field data. To make precise predictions about the impact of a vaccination program in a given community, a large amount of information

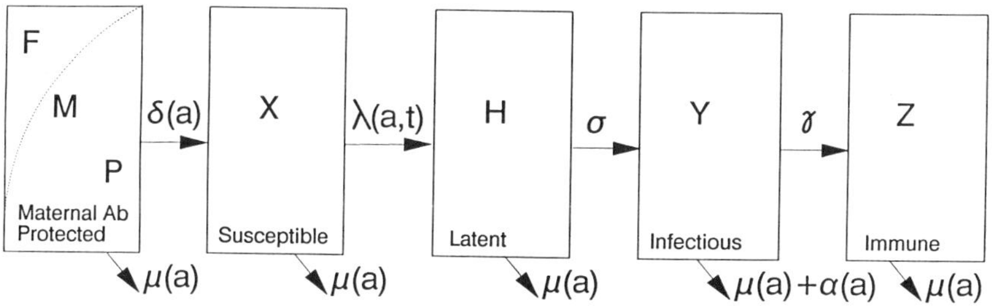

Figure 6 Structure of the more complex model used for assessing measles immunization strategies. The model follows the same overall structure as the simple model described earlier, but two groups have been added. The first of these groups contains very young children protected from infection (and unable to seroconvert after vaccination with conventional vaccines) because of their maternally derived specific antibody (M). This maternally acquired protection wanes at an age-specific rate [$\delta(a)$]. The maternal antibody protected group is subdivided further (in an age-specific way) into those who will seroconvert after vaccination with high potency E-Z vaccines (partially protected, P) and those who will not (fully protected, F). The second group contains people who have been infected but are not yet infectious (H). The latent period lasts for time $1/\sigma$. Disease-related mortality is included also, at rate $\alpha(a)$ from the infectious class, Y. The model has full age structure.

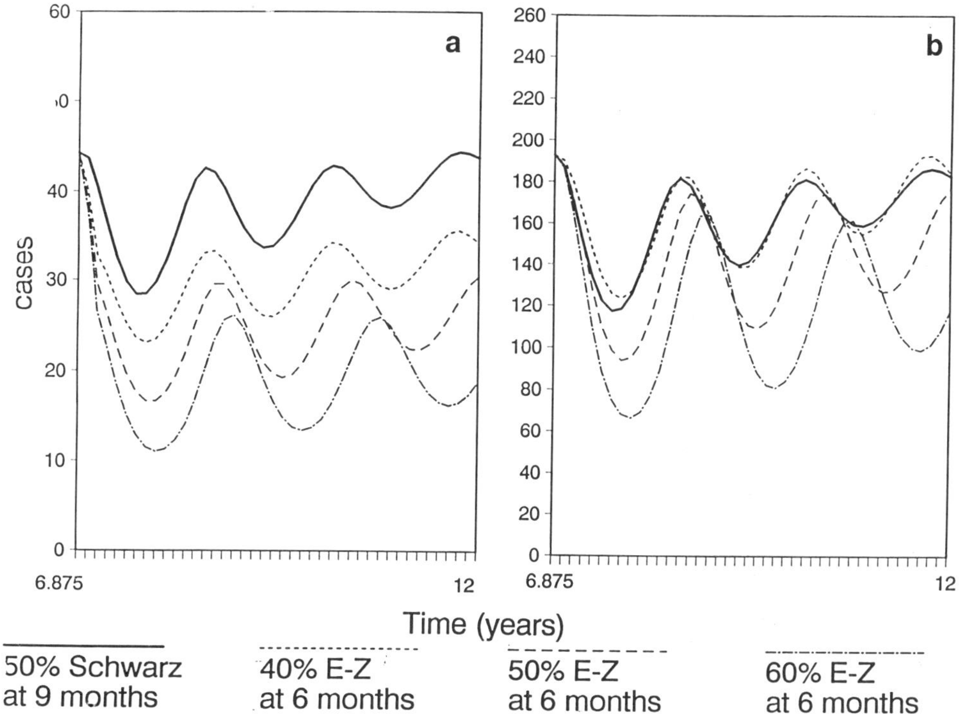

Figure 7 Comparisons of the impact of conventional vaccines given at 9 mo and E-Z vaccines given at 6 mo. (a) Incidence among those below 9 mo of age. (b) Cases over all ages. The switch in immunization policy is implemented at time 6.875 years, at the peak of an epidemic, and is introduced into a community that has been receiving 50% coverage with conventional vaccines at 9 mo of age for the previous 3 yr. The figure shows that 40% coverage with E-Z at 6 mo gives total numbers of cases comparable to a continued policy of 50% coverage with conventional vaccines at 9 mo of age. 50% Schwartz at 9 mo, ——; 40% E-Z at 6 mo, ---; 50% E-Z at 6 mo, – – –; 60% E-Z at 6 mo, –·–.

about the community is necessary. The mathematical model then allows the integration of such a body of information into a tool that can be used to make rational decisions about the best vaccination policy for that community.

IV. Future Control

Experience with the difficulties involved in controlling measles shows that the development of a safe, cheap, and effective vaccine does not herald the easy eradication of a parasite. Although such vaccines are often very effective for the individuals who receive them, the protection of whole communities requires concerted and well organized efforts to target vaccines where they will be of most benefit to the population as a whole. The planning of such intervention requires a good understanding of the interplay between the biologies of the host and the parasite. This interplay includes factors such as parasite biology, host responses, and demographic and sociological properties of the host community. Because of the complexities of this interplay between host and parasite, the outcome of a vaccination campaign will not necessarily be intuitively obvious. Theoretical work based on mathematical models can aid in the understanding of the likely impact of programs of mass vaccination. Such understanding can help in the planning of optimal strategies and in the understanding of the complex epidemiological patterns that may be observed after intervention. The existing theory for the impact of programs of mass vaccination is restricted to microparasitic diseases. If vaccines against macroparasites become available, a new body of work will be required to plan how such new vaccines will be used best.

Note added in proof: Recent findings (Garenne *et al.*, 1991) associating high child mortality with receipt of high-titre measles vaccines has led to reassessment of the WHO's recommendations.

References

Anderson, R. M., and May, R. M. (1982). Directly transmitted infectious diseases: Control by vaccination. *Science* **215,** 1053–1060.

Anderson, R. M., and May, R. M. (1983a). Vaccination against rubella and measles: Quantitative investigations of different policies. *J. Hyg.* **90,** 259–325.

Anderson, R. M., and May, R. M. (1983b). Two-stage vaccination programme against rubella. *Lancet* **ii,** 1416–1417.

Anderson, R. M., and May, R. M. (1991). "Infectious Diseases of Humans: Dynamics and Control." Oxford University Press, Oxford.

Arita, I., Wickett, J., and Fenner, F. (1986). Impact of population density on immunization programmes. *J. Hyg.* **96,** 459–466.

Basu, R. N., Jezek, Z., and Ward, N. A. (1979). "The Eradication of Smallpox from India." World Health Organization, Geneva.

Fenner, F., Henderson, D. A., Arita, I., Jezek, Z., and Ladnyi, I. D. (1988). "Smallpox and its Eradication." World Health Organization, Geneva.

Fine, P. E. M., and Clarkson, J. A. (1982). Measles in England and Wales. II. The impact of the measles vaccination programme on the distribution of immunity in the population. *Int. J. Epidemiol.* **11,** 15–25.

Foege, W., Millar, J. D., and Henderson, D. A. (1975). Smallpox eradication in West and Central Africa. *Bull. WHO* **52,** 209–222.

Foster, S. O., and Pifer, J. M. (1971). Mass measles control in West and Central Africa. *Afr. J. Med. Sci.* **2,** 151–158.

Fox, J. P., Elveback, L., Scott, W., Gatewood, L., and Ackerman, E. (1971). Herd immunity: Basic concept and relevance to public health immunization practises. *Am. J. Epidemiol.* **94,** 179–189.

Garenne, M., Leroy, Q., Bean, J. P., and Sene, I. (1991). Child mortality after high-titre measles vaccines: Prospective study in Senegal. *Lancet,* **ii,** 903–907.

McLean, A. R., and Anderson, R. M. (1988a). Measles in developing countries. I. Epidemiological parameters and patterns. *Epidemiol. Infect.* **100,** 111–133.

McLean, A. R., and Anderson, R. M. (1988b). Measles in developing countries. II. The predicted impact of mass vaccination. *Epidemiol. Infect.* **100,** 419–442.

McLean, A. R., Nokes, D. J., and Anderson, R. M. (1991). Model-based comparisons of measles immunization strategies using high dose Edmonston–Zagreb type vaccines. *Int. J. Epidemiol.* **20,** 1107–1117.

Nokes, D. J., and Anderson, R. M. (1987). Rubella vaccination policy: A note of caution. *Lancet* **i,** 1441–1442.

Taylor, W. R., KaLisa, R., Ma-Disu, M., and Weinman, J. M. (1988). Measles control efforts in urban Africa complicated by high incidence of measles in the first year of life. *Am. J. Epidemiol.* **127,** 788–794.

Graham F. Medley

Chemotherapy

12

I. Introduction

A. Parasites and Their Importance

This chapter addresses the control of parasites and parasite-induced disease by application of chemotherapy. In this context, the term "parasite" requires some definition. Overall, the macroparasites (helminths, ticks, etc.) fall within this category, but viruses do not. The bacterial and protozoan infections (e.g., malaria, tuberculosis) may be included provided that disease is related to intensity of infection, which really demands that the intensity of infection within individuals is measurable and meaningful in terms of pathogenesis of disease.

I will discuss the theory of chemotherapy application in terms of designing strategies in the most general terms. Enough similarity exists between parasites of medical and veterinary importance that this discussion can proceed without reference to specific host–parasite systems. However, some systems are better studied than others, and I will draw on the helminth parasites of humans to provide most examples, and will point out when the results derived from these parasites are not relevant to other systems.

Parasites are important causes of morbidity and mortality in human communities, and are equally important causes of production losses in animals of economic importance. For example, up to 28% of all people may be infected with *Ascaris lumbricoides,* and up to 200 million people may be harboring infections of 20 or more worms (Chan *et al.*, 1994). Several analyses have shown that prevention of veterinary disease can produce up to 1500% return on invested funds in terms of increased productivity (Morris, 1988). At first, such consequences of parasitism would appear to be sufficient evidence to warrant widespread chemotherapy against the parasites concerned, but two issues should be noted. First, the design and implementation of control strategies is not simple, and requires some understanding of the complexities of parasite population dynamics to design and a great deal of logistical planning to execute. Second, parasites exist in a world with many other competing requirements for effort and expenditure. The justification for chemotherapeutic intervention against parasites must be made with sound quantitative arguments.

Parasitic and Infectious Diseases

B. Treating Individual Hosts and Communities

Many drugs are available for chemotherapy. For human helminths, several generic families of drugs are available including ivermectin and the benzimidazoles (Gustafsson *et al.*, 1987). In comparison to other medical and veterinary treatments, these drugs are inexpensive (the cost of a full course of albendazole is less than US $1) and do not require specialist knowledge for administration. Still, parasites remain very common and virtually uncontrolled in most developing countries, providing a prime example of the problem of applying treatment to individuals, which is relatively straightforward, and preventing infection and disease in populations, which is apparently more difficult.

Theoretically, there is no reason why parasite populations cannot be controlled. The optimal control policy would be the treatment of all the hosts continuously, but this option is obviously unfeasible. Control programs do not operate outside the constraints that apply to all health care activities (Bundy and Guyatt, 1992; Guyatt and Evans, 1992), so the selection of an optimal control design will depend on the balance between program cost and effectiveness. Understanding the population biology of parasites and their interaction with hosts can lead to quantitative models that indicate how chemotherapy is administered best to achieve the desired aims of control for the least cost. I will not directly address the issues surrounding costs of programs, but I will focus on the use of population biology theory in providing insights into ways of maximizing effectiveness and briefly summarise the issues that restrict this application in practice.

C. Morbid and Mortal Consequences of Parasite Infection

Central to the population biology of helminths is their characteristic distribution in a community of hosts. (For human helminth examples, see Anderson and May, 1985a,1991.) The distribution is overdispersed, so most individuals are lightly infected (few parasites) and a few hosts harbor many parasites (see Chapter 4). This pattern has been demonstrated empirically in virtually all parasite populations studied, and has important implications for both parasites and hosts. For parasites, this distribution has the effect that the majority of the population is within relatively heavily infected hosts, which has consequences for density dependence and mating. For hosts, the distribution means that only a small proportion of the population has high parasite loads at any one time. The degree of parasite-induced morbidity and mortality within individual hosts should be related to the intensity of the parasite burden harbored by hosts, as has been demonstrated for a wide variety of infections (Greenwood *et al.*, 1987; Lwambo *et al.*, 1992; Warren *et al.*, 1993). Consequently, the heavily infected fraction of hosts attracts most of the medical or veterinary attention.

D. Parasite Life Histories

Parasites of medical and veterinary importance have life cycles drawn widely from the whole spectrum of the many and varied parasite life cycles. This chapter focuses on the stages in the human or veterinary host that are the targets for chemotherapy, and does not discuss intermediate hosts or environmental interactions, although these features are of prime importance when designing integrated control strategies that do not rely on chemotherapy alone. Parasites with reservoir hosts (i.e., hosts other than the target host that harbor the same parasite stage as the target host) are generally not controllable by chemotherapy alone. Depending on the importance of the reservoir hosts (i.e., the proportion of all parasites that are harbored by the reservoirs), chemotherapy will cure hosts that receive treatment but will not reduce transmission significantly to provide future protection. Zoonoses, which require control of parasites of the nonhuman host to reduce their human prevalence and intensity, are not considered either. Generally, the details of a control program will depend on the specific details of the parasite life cycle and its interaction with the host. For example, the population dynamics of parasites that reproduce directly on the host are very different from those of parasites that have obligatory free-living stages or intermediate hosts.

II. Processes Involved in Designing Control Programs: Theory

A. Fundamental Issues

1. Individual versus Community

A chemotherapy program against parasites must have aims that must be decided at an early stage. This step may appear to be obvious and trivial, but unless the aims are formulated explicitly, confusion can arise. One of the most basic issues that must be decided is whether the program is to be designed to maximize its impact at the individual or at the community level. Individuals suffer deleterious effects from their parasite burdens. Usually the effect is increased by increased parasite burden. Consequently, to alleviate disease in individuals, a program would aim to give chemotherapy to those individuals with the highest parasite burdens. However, as seen from the distribution of parasites among their hosts, those individuals with the highest burdens are also the least common. Figure 1 demonstrates that those individuals with intermediate parasite burdens actually may contribute more disease to the community, because their greater number outweighs their reduced disease.

Consequently, chemotherapy to the most heavily infected individuals will not reduce the number of cases of disease as much as treatment of all individuals with intermediate burdens. This pattern is the parasitological equivalent of the "prevention paradox" (Rose, 1992), which occurs in the epidemiology of nontransmissible disease: more clinical cases arise from small, more common risks (parasite burdens) than from large rare risks. Thus, the effect of parasites at the community level can be reduced most by chemotherapy that may bring only small benefits to the individual but large benefits to the population. However, transmissible diseases are dynamic by nature, so the most heavily parasitized hosts may contribute disproportionately to production of infective stages. Thus, treating this group may not cure as much disease directly, but may prevent more disease indirectly than treating moderately infected hosts.

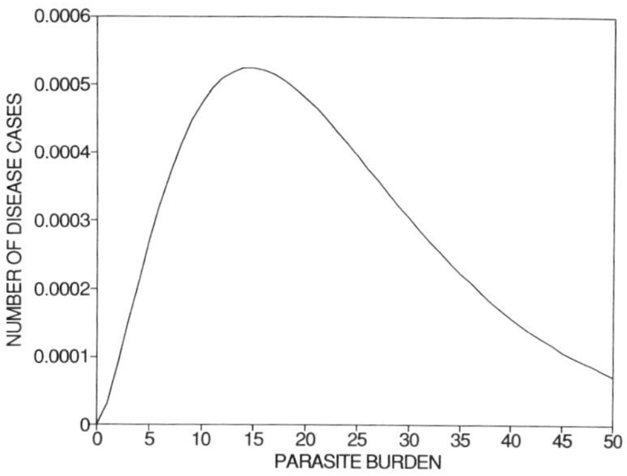

Figure 1 Amount of disease in a community as a function of the numbers of parasites within individuals. The shape of this curve comes from the assumption of an overdispersed distribution of parasites and an increasing function of disease with increased parasite burdens. The major conclusion is that hosts with intermediate burdens contribute more disease to the community than those with the highest burdens, because the greater number of hosts with intermediate burdens outweighs the greater disease in those with high burdens (Medley and Anderson, 1994; Medley and Bundy, 1994).

Until recently, the success of a chemotherapy program was measured in terms of the parasite population, that is, prevalence and intensity of infection. The current view is that programs should be aimed at reducing morbidity and mortality caused by parasite infection to some predetermined level. This change has been brought about largely by the ability to measure morbidity (for example, ultrasound investigation in human schistosome infections) and the realization that programs will be adopted only if they can be proven to be beneficial to health (Bundy, 1990a; Warren *et al.*, 1993). Eradication of parasites is achievable only by alteration of the environment to interrupt the life cycle (e.g., the installation and use of sanitation), and is thought not to be achievable by chemotherapy alone (Warren *et al.*, 1993). This consensus has been reached largely as a result of the failure of chemotherapy to eradicate parasites.

2. Ethical Issues

In the treatment of human parasites, two ethical problems often arise. The first concerns the conflict between providing treatment for the maximum benefit of individuals or for the maximum benefit of the community as a whole. Theory may indicate that most disease in a community can be prevented by treating the same people repeatedly, for example, the most heavily infected individuals, and not treating other people who may be infected. These untreated infected individuals still are suffering ill effects of parasitism, so the issue is whether the community benefit outweighs the individual benefit. This issue probably is not generally resolvable; discussion doubtless will continue on the implementation of all control programs.

The second problem revolves around the issues of side effects of drugs and the cost of diagnosing infection before prescribing treatment. On one hand, all drugs carry some risk of side effects of various degrees of seriousness. (Gustafsson *et al.*, 1987); ideally, drugs will be given to individuals only when their increased health due to removal of parasites outweighs the detrimental effects of the drug, which implies not treating uninfected people. On the other hand, diagnosing infection prior to administration is expensive; a control program that screens for infection prior to treatment may be impossible to implement or may be considerably reduced in scope compared with a program that treats a population by mass chemotherapy regardless of infection status. To some extent, this problem is resolvable by acquiring more information on risks of disease from infection and from drugs (Bundy, 1990b).

B. Population Dynamic Considerations

1. Reproduction Ratios and Density Dependence

The key concept in designing parasite control programs is that the populations involved are dynamic (see Chapter 1). Hosts and parasites are continually being removed (by death and emigration) and replaced (by birth and immigration), even when the size of the populations is constant. In this stable equilibrium state, the average number of female parasites that attain reproductive maturity produced by a female parasite during her life time is 1, that is, each female parasite replaces herself in the next generation. This quantity is termed the effective reproduction ratio, R_e. When R_e is greater than 1, the parasite population is growing; it will decline when R_e is less than 1. The effective reproduction ratio is always less than the basic reproduction ratio, R_0, which is defined as the average number of female parasites produced by one mated female in a completely susceptible host population, that is, when no other parasites are present. The difference between the effective and basic reproduction ratios is density dependence or crowding effects, which are discussed in Chapter 1. The importance of these effects to chemotherapy is that density dependence governs the rate of recovery of the parasite population following chemotherapy and interacts in complex ways with the distribution of parasites.

A relationship between the reproduction ratios can be written by considering an average density dependent effect, $f(M)$, that depends on M, the average number of para-

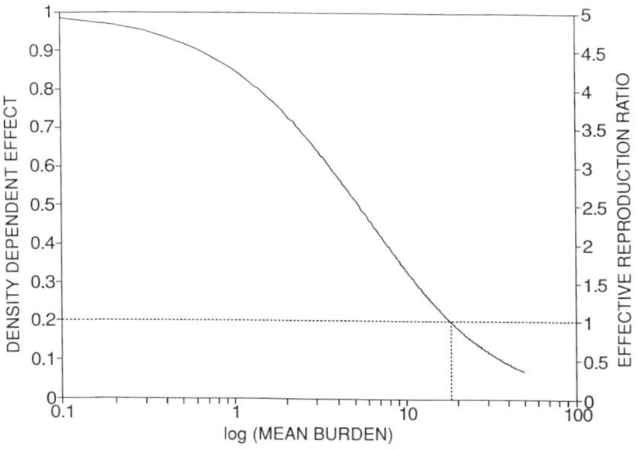

Figure 2 Relationship between the density dependent function, $f(M)$, and the effective reproduction ratio, R_e, with the mean parasite burden, M. The horizontal and vertical dashed lines indicate the mean parasite burden at which the effective reproduction ratio is unity (18.3 parasites per host), assuming that the basic reproduction ratio is 5. If the mean burden is below this value, the parasite population will grow ($R_e > 1$); if M is above this value, the parasite population will shrink ($R_e < 1$) and the equilibrium is stable. In reality, many different possible values of $f(M)$ exist for each value of M, because of the dependency of $f(M)$ on the distribution of hosts. In drawing this figure, a constant distribution of hosts has been assumed, thus giving a single value of $f(M)$ for each mean intensity.

sites per host:

$$R_e = f(M)R_0. \qquad (1)$$

The function $f(M)$ is related to the mean parasite density so $f(M) = 1$ when no density dependence is operating and $f(M) = 1/R_0$ at the equilibrium number of parasites (see Fig. 2). Note that the average density dependent effect is being considered. Those female parasites in very lightly infected hosts may leave a number of daughters very close to the maximum (R_0), whereas those in heavily infected hosts (the majority of the parasite population) will achieve considerably less reproductive success. Thus, the effect of density dependence is related intimately to the parasite distribution. The most common formulation in models of parasite populations is the assumption of some relationship for individual parasites in hosts, and then averaging over the whole parasite population using some distribution of parasites (see Anderson and May, 1982,1991; Medley *et al.*, 1993; Fig. 2).

The situation is complicated further by the fact that the distribution of parasites is itself a dynamic entity and is a function of the population dynamics of the parasites, the patterns of transmission within a community, and the perturbations induced by control interventions. Thus, the rate of infection of hosts is dependent on the distribution of parasites and the density dependence exerted as a function of the distribution, which in turn determines the rate of infection of hosts (Medley, 1992).

2. Population Consequences of Chemotherapy

Removing parasites by chemotherapy generally will reduce the average density-dependent effect; consequently, the effective reproduction ratio will increase above its equilibrium value and the parasite population will grow. This increase will cease when density dependence is again at its equilibrium value, which generally will be when the parasite population has reattained its pretreatment density. This process can be encapsulated in a simple model. Consider the rate of change of the parasite population (dM/dt): this value is negative when the population is shrinking, positive during parasite population growth, and zero at equilibrium and is measured in numbers of parasites per unit time:

$$\frac{dM}{dt} = \mu M(R_0 f(M) - 1)$$
$$= \mu M(R_e - 1). \qquad (2)$$

This equation is derived by Anderson (1982) and Anderson and May (1982,1991). The basic reproduction ratio subsumes the parasite life cycle and includes all the rates of death, reproduction, and establishment in intermediate hosts, free-living stages, and primary host. This formulation emphasizes that the death rate of adult parasites, μ, has an overall effect on the rate of recovery of a parasite population from a perturbation due to chemotherapy application. Paradoxically, the larger the death rate (the shorter the life expectancy of parasites), the faster the rate of recovery.

This relationship can be used to suggest the relative frequencies of chemotherapeutic intervention that are required to control parasite populations (Anderson and Medley, 1985). For example, consider four helminth species of humans: *A. lumbricoides, Trichuris trichiura,* hookworms, and schistosomes which are thought to have life expectancies of less than 1 yr, about 1 yr, about 4 yr, and about 5 yr, respectively. This simple relationship suggests that to maintain parasite populations at the same reduced density, treatment will have to increase in frequency with decreasing life expectancy; this conclusion is largely borne out by field experience (Warren *et al.*, 1993).

However, chemotherapy also interacts with density dependence via a more complicated route through the distribution of parasites. The posttreatment distribution of hosts is not equivalent to the pretreatment distribution; density dependence will be influenced by this change in distribution (Fig. 3). In more mathematical terms, the function $f(M)$ defined earlier is related in a complex manner to chemotherapy application and whether treatment is given at random to hosts or hosts are chosen for treatment by some criterion such as age or degree of infection. During the period of re-infection following treatment, the parasite distribution changes as treated hosts generally experience a net gain in parasites. Chemotherapy application during this period of recovery again will alter the parasite distribution and, consequently, the rate of recovery.

It is often assumed that if a parasite population is permitted to reattain precontrol levels following a period of control, the parasite distribution will be the same as the precontrol distribution. However, this need not be true, depending on the mechanisms responsible for generating the parasite distribution. Certainly it will not be true if the parasite distribution is generated by mechanisms such as acquired immunity that are dependent on the host's experience of parasites. In human helminth infections, a current theory is that individuals are predisposed to their infection levels in some way (Keymer and Pagel, 1990; Bundy and Medley, 1992). The theory does not state the mechanisms that are responsible for this predisposition—they may be genetic, behavioral, or sociological in origin. Under the theory of predisposition, individual hosts have the same relative position within the distribution (i.e., high or low burdens), but the distribution need not be the same.

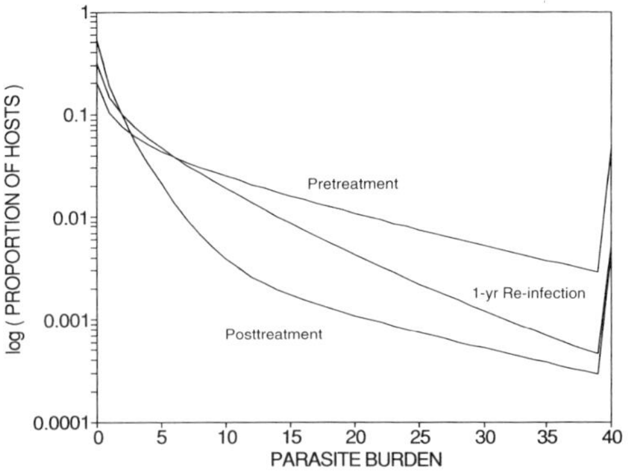

Figure 3 Effect of chemotherapy on the parasite distribution. The three curves show the expected equilibrium distribution of parasites (pretreatment) and that immediately following chemotherapy (posttreatment), giving a drug of 90% efficacy to 90% of hosts at random with respect to infection intensity. Also, the parasite distribution following 1 yr of re-infection without further treatment is shown. Details of the calculations are given by Medley *et al.* (1993).

Using theoretical models, Anderson and Medley (1985) demonstrated that the mechanism of heterogeneity generation in the parasite population can have important implications for the population dynamics of parasites and, in particular, parasite population response to treatment. Again using theoretical models, Quinnell *et al.* (1990) showed that the site of density-dependent action within the parasite life cycle (fecundity or maturation) also was of crucial importance in determining parasite population response to perturbation, and that this feature interacted with the cause of heterogeneity.

The general conclusion of these theoretical studies is that the distribution of parasites and the mechanisms that generate the distribution are significant determinants of the success of a chemotherapy program, especially when the degree and quantity of parasite-induced morbidity is determined to a large extent by the parasite distribution. Since little is understood about the heterogeneity-generating mechanisms at present (Bundy and Medley, 1992), this may appear to be an unavoidable obstacle to chemotherapy program design. However, remember that control programs will not be decided by models alone, but by field and laboratory demonstrations of effectiveness.

3. Age-Related Patterns of Infection

Parasites display a wide range of age-related patterns of prevalence and intensity of infection, which have tremendous influence on the success of a particular chemotherapy control program. If the intensity of infection is higher in some age groups than in others, those age groups with higher average burdens generally will provide the best targets for treatment. Few parasites exact the same toll on their hosts regardless of host age. Thus, the age-related intensity relationship plus the age-related morbidity per parasite determine the degree of morbidity caused by a parasite population; the age-related pattern of disease may not follow that of intensity of infection directly. Chemotherapeutic intervention will perturb this age-related pattern and, depending on the mechanisms generating the pattern, may alter it so more disease is present in the posttreatment population than in the pretreatment population. Acquired resistance to infection and disease are important components of this pattern and of the determination of the effect of chemotherapy.

4. Probability of Mating and Break Point

For dioecious parasite species, the primary host is usually the place where mating occurs. The distribution of parasites largely determines the probability of mating because parasites sharing their host with only a few others will be less likely to meet a parasite of the opposite sex than parasites in heavily infected hosts. In a comprehensive paper, May (1977) showed how the probability of mating depends on the distribution of parasites and how the relationship between mean intensity of parasites and mating probability is very nonlinear, so only at very low densities of hosts is the probability of finding a mate reduced significantly below 1. In the context of chemotherapy, the mating probability has two important effects. The first is in terms of the genetic change in the parasite population, which is discussed in relation to the development of resistance subsequently.

The second effect is the concept of the "breakpoint": the parasite density below which the probability of finding a mate is low enough to prevent the parasite population from growing. In terms of the model described earlier, $g(M)$ can be the proportion of female parasites that is mated as a function of the mean parasite burden. The effective reproduction ratio then becomes

$$R_e = f(M)g(M)R_0. \qquad (3)$$

As the mean parasite burden increases following chemotherapy, so the probability of mating increases, but the effect of density dependence increases as well, so the effective reproduction ratio is a peaked function of the mean parasite burden, as demonstrated in Fig. 4. Consequently, the rate of growth of the parasite population is fastest midway between treatment and reattainment of the equilibrium level. However, note that a second equilibrium exists in Fig. 4 compared with Fig. 2, at a mean burden of about 0.14 parasites/host. Below this level, the parasite population will decrease to zero (because the effective reproduction ratio is less than 1); this point is the breakpoint.

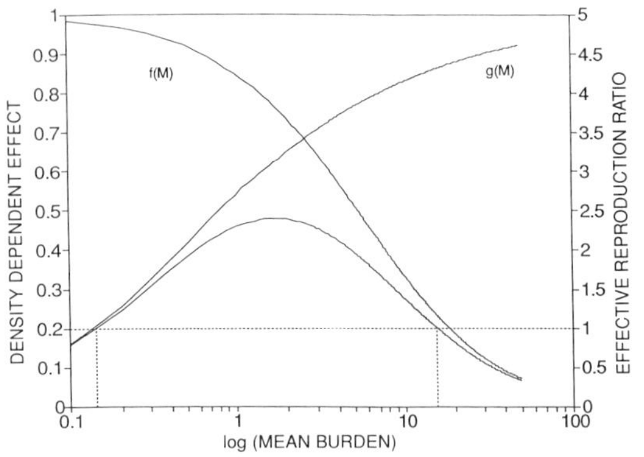

Figure 4 As in Fig. 2, but including the effect of mating probability. The probability of mating is calculated assuming a dioecious parasite from May (1977). The peaked solid curve is the product $g(M)f(M)$, which determines the relationship between the effective reproduction ratio and the mean parasite burden. Note that, compared with Fig. 2, a second equilibrium at 0.14 parasites per host occurs (see text).

However, the breakpoint has never been demonstrated to be of practical value in the field, although the decreased probability of mating could be seen in the increased proportion of unfertilized eggs of *A. lumbricoides* as overall infection declined in Japan during the 1950s and 1960s (Anderson and Medley, 1985).

5. Effects of Long-Term Chemotherapy

Until now I have considered the effect of discrete applications of chemotherapy. Once chemotherapy becomes established on a longer term, it is easier to consider the drug as acting continuously to increase parasite mortality. Drug-induced mortality then becomes part of the environment of the parasite, reducing the basic reproduction ratio, so the parasite population reattains a new lower equilibrium value (Anderson and May, 1982; Anderson and Medley, 1985; Woolhouse, 1992). It is also possible to administer antiparasite drugs so they are released into the host over a long time period, for example, the use of boluses in treating helminths of cattle. In this case, if the drug has an antilarval effect, the effect is to reduce parasite establishment as well as life expectancy. In either case, the success of the treatment program again will be determined to a large extent by the parasite distribution and the distribution of the drug within the host population.

C. Economic and Logistic Issues

1. Chemotherapy within General Health Care

The practical aspects of chemotherapy application, in terms of its position within existing health-care structures, are prominent in program design and implementation (Guyatt and Evans, 1992). Although the effects occur largely at the economic level (the cost aspect of cost–benefit analysis of chemotherapy programs), they generally are powerful enough to overturn some population dynamic considerations. For example, the treatment of school-age children with anthelmintics has obvious advantages, given the accessibility of this group to repeated health interventions, and will be an attractive policy even if evidence suggests that parasite-induced disease occurs outside this age group (Bundy, 1990b).

2. Measurement of Costs and Benefits

How is the success of a chemotherapy program best measured? The process of program design demands that one design can be compared with another quantitatively, both in terms of the benefit (alleviation of parasite-induced damage) and the economic cost. The simplest gauge of benefit, both conceptually and for measurement purposes, is the reduction in prevalence of infection at the time of treatment. However, since chemotherapy depresses the worm population for some time before recovery is complete, a better measure of the benefit of chemotherapeutic intervention is the total reduction throughout the recovery phase. Rosenfield *et al.* (1977) used prevalence changes over the recovery period as a measure of effectiveness. However, better again are measures of the proportion of hosts har-

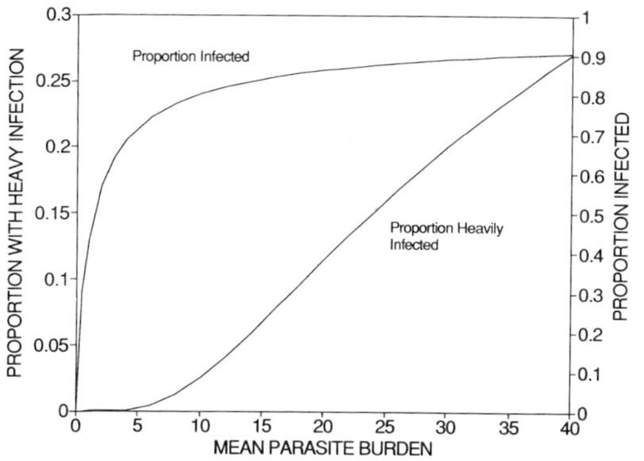

Figure 5 Relationship between the prevalence of infection and prevalence of heavy infection (greater than 50 parasites) with the mean parasite burden (Anderson and Medley, 1985). The lines were calculated using the overdispersed distribution estimated by Guyatt *et al.* (1990) with a linear decreasing function of the overdispersion parameter.

boring a high number of parasites or of the intensity of infection, because these measures are more likely to reflect disease since the level of disease in an individual is related to the number of parasites harbored (Medley *et al.*, 1993; Fig. 5).

Assuming that the benefit derived from a chemotherapy program can be measured or calculated accurately, the decision whether or not to implement the program depends on the competition for resources with other health and environmental needs such as vaccination programs and safe water provision. To have some quantitative measures by which to compare such diverse benefits, the use of QUALYs and DALYs (quality adjusted life years and disability adjusted life years, respectively) has become standard (Guyatt and Evans, 1992; Jamison, 1993). To calculate these measures, some weighting between zero and one (zero = dead and one = healthy) must be allocated to harboring a particular parasite load; then the benefit from a chemotherapy program must be recalculated in terms of the DALYs gained. The allocation of weights is necessarily arbitrary, but can be made culture specific simply by asking those people for whom the programs are intended to assign weights. Weights also can be assigned to different age groups (given the choice, most people will opt to save a mother rather than an infant) and to different times (given the choice, most people will choose to save a life now rather than at some point in the future). This second weighting method is termed "discounting" or "time-preferencing" in economics.

In veterinary situations, many of these complications do not arise since the benefits can be calculated simply in terms of increased productivity and, hence, expressed in financial terms. However, additional complications occur in the economics; for example, a program that increases productivity of milk may result in a decrease in price paid per unit due to the increased quantity.

The conclusion from this section is that a whole layer of complexity must be added to the interaction between chemotherapy and the population dynamics before a program can be assessed. Some of these complications, such as reducing the importance of future lives saved, may reduce the sensitivity of aspects of the population dynamics in projecting the effectiveness of programs, so having a complete understanding of the population biology may not be necessary before that understanding can be used to judge among chemotherapy programs. The relationships between population biology and economics have been well studied in the context of managing harvested populations (Clark, 1976), but are less well advanced in terms of controlling parasite populations. This imbalance urgently requires attention.

3. Multiple Infections

Two aspects of parasitism indicate that it may be possible, and in some cases necessary, to treat many species of parasites simultaneously in the same control program. First, because of the similarity in physiology of many parasites, antiparasite drugs usually are effective against more than one species of parasite. Second, parasites, certainly in human populations, are found in the same geographical areas (confined largely to the tropics) and are associated with poverty. A correlation between prevalence of infection within communities has been demonstrated for *A. lumbricoides* and *T. trichiura* (Booth and Bundy, 1992). On a wider geographical scale, Bundy *et al.* (1991) considered simultaneous treatment of many species. The arguments in favor of combined programs are largely economic, because the infrastructure costs of drug delivery and program administration are not duplicated; however, stronger arguments are likely to exist in terms of disease alleviation. Unfortunately, very little research has been done to understand the health implications of simultaneous infection or the interaction between different parasite populations within the same hosts.

III. Evaluation of Control Programs: Observations

A. Monitoring Control Programs

Given the complexity of host–parasite systems (and biological systems in general), monitoring the effect of a chemotherapy program is important since modeling and theoretical analysis are not guarantees of success. Also, the effectiveness of a program may change over time because of changes in both host and parasite populations (see subsequent discussion). These changes will not be detected if the host and parasite populations are not under continuous surveillance. Monitoring populations with heterogeneous distributions poses some problems. In particular, a strong nonlinear relationship exists between parasite prevalence and intensity, so large changes in intensity of infection may not be reflected in changes in prevalence (Guyatt *et al.*, 1990). Consequently, because morbidity reduction is the aim of most programs, the measure of success of a program should relate naturally to the number of heavily infected individuals. Figure 5 shows that the mean intensity of infection is related approximately linearly to the proportion of the population with a heavy infection, and is a better criterion than prevalence of infection for judging the success of a control program (Anderson and Medley, 1985; Bundy *et al.*, 1992).

B. Summary of Current Understanding of Chemotherapy Program Design: Human Helminths

1. Population Dynamics and Control

In recent years, much of the theory of chemotherapy application has revolved around attempts to include a dynamic distribution of parasites in population models (for example, see Hadeler and Deitz, 1983; Anderson and Medley, 1985, Medley *et al.*, 1993). Theory prior to this time rested largely on assumptions about the form of the parasite distribution (Anderson and May, 1982,1985a; Woolhouse, 1991,1992). The development of the theory of predisposition also led to the idea that repeated treatment of the most heavily infected individuals (selective programs) would reduce overall parasite-induced morbidity disproportionately since those individuals with low burdens were protected in some way already (Anderson and May, 1982; Anderson and Medley, 1985). However, given the difficulties of implementation of selective programs, and the fact that the theoretical results depend on the unproven assumptions concerning the causes of the parasite distribution, attention has shifted away from selective programs to mass treatment programs (Bundy, 1990a).

The inclusion of a dynamic parasite distribution has led to the quantification of the indirect benefits associated with chemotherapy, that is, benefits to those individuals who did not receive treatment. Medley *et al.* (1993) showed that, the higher the fraction of a community treated and the more effective the treatment, the greater the benefit for those who were untreated. This aspect contributes to the fact that the benefit from treatment increases nonlinearly with increased treatment effort. To allow complete cost-effectiveness evaluations to be undertaken, population dynamic models must be allied with economic frameworks (Guyatt *et al.*, 1993).

2. Control in Theory and Practice

A number of attempts have been made in recent years to quantify the effects of different chemotherapy programs in field studies (for example, see Seo *et al.*, 1980; Elkins *et al.*, 1986; Thein-Hlaing *et al.*, 1987; Bundy *et al.*, 1988). Some studies also have measured the impact that chemotherapy has on morbidity (for example, see Stephenson *et al.*, 1990; Butterworth *et al.*, 1991; Thein-Hlaing *et al.*, 1991b; Nokes *et al.*, 1992). The results of these studies have been in general agreement with current theory of the population biology of helminths. Additionally, several studies have shown that reduction of the parasite population in children reduces transmission to adults sufficiently to reduce their parasite burdens (Bundy *et al.*, 1990; Butterworth *et al.*, 1991; Thein-Hlaing *et al.*, 1991a). To some extent, attempts to measure the benefit of chemotherapy on morbidity will have underestimated the effect largely because of the sample sizes involved. As has been realized in nontransmissible disease epidemiology, treatments of common diseases that have a small but significant effect on the outcome of the disease require large trials, that is many thousands of individuals, to evaluate the program accurately (Peto, 1992).

IV. Consequences of Control Programs for Biological Interactions

A. Changes to Host Population

1. Health

A control policy is hoped to increase the health of the hosts being treated, which will have the general effect of increasing the resources required by the hosts in terms of food. Intestinal helminths tend to suppress appetite in their hosts, so removal of the parasites will increase food intake (see Chapter 9). An intimate relationship exists between parasitism and nutrition, so nutritional supplementation given simultaneously with chemotherapy may enhance the benefits of treatment (Bundy and Golden, 1987). In grazing domesticated animals, increased appetite can have repercussions on pasture quality and stocking densities (Grenfell, 1992). At the host population level, parasites are necessarily responsible for reductions in survival and reproduction, so one immediate consequence of chemotherapy is an increase in population growth rate of hosts.

The relationship between disease and parasitism has been assumed simply to be that the more parasites currently harbored by a host, the greater the risk and/or the greater the severity of disease. However, the pathogenesis of disease is rarely this uncomplicated. Disease is frequently associated with the host immune response, and may be related dynamically to the sum of exposure to parasites rather than simply to the current burden (for example, see Srividya *et al.*, 1991). In these situations, chemotherapy programs should be designed to reduce disease, even if the optimal parasite population control policy is different (Medley and Bundy, 1994).

2. Acquired Immunity

A fair generalization is that all hosts mount an immunological response to their parasites. However, in many cases, the effect the immune response has on the population dynamics of parasites and hosts is not clear, particularly for helminth infections of humans (Bundy and Medley, 1992). An acquired immune response can develop only after exposure to infection, so in any event the relationship between effective immune response and, for example, genetic determinants of responsiveness will be complicated severely by the heterogeneous patterns of exposure that occur in most parasite systems. Also, a relationship exists between nutrition and immunity so, although an effective immunity may be demonstrated for well-fed individuals, the pattern may not be reproduced in populations of hosts with varying degrees of nutritional intake (Michael and Bundy, 1992). Similarly, the interactions between the immune responses by hosts to different parasite species have received very little attention in population studies. Paradoxically, many of these complex interactions will be determined only by careful monitoring of interventions on parasite populations, that is, by chemotherapy application.

Anderson and May (1985b,1991) developed a theoretical model that assumes that the parasite population (in this case, helminth infection) is controlled by acquired immunity, that is, that immunity is the vehicle through which density dependence operates. The more parasites an individual experiences, the lower the survival or establishment rate of parasites within the host. In this case, removal of parasites by chemotherapy will have the effect of reducing exposure and, hence, reducing acquired immunity. This effect can have two important consequences. First, the distribution of parasites will not recover to its precontrol form and disease may be increased in the posttreatment compared with the pretreatment equilibrium, although the disease will have been reduced during the re-infection period immediately following chemotherapy. Second, the relationship between host age and parasite intensity will be shifted toward increased parasite numbers in older hosts; depending on how host age affects the relationship between parasite numbers and disease, this shift may result in more or less disease posttreatment.

3. Parasites as Vectors

For parasites, especially the biting arthropods and ticks, to act as vectors for other infections is not unusual. Reduction in the population of vector parasites will reduce the rate of infection of the infectious agents that they transmit. A well-established relationship exists between the density (biting rate) of vectors and the equilibrium prevalence of the transmitted infection that is nonlinear, with a significant reduction in transmitted-infection prevalence being achieved only at very low densities of vectors (Anderson and May, 1991). For chemotherapy against parasite vectors to alter the patterns of transmission of the carried infection significantly, it must be very intensive and may go beyond the levels required to reduce morbidity and mortality due to the vector parasites.

B. Changes to Parasite Population

1. Development of Genetic Resistance to Drug Action

The frequent use of a drug is a very strong selective pressure in favor of parasite genotypes expressing some resistance to the effect of the drug, especially if treatment dosing is suboptimal because of dilution of the drug, for example, since suboptimal treatment tends to give heterozygotes a selective advantage rather than kill them. Experience suggests that resistance evolves rapidly (Waller and Prichard, 1986; Anderson and May, 1991); therefore, a drug used extensively should be expected to become less effective because of genetically controlled resistance. At present, all field examples of helminth parasites with anthelmintic resistance are veterinary, largely because no useful bioassay exists for measuring resistance in human parasites. Several features may reduce the rate of development in human parasites: humans are more genetically diverse hosts than most domesticated animals, they inhabit more diverse environments, and currently drugs are used less extensively.

Resistance has several consequences, the most important of which is that only a limited number of drugs is available and, once resistance has developed to all of them, clinical disease caused by parasites becomes untreatable. This result is especially severe in the case of helminths, for which only a limited number of distinct families of drugs is available; resistance to one drug within a family generally is associated with resistance to all drugs within that family. The second consequence is that drug manufacturers market and price drugs to obtain the best economic return; the knowledge that resistance may develop in the future may be reflected in the manufacturers' policy. This possibility also has consequences in the design of policies to prevent development of resistance, because the optimal biological policy may conflict with the optimal economic policy of the manufacturer.

An important concept in population genetics of controlled populations is that of genetic refugia, that is, parasite populations that are not exposed to chemotherapy (Smith, 1990; Anderson and May, 1991). In most cases, the prominent refugia are those hosts that do not receive treatment (including reservoir hosts) and those parasite stages that are not in the treated hosts at the time of chemotherapy. The second of these is usually the most significant for medical and veterinary parasites, since at any one time most parasites are either free-living or parasitizing intermediate hosts. Refugia have two important but opposite effects that operate on development of disease resistance. On one hand, refugia provide a supply of parasites that can compete with resistant strains and may act to reduce resistance or slow its development. This possibility has been established in theory for pest (nonparasite) species (McGaughey and Whalon, 1992). On the other hand, one advantage of killing the majority of a population is the resultant loss of genetic variation (the founder effect); this benefit is lost since the unexposed parasite population harbors much genetic variability. Without the existence of refugia, the founders of the post-treatment parasite population would have a very limited genetic repertoire. Adaption of this population to changing host responses and application of other chemotherapeutics would be impaired severely.

Anderson *et al.* (1989) demonstrated the interaction between frequency of drug treatment and development of resistance, and the importance of the nonlinear relationship between mating probabilities with parasite abundance. Mating between rare genotypes occurs at a much lower frequency than that between more abundant genotypes so rare alleles, for example, those conferring drug resistance, are subject to negative selection. Drug action confers a selective advantage on resistant genotypes, but this advantage may not be sufficient to overcome the intrinsic negative selection. Discontinuous selection (as in the confinement of chemotherapy application to discrete rounds of treatment) does not necessarily result in the development of a resistant population, largely because of the negative selection between rounds, whereas continuous selection always produces a resistant population. Further, Anderson *et al.* (1989) showed that frequency of treatment is not simply related to the outcome of selection, but that various treatment regimes may result in a resistant population, a wholly susceptible population, or a mixed population and that an ordered progression from fixation of resistance to fixation of susceptibility with changing frequency and efficacy of treatment does not necessarily exist.

Theoretical studies indicate that resistance can be slowed by several management strategies. The use of multiple drugs simultaneously (i.e., in the same dose) delays but does not prevent the development of resistance relative to, for example, alternating the type of drug used (Barnes and Dobson, 1990; Smith, 1990). Two explanations for this effect are possible. The coexistence of many drug resistance genes within a single parasite is much rarer than the existence of genotypes resistant to the drugs individually. Consequently, the multiply resistant genotype must start at a much lower frequency than the singly resistant genotype. The second explanation is that multiple drugs have an increased antiparasite effect (possibly as a result of synergy or reduced resistance) which reduces the recovery time of the parasite population. The much reduced parasite population is closer to the breakpoint; consequently, a greater chance exists of single parasite infections (which cannot reproduce) and more random death (with respect to drug resistance). However, no experimental evidence supports these effects. The end result is a parasite population that is resistant to many drugs.

The second group of strategies is confined largely to the management of pasture in veterinary parasites. Grazing and treatment can be organized so parasites are divided into separate populations and are exposed to each drug only once (Barnes and Dobson, 1990). This strategy has two advantages. First, selection is confined to one generation, which is not sufficient for resistance to reach a high frequency or co-segregate with other genes. Second, genetic diversity within the parasite population is reduced significantly by the action of one drug, thus making evolution of resistance to subsequent drugs less likely and slower. These strategies are essentially ways of controlling gene flow between refugia and exposed populations.

Once resistance has developed, theoretically no reason exists to expect it to diminish in frequency, as is borne out by the field evidence. The resistant genotypes are exposed to the same evolutionary pressures as the nonresistant genotypes with respect to all aspects of survival and reproduction (e.g., host resistance and environment). Consequently, these genotypes will not be at a great selective disadvantage, if any, to those genotypes susceptible to the drug, so resistance will not be selected against greatly when drug treatment stops. It may, theoretically, be possible to (re)introduce susceptible strains that do have a selective advantage against the resistant population.

2. Adaption of Life Histories

Populations have been shown to adapt to repeated killing in both laboratory and field situations (Bennett *et al.*, 1990; Reznick *et al.*, 1990). The most frequent response to harvesting is the lowering of age of reproductive maturation. If the parasite life expectancy is reduced by chemotherapy application, those parasites that begin reproduction at an earlier age and reproduce more, even at the expense of reduced survival, will be at a selective advantage. This hypothesis has not been demonstrated in parasite populations, but is very plausible. If it occurred, such an event would make chemotherapy strategies less effective over time. Such changes would be difficult to demonstrate in human communities since demonstration would involve detecting changes in processes for which measurements, at present, are very crude or do not exist. The reduction in effectiveness of programs may be confused with development of genetic resistance to drug action. In veterinary diseases, such an event would be more detectable.

V. Summary, Discussion, and Future Prospects

Chemotherapy is a method of killing parasites that are, by definition, causing harm to their hosts. This chapter addressed how chemotherapy is distributed between hosts and through time—who gets the drugs and when—thus defining the chemotherapy program. The interactions between parasites and hosts, both at the individual and at the population level, determine the success of different programs. Given the complexity of these biological interactions, for example, the interrelationships between immunity, host and parasite genetics, nutrition, and behavior, predicting the complete consequences of chemotherapy may never be possible.

The population consequences of chemotherapy application are necessarily quantitative. Mathematical models are essential tools in developing qualitative and quantitative understanding of the consequences of particular programs. However, models generally should not attempt to provide a complete description of all the complexity of a system, but should abstract those processes that are important for understanding a particular problem. A "good" model is one in which the parameters are biologically meaningful and measurable and the results are testable in laboratory and field. The usual development cycle is for models to be tested against data and altered where necessary before testing again, continually developing understanding during this process. The consequence is that models are only as complex as required, and biological processes are only included when important.

Disrupting the parasite population by a single round of chemotherapy implies that the parasites become rarer for a time, generating repercussions for acquired immunity, age-related infection profiles, and parasite genetics as well as parasite-induced disease development. Eventually, the parasite population will recover from the perturbation, but the influence of the single treatment may be apparent for a long time if the host population has responded to the temporary reduction in parasites, for example, by loss of immunity; if the parasite genetics were altered; or if the relationship with another parasite species has been altered. The changes in parasite-induced disease also may be longer term, depending on the dynamics of disease development. Continued application of chemotherapy will emphasize these changes to parasite and host populations, and will allow the parasite population to adapt to the new circumstances of continued exposure to the drug as well.

Inclusion of these processes in mathematical frameworks has several prerequisites. At present, basic research is required in several areas before processes can be considered. First, the ecological relationship between different parasite species, their pathology, and the host responses has received little attention. Second, the population genetics of parasites and its interrelationship with host population genetics has become a possible area of study, with the development of molecular techniques to classify genes and gene products. Third, few empirical observations are available regarding the adaptive responses of host and parasite to different exposures to different drugs. Finally, the dynamics of the development of disease and the consequences of treatment are little studied, in contrast to the cellular and molecular levels of disease, which have received much attention. Many of the areas of study given here can be addressed only by considering (large) cohorts of hosts (i.e., groups of individuals) moving through time. For humans, many ethical and logistical problems are associated with conducting such studies. Instead, experiments with laboratory animals are likely to provide much useful information (Scott, 1990). One persistent problem with the monitoring of such trials is the lack of a proven methodology for assessing the intensity of parasitism without destroying the parasite burden, although such techniques are being sought.

Once an understanding of the consequences of a chemotherapy program has been developed, this program must be tested in carefully controlled trials (just as a drug is tested) before it is recommended for widespread use (Bundy, 1990a). Parasitism generally is associated with low case-fatality rates; the effects of parasites tend to be chronic rather than acute. The importance of parasites to health and productivity derives from relatively small impacts acting over a large infected population. Consequently, to ascertain the impact of chemotherapy on health and productivity improvement, trials of chemotherapy programs will have to be very large and conducted over a long period (on the same order as the life expectancy of the host).

References

Anderson, R. M. (1982). Population dynamics and control of hookworm and roundworm infections. *In* "Population Dynamics of Infectious Diseases:

Theory and Applications" (R. M. Anderson, ed.), pp. 67–108. Chapman and Hall, London.
Anderson, R. M., and May, R. M. (1982). Population dynamics of human helminth infections: Control by chemotherapy. *Nature (London)* **297,** 557–563.
Anderson, R. M., and May, R. M. (1985a). Helminth infections of humans: Mathematical models, population dynamics and control. *Adv. Parasitol.* **24,** 1–101.
Anderson, R. M., and May, R. M. (1985b). Herd immunity to helminth infection and implications for parasite control. *Nature (London)* **315,** 493–496.
Anderson, R. M., and May, R. M. (1991). "Infectious Diseases of Humans: Dynamics and Control." Oxford University Press, Oxford.
Anderson, R. M., and Medley, G. F. (1985). Community control of helminth infections of man by mass and selective chemotherapy. *Parasitology* **90,** S629–S660.
Anderson, R. M., May, R. M., and Gupta, S. (1989). Non-linear phenomena in host–parasite interactions. *Parasitology* **99,** S59–S79.
Barnes, E. H., and Dobson, R. J. (1990). Population dynamics of *Trichostrongylus colubriformis* in sheep: Computer model to simulate grazing systems and the evolution of anthelmintic resistance. *Int. J. Parasitol.* **20,** 823–831.
Bennett, A. F., Dao, K. M., and Lenski, R. E. (1990). Rapid evolution in response to high-temperature selection. *Nature (London)* **346,** 79–81.
Booth, M., and Bundy, D. A. P. (1992). Comparative prevalences of *Ascaris lumbricoides, Trichuris trichiura,* and hookworm infections and the prospects for combined control. *Parasitology* **105,** 151–157.
Bundy, D. A. P. (1990a). New initiatives in the control of helminths. *Trans. R. Soc. Trop. Med. Hyg.* **84,** 467–468.
Bundy, D. A. P. (1990b). Control of intestinal nematode infections by chemotherapy: Mass treatment versus diagnostic screening. *Trans. R. Soc. Trop. Med. Hyg.* **84,** 622–625.
Bundy, D. A. P., and Golden, M. H. N. (1987). The impact of host nutrition on gastrointestinal helminth populations. *Parasitology* **95,** 623–635.
Bundy, D. A. P., and Guyatt, H. L. (1992). Cost analysis of schistosomiasis. *Trans. R. Soc. Trop. Med. Hyg.* **86,** 646–648.
Bundy, D. A. P., and Medley, G. F. (1992). Immunoepidemiology of human geohelminthiasis: Ecological and immunological determinants of worm burden. *Parasitology* **104,** S105–S119.
Bundy, D. A. P., Cooper, E. S., Thompson, D. E., Didier, J. M., and Simmons, I. (1988). Effect of age and initial infection intensity on the rate of reinfection with *Trichuris trichiura* after treatment. *Parasitology* **97,** 469–476.
Bundy, D. A. P., Wong, M. S., Lewis, L. L., and Horton, J. (1990). Control of geohelminths by delivery of targeted chemotherapy through schools. *Trans. R. Soc. Trop. Med. Hyg.* **84,** 115–120.
Bundy, D. A. P., Chandiwana, S. K., Homeida, M. M. A., Yoon, S., and Mott, K. E. (1991). The epidemiological implications of a multiple-infection approach to the control of human helminth infections. *Trans. R. Soc. Trop. Med. Hyg.* **85,** 274–276.
Bundy, D. A. P., Hall, A., Medley, G. F., and Savioli, L. (1992). Evaluating measures to control intestinal parasitic infections. *World Health Stat. Q.* **45,** 168–179.
Butterworth, A. E., Sturrock, R. F., Ouma, J. H., Mbugua, G. G., Fulford, A. J. C., Kariuki, H. C., and Koech, D. (1991). Comparison of different chemotherapy strategies against *Schistosoma mansoni* in Machakos District, Kenya: Effects on human infection and morbidity. *Parasitology* **103,** 339–355.
Chan, M. S., Medley, G. F., Jamison, D. T., and Bundy, D. A. P. (1994). The estimation of global morbidity attributable to intestinal nematode infections. *Parasitology* (in press).
Clark, C. W. (1976). "Mathematical Bioeconomics: The Optimal Management of Renewable Resources." Wiley-Interscience, New York.
Elkins, D. B., Haswell-Elkins, M., and Anderson, R. M. (1986). The epidemiology and control of intestinal helminths in the Pulicat Lake region of Southern India. I. Study design and pre- and post-treatment observations on *Ascaris lumbricoides* infection. *Trans. R. Soc. Trop. Med. Hyg.* **80,** 774–792.
Greenwood, B. M., Bradley, A. K., Greenwood, A. M., Byass, P., Jammeh, K., Marsh, K., Tulloch, S., Oldfield, F. S. J., and Hayes, R. (1987). Mortality and morbidity from malaria among children in a rural area of The Gambia, West Africa. *Trans. R. Soc. Trop. Med. Hyg.* **81,** 478–486.
Grenfell, B. T. (1992). Parasitism and the dynamics of ungulate grazing systems. *Am. Nat.* **139,** 907–929.
Gustafsson, L. L., Beerman, B., and Abdi, Y. A. (1987). "Handbook of Drugs for Tropical Parasitic Infections." Taylor and Francis, London.
Guyatt, H. L., and Evans, D. (1992). Economic considerations for helminth control. *Parasitol. Today* **8,** 397–402.
Guyatt, H. L., Bundy, D. A. P., Medley, G. F., and Grenfell, B. T. (1990). The relationship between the frequency distribution of *Ascaris lumbricoides* and the prevalence and intensity of infection in human communities. *Parasitology* **101,** 139–143.
Guyatt, H. L., Bundy, D. A. P., and Evans, D. (1993). A population dynamic approach to the cost-effectiveness analysis of mass anthelmintic treatment: Effects of treatment frequency on *Ascaris* infection. *Trans. R. Soc. Trop. Med. Hyg.* **87,** 570–575.
Hadeler, K. P., and Deitz, K. (1983). Non-linear hyperbolic partial differential equations for the dynamics of parasite populations. *Comp. Math. Appl.* **9,** 415–430.
Jamison, D. T. (1993). Disease control priorities in developing countries: an overview. *In* Disease Control Priorities in Developing Countries" (D. T. Jamison, W. H. Mosley, A. R. Measham, and J-L.

Bobadillo, eds.). Oxford University Press, New York.

Keymer, A. E., and Pagel, M. (1990). Predisposition to hookworm infection. *In* "Hookworm Infection: Current Status and New Directions" (G. A. Schad and K. S. Warren, eds.), pp. 177–209. Taylor and Francis, London.

Lwambo, N. J. S., Bundy, D. A. P., and Medley, G. F. H. (1992). A new approach to morbidity risk assessment in hookworm endemic communities. *Epidemiol. Infect.* **108,** 469–481.

May, R. M. (1977). Togetherness among schistosomes: Its effects on the dynamics of the infection. *Math. Biosci.* **35,** 301–343.

McGaughey, W. H., and Whalon, M. E. (1992). Managing insect resistance to *Bacillus thuringiensis* toxins. *Science* **258,** 1451–1455.

Medley, G. F. (1992). Which comes first in host–parasite systems: Density dependence or parasite distribution? *Parasitol. Today* **8,** 321–322.

Medley, G. F., and Anderson, R. M. (1994). The contribution of mathematical modelling to the understanding of infectious disease. *In* "The Modelling of Vector-Borne and Other Parasitic Diseases" (B. D. Perry, ed.). International Laboratory for Research on Animal Diseases, Nairobi, Kenya (in press).

Medley, G. F., and Bundy, D. A. P. (1994). Dynamic modelling of epidemiological patterns of schistosomiasis morbidity. *Am. J. Trop. Med. Hyg.* (in press).

Medley, G. F., Guyatt, H. L., and Bundy, D. A. P. (1993). A quantitative framework for evaluating the effect of community treatment on the morbidity due to ascariasis. *Parasitology* **106,** 211–221.

Michael, E., and Bundy, D. A. P. (1992). Protein content of CBA/CA mouse diet: Relationship with host antiobody responses and the population dynamics of *Trichuris muris* (Nematoda) in repeated infection. *Parasitology* **105,** 139–150.

Morris, R. S. (1988). The effects of disease on productivity and profitability of livestock: How should it be assessed? *Proc. N. Zeal. Soc. Anim. Product.* **48,** 117–125.

Nokes, C., Grantham-McGregor, S. M., Sawyer, A. W., Cooper, E. S., Robinson, B. A., and Bundy, D. A. P. (1992). Moderate to heavy infection of *Trichuris trichiura* affect cognitive function in Jamaican school children. *Parasitology* **104,** 539–547.

Peto, R. (1992). Statistics of chronic disease control. *Nature (London)* **356,** 557–558.

Quinnell, R. J., Medley, G. F., and Keymer, A. E. (1990). The regulation of gastrointestinal helminth populations. *Phil. Trans. R. Soc. London B* **330,** 191–201.

Reznick, D. A., Bryga, H., and Endler, J. A. (1990). Experimentally induced life-history evolution in a natural population. *Nature (London)* **346,** 357–359.

Rose, G. (1992). "The Strategy of Preventive Medicine." Oxford University Press, Oxford, UK.

Rosenfield, P. L., Smith, R. A., and Wolman, M. G. (1977). Development and verification of a schistosomiasis transmission model. *Am. J. Trop. Med. Hyg.* **26,** 505–516.

Seo, B. S., Cho, S. Y., Chai, J. Y., and Hong, S. T. (1980). Comparative efficacy of various interval mass treatment on *Ascaris lumbricoides* infection in Korea. *Korean J. Parasitol.* **18,** 145–151.

Scott, M. E. (1990). An experimental and theoretical study of the dynamics of a mouse–nematode *(Heligmosomoides polygyrus)* interaction. *Parasitology* **101,** 75–92.

Smith, G. (1990). Chemotherapy: Future problems. *In* "Hookworm Disease: Current Status and New Directions" (G. A. Schad and K. S. Warren, eds.), pp. 291–303. Taylor and Francis, London.

Srividya, A., Pani, S. P., Rajagopalan, P. K., Bundy, D. A. P., and Grenfell, B. T. (1991). The dynamics of infection and disease in Bancroftian filariasis. *Trans. R. Soc. Trop. Med. Hyg.* **85,** 255–259.

Stephenson, L. S., Latham, M. C., Kinoti, S. N., Kurtz, M., and Brigham, H. (1990). Improvements in physical fitness of Kenyan schoolboys infected with hookworm, *Trichuris trichiura,* and *Ascaris lumbricoides* following a single dose of albendazole. *Trans. R. Soc. Trop. Med. Hyg.* **84,** 277–282.

Thein-Hlaing, Than-Saw, and Myint-Lwin. (1987). Reinfection of people with *Ascaris lumbricoides* following single, 6-month and 12-month interval mass chemotherapy in Okpo village, rural Burma. *Trans. R. Soc. Trop. Med. Hyg.* **81,** 140–146.

Thein-Hlaing, Than-Saw, and Myat-Lay-Kyin. (1991a). The impact of three-monthly age-targetted chemotherapy on *Ascaris lumbricoides* infection. *Trans. R. Soc. Trop. Med. Hyg.* **85,** 519–522.

Thein-Hlaing, Thane-Toe, Than-Saw, Myat-Lay-Kyin, and Myint-Lwin. (1991b). A controlled chemotherapeutic intervention trial on the relationship between *Ascaris lumbricoides* infection and malnutrition in children. *Trans. R. Soc. Trop. Med. Hyg.* **85,** 532–528.

Waller, P. J., and Prichard, R. K. (1986). Drug resistance in nematodes. *In* "Chemotherapy of Parasitic Diseases" (W. C. Campbell and R. S. Rew, eds.), pp. 339–362. Plenum Publishing, New York.

Warren, K. S., Bundy, D. A. P., Anderson, R. M., Davis, A. R., Henderson, D. A., Jamison, D. T., Prescott, N., and Senft, A. (1993). Helminth infections. *In* "Disease Control Priorities in Developing Countries" (D. T. Jamison, W. H. Mosley, A. R. Measham, and J-L. Bobadilla, eds.). Oxford University Press, New York.

Woolhouse, M. E. J. (1991). On the application of mathematical models of schistosome transmission dynamics. I. Natural transmission. *Acta Tropica* **49,** 241–270.

Woolhouse, M. E. J. (1992). On the application of mathematical models of schistosome transmission dynamics. II. Control. *Acta Tropica* **50,** 189–204.

Christopher Dye

13

Vector Control

I. Introduction

Observing a rise in the popularity of appropriate technology for vector control (Curtis, 1991), this chapter waves a flag for appropriate analysis. Its main theme is that good analysis, including good field experiments, should often lead to better strategies for vector control. I do not, of course, advocate excessively ruminative analytical work in an emergency; when an outbreak of mosquito-borne dengue fever occurs in a small tropical town, immediate fumigation with a space spray is called for (Gratz, 1991). However, even if a control method works under particular operational (rather than experimental) conditions, its general applicability will be in doubt until the success of the treatment has been explained. In any case, most vector control problems are not emergencies. As Rogers and Randolph (1985) remarked, the distribution of the tsetse fly *Glossina palpalis* in Africa has changed very little since the beginning of this century. Whatever the potential of new technology for upsetting the equilibria of the great endemic vector-borne diseases, it will never be a substitute for clear thinking.

The special feature of this chapter is that the medium of parasite control, the vector population, has a dynamic behavior all its own. The five sections that form the body of the chapter explore the dynamics of vector populations and, consequently, the dynamics of vector-borne disease in the context of particular control problems: measuring the scale of the control effort required (Section II), the significance of population regulation for genetic control (Section III) and environmental management (Section IV), competitors and predators as possible biological control agents (Section V), and the impact of vector control on disease as distinct from infection (Section VI). With the exception of Section II, the discussion that follows generally falls short of a protocol for analytical work, for ingenuity has no rules. Nor is it a comprehensive review: I mention many of the common methods by which vectors are controlled, but rather than list numerous studies under the given headings, a few examples have been selected for detailed discussion.

II. Magnitude of the Problem

The modern conceptual framework for vector control was developed by the malariologists Macdonald (1957) and Garrett-

Jones (1964). Macdonald's formula for the basic reproduction ratio (R_0) of malaria measures the number of secondary cases resulting from the introduction of one infected individual into the susceptible population (see Chapter 3) and is expressed as

$$R_0 = \frac{ma^2bp^n}{-r\ln(p)}, \qquad (1)$$

in which m is the number of adult female anopheline mosquitos per person, a is the daily biting rate of an individual female mosquito on humans (accounting for meals taken on other hosts), b is the fraction of mosquitos with infective sporozoites that actually generate human infection (and infectiousness) when biting, p is the daily survival rate, and n is the number of days between mosquito infection and the production of sporozoites in salivary glands (the so-called "extrinsic incubation period"). Human infection is summarized completely in r, which often is said to be the rate of recovery from infection but which, strictly, is the rate of recovery from infectiousness. Garrett-Jones' contribution was the isolation of the entomological components of Eq. 1 (everything except r) as the vectorial capacity, C. Vectorial capacity is the daily rate at which future inoculations arise from a currently infective case.

The qualitative value of Eq. 1 is very well known (Molineaux, 1988; Dye, 1992). Most importantly, this equation shows that C and R_0 should respond most sensitively to changes in mosquito daily survival rate, p. As DDT became available, Macdonald argued that, all other things being equal, an attack on mosquito longevity ought to be a relatively effective method of malaria control. Of course, all other things are never equal and Macdonald's argument should never be used to exclude the possibility of, for example, reducing adult population size by attacking the immature stages when they are accessible.

It pays, however, to be more circumspect about the quantitative value of Eq. 1. Vectorial capacity offers a beguilingly tangible measure of transmission rate whereas models of directly transmitted infections offer a frankly intractable β (see Chapter 4). Although Eq. 1 has provided insights into the relative importance of different parameters, it should not be viewed as a formula for estimating either C or R_0, generally for two reasons (Dye, 1992). The first reason is that Eq. 1 is an incomplete expression for R_0. For example, although b allows for the efficiency of transmission from mosquito to human, no equivalent parameter allows for the efficiency of transmission from human to mosquito. The second reason is that the components on the right side of Eq. 1 are hard to measure. For example, the problem of measuring m has been solved partially by noting that obtaining the product ma, the daily biting rate of mosquitos per person, is easier. However, this solution is only partial because ma often is measured by trained adult male bait collectors working at fixed catching stations. Mosquito biting rates are highly heterogeneous with respect to site, host age, host size, and group size, among other things. Average daily biting rate recorded by these collectors is a quantity which, at best, changes in direct proportion with ma through time.

Two questions arise: How can Eq. 1 be used quantitatively? How should R_0 for vector-borne disease be estimated? The answer to the first question is that Eq. 1 and related expressions provide foundations for devising robust comparative measures of transmission rate by vectors. Consider a program of malaria control that aims to reduce transmission by attacking adult female *Anopheles* with insecticide. By how much is the basic reproduction ratio reduced *via* the effect of insecticide on survival rate? Within the framework of Eq. 1, all aspects of vector survival are embraced by the term $p^n/-\ln(p)$. This term provides an estimate ($\hat{L}$) of the "longevity factor" L. By comparing $\hat{L}$ before and after the introduction of a control program, the success of the control effort can be estimated. The ratio $\hat{q}$ serves this purpose, where

$$\hat{q} = \frac{\hat{L}_{\text{after control}}}{\hat{L}_{\text{before control}}}. \tag{2}$$

However, the actual values of L and q will be influenced by processes not included in Eq. 1, for example, the impact of *Plasmodium* on its vector. Therefore, $\hat{L} \neq L$ and $\hat{q} \neq q$ is expected, but with a discrepancy of unknown size. Figure 1 shows the performance of the comparative index $\hat{q}$ under three different assumptions about the way in which parasites impose additional mortality on their vectors. The ratio $q/\hat{q}$ departed significantly from 1 only when p was reduced very substantially below its initial value of 0.6 (lines 2a–c). In contrast, $L/\hat{L}$ was always much less than 1, regardless of p (lines 1a–c). $\hat{L}$ consistently overestimated L, because of the substantial impact of parasites on longevity. Thus, the comparative index $\hat{q}$ performed quite well, except under extreme circumstances.

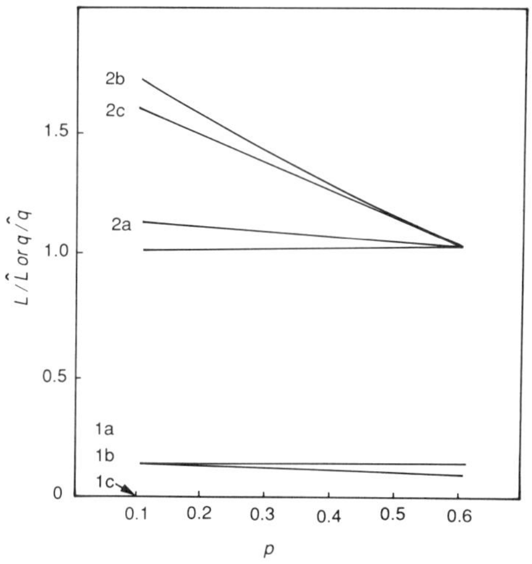

Figure 1 Performance of absolute and comparative estimates of the success of a vector control program. L and $\hat{L}$ are the actual and estimated longevity factors for a model population of *Anopheles* mosquitos before insecticidal control; the values are compared using the ratio $L/\hat{L}$ (1), over a range of values for mosquito survival, p. q and $\hat{q}$ represent the actual and estimated impacts of control on L and $\hat{L}$; the values are compared using the ratio $q/\hat{q}$ (2), assuming insecticide reduces mosquito survival to a value $p \leq 0.6$. Ideally, $L/\hat{L} = q/\hat{q} = 1$. Three different assumptions were made concerning when parasites increase the mortality rates of their vectors; (a) only during the ovarian cycle in the which the vector becomes infectious, (b) throughout infectious life, (c) throughout infected life. Reprinted with permission from Dye (1990).

Mosquito survival rate per ovarian cycle often has been estimated from the parous rate, the proportion of females that had laid eggs. This simple measure has been a mainstay of malariology for over 40 yr. Some commentators have taken the view that, because evidence now suggests that survival rate is age dependent and parasite dependent, the parous rate is defunct. However, if we accept that the absolute value of vectorial capacity is unobtainable, and that quantitative analyses in epidemiological entomology are essentially comparative analyses, then the parous rate should retain its central role in medical entomology (Dye, 1992).

Variations on the methods described in Chapter 3 are likely to be most successful in estimating R_0 for a vector-borne disease. For example, in a model of canine leishmaniasis in which phlebotomine sandflies bite domestic dogs at random (Dye *et al.*, 1992; Hasibeder *et al.*, 1992),

$$R_0 = \frac{1}{s^* u^*}. \tag{3}$$

In this simple formula, s^* and u^* are, respectively, the equilibrium proportions of dogs and sandflies that remain susceptible (uninfected). As is typical of many vector-borne diseases, a high infection rate can be maintained in the vertebrate host population while hardly any vectors are found infected. On the other hand, triatomine bugs, vectors of Chagas' disease, often have exceptionally high infection rates because they feed on blood throughout their long lives. Since in most cases $u^* \approx 1$, a good estimate of R_0 can be found without doing any entomology. This conclusion holds even when biting by sandflies is nonrandom. Sandfly preferences for particular kinds or groups of dogs (i) can be encapsulated by defining a specific force of infection (λ_i) for the proportion of dogs in each group, h_i:

$$R_0 = \frac{\Sigma h_i \lambda_i^{*2}}{u^*(\Sigma h_i \lambda_i^{*2} s_i^*)}. \tag{4}$$

Using this approach, a tentative estimate of $R_0 = 11$ has been obtained for canine leishmaniasis on the Maltese island of Gozo. Eradication of leishmaniasis from Gozo is worth contemplating because the dog population is relatively isolated, and because *Leishmania infantum* apparently has no wild animal reservoir of infection there. However, this preliminary estimate of R_0 implies that sandfly vectorial capacity would have to be reduced by more than 91% to be successful, given that the proportional reduction in vectorial capacity (P_{vc}) required for eradication is

$$P_{vc} = 1 - \frac{1}{R_0}. \tag{5}$$

III. Genetic Control of Vector Populations

"Regulation" is a word that should be reserved for those density-dependent processes that keep natural populations in the vicinity of some characteristic level of abundance, or at least help prevent extinction (see Chapter 5). "Control" should be used to refer to population suppression. For vector populations, the emphasis generally has been on control rather than regulation; and the former has sometimes failed for a lack of understanding of the latter. Although most ecologists now assume that all persistent populations are subject to density-dependent regulation of some kind for some of the time, regulation of blood-sucking insect populations is understood quite poorly. For example, practically nothing is known about density-dependent processes in *Anopheles gambiae* (s.l.), a complex of African mosquitos containing the most important malaria vectors in the world (see Christie, 1959, for the last word). Two exceptional groups of blood-sucking flies have been studied in some detail: tsetse flies and container-breeding culicine mosquitos. To illustrate the significance of density dependence here, the following discussion focuses on one well-established view of regulation, and its significance for the genetic control of *Aedes aegypti* mosquitos, the widespread vectors of yellow fever and dengue.

Because of substantial advances in transgenic technology (Crampton *et al.*, 1990; Handler and O'Brochta, 1991), the idea that genetic engineering holds promise for insect pest control has reached a wide audience (e.g., Cherfas, 1990). Remember, however, that genetically altered insects have been in existence for some time (Curtis, 1981). McDonald *et al.* (1977) created, by irradiation in the laboratory, single and double heterozygote translocations in male *A. aegypti.* The translocated males had substantially reduced fertilities of 50% and 25%, respectively. Could the release of a mixture of these aberrant males generate sufficient egg sterility to reduce the size of an adult female vector population? The answer depends on where and how much density dependence exists in an *A. aegypti* population. In this case, we need to know whether regulation acts between the point of control, the egg stage, and the offending adult stage.

Accepting the existence of regulatory processes in principle is one thing; detecting them in nature is quite another. For example, density dependence may be hard to identify if it operates with a time lag (Turchin, 1990), if it operates as a result of spatial processes within generations (Hassell, 1985, 1987), or if life table analysis is attempted with data collected over insufficient generations (Hassell *et al.,* 1989). Laboratory experiments generally are easier to carry out and interpret, but are an undesirable substitute for fieldwork because they commonly produce irrelevant results.

Aedes aegypti mosquitos offer an introduction to the study of population regulation, living as they commonly do under near-insectary conditions in the field. The larvae frequently are found in domestic water storage containers, which support few predators or competitors and in which (especially in houses) they live on an artificial food supply (e.g., maize meal accidentally dropped into the water). From an analytical perspective, these larvae come as close as any organism to living as "single populations" (May, 1981).

The best quantitative field study of *A. aegypti* was carried out by the World Health Organization's (WHO) *Aedes* Research Unit in Bangkok, the data on larval (Southwood *et al.,* 1972) and adult (Sheppard *et al.,* 1969) populations were drawn together by Dye (1984). Southwood *et al.* (1972) estimated survival rates of the different immature stages by drawing up a series of life tables after repeated cross-sectional sampling of the population over a period of about 1 yr. Sheppard *et al.* (1969) obtained estimates of adult survival rates from mark–release–recapture experiments. The important results are shown in Fig. 2. Mortality from egg to third instar was density dependent, albeit with a rather variable response to density (Fig. 2 *left*), whereas mortality from third instar to pupa was not (Fig. 2, *right*). Figure 3 describes an essentially clockwise trajectory through time for the adult population. This behavior is expected when a change in survival rate is *followed* by a change in population size. The inference is that changes in survival rate drove changes in population size, not vice versa.

The slope of the graph in Fig. 2 (*left*) is ≈ 1, implying that, over the range of egg density shown, any reduction in the egg input should be compensated almost exactly by an increase in larval survival and, consequently, should have no impact on the production of pupae and adults. Just how difficult it would be to depress the adult population by sterilizing eggs depends on the fecundity per female (F) which, for the Bangkok population, was uncertain (Fig. 4). In releasing male *A. aegypti* with translocations at a site near the Kenya coast, McDonald *et al.* (1977) were able to introduce up to 64% sterility in the egg population. In a worst-case scenario in Bangkok

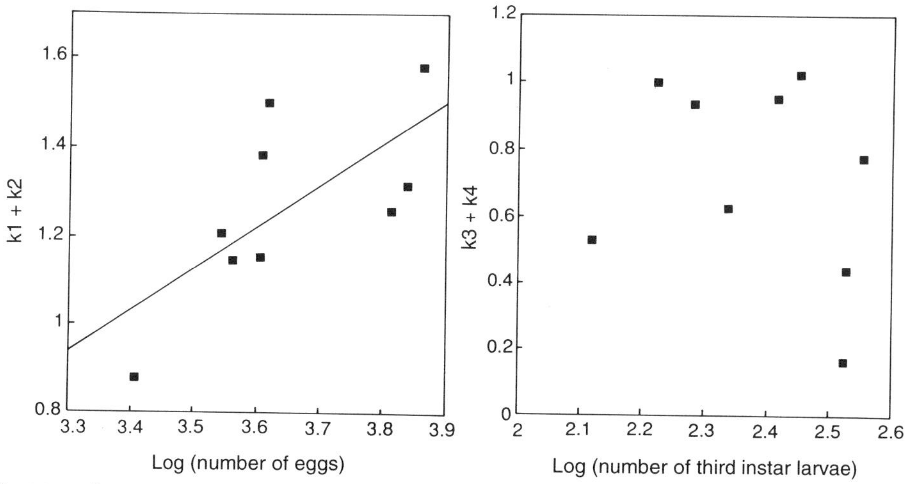

Figure 2 Mortality rates versus density for the mosquito *Aedes aegypti.* Mortalities are expressed as k values [log(number entering a stage/number surviving a stage)] where $k1 + k2$ is the mortality from egg to instar III (*left*) and $k3 + k4$ is the mortality from instar III to pupa (*right*). The solid line was obtained by linear regression. Data from Southwood *et al.* (1972).

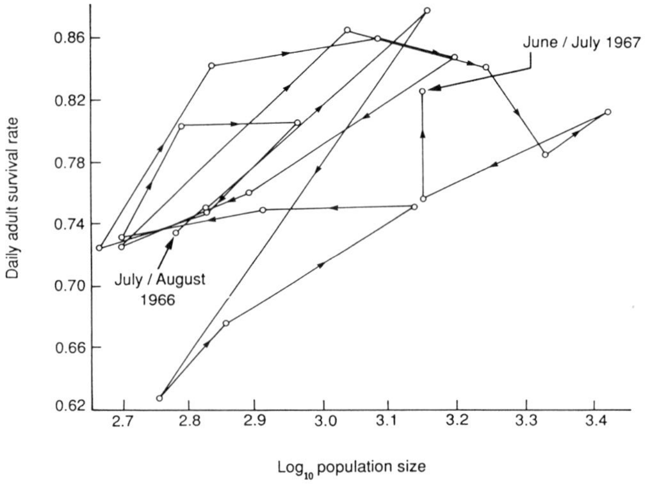

Figure 3 Daily adult survival rate of the mosquito *Aedes aegypti* related to adult population size. Both were estimated during mark–release–recapture studies carried out by Sheppard *et al.* (1969). Arrowed lines trace the trajectory through time. Reprinted with permission from Dye (1984).

($F = 58$), this effort, even if sustained, would be expected to bring no reduction to the adult population. This is essentially the result obtained by McDonald and colleagues: the rate of pupal production remained remarkably stable in the village in which the release experiment was carried out.

The new generation of transgenic insects is certain to include vectors that have been

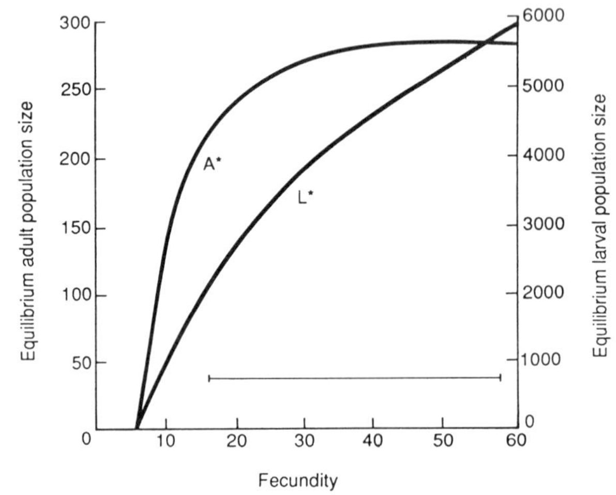

Figure 4 Variation of equilibrium adult (A*) and larval (L*) densities of the mosquito *Aedes aegypti* with female fecundity (average number of eggs laid/adult female/ovarian cycle: F), as expected from a population model described by Dye (1984). The horizontal bar reflects uncertainty about the true value of F for *A. aegypti* in Bangkok.

made refractory to parasites. At present, the most exciting advances are being made by molecular biologists and confined to laboratories. However, it is not premature for population biologists to be thinking about the consequences of introducing transgenic insects into wild vector populations. One obvious difficulty arises from the fact that only a small proportion of vectors ever becomes infected, whereas all parasites must go through vectors. In consequence, the selection pressure on parasites to overcome refractoriness will be relatively strong. Molecular biologists may have to find ways of presenting parasites with an insoluble problem, for example, by introducing two tightly linked but different refractory mechanisms into the vector genome simultaneously. Genetic engineering is still far from reaching this goal, but a natural precedent for the insoluble problem may exist. Waters *et al.* (1991) found that *Plasmodium falciparum* is related more closely to the avian malarias than to primate malarias. The former are transmitted by culicine mosquitos, the latter by anophelines. Today, culicine mosquitos feed regularly on people infected with malaria, particularly in tropical towns and cities, yet plasmodia have not taken advantage of these potential vectors (Miller, 1991).

IV. Environmental Management for Vector Control

Although regulatory processes during larval mosquito life can be demonstrated by careful field studies, they do not necessarily present an obstacle to vector control. Chubachi (1979) invented an elegant method of life table analysis for a mosquito population with partially overlapping generations, *Culex tritaeniorhynchus summorosus,* breeding in rice fields. He showed that the generation reproduction ratio, *R,* of the vector population was inversely dependent on larval density (Fig. 5). In addition, for a given larval density, Chubachi found that population growth rate was lower after the

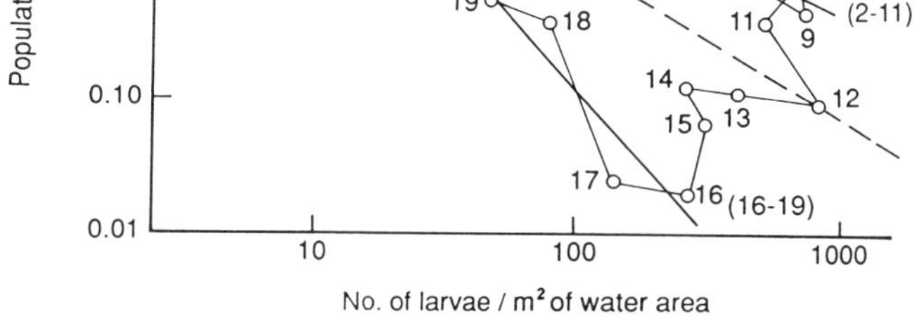

Figure 5 Population growth rate of the rice field mosquito, *Culex tritaeniorhynchus summorosus,* in relation to larval density. Points 1 – 19 represent consecutive sampling occasions during one breeding season. Solid straight lines are regression lines through these points for growth rates obtained before (*top*) and after (*bottom*) rice fields were drained (see Fig. 6). The dashed line is the best fitting regression line for the whole season. Reprinted with permission from Chubachi (1979).

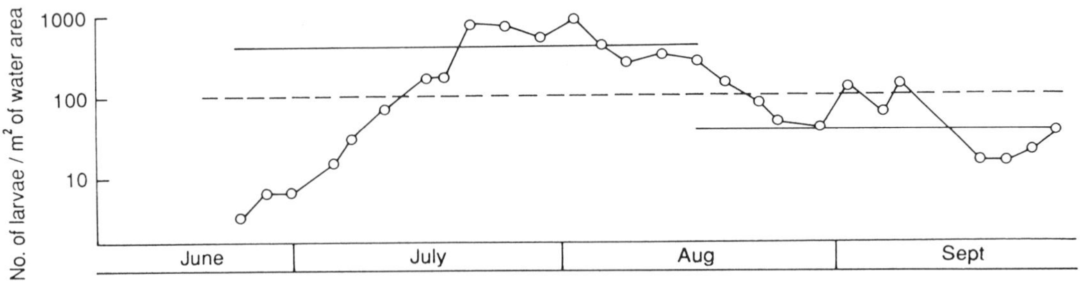

Figure 6 Dynamics of the *C. t. summorosus* larval population during the breeding season referred to in Fig. 5. Solid horizontal lines mark the estimated equilibria before (*top*) and after (*bottom*) drainage. Dashed horizontal line is the mean of the two. Reprinted with permission from Chubachi (1979).

paddy fields had been partially drained to protect the rice crop against storm damage. The equilibrium abundance of larvae was shifted downward (Fig. 6), apparently because intraspecific competition became more intense (steeper slope of the lower line in Fig. 5).

Drainage may reduce population growth rate by intensifying intraspecific competition but, better still, may be used to eradicate most mosquitos from a rice crop. Clearly, regulation during larval life will not be immediately relevant if the habitat in which larvae thrive can be eliminated (see Rajagopalan *et al.*, 1991). One of the best examples of such management is a series of trials carried out in Portugal between 1935 and 1939 (Hill and Cambournac, 1941), which aimed to reduce population densities of the malaria vector *Anopheles labranchiae. Anopheles labranchiae* develops from egg to adult in about 18 days so intermittent drainage no less frequently than every 16–17 days should be an effective control method. When small plots became completely dry in these studies, all larvae were killed, but large fields usually retained wet depressions in which some mosquitos were able to survive. Overall, about 80% larval population reduction was achieved by regular intermittent drainage. At the same time, rice yield was higher under intermittent than under continuous irrigation, grain quality was no worse, and water was saved as well. As a result, the Portuguese government passed a law in 1938 making intermittent irrigation in rice cultivation obligatory where deemed necessary by the malaria control authorities. Now, of course, the law is irrelevant because endemic malaria has been eradicated from Portugal.

V. Biological Control of Vectors

The elysium of biological control is pictured in an introduced agent—a predator, parasite, or competitor—bringing pest population size sharply downward to some lower equilibrium. If the agent were a parasite or predator, the two species of pest and control agent would, henceforth, coexist benignly below an appropriate threshold. In the case of an agricultural pest, the threshold might be defined by economics. In the case of an insect vector, the ideal threshold is defined when the basic reproduction ratio for the infection, R_0, equals 1. The insect vector persists, but insufficient vectors exist per host to ensure $R_0 > 1$; this situation would be, for example, "anophelism without malaria" (Molineaux, 1988).

Regrettably, very few instances exist to date of vectors being controlled in this way. In one of these few instances, schistosomiasis snails have been replaced by refractory species. The South American snail *Marisa cornuarietis* voraciously feeds on vegetation supporting the schistosome host, *Biomphalaria glabrata,* and also eats its eggs and hatchlings. *Biomphalaria glabrata* was eliminated from some areas of Puerto Rico after the introduction of *M. cornuarietis* and was virtually eliminated from the nearby Viegues Islands (Amin, 1983). More examples of biological control exist, some using fish as predators of mosquito larvae. *Tilapia spilurus* was successful in producing detectable reductions in *A. gambiae* mosquito populations in Somalian water storage tanks, as well as lowering the number of cases of malaria. Remember, however, that fish populations generally require careful management, including frequent restocking (Reuben *et al.,* 1991).

This is not to say, however, that the dynamics of vector biological control are usually trivial. One illuminating analysis has tackled the question of whether competitors can halt the spread of *Aedes albopictus* mosquitos in North America. *Aedes albopictus,* an important vector of dengue in Asia, first was observed in the southern United States in 1985, having been introduced in used car tires (Sprenger and Wuithiranyagool, 1986; Hawley *et al.,* 1987). *Aedes albopictus* is a container-breeding mosquito and, to become widely established in North America, must avoid or successfully compete with indigenous culicines, principally *Aedes triseriatus,* a common inhabitant of treeholes and tires throughout the eastern United States. While *A. albopictus* spreads from its point of introduction, what information can theory and experiment provide about the eventual outcome of the interaction? Livdahl and Willey (1991) tried to answer this question by carrying out replicated competition experiments in small containers with fluid from treeholes or tires. Newly hatched larvae of *A. albopictus* were reared at a range of densities (0, 7, 14, 28, and 42 per 100 ml) in the presence of the same range of densities of *A. triseriatus.* Adult females were removed as they emerged on day x (A_x); wing length (w) was used to predict the number of female offspring produced by each adult female $f(w_x)$. If D is an estimate of the delay between female emergence and first oviposition, then *per capita* population growth rates for all experimental larval cohorts can be found from

$$\hat{r} = \frac{\ln\left[\frac{1}{N_0}\sum A_x f(w_x)\right]}{D + \frac{\sum x A_x f(w_x)}{\sum A_x f(w_x)}}, \tag{6}$$

where N_0 is the initial number of females. The simplest way of expressing the impact of species j on species i, and of i on itself, is

$$\hat{r}_i \simeq \frac{dN_i}{N_i dt} = r_i + b_i N_i + b_j N_j, \tag{7}$$

where r_i is the unlimited *per capita* population growth rate of species i (in the absence of any competition) and the b's are interaction coefficients. The expected outcome of competition now can be analyzed graphically by setting $\hat{r}_i = 0$ and plotting N_j against N_i. The two resulting sets of isoclines for simulated treeholes and tires are shown in Fig. 7. This is the approach taken by classical Lotka–Volterra competition theory (Begon *et al.,* 1986); a stable equilibrium exists where the isoclines intersect. (For a more detailed description of this technique, see Chapter 5.) In model treeholes, the isoclines intersect with positive densities of both species; in model tires, *A. albopictus* is expected to win outright. These results suggest that *A. triseriatus* will not be able to prevent the spread of the invader.

The analysis concurs with the little evidence available to date from the field. *Aedes albopictus* has been numerically dominant in artificial containers since its discovery. Sprenger and Wuithiranyagool (1986) found it to be three times more abundant than *A. aegypti* and 20 times more abundant than *A. triseriatus* in tires in Texas. However, in a Louisiana forest, *A. triseriatus* and *A. albopictus* were, respectively, the first and second most abundant mosquitos colonizing glass jars attached to trees (Schreiber *et al.,* 1988).

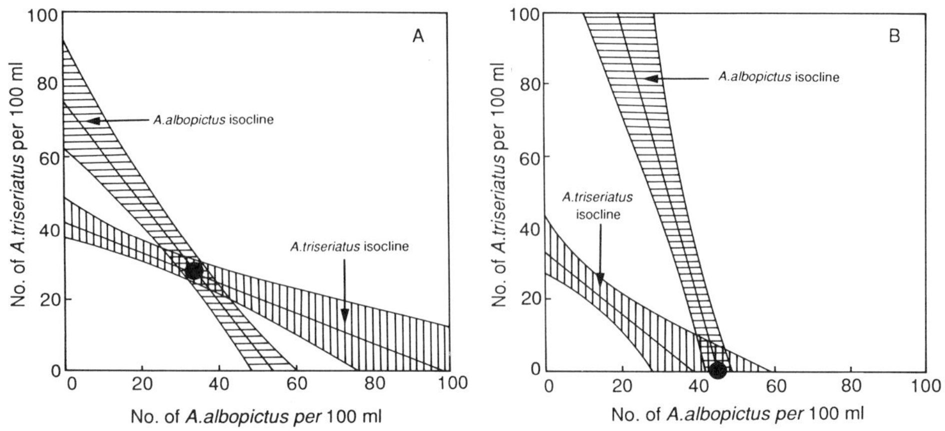

Figure 7 Expected equilibrium density combinations (circled) of the two mosquitoes *Aedes triseriatus* and *Aedes albopictus* in treeholes (*left*) and car tires (*right*). Hatched regions around each isocline define 95% confidence intervals. Reprinted with permission from Livdahl and Willey (1991).

The possibility exists that species of mosquitos other than *A. triseriatus* will influence the establishment of *A. albopictus* in sylvatic habitats. In particular, the predatory mosquito, *Toxorhynchites rutilus,* which itself has credentials as a biological control agent, can change significantly the relative abundance of emerging adults of these species. In the south-eastern United States, the presence of this predator in treeholes favors *Orthopodomyia signifera* and *Anopheles barberi,* and disfavors *A. triseriatus* (Brad-

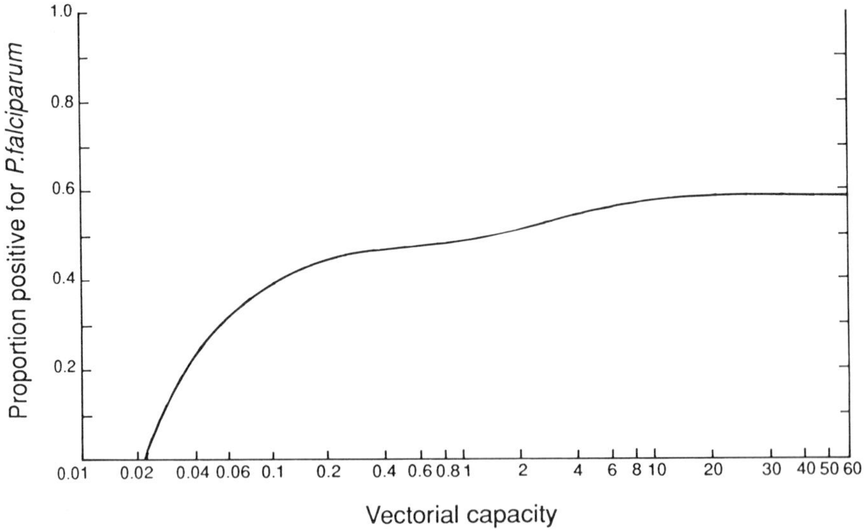

Figure 8 Expected proportion of persons positive for malaria (*Plasmodium falciparum*) infection as a function of mosquito vectorial capacity, according to the Garki model of malaria. Adapted with permission from Molineaux and Gramiccia (1980).

shaw and Holzapfel, 1983,1988). The precise effects of the expansion of the *A. albopictus* range from its point of invasion in Texas remain to be seen. Additional analysis should go hand in hand with additional observation.

VI. Vector Control, Infection, and Disease

Because malaria is the most important of all insect-borne parasitic diseases, it has received more interest from quantitative epidemiologists than any other. However, until recently, attention has been restricted to infection rather than disease, the former being a necessary but not a sufficient condition for the latter. Accordingly, mathematical models have been almost exclusively compartmental models, describing only the prevalence of infection in humans and anopheline mosquitos.

The distinction between infection and disease is most important in areas endemic for malaria, particularly sub-Saharan Africa, where investigators long have realized that vector control must be very effective to have any significant impact on the prevalence of infection. The point is illustrated nicely with the prediction made by the Garki model, the most detailed malaria model ever constructed (Molineaux and Gramiccia, 1980). In one village highly endemic for malaria in this area of northern Nigeria, an index related to vectorial capacity (containing some of but not all the entomological components of Eq. 1) varied from dry to wet season, ranging typically between 0.2 and 20 per day. Figure 8 shows that the equilibrium prevalence of human infection established anywhere within this large range is expected to be about 0.5. If the true vectorial capacity had an equilibrium level of 4 per day, the abundance of female *Anopheles* would have to be reduced by about two orders of magnitude to produce any significant reduction in human prevalence.

Among the ≈50% of persons infected in the endemic steady state, only some will be found suffering from malaria, the disease. If prevalence in the whole population cannot be reduced, could the incidence of severe malaria (cerebral malaria or malaria anemia) that follows perhaps 2% of all clinical attacks be reduced (Greenwood *et al.*, 1991)? Some field experiments with insecticide-impregnated bednets in The Gambia show that it is. First, Snow *et al.* (1987) showed that, although bednets had a slight but insignificant impact on the percentage of persons giving positive blood films, they did reduce the number of febrile episodes significantly, both for all cases of parasitemia and for those showing high parasitemia (Fig. 9). They confirmed these results in a follow-up study Snow *et al.* (1988), which also demonstrated that bednets could be effective in reducing spleen size and anemia. Alonso *et al.* (1991) went further, showing that treated nets could have a profound effect on childhood mortality. Figure 10 illus-

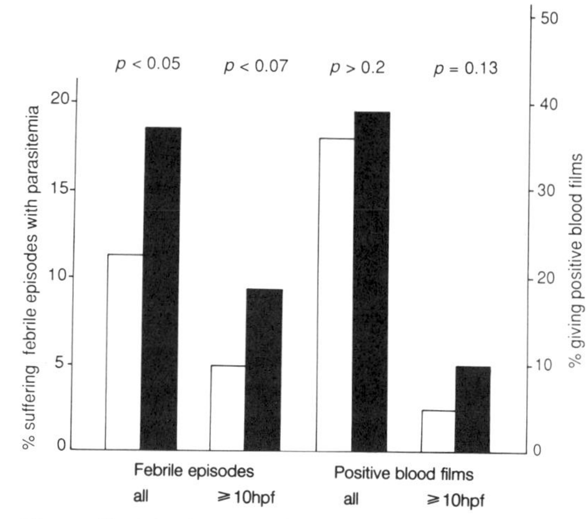

Figure 9 The impact of bednets with (open bars) or without (solid bars) permethrin on malaria infection and disease. Infection is measured by the proportion of persons giving positive blood films (all positive films, or high density films with ≥10 parasites/high power field). Disease is measured by the percentage of persons suffering febrile episodes with parasitemia. p values above the bars indicate probabilities that the difference between solid and open bars is due to chance. Data from Snow *et al.* (1987).

trates the results of the field trial that found a 30% reduction in mortality attributable to malaria among Gambian children.

The impact of bednets on malaria morbidity in endemic areas cannot yet be fully accounted for, though it is more persuasively explained in terms of strain-specific immunity (Lines and Armstrong, 1992), or strain-specific virulence, than in terms of parasite dosage (Jeffrey *et al.,* 1959). The key point is that malaria sporozoites inoculated into hosts in highly endemic areas will often be greatly outnumbered by resident parasites, which have multiplied in the liver and blood. A few more parasites could not easily compete quantitatively, and it is easier to imagine a new episode of illness arising from qualitatively (antigenically) different parasites.

The quantitative relationship between vector dynamics, infection, and disease is better understood for onchocerciasis (river blindness), for which the risk of becoming blind is a reasonably well-defined function of microfilarial load accumulated over time. Plaisier *et al.* (1990,1991) have incorporated the risk of blindness into ONCHOSIM, a simulation model developed in collaboration with the Onchocerciasis Control Programme (OCP) in West Africa (for further discussion, see Chapter 7). Figure 11 shows how the model already has made very good predictions about microfilarial distribution in the human population before and after 12 yr of larviciding against *Simulium damnosum* blackflies. Since vector control has been successful to the point that transmission has ceased in the center of the OCP

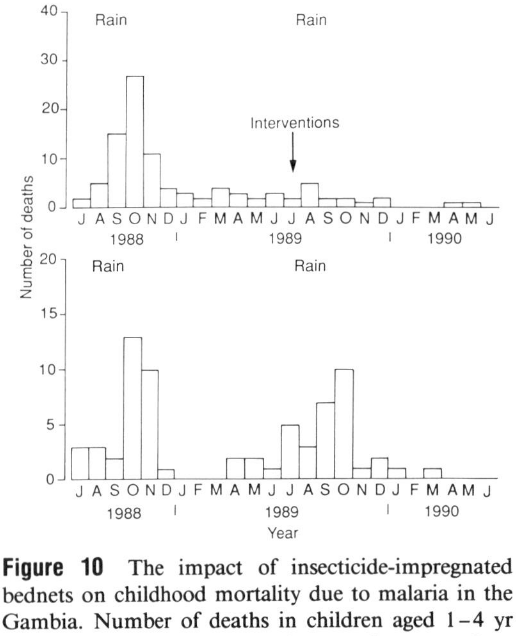

Figure 10 The impact of insecticide-impregnated bednets on childhood mortality due to malaria in the Gambia. Number of deaths in children aged 1–4 yr were recorded in pre- and postintervention years. Over 90% of children in village A (*top*) slept under bednets from July 1989 onwards. Reprinted with permission from Alonso *et al.* (1991). The effect of insecticide-treated bed nets on mortality of Gambian children. *Lancet.* 337, 1499–1502, © by The Lancet Ltd.

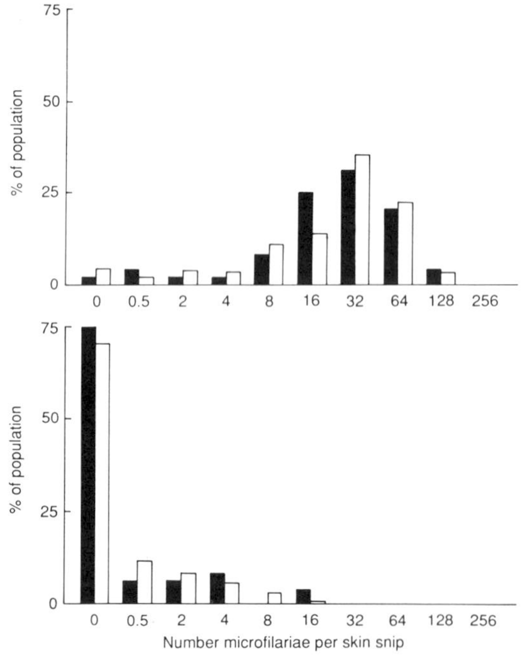

Figure 11 Observed (solid bars) and expected (open bars) frequency distributions of *Onchocerca volvulus* microfilarial load before (*top*) and after (*bottom*) 12 yr of larvicidal control in the Onchocerciasis Control Programme. Numbers on the x axis are the lower limits of the categories used. Reprinted with permission from Habbema *et al.* (1992).

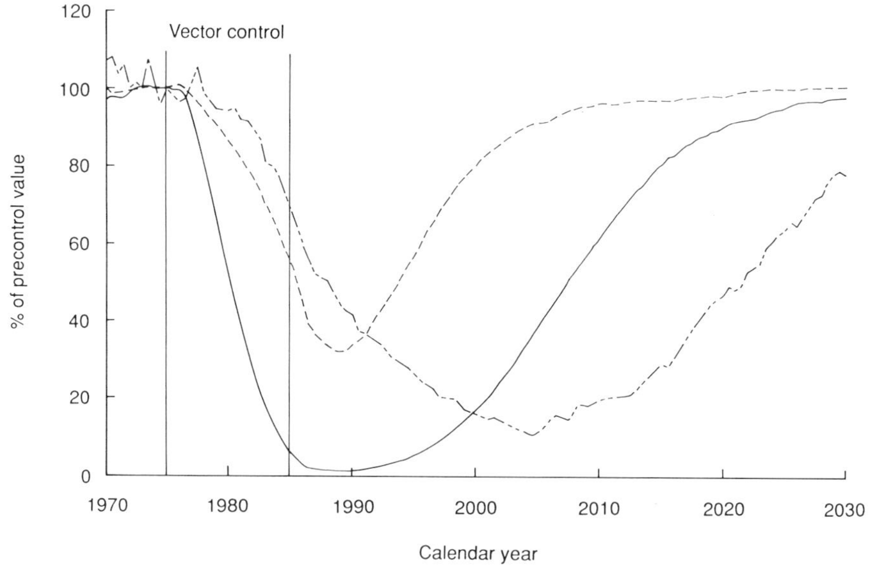

Figure 12 The dynamics of *Onchocerca* and onchocerciasis during and after 10 yr of vector control in West Africa. Three trajectories are shown for the prevalence of microfilaremia (- - -), the prevalence of blindness (——-- -—), and the community microfilarial load (CMFL, geometric mean skin snip count for adults; ———). Reprinted with permission from Habbema *et al.* (1992).

area, the rate of decline of the prevalence of blindness is governed principally by human birth and death rates, and should be relatively easy to predict (Fig. 12). However, predicting rates of increase of blindness, and of lesser ocular disabilities, after termination of vector control should be more difficult since these events will depend on vector and parasite population dynamics, as well as on human demography. ONCHOSIM has yet to be tested under these more demanding circumstances (Habbema *et al.*, 1992).

VII. Conclusions

The examples provided in this chapter illustrate some of the approaches to, and current issues in, vector control. The particulars are summarized easily, and some general points emerge from them. The concept of vectorial capacity long has been used in malariology and now is being applied to other vector-borne diseases. However, the formula for vectorial capacity is not a recipe for estimating absolute transmission rates, including the basic reproduction ratio, R_0. The principal quantitative value of vectorial capacity lies in providing a framework for comparative analysis. R_0 for vector-borne diseases is best found using variations on the methods devised for directly transmitted infections.

The distinction between infection and disease is becoming increasingly important in studies of parasite epidemiology (see Chapter 11). Notwithstanding the progress made for onchocerciasis, the role of quantitative analytical work in the near future is likely to be in helping investigators to understand the relationships between infection and disease, rather than in making models that attempt to predict in detail the impact of vector control on disease.

Regulatory or density-dependent processes in vector populations have been studied infrequently, although evidence suggests that these processes could be major hindrances to control. The most effective method of vector population control will act after the point of compensation but before the point of transmission. Environmental management, such as draining mosquito larval habitats (see Chapter 7), may do this effectively. If all regulation occurs during larval life, no compensation will be possible. The reductions in pupal and adult numbers should be related linearly. As intellectually enticing as these arguments about nonlinearities are, they must be kept in perspective: compensation will be irrelevant in the force of massive density-independent mortality.

The ideal of biological control has been met rarely, but the few good examples that do exist encourage additional careful evaluation of promising candidates. That populations of the control agent need to be managed closely is not necessarily a disadvantage. The philosophy of vector control in many developing countries is shifting from vertical to horizontal, from nationally organized top-down programs to community participation. Local management means local acceptance for the control method (by no means a formality) and often provides an effective, albeit unskilled, monitoring service. Clearly, the same rules apply to some kinds of environmental management, such as keeping the peridomestic arena free from tires, tin cans, and the like, which are excellent breeding sites for *Aedes* mosquitos.

That said, it should be emphasized (if not already clear from the diversity of examples) the "high tech/low tech," vertical/horizontal, and secondary/primary health-care dichotomies should not be seen as bad or good, but as alternatives applicable under different circumstances. There are, in fact, neither philosophical nor technical panaceas for vector control.

Acknowledgment

I thank Jo Lines for helpful comments.

References

Alonso, P. L., Lindsay, S. W., Armstrong, J. R. M., Conteh, M., Hill, A. G., David, P. H., Fegan, G., de Francisco, A., Hall, A. J., Shenton, F. C., Cham, K., and Greenwood, B. M. (1991). The effect of insecticide-treated bed nets on mortality of Gambian children. *Lancet* **337,** 1499–1502.

Amin, A. (1983). Control of snail intermediate hosts of schistosomiasis. *In* "Pest and Vector Management in the Tropics" (A. Youdeowei and M. W. Service, eds.), pp. 318–323. Longman, London.

Begon, M., Harper, J. L., and Townsend, C. R. (1986). "Ecology: Individuals, Populations and Communities." Blackwell Scientific Publications, Oxford.

Bradshaw, W. E., and Holzapfel, C. M. (1983). Predator-mediated, non-equilibrium coexistence of tree-hole mosquitoes in southeastern North America. *Oecologia* **57,** 239–256.

Bradshaw, W. E., and Holzapfel, C. M. (1988). Drought and the organization of tree-hole mosquito communities. *Oecologia* **74,** 507–514.

Cherfas, J. (1990). Malaria research—What next? *Science,* **247,** 399–403.

Christie, M. (1959). A critical review of the role of the immature stages of anopheline mosquitoes in the regulation of adult numbers, with particular reference to *Anopheles gambiae. Trop. Dis. Bull.* **56,** 385–389.

Chubachi, R. (1979). An analysis of the generation mean life table of the mosquito *Culex tritaeniorhynchus summorosus,* with particular reference to population regulation. *J. Anim. Ecol.* **48,** 681–702.

Crampton, J., Morris, A., Lycett, G., Warren, A., and Eggleston, P. (1990). Transgenic mosquitoes: A future vector control strategy? *Parasitol. Today* **6,** 31–36.

Curtis, C. F. (1981). Chromosome manipulations for insect pest control. *Chrom. Today* **7,** 138–147.

Curtis, C. F. (ed.) (1991). "Control of Disease Vectors in the Community." Wolfe, London.

Dye, C. (1984). Models for the population dynamics of the yellow fever mosquito, *Aedes aegypti. J. Anim. Ecol.* **53,** 247–268.

Dye, C. (1990). Epidemiological significance of vector–parasite interactions. *Parasitology* **101,** 409–415.

Dye, C. (1992). The analysis of parasite transmission by blood-sucking insects. *Ann. Rev. Entomol.* **37,** 1–19.

Dye, C., Killick-Kendrick, R., Vitutia, M. M., Walton, R., Killick-Kendrick, M., Harith, A. E., Guy, M. W., Canavate, M. C., and Hasibeder, G. (1992). Epidemiology of canine leishmaniasis: Prevalence, incidence and basic reproduction number calculated from a cross-sectional serological survey on the Island of Gozo, Malta. *Parasitology* **105,** 35–41.

Garrett-Jones, C. (1964). The human blood index of malaria vectors in relation to epidemiological assessment. *Bull. W. H. O.* **30,** 241-261.
Gratz, N. G. (1991). Emergency control of *Aedes aegypti* as a disease vector in urban areas. *J. Am. Mosquito Control Assoc.* **7,** 353-365.
Greenwood, B., Marsh, K., and Snow, R. (1991). Why do some African children develop severe malaria? *Parasitol. Today* **7,** 277-281.
Habbema, J. D. F., Alley, E. S., Plaisier, A. P., van Oortmarssen, G. J., and Remme, J. H. F. (1992). Epidemiological modelling for onchocerciasis control. *Parasitol. Today* **8,** 99-103.
Handler, A. M., and O'Brochta, D. A. (1991). Prospects for gene transformation in insects. *Annu. Rev. Entomol.* **36,** 159-183.
Hasibeder, G., Dye, C., and Carpenter, J. (1992). Mathematical modelling and theory for estimating the basic reproduction number of canine leishmaniasis. *Parasitology* **105,** 43-53.
Hassell, M. P. (1985). Insect natural enemies as regulating factors. *J. Anim. Ecol.* **54,** 323-334.
Hassell, M. P. (1987). Detecting regulation in patchily distributed animal populations. *J. Anim. Ecol.* **56,** 705-713.
Hassell, M. P., Latto, J., and May, R. M. (1989). Seeing the wood for the trees: Detecting density dependence from existing life table studies. *J. Anim. Ecol.* **58,** 883-892.
Hawley, W. A., Reiter, P., Copeland, R. S., Pumpuni, C. B., and Craig, G. B. Jr. (1987). *Aedes albopictus* in North America: Probable introduction in used tires from Northern Asia. *Science* **236,** 1114-1115.
Hill, R. B., and Cambournac, F. J. C. (1941). Intermittent irrigation in rice cultivation, and its effect on yield, water consumption and *Anopheles* production. *Am. J. Trop. Med. Hyg.* **21,** 123.
Jeffery, G. M., Young, M. D., Burgess, R. W., and Eyles, D. E. (1959). Early activity in sporozoite-induced *Plasmodium falciparum* infections. *Ann. Trop. Med. Parasitol.* **53,** 51-58.
Lines, J., and Armstrong, J. R. M. (1992). For a few parasites more: inoculum size, vector control and strain—specific immunity to malaria. *Parasitol. Today* **8,** 381-383.
Livdahl, T. P., and Willey, M. S. (1991). Prospects for an invasion: Competition between *Aedes albopictus* and native *Aedes triseriatus. Science* **253,** 189-191.
Macdonald, G. (1957). "The Epidemiology and Control of Malaria." Oxford University Press, London.
May, R. M. (1981). Models for single populations. *In* "Theoretical Ecology: Principles and Applications" (R. M. May, ed.), pp. 5-29. Blackwell Scientific Publications, Oxford.
McDonald, P. T., Hausermann, W., and Lorimer, N. (1977). Sterility introduced by release of genetically altered males to a domestic population of *Aedes aegypti* at the Kenya coast. *Am. J. Trop. Med. Hyg.* **26,** 553-561.
Miller, L. H. (1991). Postscript. *In* "Malaria: Waiting for the Vaccine" (G. A. Targett, ed.), pp. 221-224. Wiley and Sons, Chichester.
Molineaux, L. (1988). The epidemiology of human malaria as an explanation of its distribution, including some implications for its control. *In* "Malaria: Principles and Practice of Malariology" (W. Wernsdorfer and I. McGregor, eds.), pp. 913-998. Churchill Livingstone, Edinburgh.
Molineaux, L., and Gramiccia, G. (1980). "The Garki Project." World Health Organization, Geneva.
Plaisier, A. P., van Oortmarssen, G. J., Habbema, J. D. F., Remme, J., and Alley, E. S. (1990). ONCHOSIM: A model and computer simulation program for the transmission and control of onchocerciasis. *Computer Models Programs Biomed.* **31,** 43-56.
Plaisier, A. P., van Oortmarssen, G. J., Remme, J., Alley, E. S., and Habbema, J. D. F. (1991). The risk and dynamics of onchocerciacis recrudescence after cessation of vector control. *Bull. W. H. O.* **69,** 169-178.
Rajagopalan, P. K., Das, P. K., Panicker, K. N., Reuben, R., Raghuatha Rao, D., Self, L. S., and Lines, J. D. (1991). Environmental and water management for mosquito control. *In* "Control of Disease Vectors in the Community" (C. F. Curtis, ed.), pp. 121-138. Wolfe, London.
Reuben, R., Raghunatha Rao, D., Sebastian, A., Corbet, P. S., Wu, N., and Liao, G. (1991). Biological control methods suitable for community use. *In* "Control of Disease Vectors in the Community" (C. F. Curtis, ed.), pp. 139-158. Wolfe, London.
Rogers, D. J., and Randolph, S. E. (1985). Population ecology of tsetse. *Annu. Rev. Entomol.* **30,** 197-216.
Schreiber, E. T., Meek, C. L., and Yates, M. M. (1988). Vertical distribution and species coexistence of treehole mosquitoes in Louisiana. *J. Am. Mosquito Control Assoc.* **4,** 9-14.
Sheppard, P. M., Macdonald, W. W., Tonn, R. J., and Grab, B. (1969). The dynamics of an adult population of *Aedes aegypti* in relation to dengue haemorrhagic fever in Bangkok. *J. Anim. Ecol.* **38,** 661-702.
Snow, R. W., Rowan, K. M., and Greenwood, B. M. (1987). A trial of permethrin-treated bed nets in the prevention of malaria in Gambian children. *Trans. R. Soc. Trop. Hyg.* **81,** 563-567.
Snow, R. W., Lindsay, S. W., Hayes, R. J., and Greenwood, B. M. (1988). Permethrin-treated bed nets (mosquito nets) prevent malaria in Gambian children. *Trans. R. Soc. Trop. Med. Hyg.* **82,** 838-842.
Southwood, T. R. E., Murdie, G., Yasuno, M., Tonn, R. J., and Reader, P. M. (1972). Studies on the life

budget of *Aedes aegypti* in Wat Samphaya, Bangkok, Thailand. *Bull. W. H. O.* **46,** 211–226.

Sprenger, D., and Wuithiranyagool, T. (1986). The discovery and distribution of *Aedes albopictus* in Harris County, Texas. *J. Am. Mosquito Control Assoc.* **2,** 217–219.

Turchin, P. (1990). Rarity of density dependence or population regulation with lags? *Nature (London)* **344,** 660–663.

Waters, A. P., Higgins, D. G., and McCutchan, T. F. (1991). *Plasmodium falciparum* appears to have arisen as a result of lateral transfer between avian and human hosts. *Proc. Nat. Acad. Sci. USA* **88,** 3140–3144.

D. W. T. Crompton

14

Ascaris lumbricoides

I. Introduction

According to Yamaguti (1961), 16 species of *Ascaris* are known from mammalian hosts. *Ascaris lumbricoides* in humans and *Ascaris suum* in pigs are by far the most common and best known members of the genus. Some uncertainty remains about whether *A. lumbricoides* and *A. suum* are valid separate species (Lýsek, 1967; Gibson, 1983), but in this chapter *A. lumbricoides* has been assumed to be a valid species in its own right.

Ascaris lumbricoides is a remarkably infectious and persistent parasite. Stoll (1947) estimated that roughly one-fifth of the world's population was infected with *A. lumbricoides.* The situation appears not to have changed; about one billion people currently are believed to harbor this worm (Crompton, 1989a). Unfortunately, the distribution and abundance of *A. lumbricoides* reflect the extent and severity of human deprivation, poverty, undernutrition, and underdevelopment. Sustained relief from ascariasis may seem not to be possible until the populations of developing countries have access to the same standards of sanitation, water supply, services, and economic growth that now are provided in industrialized countries, where ascariasis is found rarely (Yokogawa, 1985; Crompton, 1989a). Ascariasis is the name given to the spectrum of disease symptoms that present in some infected humans during the course of an infection with *A. lumbricoides.* Both adults and children experience acute life-threatening ascariasis, in which intestinal obstructions and biliary complications predominate (Maki, 1972; Pinus, 1985; Khuroo *et al.,* 1990; Thein-Hlaing *et al.,* 1990; Chai *et al.,* 1991). As a result, perhaps as many as 100,000 people die annually (Pawlowski and Davis, 1989). Pneumonitis and respiratory disturbance occur in some cases of ascariasis (Pawlowski, 1987) and allergic/atopic reactivity induced by the synthesis of IgE in response to *A. lumbricoides* may be observed (Pritchard, 1992). Chronic ascariasis is one of the many factors that impair the growth and healthy development of children (Stephenson, 1987; Tomkins and Watson, 1989; Crompton, 1992). In this chapter, aspects of the epidemiology and population biology of *A. lumbricoides* are discussed in conjunction with the results of theoretical and operational research into methods for control of the infection.

Parasitic and Infectious Diseases

Table 1 Features of the Life History of *Ascaris lumbricoides*[a]

Event	Observation	Reference
Embryonation	10–14 days at 30 ± 2°C 45–55 days at 17 ± 1°C	Pawlowski and Arfaa (1984) Pawlowski and Arfaa (1984)
Tissue migration (using experimental infections)		
Small intestine to liver	4 (2–8) days pi	Yoshida (1918)
Liver to lungs	7 (14) days pi	Yoshida (1918)
Lungs to small intestine	14–20 days pi	Nichols (1956)
Normal location of adult worms	Mainly jejunum	Makidono (1956)
Prepatent period (from time of infection to first detection of eggs in stools)	67–76 days 67 days in children <4 yr	Takata (1951) Akamatsu (1959)
Fecundity (estimate of egg production/day per female worm)	238,722 (134,462–358,750)	Sinniah (1982)
Life span	1–2 yr	Anderson (1982)
Egg survival (retention of infectivity in host environment)	Variable up to 15 yr	Krasnonos (1978); see WHO (1967), Storey and Phillips (1985)

[a] Reproduced with permission from Crompton (1989b).

II. Life History

A. General Features

The life-history pattern of *A. lumbricoides* is direct; the parasite is passed from human to human by the ingestion of eggs containing infective second-stage larvae (Crompton, 1989b). Adult male and female worms living in the small intestine copulate, eggs are released in the stools, and, in communities in which sanitation is inadequate, the environment quickly becomes contaminated. A female worm releases over 200,000 eggs per day, perhaps for about 1 yr (Table 1); although many of these eggs must perish, many will embryonate to produce second-stage larvae if given moisture, oxygen, and shade. Additional details of developmental events and life-history features are given in Table 1.

Two populations and various rate processes are contained in the life history of *A. lumbricoides* (Fig. 1). The first population is the number of sexually mature *A. lumbricoides* in the host population; the second is the number of viable infective stages free in the environment of the host population. The population of mature worms gains new members through the natural route of infection, while losses occur as the worms or the infected hosts die. The adult worms do not

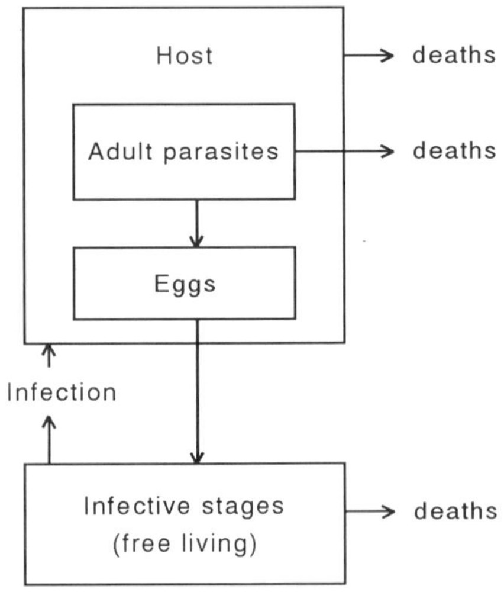

Figure 1 Flow chart representing the life history of *Ascaris lumbricoides*.

Table 2 Observations on Nonhuman Hosts of *Ascaris lumbricoides*[a]

Host	Stage	Reference
Brown bear (n)	Stage unknown	Yamaguti (1961)
Cat (e)	Larvae in lungs	Yoshida (1918)
Chimpanzee (n)	Mature adults	Pillers (1924), Orihel (1970)
Dog (e)	Larvae in lungs	Gaur and Deo (1972)
Gibbon (n)	Mature adults	Dunn and Greer (1962), Orihel (1970)
Goat (e)	Larvae in lungs	Gaur and Deo (1972)
Gorilla (n)	Mature adults	Orihel (1970)
Guinea pig (e)	Larvae in lungs (7 days pi), intestine (8 days pi)	Yoshida (1919)
Lamb (e)	Larvae in lungs	Gaur and Deo (1972)
Monkey (e)	Larvae in lungs	Yoshida (1918), Gaur and Deo (1972)
Cynomolgus (n)	Mature adults	Habermann and Williams (1957)
Rhesus (n)	Mature adults	Habermann and Williams (1957), Remfry (1982)
Mouse (e)	Larvae in lungs	Stewart (1917), Nichols (1956)
Orangutan (n)	Mature adults	Orihel (1970)
Pig (e)	Mature adults	Galvin (1968)
Rabbit (e)	Larvae in intestine	Yoshida (1919)
	Larvae in lungs	Galvin (1968)
Rat (e)	Larvae in lungs	Stewart (1917)
Squirrel (n)	Immature stages?	Rausch and Tiner (1948)

[a] Adapted with permission from Crompton (1989b).
[b] n, Natural infection; e, experimental; pi, post infection.

replicate within the host. The population of free-living infective stages (Fig. 1) gains recruits from the proportion of eggs that develops following expulsion from the host, while losses occur as infective stages perish from natural causes or are removed during successful transmission.

B. Host Specificity

Available evidence suggests that *A. lumbricoides* is highly specific for its human host. Sexually mature worms identified as *A. lumbricoides* have been recovered from a few species of nonhuman primate, and developmental stages can be retrieved from the lungs and intestines of a range of experimentally infected mammalian hosts (Table 2). At present, no evidence substantiates the idea that reservoir hosts play a significant role in the epidemiology of *A. lumbricoides* in endemic areas. Pigs are infected readily with *Ascaris* of human origin and vice versa (see Crompton, 1989b) and must be considered sources of *Ascaris* infection for humans in some places.

C. Transmission

Infections of *A. lumbricoides* are transmitted to humans readily through the fecooral route. Many surveys have detected eggs of *Ascaris* spp. in the soil around houses, schools, and public places (Yadav and Tandon, 1989; Wong and Bundy, 1990). In rural areas of developing countries, young children crawling and playing on contaminated soil are particularly at risk from *A. lumbricoides* infection. Finding as many as 100 eggs of *A. lumbricoides* per g soil in an area of endemic ascariasis is not unusual (Morishita, 1972). The structure and composition of the soil are important factors in the survival and transmission of *A. lumbricoides;* the more clay in a soil, as opposed to sand, the better the survival (Janssens, 1985). The structure of the soil also influences the rate at which the eggs become buried and removed from contact with hosts (Storey and Phillips, 1985), even if infectivity may be retained for years (Müller, 1953; Krasnonos, 1978).

Widespread contamination of the household environment occurs when sanitation and sewage treatment are lacking or inadequate and when "night soil" is used as a fertilizer for crops. The eggs of *A. lumbricoides* are invariably sticky and, in areas of endemic ascariasis, may be found adhering to utensils, furniture, money, fruit, vegetables, door handles, and fingers (Kagei, 1983). Since eggs also occur in the air and household dust, the possibility exists of their being inhaled and swallowed (WHO, 1967; Bidinger *et al.*, 1981; Kroeger *et al.*, 1992). Under experimental conditions, larval *Ascaris* can be transferred between hosts (Yoshida, 1918; Stephenson *et al.*, 1977), can enter the body through skin lesions (Urabe, 1957), and may even pass from the mother to the unborn child (Chu *et al.*, 1972; Rathi *et al.*, 1981).

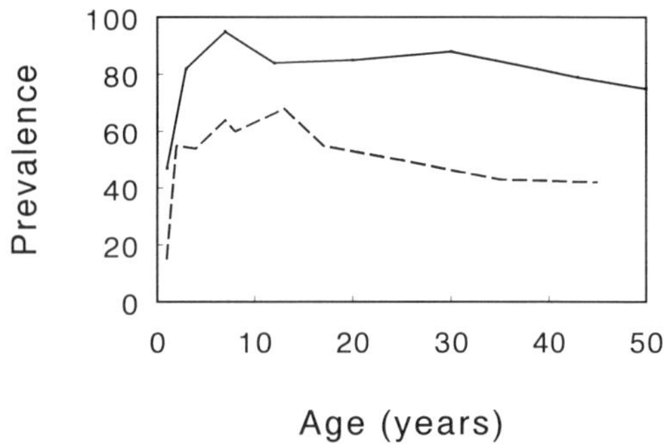

Figure 2 Relationship between prevalence of *Ascaris lumbricoides* and host age. Redrawn with permission from Elkins *et al.* (1986) (——) and Forrester *et al.* (1988) (- - -).

III. Prevalence

Estimates of the prevalence of ascariasis usually are made by determining the number of egg-positive cases in a sample of people from the population of interest. Methods for obtaining population samples have been proposed and discussed by the World Health Organization (WHO, 1986) and by Thein-Hlaing (1989). Diagnosis usually is made by means of stool examination (Thienpont *et al.*, 1986), the Kato Katz procedure (WHO, 1992) being the method of choice. This procedure is simple to carry out, is relatively quick and inexpensive, is found to be reproducible when used by different technicians in different places, and consequently provides useful comparative results. The Kato Katz procedure also gives quantitative results; egg counts can be undertaken and expressed as eggs per gram (epg; see Section IV). Some problems must be kept in mind when estimating prevalence and intensity on the basis of stool examination (Hall, 1982). For example, infections consisting of male *A. lumbricoides* alone will not be detected (Seo *et al.*, 1979b). Despite the tedious nature of stool sampling and examination, as yet no better method exists for the standard investigation of the epidemiology of *A. lumbricoides* in humans.

A. Prevalence and Host Age

Overall estimates of prevalence for countries or regions are useful because they help define the public health significance of a problem and so contribute to the determination of health priorities (Walsh and Warren, 1979). A more useful approach to understanding the prevalence of ascariasis and population biology of *A. lumbricoides* is examining the relationship between prevalence and host age. Graphical presentations of this relationship are shown in Fig. 2, the curves having been constructed from the results of cross-sectional surveys. Additional examples are given by Anderson and May (1985). The age–prevalence patterns shown in the graphs are believed to be widely typical; infections first are acquired when children are in their second year of life and are spending time away from their mothers, crawling around in and near their homes. Maximum prevalence values usually are observed when children are 5–10 yr old. A very gradual decline with age is also common. The biological significance of the gradual decline is discussed in Section IV.

B. Prevalence and Host Sex

A range of surveys has found that, when the prevalence of *A. lumbricoides* infection in a given community tends to vary by sex, more females tend to be infected than males. This relationship appears to hold independently of age, when similar numbers of both sexes were examined. The effect is quite widespread, having been observed in studies in Tanzania (McCullough, 1974), Indonesia (Cross *et al.,* 1975), Iran (Arfaa and Ghadirian, 1977), Papua New Guinea (Shield *et al.,* 1980), the Philippines (Cabrera and Valeza, 1980), and Thailand (Harinasuta and Charoenlarp, 1980). An estimation based on the results of pooled samples from surveys conducted throughout India showed that 5.7% of 30,824 males and 10.4% of 23,490 females were infected with *A. lumbricoides* (Crompton, 1989a). This sex effect has not been explained and is not always observed.

C. Prevalence and Environment

Climatic factors should be expected to influence the distribution and prevalence of *A. lumbricoides* because of their role in embryonation and in the survival of infective eggs. In Africa, the prevalence is generally low in countries with arid climates and high in those with consistent wet weather and warmth (Crompton and Tulley, 1987; Prost, 1987). For example, in Chad and Mali, overall prevalences of 3.8% and 0.1% can be calculated, compared with 38.3% and 57.3% for Kenya and Zaire, respectively, both of which have large areas with levels of rainfall that attract human settlements (Crompton, 1989a). The distribution of *A. lumbricoides* in any country is bound to reflect the distribution of human communities and their dependence on rainfall and water supply. During 1988–1990, Ratard *et al.* (1991) conducted a national survey of *A. lumbricoides* involving 22,166 children in Cameroon, tropical West Africa. The results are displayed graphically in Fig. 3, which shows how prevalence can be related to climate. During a survey of *A. lumbricoides* infection in Thailand (Harinasuta and Charoenlarp, 1980), the prevalence was found to be 7.5% in the dry northeast and 70.5% in the south of the country, which usually has no dry season.

Seasonal effects on the prevalence of *A. lumbricoides* infection also reflect variations or patterns in the climate. In Saudi Arabia, Gelphi and Mustafa (1967) studied the etiology of pneumonitis that appeared annually for 3 mo from March to May. The syndrome was considered to be linked to the larval migration of *A. lumbricoides* associated with the optimal conditions for embryonation that follow a brief rainy season. Seasonal effects were detected by Seo *et al.* (1979a) during a study of reinfection with *A. lumbricoides* in rural people in South Korea, following mass chemotherapy. During the course of 1 yr, peaks in egg-positive rates and in the expulsion of adult worms and of immature worms were found in February, March, August, and September (Fig. 4). These peaks can be related to earlier periods of seasonal peak transmission, that is, November to December for the larger peak and May to June for the smaller. As Seo *et al.* (1979a) pointed out, the seasonal effect may be indirect; apparently at the time of the study, the traditional consumption of pickles during November and December was a likely source of infection.

The type of community and the nature of its housing provide a variety of factors that influence the prevalence and distribution of *A. lumbricoides* infection. Generally, the poorer the quality of housing and community services, the more likely *A. lumbricoides* is to persist and flourish (Annan *et al.,* 1986; Holland *et al.,* 1988). Until recently, more cases of *A. lumbricoides* infection were considered to occur in rural than

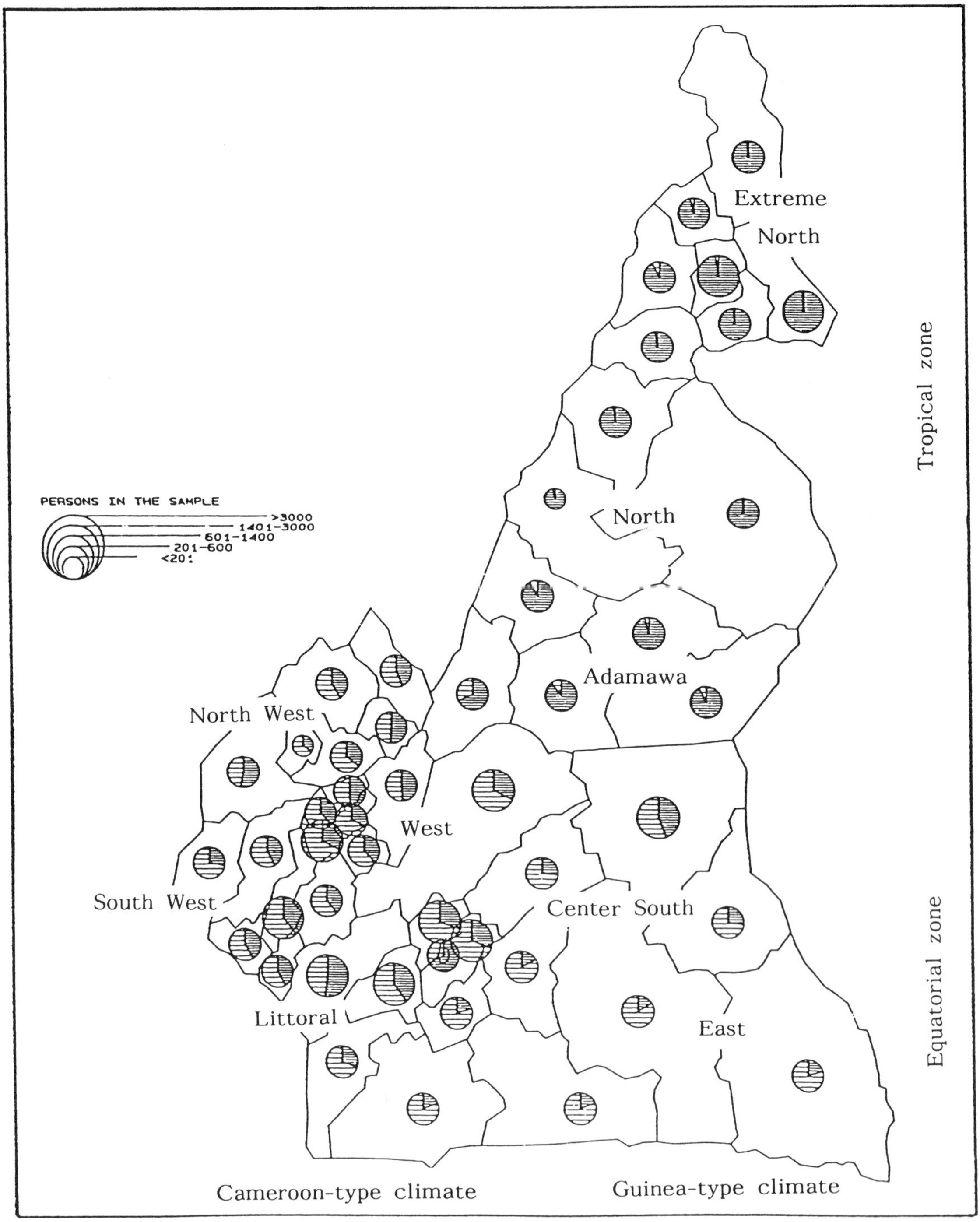

Figure 3 Cartographic display of the prevalence of *Ascaris lumbricoides* in children in Cameroon. The data were reported by Ratard *et al.* (1991). The map was produced using PARADISMAP (Coombs *et al.*, 1991). The Extreme North, North, and most of the Adamawa province are located in the tropical zone and experience a short rainy season (500–1,000 mm annually on 60–75 days) followed by a long dry season. Sunshine averages 67% daylight and the mean annual temperature is 28°C. Much of the remainder of the country is in the equatorial zone, where rainfall occurs monthly (150–260 days per year). The Cameroon-type climate has a long rainy season with a rainfall of 2,000–10,000 mm annually. On average, sunshine occurs during 22–43% daylight and mean annual temperatures are 24–26°C. Infected population, filled; uninfected population, hatched.

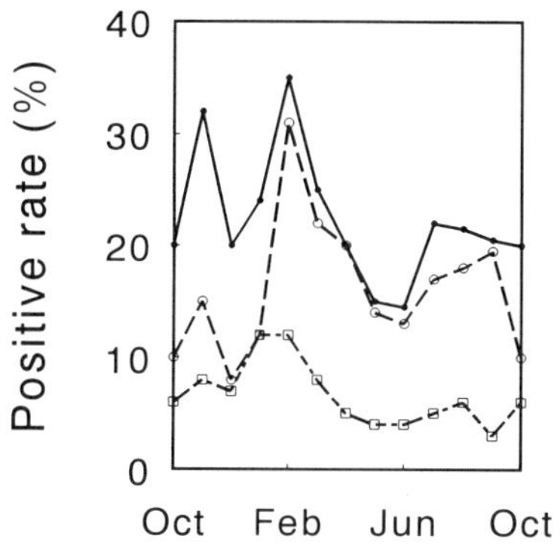

Figure 4 Evidence for the seasonal fluctuation of infection rates with *Ascaris lumbricoides* in South Korea, based on the detection of worms and eggs in stool samples. Adult worms (●); immature worms (□); eggs (○). Redrawn with permission from Seo *et al.* (1979a).

in urban communities. Most people in developing countries appeared to live in rural villages, where the lifestyle was assumed to involve less population mobility, thereby increasing the risk of contact with contaminated soil. For example, Seo (1980) analyzed the results of a survey of 24,886 people living in South Korea in 1971 and found the prevalence of *A. lumbricoides* infection to be 59.7% in rural dwellers and 42.6% in urban dwellers.

The situation may be changing, however, as a consequence of urbanization (Crompton and Savioli, 1993), which has been described as the unplanned, uncontrolled migration of people from rural villages to urban settlements. By the year 2025, about 57% of the population of developing countries will be living in large urban conurbations if current trends continue (Harpham and Stephens, 1991), with the implication that fewer cases of *A. lumbricoides* infection will occur in rural areas.

Another environmental influence on the prevalence of *A. lumbricoides* infection is that provided by the family itself. Williams *et al.* (1974) observed that *A. lumbricoides* infections tend to be distributed in clusters in families; when one child in a family was found to be infected, the siblings were highly likely also to be infected. Forrester *et al.* (1988) undertook a survey of 428 households in a shanty town in Coatzacoalcos, Mexico. Altogether, 2098 people took part in the study; the number of individuals per household ranged from 1 to 26. Stool specimens were collected and investigated by means of the Kato Katz procedure. The distribution of the individuals with high egg counts was found not to be random. Heavily infected individuals were clustered together in households of 6 or more residents. Two explanations may be offered for this result. First, more eggs would be expected in the vicinity of a household with heavily infected individuals, and such households would function as transmission foci. Second, any genetic predisposition to harbor heavy worm burdens (Section IV,D) might be expected to express itself among members of the same family group living together.

IV. Intensity

The intensity of an *A. lumbricoides* infection is defined as the mean number of worms per person. Knowledge of the intensity is crucial for understanding the population biology and transmission dynamics of *A. lumbricoides* (Anderson and May, 1985,1991; Anderson, 1986); for evaluating its public health significance, since morbidity and mortality increase as worm burden increases (Pawlowski and Davis, 1989); and for implementing and monitoring control programs (WHO, 1987). Intensity can be measured directly by counting the number of worms passed after anthelmintic chemotherapy (Croll and Ghadirian, 1981) or indirectly by counting eggs in a stool sample (WHO, 1992).

A greater fecal egg count is assumed to indicate a greater number of sexually mature female *A. lumbricoides* present in the infected individual. Care is needed when interpreting fecal egg counts since density-dependent constraints on fecundity may mask the number of worms present (Hall, 1982;

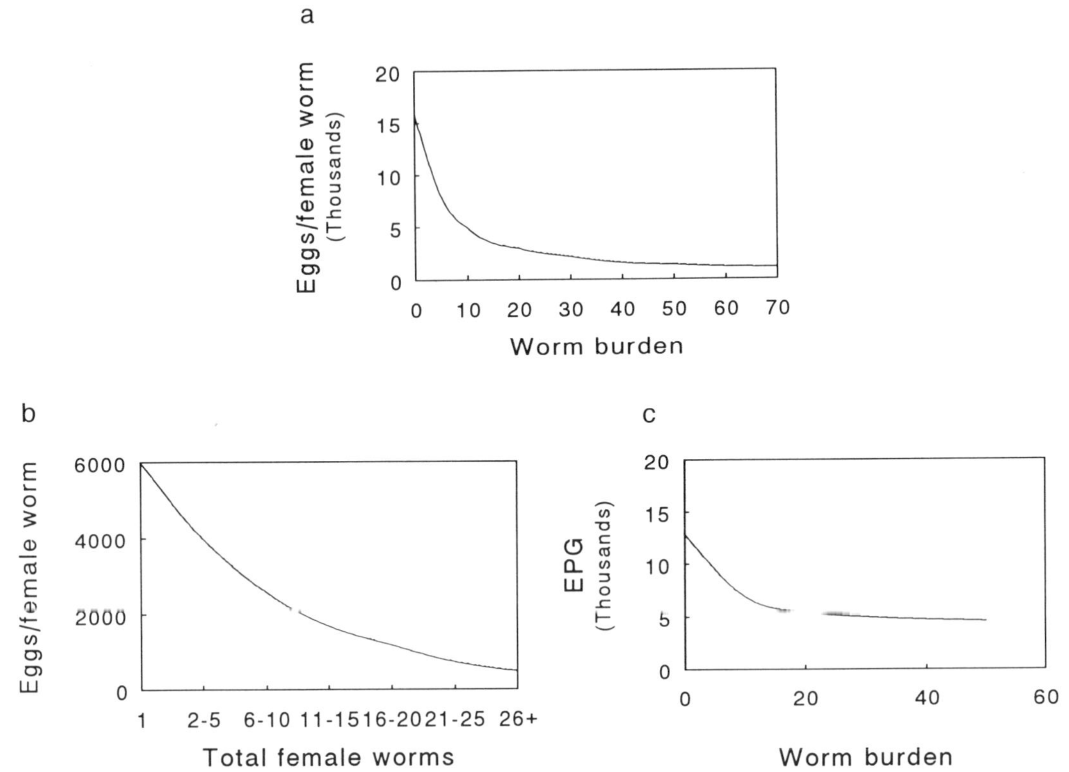

Figure 5 Evidence for density-dependent egg production of *Ascaris lumbricoides,* as shown by the relationship between fecundity of individual female worms and the worm burden in the host. Redrawn with permission from (a) Thein-Hlaing *et al.* (1984), (b) Holland *et al.* (1989), and (c) Forrester and Scott (1990).

Keymer, 1982). Thein-Hlaing *et al.* (1984), Holland *et al.* (1989), and Forrester and Scott (1990) demonstrated the actual compared with the theoretical existence of density-dependent effects on the fecundity of female *A. lumbricoides* in Myanmar, Nigeria, and Mexico, respectively (Fig. 5). However, in general terms, egg counts do tend to reflect worm burden (Forrester and Scott, 1990; Fig. 6).

Anecdotal evidence suggests that, in endemic areas of *A. lumbricoides* infection where the prevalence is high (>50%), more heavily infected individuals will be present. The relationship between prevalence and intensity has been investigated by Guyatt *et al.* (1990), using data from 13 epidemiological surveys in which the worms passed by infected individuals after anthelmintic chemotherapy were collected and counted. Prevalence and mean worm burden were found to have a nonlinear relationship (Fig. 7) that is well described by the negative binomial distribution (see Section IV,D).

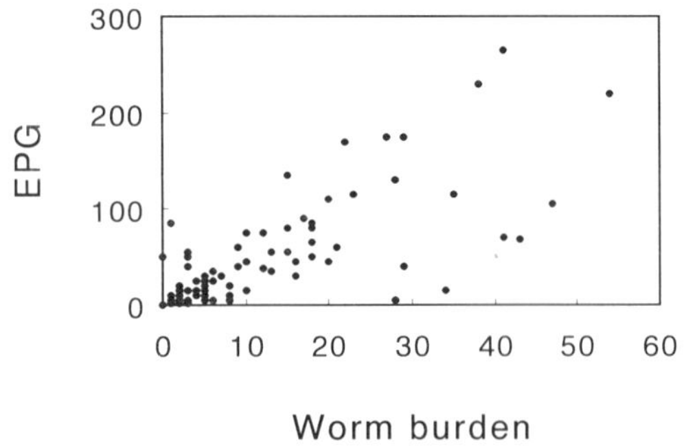

Figure 6 The relationship between fecal egg counts (epg) and numbers of *Ascaris lumbricoides* per host. Redrawn with permission from Forrester and Scott (1990).

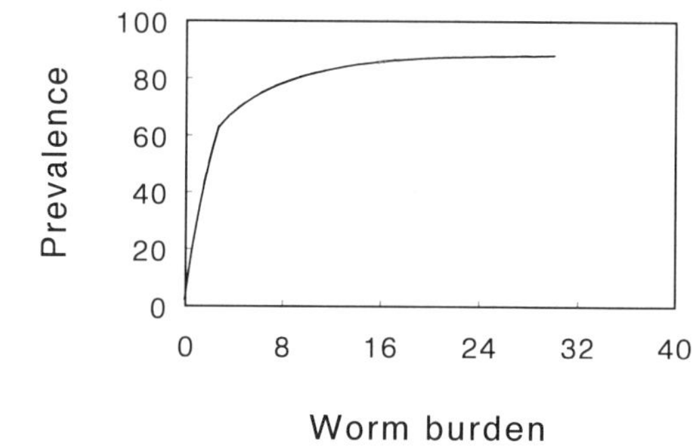

Figure 7 Results of an investigation into the relationship between the prevalence of *Ascaris lumbricoides* and the mean worm burden, assuming that the parasites are distributed according to the negative binomial. The curve represents the relationship when $k = 0.543$. Data from 13 field studies were used to construct the curve. Redrawn with permission from Guyatt *et al.* (1990).

Guyatt *et al.* (1990) suggested several population processes to account for this relationship, ranging from density-dependent effects and human behavior to parasite-induced host mortality. Whatever the mechanisms, they are highly consistent since they are based on data collected independently from different countries, continents, climates, and cultures.

A. Intensity and Host Age

Probably the most useful baseline for studying infections of *A. lumbricoides* in a community is the investigation of the relationship between intensity and host age (Fig. 8). Typically, the intensity is observed to rise rapidly during childhood and to fall away with increasing age (Fig. 8a). Two of the exceptions to this pattern that have been found were in Iran (Fig. 8b), found by Arfaa and Ghadirian (1977), and in the Dominican Republic, found by Beattie *et al.* (1990).

During childhood, the intensity profile (Fig. 8a) appears not unlike the prevalence profile (Fig. 2), but intensity usually declines markedly with age, indicating that although older people may remain infected, they do not harbor as many worms as children. An immunological and a socioenvironmental hypothesis have been proposed to explain the decline in intensity with age. Exposure to repeated infections with *A. lumbricoides* during early life might induce some level of protective immunity. Pigs develop some immunity to *A. suum* (Lloyd and Soulsby, 1985; Eriksen, 1981) and human hosts display complex allergic reactivity (see Kennedy, 1992; Pritchard, 1992), secreting antibody in response to antigens of

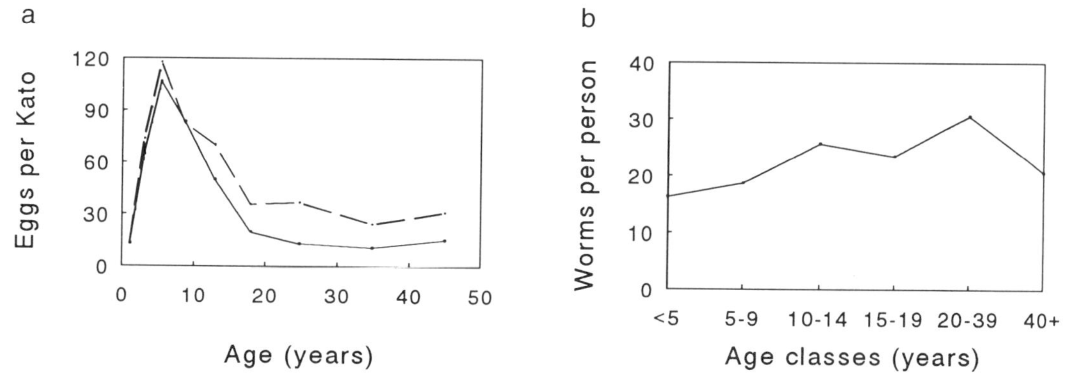

Figure 8 Relationship between intensity of *Ascaris lumbricoides* and host age. Intensity was measured indirectly by counting eggs in stool samples using the Kato technique (a) and directly by counting worms after expulsion chemotherapy (b). Redrawn with permission from (a) Forrester *et al.* (1988) and from data tabulated by (b) Arfaa and Ghadirian (1977).

A. lumbricoides (Jones, 1977; Haswell-Elkins *et al.*, 1992). Evidence from re-infection studies indicates that, when people have been kept free of *A. lumbricoides* through regular use of anthelmintic drugs, the prevalence of the infection increases above the pretreatment value when the treatment is stopped, perhaps because resistance to the infection has been weakened because of the lack of stimulus from resident worms.

The socioenvironmental explanation for higher intensities of *A. lumbricoides* infection in children relative to adults (Fig. 8a) depends largely on differences in personal hygiene and behavior as individuals interact with their surroundings. Young children suck their fingers, crawl on the floor and ground around the home, and are expected to be exposed to more infective eggs than adults, especially in areas in which sanitation is lacking or inadequate. Progress in understanding the nature of the human immune response to *A. lumbricoides* seems more likely to provide a test of the immune hypothesis for explaining the decline in intensity with age. Unraveling social and behavioral factors in a complex environment seems more challenging. Perhaps aspects of both hypotheses will contribute to the explanation.

B. Intensity and Host Sex

The results of a survey in India (Elkins *et al.*, 1986) showed that the intensity of *A. lumbricoides* infection tends to be higher in females of most age classes than in males; the difference is more obvious after the early years of childhood. The explanation for this observation is not yet understood, and the observation is not universal (Asaolu *et al.*, 1992).

C. Frequency Distribution of Numbers of Worms per Host

When the frequency distribution, numbers of *A. lumbricoides* per host, is investigated, most hosts are observed to harbor a few worms whereas a few hosts harbor many worms (Fig. 9). In other words, the distribution is aggregated or overdispersed rather than random. The observations are found to be described best by the negative binomial distribution (Bliss and Fisher, 1953; Seo *et al.*, 1979b; Anderson, 1986). When the negative binomial distribution is applied to observations on numbers of worms per host, values for the parameter k can be calculated. This parameter provides an inverse measure of the degree of aggregation or overdispersion; the greater the aggregation, the smaller the value of k. Guyatt *et al.* (1990) summarized data from 13 studies that showed the value of k for the distribution of numbers of *A. lumbricoides* in human populations to vary from 0.36 to 0.81. The value of k also appears to vary between host age classes. For example, for children aged 2–4 yr, the value was found to be 0.85; in those aged 10–14 yr, the value was 1.73 (Elkins *et al.*, 1986). Anderson and May (1991) identified four consequences of aggregation that are important for the understanding of the population biology, epidemiology, morbidity, and control prospects of *A. lumbricoides* infections in humans. First, aggregation influences the mate-finding chances of dioecious worms such as *A. lumbricoides,* thereby enhancing overall egg production and transmission in the community. Second, aggregation increases regulation of populations of *A. lumbricoides* through density-dependent effects. Third, aggregation leads to relatively few individuals harboring sufficient worms to precipitate life-threatening or severe morbidity (Section IV,E). Fourth, knowledge of aggregation must not be ignored when control strategies are designed (Section VII).

D. Predisposition

According to the Oxford Dictionary, the word "predisposition" should be used to describe a condition or situation that has been influenced in advance or inclined beforehand. The term has been used by parasitologists for the observation made during re-

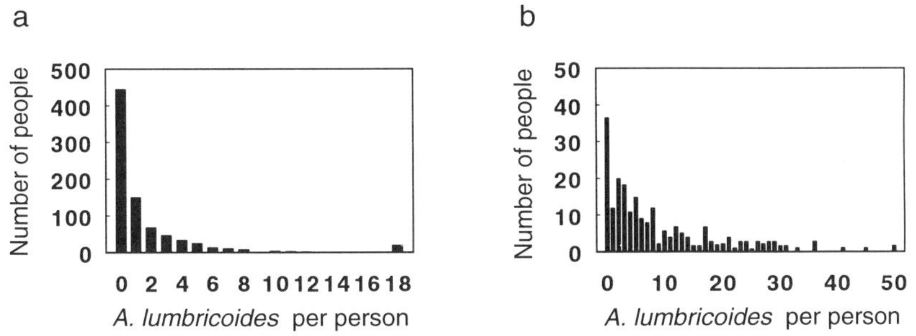

Figure 9 Frequency distribution of numbers of *Ascaris lumbricoides* per host. Redrawn with permission from (a) Seo *et al.* (1979b) and (b) Elkins *et al.* (1986).

infection studies of *A. lumbricoides,* in which individuals after anthelmintic chemotherapy seem to acquire worm burdens similar to those they harbored before treatment. Worm burdens of individuals, or egg counts, are measured and compared before and after treatment; invariably, the same individuals consistently are found to be identified with a particular intensity of infection. This result is statistically significant and has been demonstrated in studies in India (Haswell-Elkins *et al.,* 1987), Myanmar (Thein-Hlaing *et al.,* 1987), Nigeria (Holland *et al.,* 1989), Mexico (Forrester *et al.,* 1990), Malaysia (Chan *et al.,* 1992), and Bangladesh (Hall *et al.,* 1992), but not in Iran (Croll and Ghadirian, 1981).

The relationship between individuals and their intensity of *A. lumbricoides* infection was reported to continue after three rounds of expulsion chemotherapy (Bundy *et al.,* 1987; Holland *et al.,* 1989; Hall *et al.,* 1992). In the Nigerian study, worms were collected on three occasions (Table 3; phases I, II, and III). The numbers retrieved were investigated by host age and sex. On the basis of an arbitrary classification of worm burdens into heavy, light, and zero, use of a transition matrix revealed that 51% of individuals judged to be uninfected at the start of the study remained in that state, whereas 31% of the individuals judged to be heavily infected at the start consistently were found to be heavily infected. This result highlights the fact that, although a statistically significant relationship between pre- and posttreatment intensities exists at a population level, not all individuals return to the same infection intensity after treatment.

Immunological and socioenvironmental hypotheses (Section II,A) might be proposed to explain predisposition. Holland *et al.* (1992) continued their earlier work (Holland *et al.,* 1989) by investigating Class I HLAs in a sample of subjects as well as *Ascaris* egg counts in the soil around their homes. On the basis of detection of eggs, all subjects were likely to have been exposed repeatedly to infection, but some individuals remained uninfected. If these individuals are predisposed to remain uninfected, then inherent (presumably genetically based) host features (Keymer and Pagel, 1990) appear to be involved. Interestingly, Holland *et al.* (1992) found that the individuals who remained uninfected despite exposure to eggs of *A. lumbricoides* lacked A30/31 antigens. The relationship between HLA and susceptibility to helminth infection is now beginning to receive more attention (Chan *et al.,* 1984; Hirayama *et al.,* 1987; Bundy, 1988), but the work is difficult to undertake in rural communities in developing countries. In their study of the intensity of re-infection in an urban community in Bangladesh, Hall *et al.* (1992) drew attention to the fact that, despite the prevalence being 89%

Table 3 Results from an Analysis of Numbers of *Ascaris lumbricoides* Passed by Nigerian Children after Repeated Anthelmintic Chemotherapy[a,b]

	Phases I and II		Phases II and III	
Group	Sample size	Correlation coefficient	Sampled	Correlation
Total	756	0.1873****	580	0.2781****
males	390	0.1750****	298	0.3166****
females	366	0.2038****	282	0.2341****
5–6 yr	162	0.1892****	148	0.3084****
males	91	0.1876**	83	0.3295****
females	71	0.1810*	65	0.2871****
7–8 yr	282	0.1906****	252	0.2510****
males	138	0.1842***	120	0.3334****
females	144	0.2180****	132	0.1613**
9–10 yr	142	0.1667**	119	0.3164****
males	69	0.1247**	58	0.3273****
females	73	0.1170NS	61	0.3134***
11–12 yr	99	0.1751**	51	0.2019*
males	56	0.0899NS	30	0.1859NS
females	43	0.3058**	21	0.2251NS
13–16 yr	71	0.3029****	—	—
males	36	0.3219**	—	—
females	35	0.3293**	—	—

[a] Reproduced with permission from Holland *et al.* (1989).
[b] Kendall's rank correlation coefficient for worm numbers for the three phases (see text). * P ≤ 0.05, ** P ≤ 0.01, *** P ≤ 0.001, **** P ≤ 0.0001, NS, not significant.

and the mean worm burden being 18.5 at one stage, some of the 880 people who participated appeared not to be susceptible to infection with *A. lumbricoides*.

E. Intensity and Disease

The state of ill health that can be attributed to the presence of *A. lumbricoides* is known as ascariasis (WHO, 1967). Acute ascariasis presents as allergic and respiratory problems associated with migrating larval stages or a variety of obstructions and surgical complications caused by the adult worms. Chronic ascariasis also involves the adult worms and now is widely accepted as a contributory factor in the etiology and persistence of impaired nutritional status during childhood (Crompton, 1992). In cases of ascariasis involving adult worms, the severity of the disease is expected to be related to the intensity of infection.

An illustration of the complexity of acute ascariasis caused by adult *A. lumbricoides* is given in Table 4. Among children, intestinal obstructions appear to be the most common manifestation of acute ascariasis; the greater the number of worms present, the greater the risk from life-threatening obstruction. Additional analysis of these data suggests that intestinal obstructions may occur more frequently in girls than in boys; the same pattern is seen for biliary obstructions. No doubt many worms are needed to obstruct the lumen of the small intestine, although on one occasion a single *A. lumbricoides* surrounded by multiple adhesions was judged to be the cause of an intestinal obstruction (Ghawss and Willan, 1990). Reports of biliary complications invariably state that one worm was removed by surgical procedure from the bile duct; since patients not infrequently die from acute biliary ascariasis (Table 4), investigators have

Table 4 Complications of Acute Ascariasis Related to Adult *Ascaris lumbricoides* in Relation to Patient Age and Gender[a,b]

		Age		Gender		Deaths	
	All cases	Child	Adult	Male	Female	Child	Adult
Biliary system	1124	380	374	92	117	4	23
Gastrointestinal tract	3408	2149	73	572	716	85	9
Hepatic abscess	100	52	6	17	15	13	1
Pancreatitis	67	28	26	11	15	1	2
Miscellaneous complications	94	77	15	50	41	9	3
Totals	4793	2686	494	742	964	112	38

[a] D. W. T. Crompton and N. R. Fraser, unpublished observations.
[b] The cases recorded in this table were abstracted from 230 reports published between 1971 and 1992. The cases described were from 54 countries. Often reports were found to be incomplete, with no information given about the age or gender of the patients.

claimed that one worm can kill. Regrettably, published cases about biliary ascariasis rarely carry information about the intensity of infection. Various chemical agents, from anesthetics to pepper, have been suggested as stimuli that induce *A. lumbricoides* to migrate from its usual microhabitat in the small intestine; migration from an overcrowded habitat seems equally plausible. Data assembled by Chai *et al.* (1991) strongly support the contention that morbidity due to acute ascariasis is related closely to the intensity of infection. During the past 25 yr, the overall prevalence of *A. lumbricoides* infection in the population of South Korea has fallen from over 60% to less than 1%; the numbers of acute ascariasis cases and *Ascaris*-related deaths have decreased in proportion. By crude extrapolation from statistics given by Seo (1990), the mean number of worms per infected person in areas of high endemicity fell from about 10 to about 2 worms during the period of investigation. Interestingly, over 90% of surgical ascariasis cases that were reported in South Korea from 1955 to 1989 concerned the biliary system. Given the high prevalence and intensity of *A. lumbricoides* at the start of this period, biliary ascariasis seems more likely to develop when one worm leaves a larger subpopulation.

Chronic ascariasis, as defined earlier, is related to the intensity of infection as well as to other variables such as host age, nutritional status, and socioeconomic factors (WHO, 1987; Holland, 1989). For example, the impact on host nutritional status of 10 adult *A. lumbricoides* in the lumen of the small intestine of a 3-yr-old child may well be greater than their impact on a 9-yr-old child (Seo and Chai, 1980), given the large size of adult *A. lumbricoides.* When intensity is reduced through intervention with anthelmintic chemotherapy, the nutritional status of previously infected children improves significantly (Thein-Hlaing *et al.*, 1991).

V. Concurrent Infections

Infections of *A. lumbricoides* rarely are found alone in human communities. The commonly used techniques for the examination of feces nearly always reveal the presence of other soil-transmitted helminth species and protozoan infections. For example, in a survey to identify intestinal parasitic infections in a sample of 422 preschool children from villages in rural Ghana, Annan *et al.* (1986) found *Chilomastrix mesnili, Entamoeba* spp., *Giardia intestinalis, A. lumbricoides, Dicrocoelium hospes, Enterobius vermicularis,* hookworm (probably *Necator americanus), Schistosoma haematobium, Schistosoma mansoni, Strongyloides stercoralis,* and *Trichuris trichiura.* In one village, the average number of infections per child examined was 1.97, with one child harboring five parasites concurrently. Annan *et al.* (1986) investigated the data obtained for the six common infections to see whether some species of parasite typically occur together or avoid each other, or whether the relationships are generated through random events. The chi-squared statistic was chosen to assess putative associations between infections after comparing the expected with the observed frequencies for pairs of infections. *Ascaris lumbricoides* was found to be associated with the presence of both *T. trichiura* and hookworm, but a noticeable lack of association existed between *G. intestinalis* and the presence of the other parasites. Polyparasitism is one of the problems that has confounded investigations into the contribution of ascariasis to childhood malnutrition (Keusch and Migasena, 1982).

In making the case for control strategies that might be used simultaneously against the common species of soil-transmitted helminths, Booth and Bundy (1992) observed that not only do *A. lumbricoides* and *T. trichiura* have closely related distributions, but they also share a remarkably consistent degree of prevalence in a given community. Booth and Bundy abstracted data from the literature and found statistically significant correlations ($P < 0.0001$; Spearman Rank Correlation Coefficient) between the prevalences of *A. lumbricoides* and *T. trichiura* in communities in India, Indonesia, Brazil, Chile, and Cameroon. Relationships between species of hookworm and the other soil-transmitted helminth species were less clear and more variable. The feco-oral route of transmission of both *A. lumbricoides* and *T. trichiura* is fundamentally different from obligatory skin penetration that leads to the establishment of infections with *N. americanus.* This difference is suggested to account for the weaker correlations between hookworm and *A. lumbricoides* or *T. trichiura* infection.

The intensities of infection of *A. lumbri-*

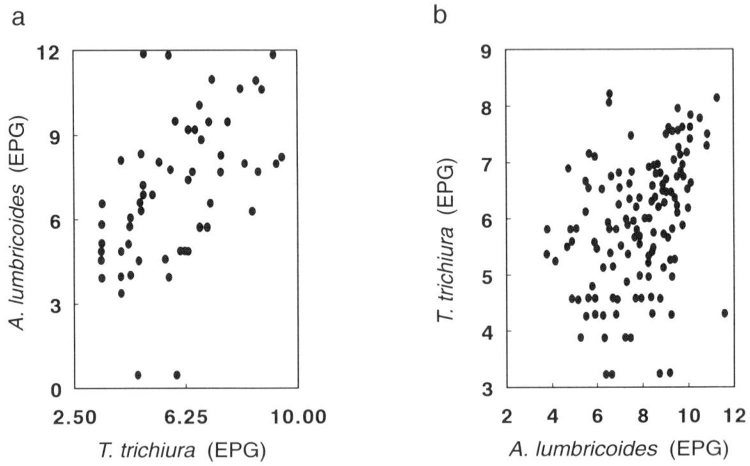

Figure 10 Relationship between intensities of *Ascaris lumbricoides* and *Trichuris trichiura* in individual children living (a) in Panama and (b) in Sierra Leone. In both studies, intensity was measured indirectly as eggs per gram feces (epg). The relationship for both data sets was found to be statistically significant using Kendall's coefficient [reproduced with permission from (a) Robertson *et al.*, 1989 and (b) Webster *et al.*, 1990] of rank correlation [(a) $P < 0.001$; (b) $P < 0.01$].

coides and *T. trichiura* have been found to be significantly related in studies of children in Panama (Robertson *et al.*, 1989), Nigeria (Holland *et al.*, 1989), and Sierra Leone (Webster *et al.*, 1990). Results from two of these studies, each of which assessed intensity indirectly by means of egg counts, are illustrated in Fig. 10.

VI. Mathematical Framework

Beginning with information of the type summarized in Table 1 and Fig. 1, and the results of epidemiological surveys that covered large samples across a wide age range, Anderson and May (1985, 1991) applied mathematical techniques to elucidate how populations of *A. lumbricoides* interact with populations of hosts. Models were developed to help explain how infections of *A. lumbricoides* are maintained with such stability in endemic areas and how transmission is achieved. Perhaps the key feature that emerged from this body of work is the emphasis placed on the significance of the distribution of worm burdens in human populations.

Most importantly for individuals planning control programs for ascariasis, mathematical models can indicate how best to use available resources and can identify achievable objectives that can be monitored as the control program unfolds. A model also can provide intellectually satisfactory explanations for certain observations. In practice, events leading to the continuation of an infection such as *A. lumbricoides* cannot be measured accurately. The model framework, however, has produced the concept of the basic reproduction ratio (R_0) of infection, defined as the number of female offspring produced by a sexually mature female worm that, in the absence of density-dependent constraints, survive to reproduce (Anderson, 1989). When R_0 equals 1, every mature female worm will be replaced by another female worm. Successful control measures aim to reduce R_0, ideally to a value below 1, at which time the population of *A. lumbricoides* in the community will decline. Estimates of R_0 require knowledge of the average intensity of infection in the community, the degree of aggregation of worms in the population, and the severity of density-dependent constraints on the parasite population (Anderson and May, 1985, 1991). Although *A. lumbricoides* has a seemingly simple life history (Fig. 1), the interactions of humans with their environment that lead to heterogeneous exposure to infection, in conjunction with equally variable susceptibility between individuals, means that the life history of the worm is actually rather complex. Anderson and May (1985) recognized the pitfall of constructing and using an oversimplified model. However, such a model may be no less risky than interpreting data by eye or relying on intuition or common sense to understand the population biology of *A. lumbricoides.*

VII. Control Strategies

Prospects for the control of ascariasis are better now than ever before, at least to the extent of developing strategies that are in theory realistic and feasible. Four lines of work have led to this position. First, the mathematical approach pioneered and disseminated over the past 15 yr by Anderson and May (1991) has identified gaps in the knowledge needed to understand how *A. lumbricoides* persists with such stability in human communities. The application of this work has stimulated much high quality epidemiological research with clear objectives, leading to many of the results summarized in this review. Second, the activities of the Japanese Organization for International Co-operation in Family Planning and the Asian Parasite Control Organization have produced a wealth of information about how to integrate programs for family planning, nutrition, and parasite control (Trainer, 1985,1989; Yokogawa, 1985; Seo, 1990). Third, many developing countries, the populations of which endure most of the morbidity associated with ascariasis, have established the infrastructure needed to promote primary health care. Control measures for ascariasis are relatively simple. People can do much to help themselves as health awareness develops in response to health education (Jancloes and Cornet, 1980; Jancloes and Jancloes-Diepart, 1981). Fourth, anthelmintic drugs developed by the international research-based pharmaceutical industry are widely available, highly efficacious, and easy to use (WHO, 1990). Anthelmintic drugs provide the most rapid means of reducing the intensity of infection, so morbidity declines to tolerable levels. Then sustaining the progress made becomes the challenge, given the survival characteristics of the infective eggs of *A. lumbricoides* in the environment (Section II,C).

After consideration of the factors influencing the distribution of *A. lumbricoides* in human populations and the significance of the observed aggregation, Anderson (1989) proposed three forms of anthelmintic drug administration in the community, in addition to the treatment of infected individuals who present themselves to doctors or health workers. Mass treatment can be offered to all individuals in an area of high endemicity. The drugs have such a good record that the technical costs and time involved in diagnosis seem unnecessary (WHO, 1990). Targeted treatment is offered to a group within the community, for example, children of primary-school age. Treatment targeted at children in this way is particularly attractive because the school can serve as the center for drug delivery and administration, and because modern health education can be developed and integrated with the chemotherapy (Stephenson *et al.*, 1983; Bundy *et al.*, 1990; Asaolu *et al.*, 1991). With *A. lumbricoides*, children experience the most morbidity because they harbor more worms, on average, than do adults (Fig. 8). Selective treatment is particularly attractive on theoretical grounds because anthelmintic drugs should be given only to those who are identified as heavily infected. The use of relatively few doses of drug reduces costs and perhaps helps delay or avoid conditions that might lead to the development of drug resistance. Some of the practical difficulties of applying selective treatment have been discussed by Hall *et al.* (1992).

The three forms of community anthelmintic drug application proposed by Anderson (1989) were tested in rural villages in Nigeria by Asaolu *et al.* (1991). The results of this study are summarized in general terms in Table 5. The most satisfactory result was the finding that, in the community in which only primary-school children were treated, the intensity of infection dropped significantly in the other untreated residents. The explanation for this decline was assumed to be the reduction in the number of eggs being passed by children into the community; no evidence was obtained to suggest that other villagers were acquiring anthelmintic drugs from elsewhere. Under conditions of true primary health care, however, the independent pursuit of anthelmintic drugs by villagers in response to information spreading from the school would be seen as evidence of a successful control program. A similar result was obtained by Bundy *et al.* (1990), who provided four cycles of treatment to children aged 2–15 yr on the island of Montserrat. Not only did the prevalence and intensity of *A. lumbricoides* and *T. trichiura* decline in the treated age group, but a decline also was detected in the 16- to 25-yr age group that received no treatment.

Since *A. lumbricoides* infections usually coexist with other intestinal infections (Section V), compelling reasons exist for promoting control programs aimed at soil-transmitted helminthiases generally, although *A. lumbricoides* is most vulnerable because it is expelled most readily in response to a single oral dose of anthelmintic drug. Savioli *et al.* (1992) summarized sev-

Table 5 General Effects of Chemotherapy Strategies on the Intensity of *Ascaris lumbricoides* in Rural Communities in Oyo State, Nigeria[a,b]

Village	Participants[c]	Chemotherapy[d]	Mean e.p.g.[e] Before	Mean e.p.g.[e] After	Significance[f]
1	288	Control	7500	4700	NS[f]
2	185	Selective[g]	6800	4250	NS
3	211	Targeted[h]	9000	2600	$P < 0.0001$
4	224	Mass[i]	12000	1500	$P < 0.0001$

[a] Reproduced with permission from Crompton (1991).
[b] Full details given by Asaolu *et al.* (1991).
[c] Number of subjects providing stools before and after treatment.
[d] Chemotherapy consisted of four doses of levamisole (Ketrax, ICI Pharmaceuticals, Macclesfield, U. K.) given at 3-mo intervals.
[e] Intensity was measured indirectly by egg counts; the values have been rounded up or down and standard deviations omitted for clarity.
[f] NS, Not significant.
[g] Treatment provided for the individuals constituting the 20% with the heaviest infections.
[h] Treatment provided for children aged 2–15 yr.
[i] Treatment provided for any willing individual, but not for pregnant women or children under 2 yr of age.

eral of the benefits that should accrue to communities that are helped to adopt this approach. Despite the costs, which must seem to be immense for developing countries in which many health priorities compete for a share of a small budget, concern to bring ascariasis and other soil-transmitted helminthiases under control is justified.

Acknowledgments

It is a pleasure to thank S. O. Asaolu, I. Coombs, A. Davis, M. Hodges, C. V. Holland, Z. S. Pawlowski, L. J. Robertson, L. Savioli, R. C. Stoddart, and D. E. Walters for many helpful discussions. Thanks also to P. Johnston and M. McCullogh for their preparation of the manuscript.

References

Akamatsu, T. (1959). Study of the experimental infection with *Ascaris* in babies and infants. *Nippon Shonika Gakkai Zasshi* **62,** 1584–1592.

Anderson, R. M. (1982). The population dynamics and control of hookworm and roundworm infections. *In* "Population Dynamics of Infectious Diseases: Theory and Applications" (R. M. Anderson, ed.), pp. 67–108. Chapman and Hall, London.

Anderson, R. M. (1986). The population dynamics and epidemiology of intestinal nematode infections. *Trans. R. Soc. Trop. Med. Hyg.* **80,** 686–696.

Anderson, R. M. (1989). Transmission dynamics of *Ascaris lumbricoides* and the impact of chemotherapy. *In* "Ascariasis and Its Prevention and Control" (D. W. T. Crompton, M. C. Nesheim, and Z. S. Pawlowski, eds.), p. 253–273. Taylor and Francis, London.

Anderson, R. M., and May, R. M. (1985). Helminth infections of humans: Mathematical models, population dynamics, and control. *Adv. Parasitol.* **24,** 1–101.

Anderson, R. M., and May, R. M. (1991). "Infectious Diseases of Humans. Dynamics and Control." Oxford University Press, Oxford.

Annan, A., Crompton, D. W. T., Walters, D. E., and Arnold, S. E. (1986). An investigation of the prevalence of intestinal parasites in pre-school children in Ghana. *Parasitology* **92,** 209–217.

Arfaa, F., and Ghadirian, E. (1977). Epidemiology and mass-treatment of ascariasis in six rural communities in central Iran. *Am. J. Trop. Med. Hyg.* **26,** 866–871.

Asaolu, S. O., Holland, C. V., and Crompton, D. W. T. (1991). Community control of *Ascaris lumbricoides* in rural Oyo State, Nigeria: Mass, targeted and selective treatment with levamisole. *Parasitology* **103,** 291–298.

Asaolu, S. O., Holland, C. V., Jegede, J. O., Fraser,

N. R., Stoddard, R. C., and Crompton, D. W. T. (1992). The prevalence and intensity of soil-transmitted helminthiases in rural communities in Southern Nigeria. *Ann. Trop. Med. Parasitol.* **86,** 279–287.

Beattie, P., Jeavons, L., and Crompton, D. W. T. (1990). *Ascaris lumbricoides* and *Trichuris trichiura* in communities near San Jose de la Matas, Republica Dominicana. *Méd. Chirurg. Digest.* **19,** 437–439.

Bidinger, P. D., Crompton, D. W. T., and Arnold, S. (1981). Aspects of intestinal parasitism in villages from rural peninsular India. *Parasitology* **83,** 373–380.

Bliss, C. I., and Fisher, R. A. (1953). Fitting the negative binomial distribution to biological data and a note on the efficient fitting of the negative binomial. *Biometrics* **9,** 176–200.

Booth, M., and Bundy, D. A. P. (1992). Comparative prevalences of *Ascaris lumbricoides, Trichuris trichiura,* and hookworm infections and the prospects for combined control. *Parasitology* **105,** 151–157.

Bundy, D. A. P. (1988). Population ecology of intestinal helminth infections in human communities. *Phil. Trans. R. Soc. London B* **321,** 405–420.

Bundy, D. A. P., Cooper, E. S., Thompson, D. E., Didier, J. M., and Simmons, I. (1987). Epidemiology and population dynamics of *Ascaris lumbricoides* and *Trichuris trichiura* infection in the same community. *Trans. R. Soc. Trop. Med. Hyg.* **81,** 987–993.

Bundy, D. A. P., Wong, M. S., Lewis, L. L., and Horton, J. (1990). Control of geohelminths by delivery of targeted chemotherapy through schools. *Trans. R. Soc. Trop. Med. Hyg.* **84,** 115–120.

Cabrera, B. D., and Valeza, F. (1980). The reinfection rate of soil-transmitted helminths in the pilot areas after treatment. *In* "Collected Papers on the Control of Soil-Transmitted Helminthiases" (M. Yokogawa, S. Hayashi, A. Kobayashi, N. Kagei, N. Suzuki, and C. Kunii, eds.), Vol. 1, pp. 141–149. Asian Parasite Control Organization, Tokyo.

Chai, J. Y., Cho, S. Y., Lee, S. H., and Seo, B. S. (1991). Reduction in the incidence of biliary and other surgical complications of ascariasis according to the decrease of its national egg prevalence in Korea. *Korean J. Parasitol.* **29,** 101–111.

Chan, L., Kan, S. P., and Bundy, D. A. P. (1992). The effect of repeated chemotherapy on age-related predisposition to *Ascaris lumbricoides* and *Trichuris trichiura. Parasitology* **104,** 371–377.

Chan, S. H., Dissanayake, S., Mak, J. W., Ismail, M. M., Wee, G. B., Srinivasan, N., Soo, B. H., and Zaman, V. (1984). HLA and filariasis in Sri Lankans and Indians. *Southeast Asian J. Trop. Med. Public Health* **15,** 281–286.

Chu, W. G., Chen, P. M., Huang, C. C., and Hsu, C. T. (1972). Neonatal ascariasis. *J. Pediatr.* **81,** 783–785.

Coombs, I., Crompton, D. W. T., Mears, K., and Petrie, G. (1991). Mapping the distribution and abundance of ascariasis. *Trans. R. Soc. Trop. Med. Hyg.* **85,** 839.

Croll, N. A., and Ghadirian, F. (1981). Wormy persons: Contributions to the nature and patterns of overdispersion with *Ascaris lumbricoides, Ancylostoma duodenale, Necator americanus* and *Trichuris trichiura. Trop. Geograph. Med.* **33,** 241–248.

Crompton, D. W. T. (1989a). Prevalence of ascariasis. *In* "Ascariasis and Its Prevention and Control" (D. W. T. Crompton, M. C. Nesheim, and Z. S. Pawlowski, eds.), pp. 45–69. Taylor and Francis, London.

Crompton, D. W. T. (1989b). Biology of *Ascaris lumbricoides. In* "Ascariasis and its Prevention and Control" (D. W. T. Crompton, M. C. Nesheim and Z. S. Pawlowski, eds.), pp. 9–44. Taylor and Francis, London.

Crompton, D. W. T. (1991). The challenge of parasitic worms. *Trans. Nebraska Acad. Sci.* **18,** 73–86.

Crompton, D. W. T. (1992). Ascariasis and childhood malnutrition. *Trans. R. Soc. Trop. Med. Hyg.* **86,** 577–579.

Crompton, D. W. T., and Savioli, L. (1993). Intestinal parasitic infections and urbanization. *Bull. WHO* **71,** 1–7.

Crompton, D. W. T., and Tulley, J. J. (1987). How much ascariasis is there in Africa? *Parasitol. Today* **3,** 123–127.

Cross, J. H., Clarke, M. D., Durfee, P. T., Irving, G. S., Taylor, J., Partono, F., Joesoef, A., Hudojo, and Oemijati, S. (1975). Parasitology survey and seroepidemiology of amoebiasis in South Kalimantan (Borneo), Indonesia. *Southeast Asian J. Trop. Med. Public Health* **6,** 52–60.

Dunn, F. L., and Greer, W. E. (1962). Nematodes resembling *Ascaris lumbricoides* L. 1758, from a Malayan gibbon *Hylobates agilis* F. Cuvier, 1821. *J. Parasitology* **48,** 150.

Elkins, D. B., Haswell-Elkins, M. R., and Anderson, R. M. (1986). The epidemiology and control of intestinal helminths in the Pulicat Lake region of Southern India. I. Study design and pre- and post-treatment observations on *Ascaris lumbricoides* infection. *Trans. R. Soc. Trop. Med. Hyg.* **80,** 774–792.

Eriksen, L. (1981). "Host–Parasite Relations in *Ascaris suum* Infection in Pigs and Mice." Institute of Internal Medicine, Royal Veterinary and Agricultural University, Copenhagen.

Forrester, J. E., and Scott, M. E. (1990). Measurement of *Ascaris lumbricoides* infection intensity and the dynamics of expulsion following treatment with mebendazole. *Parasitology* **100,** 303–308.

Forrester, J. E., Scott, M. E., Bundy, D. A. P., and Golden, M. H. N. (1988). Clustering of *Ascaris lumbricoides* and *Trichuris trichiura* infections within households. *Trans. R. Soc. Trop. Med. Hyg.* **82,** 282–288.

Forrester, J. E., Scott, M. E., Bundy, D. A. P., and Golden, M. H. N. (1990). Predisposition of individuals and families in Mexico to heavy infection with *Ascaris lumbricoides* and *Trichuris trichiura*. *Trans. R. Soc. Trop. Med. Hyg.* **84,** 272-276.

Galvin, T. J. (1968). Development of human and pig *Ascaris* in the pig and rabbit. *J. Parasitol.* **54,** 1085-1091.

Gaur, S. N. S., and Deo, P. G. (1972). Observations on the transmission of *Ascaris lumbricoides* (pig and human strains) in certain heterologous hosts. *Ind. J. Anim. Sci.* **42,** 281-284.

Gelphi, A. P., and Mustafa, A. (1967). Seasonal pneumonitis with eosinophilia: A study of larval ascariasis in Saudi Arabs. *Am. J. Trop. Med. Hyg.* **16,** 646-657.

Ghawss, M. I. M., and Willan, P. L. T. (1990). Subacute non-bolus intestinal obstruction caused by *Ascaris lumbricoides*. *Br. J. Clin. Pract.* **44,** 243-244.

Gibson, D. I. (1983). The systematics of ascaridoid nematodes—A current assessment. *In* "Concepts in Nematode Systematics" (A. R. Stone, H. M. Platt, and L. F. Khalil, eds.), pp. 321-338. Academic Press, London.

Guyatt, H. L., Bundy, D. A. P., Medley, G. F., and Grenfell, B. T. (1990). The relationship between the frequency distribution of *Ascaris lumbricoides* and the prevalence and intensity of infection in human communities. *Parasitology* **101,** 139-143.

Habermann, R. T., and Williams, F. P. (1957). Diseases seen at necropsy of 708 *Macaca mulatta* (Rhesus monkey) and *Macaca philippinensis* (Cynomolgus monkey). *Am. J. Vet. Res.* **18,** 419-426.

Hall, A. (1982). Intestinal helminths of man: The interpretation of egg counts. *Parasitology* **85,** 605-613.

Hall, A., Anwar, K. S., and Tomkins, A. (1992). Intensity of reinfection with *Ascaris lumbricoides* and its implications for parasite control. *Lancet* **339,** 1253-1257.

Harinasuta, T., and Charoenlarp, P. (1980). The nematode infections in Thailand. *In* "Collected Papers on the Control of Soil-Transmitted Helminthiases" (M. Yokogawa, S. Hayashi, A. Kobayashi, N. Kagei, N. Suzuki and C. Kunii, eds.), Vol. 1, pp. 87-93. Asian Parasite Control Organization, Tokyo.

Harpham, T., and Stephens, C. (1991). Urbanization and health in developing countries. *World Health Stat. Q.* **44,** 62-69.

Haswell-Elkins, M. R., Elkins, D. B., and Anderson, R. M. (1987). Evidence for predisposition in humans to infection with *Ascaris,* hookworm, *Enterobius* and *Trichuris* in a South Indian fishing community. *Parasitology* **95,** 323-337.

Haswell-Elkins, M. R., Leonard, H., Kennedy, M. W., Elkins, D. B., and Maizels, R. M. (1992). Immunoepidemiology of *Ascaris lumbricoides:* Relationships between antibody specificities, exposure and infection in a human community. *Parasitology* **104,** 153-159.

Hirayama, K., Matsushita, S., Kikuchi, I., Iuchi, M., Ohta, N., and Sasazuki, T. (1987). HLA-DQ is epistatic to HLA-DR in controlling the immune response to schistosomal antigen in humans. *Nature (London)* **327,** 426-430.

Holland, C. V. (1989). Man and his parasites: Integration of biomedical and social approaches to transmission and control. *Social Sci. Med.* **29,** 403-411.

Holland, C. V., Taren, D. L., Crompton, D. W. T., Nesheim, M. C., Sanjur, D., Barbeau, I., Tucker, K., Tiffany, J., and Rivera, G. (1988). Intestinal helminthiases in relation to the socioeconomic environment of Panamanian children. *Social Sci. Med.* **26,** 209-213.

Holland, C. V., Asaolu, S. O., Crompton, D. W. T., Stoddart, R. C., Macdonald, R., and Torimiro, S. E. A. (1989). The epidemiology of *Ascaris lumbricoides* and other soil-transmitted helminths in primary school children from Ile-Ife, Nigeria. *Parasitology* **99,** 275-285.

Holland, C. V., Crompton, D. W. T., Asaolu, S. O., Crichton, W. B., Torimiro, S. E. A., and Walters, D. E. (1992). A possible genetic factor influencing protection from infection with *Ascaris lumbricoides* in Nigerian children. *J. Parasitol.* **78,** 915-916.

Jancloes, M. F., and Cornet, P. (1980). Contrôle épidémiologique des nématodoses intestinales dans une zone rurale au Zaïre. *Rev. Epidémiol. Santé Publ.* **28,** 89-103.

Jancloes, M., and Jancloes-Diepart, M. (1981). Campagnes périodiques d'assainissement et de chimiothérapie de masse contre les nématodes intestinaux appliquées isolement et en combinaison, au bas-Zaïre. *Ann. Soc. Belge Méd. Trop.* **61,** 111-118.

Janssens, P. G. (1985). Chemotherapy of gastrointestinal nematodiasis in man. *In* "Chemotherapy of Gastrointestinal Helminths" (H. Vanden Bossche, D. Thienpont, and P. G. Jassens, eds.), Handbook of Experimental Pharmacology, Vol. 77, pp. 183-406. Springer-Verlag, Berlin.

Jones, H. I. (1977). Haemagglutination tests in the study of *Ascaris* epidemiology. *Ann. Trop. Med. Parasitol.* **71,** 219-226.

Kagei, N. (1983). Techniques for the measurement of environmental pollution by infective stage of soil-transmitted helminths. *In* "Collected Papers on the Control of Soil-Transmitted Helminthiases" (M. Yokogawa, S. Hayashi, A. Kobayashi, N. Kagei, N. Suzuki, and C. Kunii, eds.), Vol. 2, pp. 27-46. Asian Parasite Control Organization, Tokyo.

Kennedy, M. W. (1992). Genetic control of the antibody response to parasite allergens. *In* "Allergy and Immunity to Helminths" (R. Moqbel, ed.), pp. 63-80. Taylor and Francis, London.

Keusch, G. T., and Migasena, P. (1982). Biological implications of polyparasitism. *Rev. Infect. Dis.* **4,** 880-882.

Keymer, A. (1982). Density-dependent mechanisms in

the regulation of intestinal helminth populations. *Parasitology* **84,** 573-587.
Keymer, A., and Pagel, M. (1990). Predisposition to helminth infection. *In* "Hookworm Disease: Current Status and New Directions" (G. A. Schad and K. S. Warren, eds.), pp. 177-209. Taylor and Francis, London.
Khuroo, M. S., Zargar, S. A., and Mahajan, R. (1990). Hepatobiliary and pancreatic ascariasis in India. *The Lancet* 1503-1506.
Krasnonos, L. N. (1978). Prolonged survival of *Ascaris lumbricoides* L., 1758 ova in the soil in Samarkand. *Med. Parazitol. Parazit. Bolezni* **47,** 103-105. (Cited in *Helminthol. Abstr.* 1979, 411.)
Kroeger, A., Schulz, B., Witte, B., Skewes-Ramm, R., and Etzler, A. (1992). Helminthiasis and cultural change in the Peruvian rainforest. *J. Trop. Med. Hyg.* **95,** 104-113.
Lloyd, S., and Soulsby, E. J. L. (1985). Ascariasis in animals. *In* "Ascariasis and Its Public Health Significance" (D. W. T. Crompton, M. C. Nesheim, and Z. S. Pawlowski, eds.), pp. 25-36. Taylor and Francis, Basingstoke, England.
Lýsek, H. (1967). On the host specificity of ascarids of human and pig origin. *Helminthologia* **8,** 309-312.
Maki, T. (1972). Surgical diseases due to *Ascaris lumbricoides. In* "Progress of Medical Parasitology in Japan" (K. Morishita, Y. Komiya, and H. Matsubayashi, eds.), Vol. 4, pp. 221-270. Meguro Parasitological Museum, Tokyo.
Makidono, J. (1956). Observations on *Ascaris* during fluoroscopy. *Am. J. Trop. Med. Hyg.* **5,** 699-702.
McCullough, F. (1974). Observations on *Ascaris lumbricoides* infection in Mwanza, Tanzania. *In* "Parasites in Man and Animals in Africa" (C. Anderson and W. L. Kilama, eds.), pp. 359-385. East African Literature Bureau, Nairobi.
Morishita, K. (1972). Studies on the epidemiological aspects of ascariasis in Japan and basic knowledge concerning its control. *In* "Progress of Medical Parasitology in Japan" (K. Morishita, Y. Komiya, and H. Matsubayashi, eds.), Vol. 4, pp. 1-153. Meguro Parasitological Museum, Tokyo.
Müller, G. (1953). Untersuchungen über die Lebensdauer von Askarideneiern in Gartenerde. *Zbl. Bakteriol. Parasitenk. Infektionskrankh. Hyg.* **159,** 377-379.
Nichols, R. L. (1956). The etiology of visceral larva migrans. II. Comparative larval morphology of *Ascaris lumbricoides, Necator americanus, Strongyloides stercoralis* and *Ancylostoma caninum. J. Parasitol.* **42,** 363-399.
Orihel, T. C. (1970). The helminth parasites of nonhuman primates and man. *Lab. Anim. Care* **20,** 395-401.
Pawlowski, Z. S. (1987). Ascariasis. *In* "Intestinal Helminthic Infections" (Z. S. Pawlowski, ed.), pp. 595-615. Bailliere Tindall, London.
Pawlowski, Z. S., and Arfaa, F. (1984). Ascariasis. *In* "Tropical and Geographical Medicine" (K. S. Warren and A. A. F. Mahmoud, eds.), pp. 347-358. McGraw-Hill, New York.
Pawlowski, Z. S., and Davis, A. (1989). Morbidity and mortality in ascariasis. *In* "Ascariasis and Its Prevention and Control" (D. W. T. Crompton, M. C. Nesheim, and Z. S. Pawlowski, eds.), pp. 71-86. Taylor and Francis, London.
Pillers, A. W. M. (1924). *Ascaris lumbricoides* causing fatal lesions in a chimpanzee. *Ann. Trop. Med. Parasitol.* **18,** 101-102.
Pinus, J. (1985). Surgical complications of ascariasis in Brazil. *In* "Ascariasis and Its Public Health Significance" (D. W. T. Crompton, M. C. Nesheim, and Z. S. Pawlowski, eds.), pp. 161-166. Taylor and Francis, Basingstoke, England.
Pritchard, D. J. (1992). Parasites and allergic disease. *In* "Allergy and Immunity to Helminths" (R. Moqbel, ed.), pp. 38-50. Taylor and Francis, London.
Prost, A. (1987). L'ascaridiose en Afrique de l'ouest. Revue épidémiologique. *Ann. Parasitol. Hum. Comp.* **62,** 131-155.
Ratard, R. C., Kouemeni, L. E., Ekani Bessala, M. M., Ndamkou, C. N., Sama, M. T., and Cline, B. L. (1991). Ascariasis and trichuriasis in Cameroon. *Trans. R. Soc. Trop. Med. Hyg.* **85,** 84-88.
Rathi, A. K., Batra, S., Chandraa, J., and Mehrotra, S. N. (1981). Ascariasis causing intestinal obstruction in a 45-day-old infant. *Ind. Pediatr.* **18,** 751-752.
Rausch, R., and Tiner, J. D. (1948). Studies on the parasitic helminths of the North Central States. I. Helminths of Sciuridae. *Am. Midl. Nat.* **39,** 728-747.
Remfry, J. (1982). The endoparasites of rhesus monkeys *(Macaca mulatta)* before and after capture. *Microbiologica* **5,** 143-147.
Robertson, L. J., Crompton, D. W. T., Walters, D. E., Nesheim, M. C., Sanjur, D., and Walsh, E. A. (1989). Soil-transmitted helminth infections in school children from Cocle Province, Republic of Panama. *Parasitology* **99,** 287-292.
Savioli, L., Bundy, D., and Tomkins, A. (1992). Intestinal parasitic infections: a soluble public health problem. *Trans. R. S. Trop. Med. Hyg.* **86,** 353-354.
Seo, B. S. (1980). Control problems of ascariasis in Korea with special reference on the related biology and epidemiology. *In* "Collected Papers on the Control of Soil-Transmitted Helminthiases" (M. Yokogawa, S. Hayashi, A. Kobayashi, N. Kagei, N. Suzuki, and C. Kunii, eds.), Vol. 2, pp. 194-216. Asian Parasite Control Organization, Tokyo.
Seo, B. S. (1990). Epidemiology and control of ascariasis in Kenya. *Korean J. Parasitol.* **28,** 49-61.
Seo, B. S., and Chai, J. Y. (1980). Chronologic growth pattern of *Ascaris lumbricoides. Korean J. Parasitol.* **18,** 164-170.

Seo, B. S., Cho, S. Y., and Chai, J. Y. (1979a). Seasonal fluctuation of *Ascaris* reinfection incidences in a rural Korean population. *Korean J. Parasitol.* **17,** 11–18.

Seo, B. S., Cho, S. Y., and Chai, J. Y. (1979b). Frequency distribution of *Ascaris lumbricoides* in rural Koreans with special reference on the effect of changing endemicity. *Korean J. Parasitol.* **17,** 105–113.

Shield, J. M., Scrimgeour, E. M., and Vaterlaws, A. L. (1980). Intestinal helminths in an adult hospital population in the Eastern Highlands of Papua New Guinea: Relationship with anaemia, eosinophilia and asthma. *Papua New Guinea Med. J.* **23,** 157–164.

Sinniah, B. (1982). Daily egg production of *Ascaris lumbricoides:* The distribution of eggs in the faeces and the variability of egg counts. *Parasitology* **84,** 167–175.

Stephenson, L. S. (1987). "The Impact of Helminth Infections on Human Nutrition." Taylor and Francis, London.

Stephenson, L. S., Georgi, J. R., and Cleveland, D. J. (1977). Infection of weanling pigs with known numbers of *Ascaris suum* fourth stage larvae. *Cornell Vet.* **67,** 92–102.

Stephenson, L. S., Crompton, D. W. T., Latham, M. C., Arnold, S. E., and Jansen, A. A. J. (1983). Evaluation of a four year project to control *Ascaris* infection in children in two Kenyan villages. *J. Trop. Pediatr.* **29,** 175–184.

Stewart, F. H. (1917). On the development of *Ascaris lumbricoides* Lin. and *Ascaris suilla* Duj. in the rat and mouse. *Parasitology* **9,** 213–227.

Stoll, N. R. (1947). This wormy world. *J. Parasitol.* **33,** 1–18.

Storey, G. W., and Phillips, R. A. (1985). The survival of parasite eggs throughout the soil profile. *Parasitology* **91,** 585–590.

Takata, I. (1951). Experimental infection of man with *Ascaris* of man and the pig. *Kitasato Arch. Exp. Med.* **23,** 49–59.

Thein-Hlaing (1989). Epidemiological basis of survey design, methodology and data analysis for ascariasis. *In* "Ascariasis and Its Prevention and Control" (D. W. T. Crompton, M. C. Nesheim, and Z. S. Pawlowski, eds.), pp. 351–368. Taylor and Francis, London.

Thein-Hlaing, Thane-Toe, Than-Saw, Myat-Lay-Kyin, and Myint-Lwin. (1991). A controlled chemotherapeutic intervention trial on the relationship between *Ascaris lumbricoides* infection and malnutrition in children. *Trans. R. Soc. Trop. Med. Hyg.* **85,** 523–528.

Thein-Hlaing, Than-Saw, Htay-Htay-Aye, Myint-Lwin, and Thein-Maung-Myint. (1984). Epidemiology and transmission dynamics of *Ascaris lumbricoides* in Okpo village, rural Burma. *Trans. R. Soc. Trop. Med. Hyg.* **78,** 497–504.

Thein-Hlaing, Than-Saw, and Myint-Lwin (1987). Reinfection of people with *Ascaris lumbricoides* following single 6-month and 12-month interval mass chemotherapy in Opko village, rural Burma. *Trans. R. Soc. Trop. Med. Hyg.* **81,** 140–146.

Thein-Hlaing, Myat-Lay-Kyin, Hlaing-Mya, and Maung-Maung. (1990). Role of ascariasis in surgical abdominal emergencies in the Rangoon Children's Hospital, Burma. *Ann. Trop. Pediatr.* **10,** 53–60.

Thienpont, D., Rochette, F., and Vanparijs, O. F. J. (1986). "Diagnosing Helminthiasis by Coprological Examination," 2d Ed. Janssen Research Foundation, Beerse, Belgium.

Tomkins, A., and Watson, F. (1989). "Malnutrition and Infection." World Health Organization, Geneva.

Trainer, E. S. (1985). Mass parasite control: A good beginning. *World Health Forum* **6,** 248–253.

Trainer, E. S. (1989). An alternative strategy for promoting ascariasis control. *In* "Ascariasis and Its Prevention and Control" (D. W. T. Crompton, M. C. Nesheim, and Z. S. Pawlowski, eds.), pp. 321–330. Taylor and Francis, London.

Urabe, K. (1957). Human experiment on percutaneous infection of *Ascaris. Igaku Kenkyu* **27,** 277–281.

Walsh, J. A., and Warren, K. S. (1979). Selective primary health care. *N. Engl. J. Med.* **301,** 967–974.

Webster, J., Hodges, M., Crompton, D. W. T., and Walters, D. E. (1990). Intestinal parasitic infections in children from Freetown, Sierra Leone. *J. Sierra Leone Med. Dent. Assoc.* **5,** 144–155.

Williams, D., Burke, G., and Owen Hendley, J. (1974). Ascariasis: A family disease. *J. Pediatr.* **84,** 853–854.

Wong, M. S., and Bundy, D. A. P. (1990). Quantitative assessment of contamination of soil by the eggs of *Ascaris lumbricoides* and *Trichuris trichiura. Trans. R. Soc. Trop. Med. Hyg.* **84,** 567–570.

World Health Organization (1967). "Control of Ascariasis." WHO Tech. Rep. 379, World Health Organization, Geneva.

World Health Organization (1986). "Sample Size Determination." Document WHO/HST/ESM/86/I. World Health Organization, Geneva.

World Health Organization (1987). "Prevention and Control of Intestinal Parasitic Infections." WHO Tech. Rep. 749, World Health Organization, Geneva.

World Health Organization (1990). "WHO Model Prescribing Information: Drugs Used in Parasitic Diseases." World Health Organization, Geneva.

World Health Organization (1992). "Bench Aids for the Diagnosis of Intestinal Helminths." WHO/CDS/IPI. World Health Organization, Geneva.

Yadav, A. K., and Tandon, V. (1989). Prevalence of nematode eggs in the urban area of the city of Shillong, India—A public health problem. *Health Hyg.* **10,** 158–161.

Yamaguti, S. (1961). "Systema Helminthum—The Nematodes of Vertebrates," Vol. III, Part 1. Interscience Publishers, New York.
Yokogawa, M. (1985). JOICFP's experience in the control of ascariasis within an integrated programme. *In* "Ascariasis and Its Public Health Significance" (D. W. T. Crompton, M. C. Nesheim, and Z. S. Pawlowski, eds.), pp. 265–277. Taylor and Francis, Basingstoke, England.
Yoshida, S. (1919). On the development of *Ascaris lumbricoides* L. *J. Parasitol.* **5**, 105–115.

Mark E. J. Woolhouse

Epidemiology of Human Schistosomes

15

I. Background

Schistosomes are digenetic trematodes that first were described by Bilharz (1852). The adult parasites are up to 30 mm long and occur as male–female pairs in the abdominal blood vessels of mammalian hosts. Three main species infect humans, causing the disease schistosomiasis. *Schistosoma haematobium* is associated with the blood vessels of the bladder and kidneys and is endemic in much of Africa and Arabia. *Schistosoma mansoni* and *Schistosoma japonicum* are associated with the intestinal blood vessels. *Schistosoma mansoni* is endemic in Africa, Arabia, and South and Central America. *Schistosoma japonicum* occurs in Asia.

The schistosome life cycle (Fig. 1) first was described early in the twentieth century. Eggs are released from the human host via the urine or feces. If deposited in fresh water, an egg hatches to produce a free-swimming ciliated larva, a miracidium, approximately 0.2 mm long. The miracidium penetrates the tissues of the intermediate host, a freshwater snail, and transforms into a sporocyst. The sporocyst undergoes several rounds of asexual reproduction within the snail, producing thousands of daughter sporocysts. These sporocysts mature and emerge from the snail as a second free-swimming larval stage, a cercaria, approximately 1 mm long with a characteristic forked tail. The cercaria penetrates the skin of the human host, loses its tail, and enters the blood vessels as a schistosomulum. All human schistosomes are capable of infecting other mammals and, especially for *S. japonicum,* nonhuman hosts can provide important reservoirs of infection. The schistosomulae migrate to the blood vessels of the lung where they mature before transferring to the abdomen, as mated adult schistosomes. Each schistosome pair produces hundreds to thousands of eggs per day. The eggs are spined and are worked passively through the tissues to the excretory tract. Most of the pathology associated with infection is due to immune responses to eggs that become trapped in the host tissues.

Current estimates suggest that up to 200 million people are infected with schistosomes [World Health Organization (WHO), 1990a]. A small proportion of these individuals experienced morbidity as a result of infection. Schistosomiasis is essentially a chronic illness, the severity of which is a function of the intensity and duration of infection (von Lichtenberg, 1987; Gryseels and Polderman, 1991).

Parasitic and Infectious Diseases

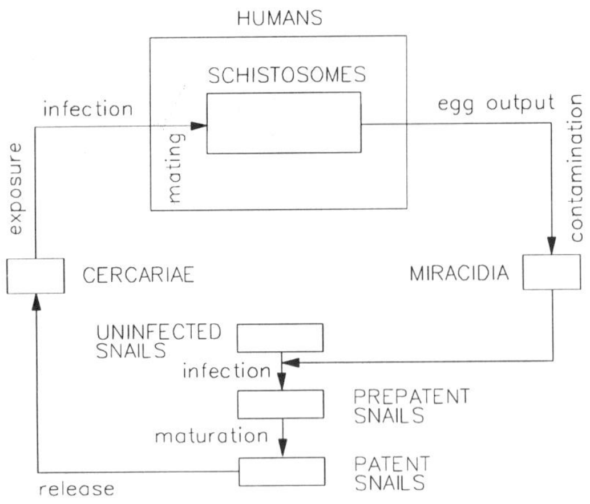

Figure 1 Diagrammatic representation of the life cycle of human schistosomes.

II. Basic Theory

Schistosomes do not reproduce directly within the human host. An infected individual's contribution to transmission is related quantitatively to parasite burden (as is the level of morbidity). A model of schistosome transmission dynamics must, therefore, consider the intensity of infection in the human population. In contrast, schistosomes undergo extensive multiplication within the snail host. An infected snail's contribution to transmission is independent of the number of miracidia penetrating it (Christie and Prentice, 1978; Touassem and Theron, 1989). Therefore, a model need consider only the prevalence of infection in the snail population.

Mathematical models can be simplified by allowing for the different turnover rates of the host and parasite populations (Table 1). Human abundances change relatively slowly and, for most purposes, can be assumed to be constant. At the other extreme, the turnover of cercariae and miracidia populations is so rapid that larval stage densities can be assumed always to be at equilibrium for given densities of patent infected snails and adult schistosomes, respectively. Larval stage dynamics do not, therefore, have to be considered explicitly. These simplifications allow a much more manageable model than inspection of Fig. 1 would suggest.

Consider a human population of H individuals that makes use of a single water body of unit dimensions containing N intermediate host snails. Levels of schistosome infection are represented by the following variables: m, the mean number of schistosome pairs per person; z, the proportion of snails supporting prepatent infections; y, the proportion of snails supporting patent infections; and x, the proportion of uninfected snails, where $x = 1 - z - y$. The dynamics of this system can be expressed by the model

$$\frac{dm}{dt} = \alpha\eta N y - \gamma m$$

$$\frac{dz}{dt} = \beta\kappa H m x - \sigma z - \mu' z \qquad (1)$$

$$\frac{dy}{dt} = \sigma z - \mu'' y,$$

where η is the rate of exposure per person due to behavior, α is the rate of infection with schistosome pairs per person per patently infected snail per unit exposure (also referred to as susceptibility to infection), γ is the *per capita* mortality rate of schistosome pairs (the mean life expectancy is $1/\gamma$), κ is the rate of contamination with excreta per person, β is the *per capita* rate of infection of snails per schistosome pair per unit contamination, σ is the rate of maturation of prepatent infections of snails (the mean prepatent period is $1/\sigma$), and μ' and μ'' are the *per capita* mortality rates of prepatent and pat-

Table 1 Life Expectancies Relevant to Schistosome Transmission

Human	60 yr
Mated female schistosome	3–5 yr
Snail	4–12 wk
Cercaria	<24 hr
Miracidium	<12 hr

ently infected snails, respectively (with mean life expectancies $1/\mu'$ and $1/\mu''$, respectively). Equation 1 captures the spirit of a wide range of more detailed and complicated models (see Cohen, 1977; Anderson and May, 1991; Woolhouse, 1991).

Equation 1 generates the following expression for R_0, the basic reproduction ratio:

$$R_0 = \frac{\alpha\eta N\beta\kappa H\sigma}{\gamma\mu''(\sigma + \mu')}. \tag{2}$$

Equilibrium values for levels of infection are denoted m^*, z^*, and y^*. If $R_0 < 1$, the only stable solution lies at the origin:

$$\begin{aligned} m^* &= 0 \\ z^* &= 0 \\ y^* &= 0. \end{aligned} \tag{3}$$

If $R_0 > 1$, the solution at the origin is unstable and the model generates the following equilibrium levels of infection in the human and snail populations:

$$\begin{aligned} m^* &= \frac{\alpha\eta N\sigma}{\gamma(\sigma + \mu'')} - \frac{\mu''(\sigma + \mu')}{\beta\kappa H(\sigma + \mu'')} \\ y^* &= \frac{\sigma}{\sigma + \mu''}\left(1 - \frac{1}{R_0}\right). \end{aligned} \tag{4}$$

A complication arises if the model is expanded to consider schistosome mating. In this case, a nonzero equilibrium can be attained only if initial m, z, and y exceed a combination of values, referred to as the breakpoint, at which the tendency for levels of infection to increase is balanced by a low probability that a given female schistosome will be mated, which decreases as m decreases (May, 1977). In practice, the distribution of schistosomes among human hosts appears to be sufficiently overdispersed that most schistosomes are found in a few hosts, so even at low m most female schistosomes are (potentially) mated. Consequently, breakpoint levels of infection are too close to zero to be of epidemiological significance.

III. Epidemiology

A. Infection Intensity

A major limitation for epidemiological studies of human schistosomes is that the infection intensity, m, can be estimated only indirectly. Most field studies describe the infection intensity in terms of egg excretion rates. Advances in the detection of adult schistosome antigens may provide a quantitative index of adult schistosome burdens that is independent of egg production (de Jonge *et al.*, 1989). The only direct estimates of adult schistosome burdens come from autopsy data (Cheever, 1968; Cheever *et al.*, 1977). These data are valuable but are available only from small and nonrandom samples of human populations.

Infection intensities in a human community typically are overdispersed, that is, most individuals release few or no eggs, whereas a few individuals release very many eggs. For example, a study of *S. haematobium* infection in Zimbabwe (Chandiwana *et al.*, 1991) found that just 3% of the population was responsible for 50% of egg output (Fig. 2a). Infection intensities also show a characteristic pattern of variation with age, typically (although not invariably) rising to a peak in older children and declining in adults (Figs. 2b–d). This decline often is observed to be more severe for infection with *S. haematobium.* An important question is whether the decline is the result of age-related changes in exposure or whether it reflects resistance to infection in older age groups, possibly as a result of acquired immunity (Warren, 1973).

B. Host Behavior

Several studies have attempted to quantify human exposure to infection by observing water contact behavior in a community (e.g., Dalton and Pole, 1978). The factors considered to determine relative exposure are summarized in Table 2. The functional relationships between behavior and expo-

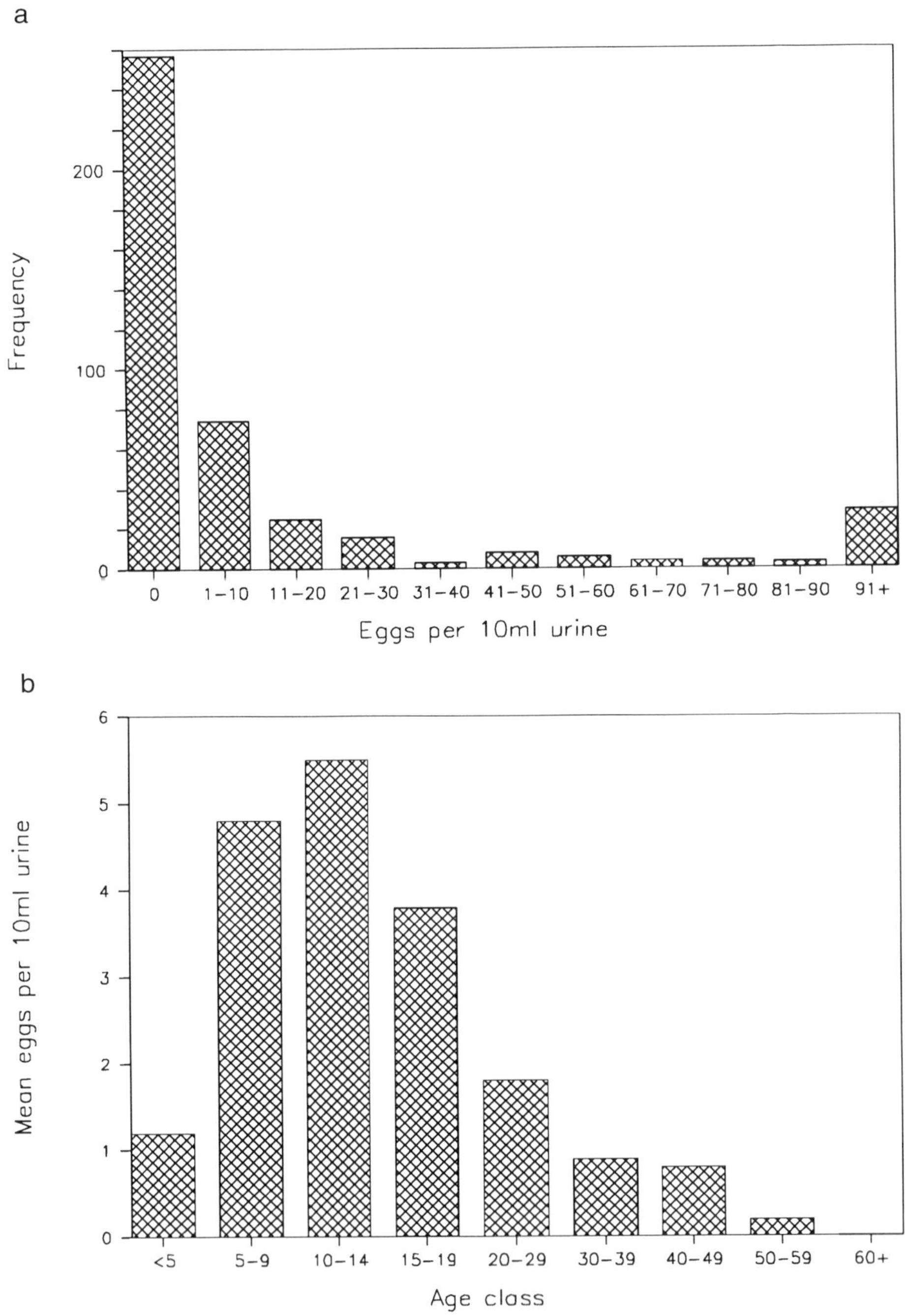

Figure 2 (a) Frequency distribution of egg count data for *Schistosoma haematobium* infections in a community in Zimbabwe. Data from Chandiwana *et al.* (1991). The data are highly overdispersed with a variance-to-mean ratio of 283. (b) Relationship between geometric mean egg counts and age (same data). (c) Relationship between geometric mean egg counts and age for *Schistosoma mansoni* infections in a community in St. Lucia. Data from Jordan and Webbe (1982). (d) Relationship between geometric mean egg counts and age for *Schistosoma japonicum* infections in a community in the Philippines. Data from Olveda *et al.* (1983).

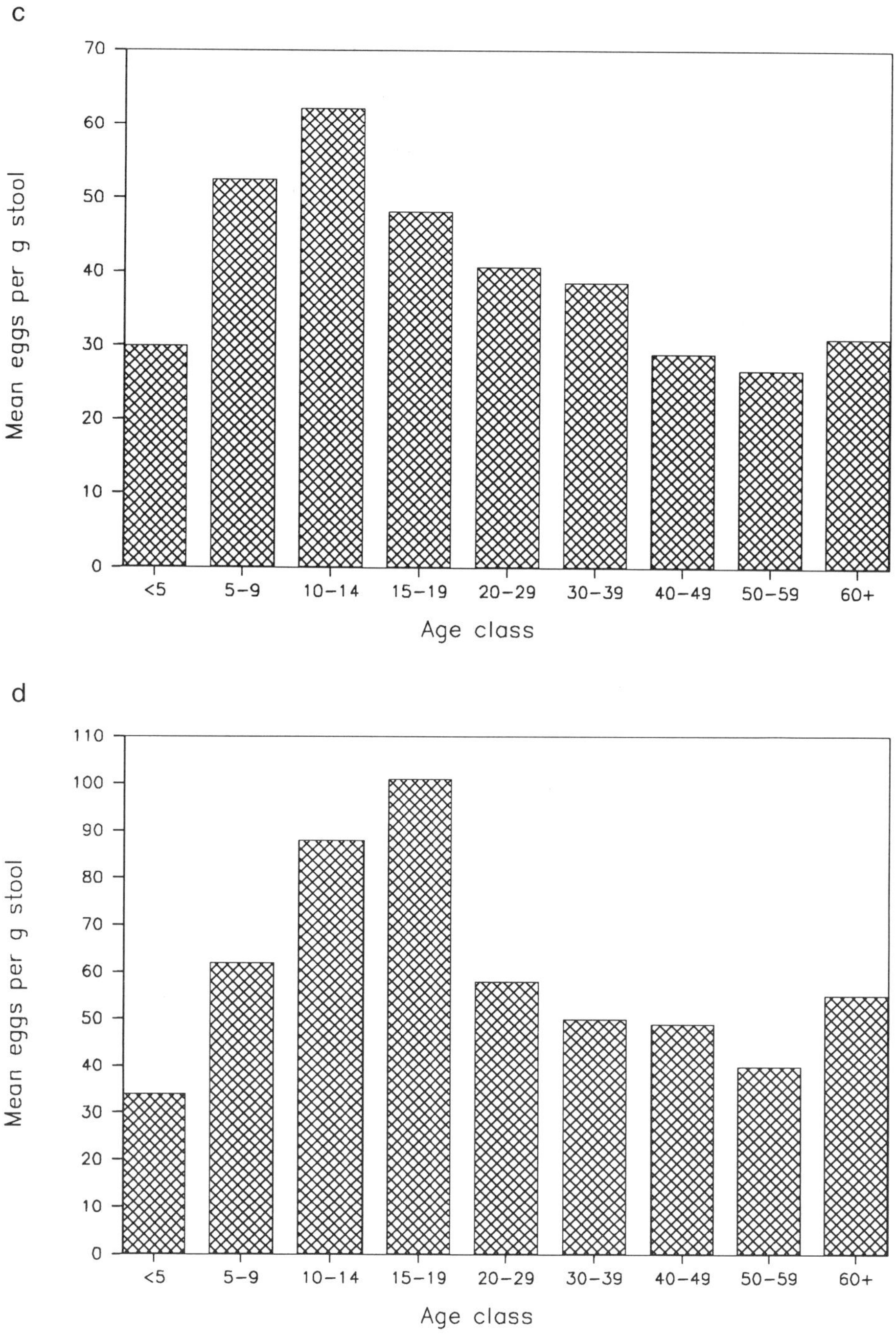

Figure 2 *(Continued)*

Table 2 Components of Exposure Index

Each contact weighted according to
duration
activity related to
area of skin surface exposed
use of soap
cercarial density related to
time of day
density of patently infected snails which varies with
site
time (within and between years)

sure must, for obvious ethical reasons, be based on deduction rather than experiment, although some support for particular exposure indices comes from treatment–re-infection studies (see subsequent discussion). Exposure commonly is assumed to be a linear function of the duration of contact, the fraction of skin surface exposed, and cercarial density. Cercarial density is expected to reflect variation in the density of patently infected snails and diurnal patterns in cercarial emergence (Fig. 3). Distinguishing the behavioral components of exposure (number of contacts, duration, activity carried out, time of day) from the biological components (density of patently infected snails) is useful. These factors are represented in the basic model by η and Ny, respectively.

The behavioral components of exposure are subject to enormous variation among populations, since water contact behavior reflects local customs and traditions, water requirements, and available water supplies. However, some general patterns are apparent. One such pattern is that rates of exposure are overdispersed. For example, in the community in Zimbabwe discussed earlier, 12% of the population is responsible for 50% of water contact (Fig. 4a). Another pattern is that rates of exposure change with age; several studies have reported higher exposure among older children than among adults (Figs. 4b–d).

Much less information is available regarding contamination of water bodies with excreta. Fecal contamination in some African communities tends to be associated with bathing and laundry rather than with direct defecation into the water (Chandiwana, 1986; Ouma, 1988), but this situation may not exist elsewhere (Nelson, 1990). Chil-

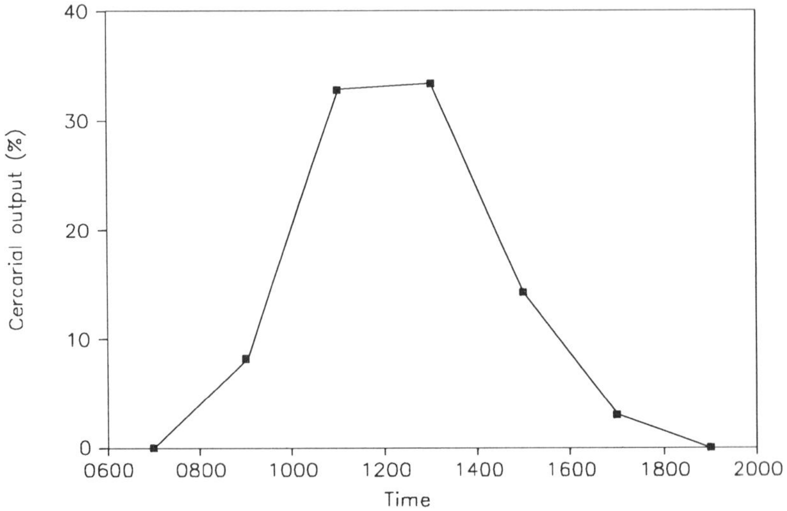

Figure 3 Circadian rhythm in output of *Schistosoma haematobium* cercariae from *Bulinus globosus* snails. Data from Pitchford *et al.* (1969).

dren, who tend to have high rates of both egg output and water contact, are likely to be disproportionately responsible for transmission from humans to snails.

C. Rates of Infection and Loss of Infection

Information on the rates of infection of humans comes from treatment-re-infection studies. In such studies, a subset of the human population is cured of infection using anthelmintic drugs; the rate at which this group becomes re-infected is monitored over the following months. The rate of infection is represented by the term $\alpha\eta Ny$. This term can be dissected into susceptibility, α, and the behavioral and biological components of exposure, η and Ny, respectively. Variations in rates of infection among individuals or among groups imply variations in one or more of these parameters.

One result from these studies is that significant predisposition to infection exists; individuals with higher egg outputs before treatment tend to become re-infected more rapidly (Bensted-Smith *et al.*, 1987; Tingley *et al.*, 1988; Etard *et al.*, 1990). This result precludes short-term variation in susceptibility or exposure as the only determinants of variation in levels of infection.

Rates of re-infection have been demonstrated to be correlated positively with rates of exposure for *S. mansoni* in Kenya (Butterworth *et al.*, 1985) and for *S. haematobium* in both The Gambia (Wilkins *et al.*, 1987) and Zimbabwe (Chandiwana *et al.*, 1991), confirming that the exposure indices have some epidemiological significance. All these studies also found independent effects of age on re-infection rates. The Gambian study also demonstrated a negative relationship between re-infection rates and levels of anti-adult and anti-egg IgE antibody responses (see subsequent discussion), in addition to effects of exposure and age (Hagan *et al.*, 1991). This result suggests that susceptibility to infection, α, is a function of acquired immunity.

The only estimates of schistosome mortality rates in the human host derive from data on the decline in egg output rates in areas where transmission has been interrupted. These data (assuming that the declines are the result of mortality of schistosome pairs) give estimates of mean life expectancies, $1/\gamma$, for *S. mansoni* and *S. haematobium* of 3-5 yr (Goddard and Jordan, 1980; Wilkins *et al.*, 1984) or more (Vermund *et al.*, 1983). Given these mortality rates, up to 2% of schistosomes would remain alive after 20-30 yr, and some initially heavily infected individuals still would be excreting eggs after this period, as has been reported (Chabasse *et al.*, 1985).

D. Intermediate Host

The intermediate hosts for *S. mansoni, S. haematobium,* and *S. japonicum* are freshwater snails of the genera *Biomphalaria, Bulinus,* and *Oncomelania,* respectively. These snails occur in a wide range of habitats, from seasonal streams to large lakes and including artificial habitats such as irrigation systems and rice fields. Substantial information is available from both laboratory and field studies on the population dynamics of the snails and the epidemiology of their infection with schistosomes (see Anderson and May, 1979; Woolhouse and Chandiwana, 1990a).

Snail population densities tend to be highly variable over spatial scales of a few meters and temporal scales of a few weeks (Fig. 5). The prevalence of patent infections is similarly patchy in both space and time (Fig. 5). Prevalences of patent infections are typically quite low (less than 10%), largely because the prepatent period tends to be long (4 wk or more) in comparison with snail life expectancy (3-12 wk) and because infection substantially reduces snail life expectancy (to 2-6 wk). Still, over small distances over which extensive human water contact occurs, the rate of contamination with egg-bearing excreta may be sufficient to generate prevalences of patent infections in

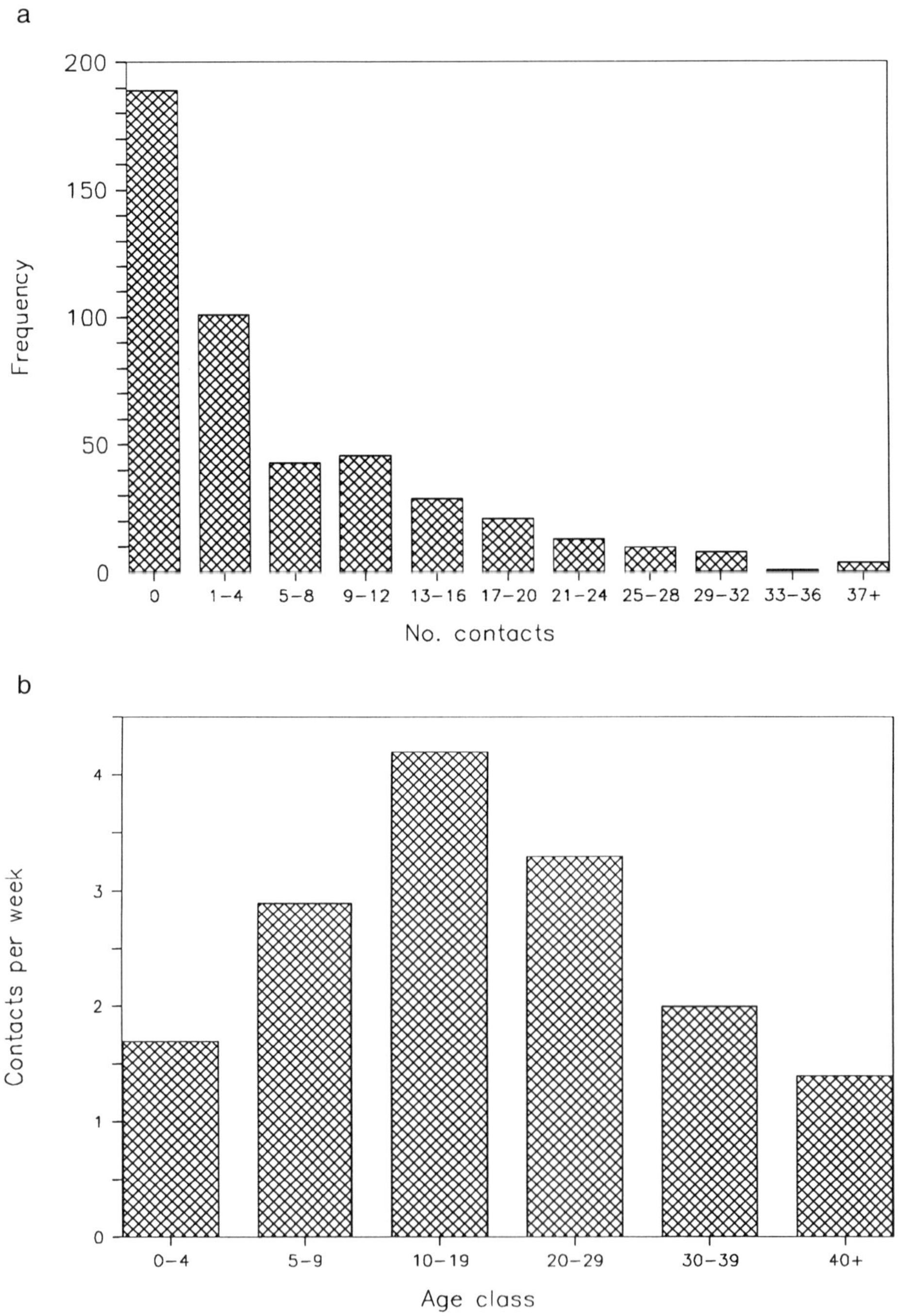

Figure 4 (a) Frequency distribution of water contacts observed over 2 wk for a rural community in Zimbabwe. Data from Chandiwana and Woolhouse (1991). The data are overdispersed with a variance-to-mean ratio of 12.1. (b) Relationship between water contact rates and age (same data). More detailed indices of exposure (see Table 2) show similar age-related variation. (c) Relationship between water contact rates and age for a fishing community in Ghana. Data from Dalton and Pole (1978). (d) Relationship between net duration of water contact and age for a rural community in Kenya. Data from Butterworth *et al.* (1984).

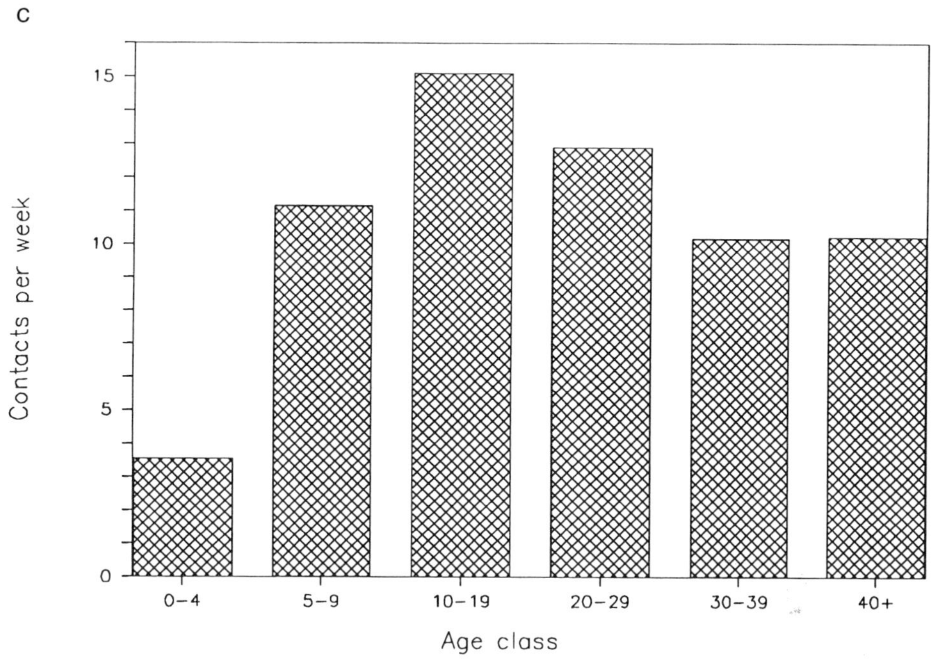

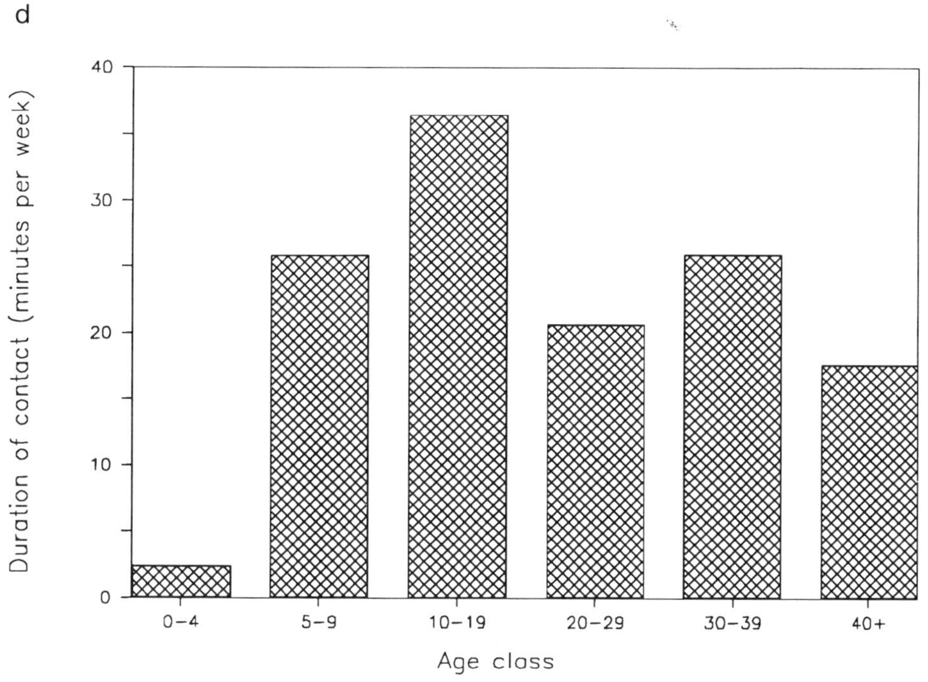

Figure 4 *(Continued)*

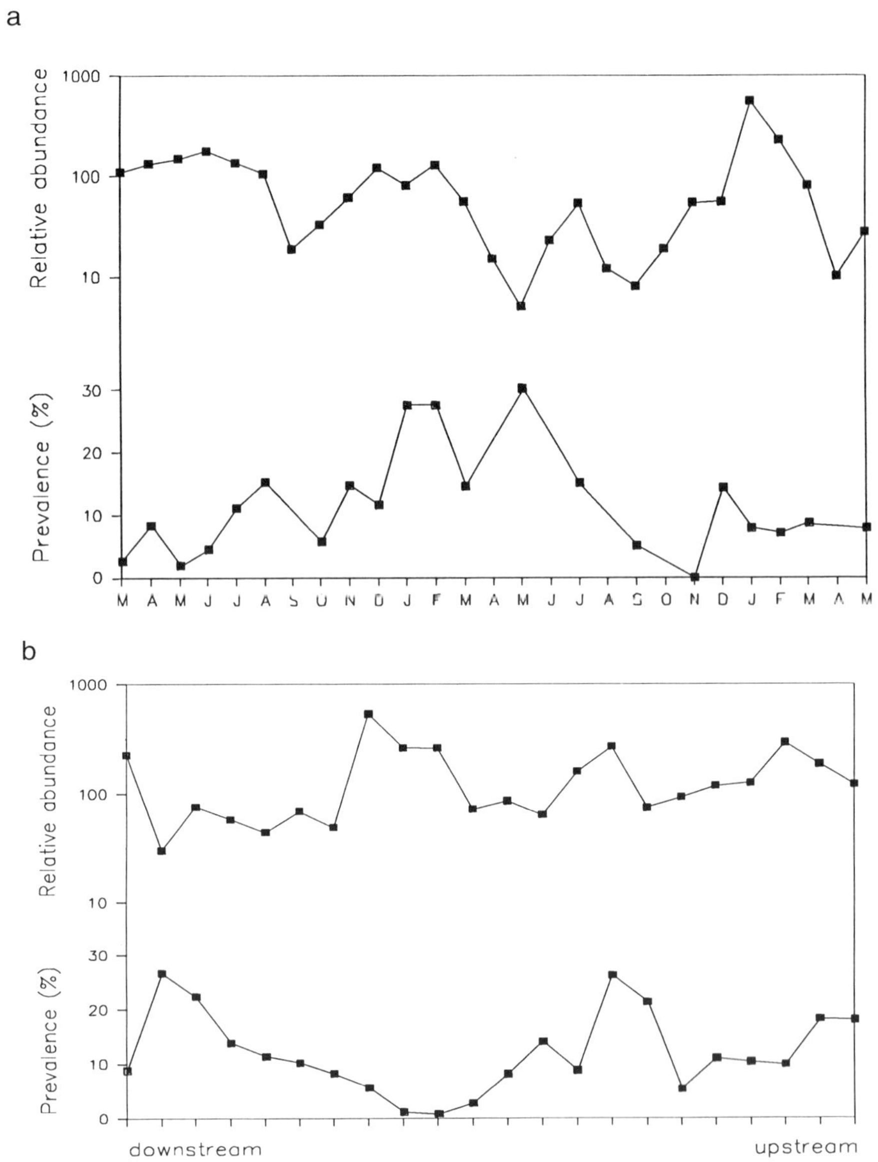

Figure 5 (a) Monthly variation in relative snail abundance (as indexed by fractional sampling) and the prevalence of patent *Schistosoma haematobium* infections for *Bulinus rohlfsi* in the Volta Lake. Data from Klumpp and Chu (1977). (b) Spatial variation in relative snail abundance and the prevalence of patent schistosome infections for *Bulinus globosus* in a river habitat in Zimbabwe. Each division represents a contiguous 40-m length of river. Data pooled over 18 mo, see Woolhouse and Chandiwana (1990a,b).

excess of 50% (Udonsi, 1990; Woolhouse and Chandiwana, 1990a). Temporal, especially seasonal, variations in prevalence reflect the effects of temperature on the prepatent period (which may be very long at lower temperatures; Shiff *et al.*, 1975), changes in the age structure of the snail population (due to seasonal variation in fecundity and mortality rates; Sturrock, 1973), and changes in human water contact rates.

The *per capita* rate of infection of snails has been estimated by a technique known as catalytic curve analysis. If the *per capita* rate of infection of uninfected snails (represented by λ where λ is equivalent to the term $\beta\kappa Hm$ in the basic model), snail mortality rates, and the prepatent period are constant, then Eq. 1 can be solved to give x, y, and z as functions of time. For a given cohort of snails, given $x = 1$ and $z = y = 0$ at time $t = 0$, time can be equated with age. Catalytic models have been applied to age–prevalence data from the field (using snail size to estimate age) under approximately steady-state conditions, when all parameters were known except λ (Woolhouse and Chandiwana, 1990b,1992). The models provided good fits to the data (Fig. 6a,b) and suggested that up to 5% of snails acquire infection per week. Miracidia do not infect snails at random (Woolhouse *et al.*, 1990), so the rate of infection may be higher than this among subpopulations of snails. Nonrandom infection may be the result of patchiness in the contamination of the water with excreta as well as differences between snails in their susceptibility to infection (Richards, 1984). Loss of infection in snails has been reported (Sturrock, 1967), but does not occur at rates that are likely to be epidemiologically significant (Woolhouse, 1989).

E. Density Dependence

One constraint on schistosome population growth is that the rate at which a patently infected snail releases cercariae is independent of the number of miracidia with which it is infected. This constraint is the only density-dependent constraint incorporated in the model of Eq. 1. However, density-dependent processes also may operate in the human host, and could act to reduce schistosome egg output (a component of β), to increase schistosome mortality (γ), and/or to reduce the rate of infection (α).

Direct evidence exists only for the first of these effects. Comparison of egg output rates with postmortem counts of parasite burdens suggests that more heavily infected individuals may have lower rates of egg output per schistosome pair. The relationship is not marked for *S. mansoni,* and statistical analyses of these data have given equivocal results (Cheever, 1968; Medley and Anderson, 1985; Keymer and Slater, 1987). In contrast, the situation is far more dramatic for *S. haematobium,* for which the data suggest that more heavily infected individuals have lower net egg output rates (Cheever *et al.,* 1977; Fig. 7). A similar effect has been demonstrated clearly for *Schistosoma spindale* infections of cattle (De Bont *et al.,* 1991).

F. Acquired Immunity

Various cellular and antibody responses to schistosome infection have been reported in humans (see Hagan, 1987), including responses to schistosomulae, adult schistosomes, and eggs. In principle, such responses could produce density-dependent effects on the parameters α, γ, and β

Egg output, a component of β, reflects both schistosome fecundity and the rate of passage of eggs through host tissues. Some evidence suggests that immune responses have an effect on schistosome fecundity. Immune responses to eggs in the tissues typically result in the formation of granulomas (which are responsible for the pathological effects of infection). However, in experimentally infected mice, evidence exists that acquired immune responses act to *increase* the rate of egg output (Doenhoff *et al.,* 1978). At present, whether acquired immunity has a significant effect on egg output in humans is not clear. The observed density dependence of egg output per schistosome pair (see previous discussion) may be the result of physical effects such as crowding.

Immune-mediated killing of schistosomulae has been demonstrated *in vitro.* Studies of experimentally infected animals indi-

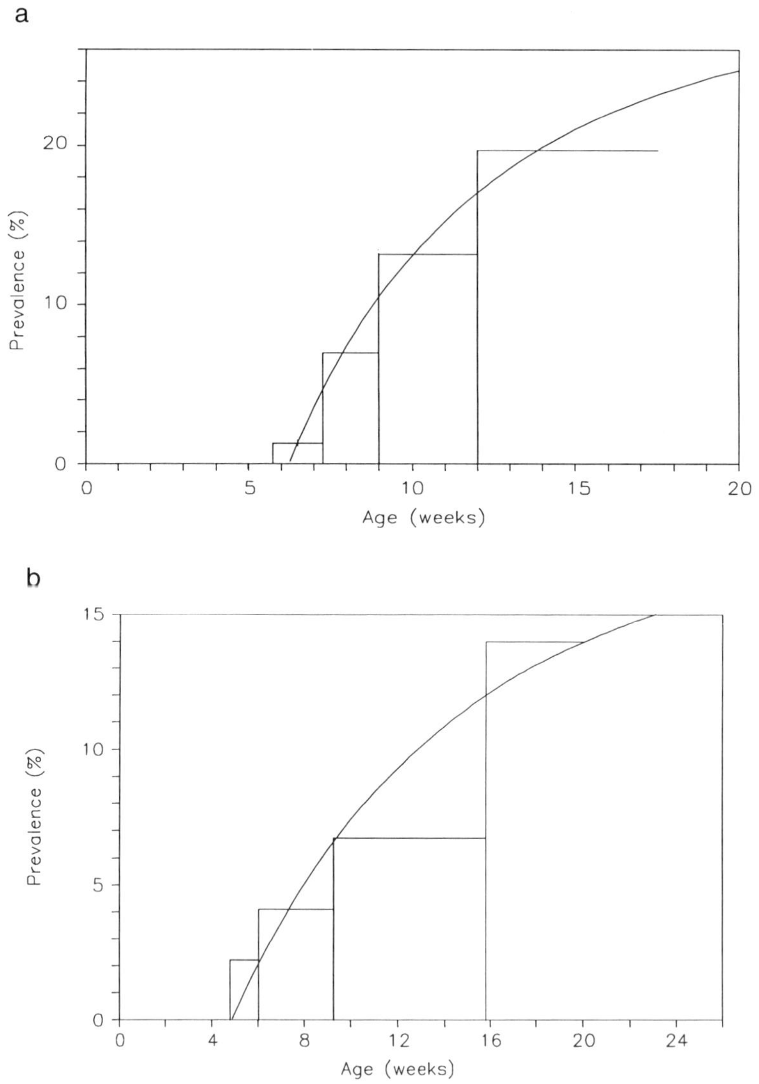

Figure 6 (a) Age–prevalence relationships for patent schistosome infections of *Bulinus globosus* snails. Field data (histogram) are compared with a catalytic model (line) with fitted $\lambda = 0.050$ (see text). Reprinted with permission from Woolhouse and Chandiwana (1990a). (b) Age–prevalence relationships for *Schistosoma mansoni* infections of *Biomphalaria pfeifferi*. Fitted $\lambda = 0.034$. Reprinted with permission from Woolhouse and Chandiwana (1992) and Cambridge University Press.

cate that these responses are protective *in vivo* (see McLaren and Smithers, 1987). In humans, both cellular (especially eosinophil) and antibody (especially IgE) responses have been implicated in protection (Butterworth and Hagan, 1987). Acquired immunity against schistosomulae is associated with the presence of mature infections, a phenomenon referred to as concomitant immunity, which may imply cross-reactivity between schistosomulum antigens and adult and/or egg antigens (Butterworth, 1987).

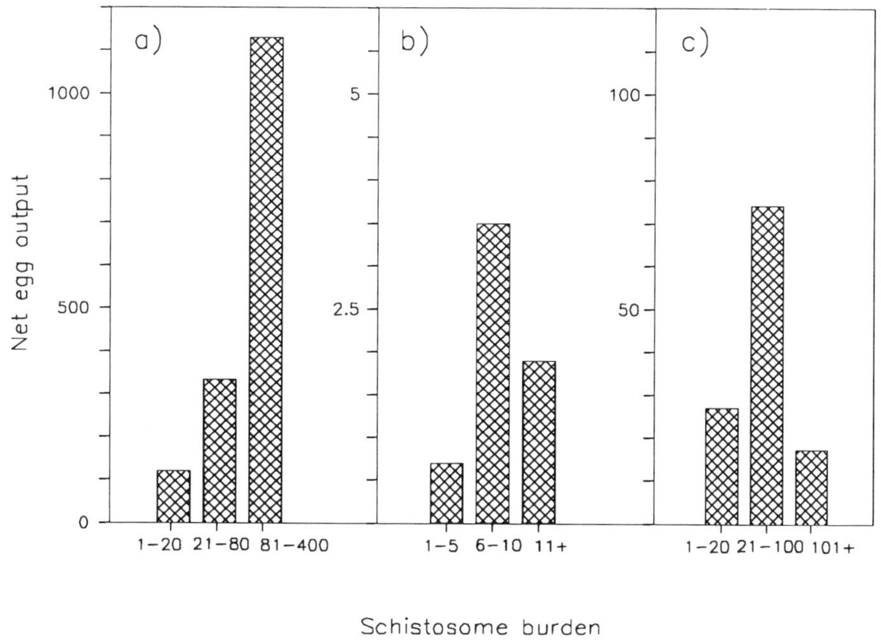

Figure 7 Relationship between net egg output and schistosome burden. (a) *Schistosoma mansoni* infections of humans in Brazil. Mean eggs per g feces compared with number of female worms. Data from Cheever (1968). (b) *Schistosoma haematobium* infections of humans in Egypt. Geometric mean eggs per 10 ml urine compared with number of female worms. Data from Cheever *et al.* (1977). (c) *Schistosoma spindale* infections of cattle in Sri Lanka. Mean miracidia per 100 g feces compared with number of schistosome pairs. Data from De Bont *et al.* (1991).

These responses would result in increased resistance (reduced susceptibility) to infection as a function of past history of infection, that is, a density-dependent effect on the parameter α. The situation may be complicated further by the existence of blocking antibodies, immune responses that inhibit any protective response and delay the development of acquired resistance (Butterworth *et al.*, 1988).

Currently, no good evidence is available that immune responses have significant effects on adult schistosome mortality rates, γ, *in vivo*.

Under equilibrium conditions, with $y = y^*$, Eq. 1 of the basic model can be solved to give m as a function of time. For a given cohort of humans, given $m = 0$ at $t = 0$, time can be equated with age. The model predicts a monotonic rise in infection intensity with age (see Woolhouse, 1991). The effect of acquired immunity can be incorporated by making α a decreasing function of cumulative infection intensity or cumulative exposure (Anderson and May, 1991). Infection intensity now rises to a peak and subsequently declines. This pattern also can be obtained by allowing acquired immunity to act on schistosome mortality rate, γ. Alternatively, the same pattern can be generated, in the absence of acquired immunity, by allowing the rate of exposure, η, to decline with age.

However, models incorporating acquired immunity, in contrast to those allowing only age-dependent exposure, predict systematic variation in the shape of the age–intensity relationship between populations subject to high and low transmission (Fig. 8a). One predicted pattern is that populations subject to high transmission should show higher peak levels of infection occurring in younger age classes. This prediction is supported by field data on *S. haematobium* infections of school children (Fig. 8b), a result that is consistent with a significant

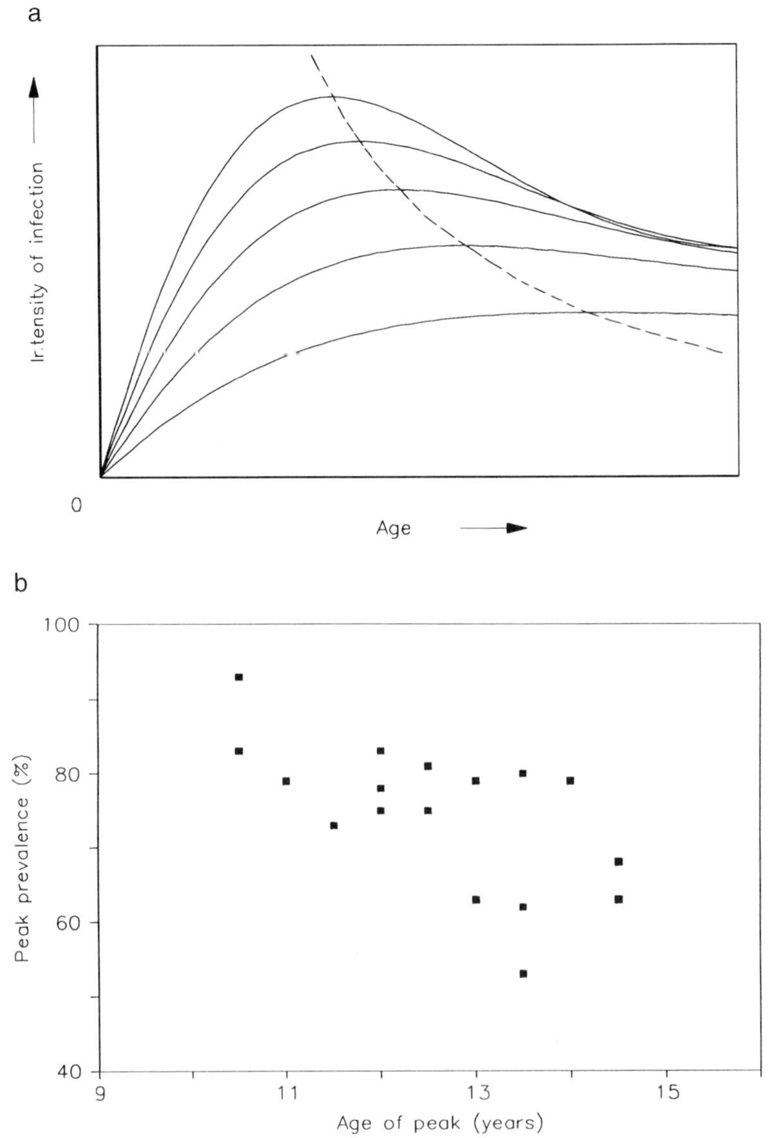

Figure 8 (a) Age–intensity curves for populations subject to different strengths of transmission (solid lines) as predicted by a model in which acquired immunity acts to reduce susceptibility to infection (see text). The predicted relationship between the peak intensity of infection and the age at which peak intensity occurs is shown by the dashed line. (b) Observed relationship between the peak prevalence of *Schistosoma haematobium* infection and the age at which peak prevalence occurs for 17 primary schools in Zimbabwe. From Woolhouse *et al.* (1991a). Reprinted with permission from *Nature (London)*, MacMillan Magazines, Ltd.

effect of acquired immunity (Woolhouse *et al.,* 1991a).

G. Heterogeneous Transmission

Heterogeneities in rates of transmission may arise through variation in patterns of exposure and contamination, as well as through variations in susceptibility to infection due to acquired immunity and other factors. Heterogeneity in water contact behavior can be partitioned as variation in individual net contact rates, variation in the net contact rate at different sites, and variation in the distributions of individual contacts among sites, referred to as the contact pattern (Fig. 9). Variation in individual net contact rates often is substantial (Fig. 2a) and reflects differences in factors such as age, sex, occupation, religion, socioeconomic status, and proximity to contact sites (Farooq and Mallah, 1966; Dalton and Pole, 1978; de Lima e Costa *et al.,* 1985). Variation between sites also may be significant; in many communities, most contacts are confined to relatively few, more-or-less discrete sites (Dalton and Pole, 1978; Butterworth *et al.,* 1984; Chandiwana and Woolhouse, 1991). Variation in contact pattern has received little attention, but individuals typically concentrate their activities at a subset of the sites available. The partitioning of sites between the sexes is particularly important, although children are less discriminating in this respect than adults (Butterworth *et al.,* 1984; Chandiwana and Woolhouse, 1991).

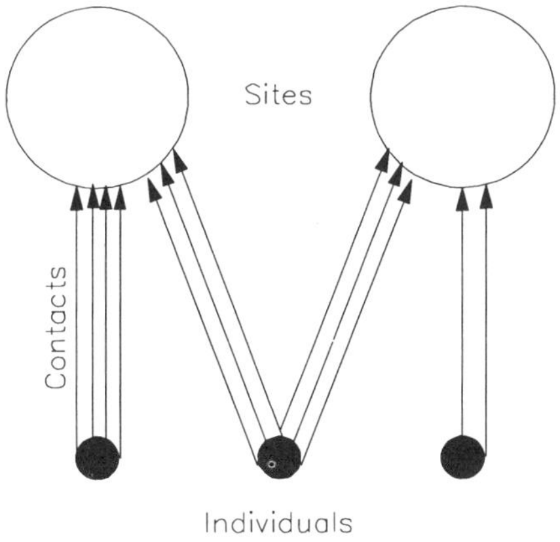

Figure 9 Diagrammatic representation of heterogeneities in water contact rates. Each arrow represents one contact by one individual at one site per unit time. Three types of heterogeneity are illustrated: (1) different net contact rates of individuals; (2) different net contact rates at sites; and (3) differences in the contact pattern, the way in which individuals distribute their contacts among sites.

The epidemiological significance of heterogeneities in contact behavior has been explored using mathematical models (Barbour, 1978; Woolhouse *et al.,* 1991b). The model of Eq. 1 can be modified to consider the dynamics of the variables M_i, the number of schistosome pairs in person i for $i = 1$ to H, and z_j and y_j, the proportion of prepatent and patent snails, respectively, in site j for $j = 1$ to L where L is the number of sites and each site supports N snails. The model now becomes

$$\frac{dM_i}{dt} = \alpha\left(\sum_j \eta_{ij} N y_j\right) - \gamma M_i$$

$$\frac{dz_j}{dt} = \beta\left(\sum_i \kappa_{ij} M_i\right) x_j - \sigma z_j - \mu' z_j \quad (5)$$

$$\frac{dy_j}{dt} = \sigma z_j - \mu'' y_j,$$

where η_{ij} and κ_{ij} are, respectively, the rates of exposure and contamination by individual i at site j. For this model, R_0 is given by

$$R_0 = \frac{\alpha N \beta \sigma}{\gamma \mu''(\sigma + \mu')} \epsilon, \quad (6)$$

where ϵ is the dominant eigenvalue of the $L \times L$ matrix with elements $\mu_{jk} = \Sigma_i(\eta_{ij}\kappa_{jk})$. An estimate of ϵ has been made from water contact observations of a community in Zimbabwe, for which the calculated effect was to increase R_0 substantially, by a factor of 4.4, largely because of variation in individual net contact rates (Woolhouse *et al.,* 1991b). This estimate was an upper limit because it assumed that both exposure and contamination rates were proportional to contact rates. For the special case of $L = 1$, a more general expression for ϵ (see Dietz, 1980) is

$$\epsilon = H(\bar{\eta}\bar{\kappa} + \mathrm{SD}(\eta)\mathrm{SD}(\kappa)r_{\eta\kappa}), \quad (7)$$

where SD represents standard deviation and $r_{\eta\kappa}$ is the linear correlation between individual rates of exposure and contamination. This correlation has yet to be estimated in the field.

Other heterogeneities also may contribute to effects on R_0. Sites are likely to vary in the density of snails, N, as well as in the susceptibility of snails to infection, a component of β. Individuals may vary in their innate susceptibility to infection, affecting α.

The effects of heterogeneities in transmission on equilibrium levels of infection have not been analyzed. Such analysis also would have to consider any individual variation in the ability to develop acquired immunity caused, for example, by genetic or nutritional factors (see Wakelin, 1988).

IV. Control

The most important methods for the control of schistosome infection are chemotherapy, snail control, and improved water supplies and sanitation.

Many control programs and trials have involved chemotherapy (see Mott, 1987). The anthelmintic of choice is the drug praziquantel, which is highly efficacious in a single dose. The important considerations are the frequency of treatment and the choice of mass, selective, or targeted treatment.

Also, many attempts have been made at snail control using molluscicides, biological controls, or habitat alteration. Molluscicides may be synthetic or natural (derived from plants such as the Ethiopian soap berry, *Phytolacca dodecandra*). The important considerations are the frequency and timing of application and the choice of blanket or focal (confined to selected sites) application. Biological control has been attempted using both predators and competitors; this approach has been most successful in artificial habitats such as irrigation systems (see Madsen, 1990). Habitat alteration includes attempts to design irrigation systems to provide unsuitable environments for snails (Pike, 1987).

Improved water supplies and sanitation programs have been included in several control programs (e.g., Jordan, 1985). The effect of these approaches is modifying behavior, decreasing rates of exposure and contamination of the water with excreta. Modifications of behavior can be reinforced by health education (WHO, 1990b).

The expected effects of different control measures can be explored theoretically (see Woolhouse, 1992). For example, regular chemotherapy can be incorporated into mathematical models as a decrease in schistosome life expectancy, $1/\gamma$; sustained snail control as a decrease in snail density, N; and improved sanitation as a decrease in the rate of contamination, κ. An important theoretical conclusion is that changes in different parameters can have different effects on equilibrium levels of infection. These effects can be illustrated using the model of Eq. 4 (which ignores such complications as density dependence and acquired immunity). For certain parameter values, a reduction in κ is less effective than equivalent reductions in N or $1/\gamma$ (Fig. 10), implying that sanitation may be relatively inefficient compared with chemotherapy or snail control. This result, first proposed by Macdonald (1965), remains controversial but is broadly consistent with field experience. If parameter values are reduced sufficiently so that R_0 falls below 1 the model predicts the eradication of infection. The degree of effort required to achieve eradication is related to the precontrol value of R_0 (Anderson and May, 1991).

In practice, eradication of schistosome infection is extremely difficult, which may imply that, typically, $R_0 \gg 1$. Unfortunately, no good estimates of R_0 for schisto-

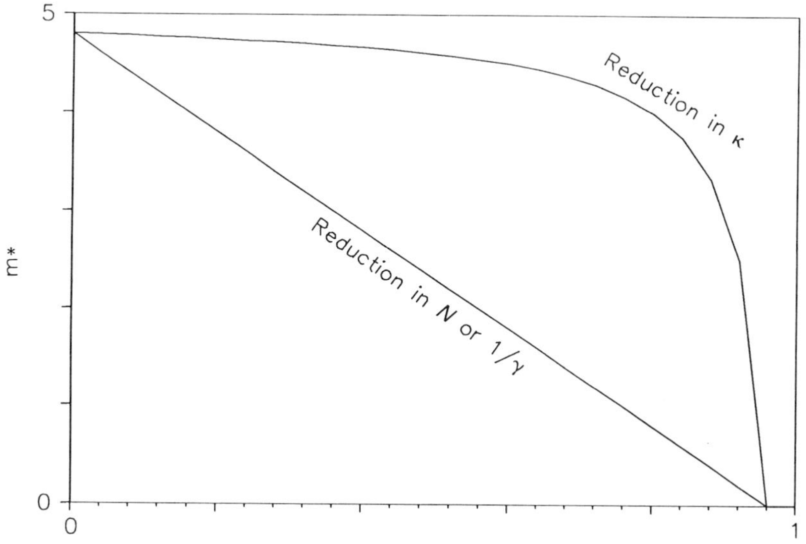

Figure 10 Relationship between equilibrium mean number of schistosome pairs per person (m^*) and schistosome life expectancy ($1/\gamma$), snail population density (N), and the rate of contamination (κ), as predicted by Eq. 4. Values of parameter combinations in the absence of control are ($\alpha\eta N/\gamma = 5$ and ($\beta\kappa H/\mu'') = 5$ with $\sigma \gg \mu', \mu''$. The expected effect on m^* is identical for reductions in $1/\gamma$ resulting, for example, from regular mass chemotherapy, or in N resulting, for example, from snail control. Different effects are expected for reductions in κ resulting, for example, from improved sanitation, although the details of this comparison depend on the parameter values used. Threshold parameter values exist below which $R_0 < 1$, so $m^* = 0$.

some infection are available because of the difficulty in measuring the parameters of Eq. 2, especially α and β, and the uncertainty about density-dependent effects, especially those that result from acquired immunity. Rates of re-infection following mass chemotherapy also tend to be high; pretreatment levels of infection are reached in a few years (see Wilkins, 1989). This result also is consistent with a high R_0, although the interpretation of these data often is complicated by incomplete coverage of chemotherapy and by the immigration of infected individuals.

Because eradication is rarely possible, attention has shifted to sustained reductions in levels of infection or reductions in morbidity due to infection (e.g., see Gryseels and Polderman, 1991). This approach favors concentrating control efforts on heavily exposed and susceptible individuals, particularly children. Unfortunately, field studies rarely monitor the effects of control for periods longer than 5 yr, making assessment of the long-term efficacy of control difficult, particularly since sustained control will affect the development of acquired immunity in the whole population.

In many countries where schistosome infection is endemic, severe logistic and financial constraints exist on the implementation of control measures. Maximizing the cost effectiveness of control programs is essential. Costs can be incorporated into mathematical models. Some attempts have been made to integrate theoretical analyses with data recording the outcome of control trials; such studies have led to useful recommendations for the design of control programs (Rosenfield *et al.*, 1977; Wiemer, 1987). However, geographic variation in aspects of schistosome epidemiology (affecting the parasite and the snail and human hosts) and control (health care priorities and resources) may compromise global solutions to the problem of schistosomiasis.

V. Conclusions and Future Work

Field studies of human schistosome epidemiology show some clear and consistent patterns. The most important of these are (1) the characteristic relationship between infection intensity (as indexed by egg output) and age, with a peak in older children or young adults, and (2) large variation in infection intensities, with a markedly overdispersed distribution of egg outputs (both overall and within age groups). At any one time, a very small fraction of the population is likely to be responsible for the bulk of schistosome transmission. Although much progress has been made in interpreting and explaining these patterns, several important questions remain. Answers to these questions will require long-term, detailed, and integrated field studies of individuals in endemic areas.

Developing meaningful indices of exposure to infection has proven possible, although the gathering and processing of data on human water contact behavior is a difficult and time-consuming task. Age-dependent variation in exposure clearly contributes to age-dependent variation in infection intensity, but other factors, notably acquired immunity, also appear to be important. The contribution of variation in exposure to age-independent variation in infection intensity is less well understood. A major unresolved problem concerns the long-term stability of exposure. Whether individuals tend to maintain high or low rates of exposure (relative to their age group) over periods of a decade or more is not known. The same uncertainty applies to the long-term stability of the spatial distribution of water contacts. Long-term studies of patterns of exposure are needed. Further, the contamination of water bodies with egg-bearing excreta has received far too little attention, despite the obvious relevance to patterns of schistosome transmission.

Other factors that may contribute to age-independent variation in infection intensity are poorly understood. Variation among humans (possibly genetically determined) is likely to exist, both in innate susceptibility and in the ability to mount a protective immune response, but the magnitudes of any such effects are unknown. With respect to acquired immunity, a formidable gap remains between characterizing antischistosomal effects *in vitro* on the one hand and the epidemiological effects of immune responses at the population level on the other hand. Field studies are needed that relate changes in individual infection intensities to exposure and immunological assays, and examine the long-term dynamics of the interrelationships between these variables in populations subject to uninterrupted transmission. A problem in interpreting field data is the relationship between schistosome burden and egg output, which is complicated by the action of density-dependent processes, including any effects of acquired immunity. Indices of the schistosome burden based on antigen production may help clarify this relationship.

The abundance and distribution of infected snails are highly variable and must be monitored closely to understand patterns of infection within the human population. Although good information is available on short-term variability in the abundance of intermediate host snails and of infected snails, data are needed on variability over time scales relevant to human infection (up to 10 yr).

Understanding the complex population dynamics of human schistosomes has benefited from the integration of field work, laboratory experiments, and theoretical analysis. Mathematical models will continue to prove powerful aids to interpreting the results of both laboratory and field studies. Improvements in the understanding of schistosome epidemiology, especially the long-term consequences of control measures, are hoped to lead to further improvements in the design of control programs

References

Anderson, R. M., and May, R. M. (1979). Prevalence of schistosome infections within molluscan populations: Observed patterns and theoretical predictions. *Parasitology* **79,** 63–94.

Anderson R. M., and May, R. M. (1985). Helminth infections of humans: Mathematical models, population dynamics, and control. *Adv. Parasitol.* **24,** 1–101.

Anderson R. M., and May, R. M. (1991). "Infectious Diseases of Humans: Dynamics and Control." Oxford University Press, Oxford.

Barbour, A. (1978). Macdonald's model and the transmission of bilharzia. *Trans. R. Soc. Trop. Med. Hyg.* **72,** 6–15.

Bensted-Smith R., Anderson, R. M., Butterworth, A. E., Dalton, P. R., Kariuki, H. C., Koech, D., Mugambi, M., Ouma, J. H., Arap Siongok, T. K., and Sturrock, R. F. (1987). Evidence for predisposition of individual patients to reinfection with *Schistosome mansoni* after treatment. *Trans. R. Soc. Trop. Med. Hyg.* **81,** 651–654.

Bilharz, T. M. (1852). Fernere Beobachtungen uber das die Pfortader des Menschen bewohnende *Distomum haematobium* und sein verhaltniss zu gewissen pathologischen Bildungen aus brieflichen Mitheilungen an Professor v. Siebold vom 29. Marz 1852. *Zeitschrift für Wissenschaftliche Zoologie,* Liepzig **4,** 72–76.

Butterworth, A. E. (1987). Immunity in human schistosomiasis. *Acta Tropica* **44,** (Suppl. 12) 31–40.

Butterworth, A. E., and Hagan, P. (1987). Immunity in human schistosomiasis. *Parasitol. Today* **3,** 11–16.

Butterworth, A. E., Dalton, P. R., Dunne, D. W., Mugambi, M., Ouma, J. H., Richardson, B. A., Arap Siongok, T. K., and Sturrock, R. F. (1984). Immunity after treatment of human schistosomiasis mansoni. I. Study design, pretreatment observations and the results of treatment. *Trans. R. Soc. Trop. Med. Hyg.* **78,** 108–123.

Butterworth, A. E., Bensted-Smith, R., Capron, A., Capron, M., Dalton, P. R., Dunne, D. W., Grzych, J. M., Kariuki, H. C., Khalife, J., Koech, D., Mugambi, M., Ouma, J. H., Arap Siongok, T. K., and Sturrock, R. F. (1988). Immunity in human schistosomiasis mansoni: Prevention by blocking antibodies of the expression of immunity in young children. *Parasitology* **94,** 281–300.

Butterworth, A. E., Capron, M., Cordingley, J. S., Dalton, P. R., Dunne, D. W., Kariuki, H. C., Kimani, G., Koech, D., Mugambi, M., Ouma, J. H., Prentice, M. A., Richardson, B. A., Arap Siongok, T. K., Sturrock, R. F., and Taylor, D. W. (1985). Immunity after treatment of human schistosomiasis mansoni. II. Identification of resistant individuals, and analysis of their immune responses. *Trans. R. Soc. Trop. Med. Hyg.* **79,** 393–408.

Chabasse, D., Bertrand, G., Leroux, J. P., Gouthery, N., and Hocquet, P. (1985). Bilharziose a *Schistosoma mansoni* évolutive découverte 37 ans après l'infestation. *Bull. Soc. Pathol. Exotique* **78,** 643–647.

Chandiwana, S. K. (1986). How *Schistosoma mansoni* eggs reach natural waterbodies. *Trans. R. Soc. Trop. Med. Hyg.* **80,** 963–964.

Chandiwana, S. K., and Woolhouse, M. E. J. (1991). Heterogeneities in water contact patterns and the epidemiology of *Schistosoma haematobium. Parasitology* **103,** 363–370.

Chandiwana, S. K., Woolhouse, M. E. J., and Bradley, M. (1991). Factors affecting the intensity of reinfection with *Schistosoma haematobium* following treatment with praziquantel. *Parasitology* **102,** 73–83.

Cheever, A. W. (1968). A quantitative post-mortem study of schistosomiasis mansoni in man. *Am. J. Trop. Med. Hyg.* **17,** 38–64.

Cheever, A. W., Kamel, I. A., Elwi, A. M., Mosimann, J. E., and Danner, R. (1977). *Schistosoma mansoni* and *S. haematobium* infections in Egypt. II. Quantitative parasitological findings at necropsy. *Am. J. Trop. Med. Hyg.* **26,** 702–716.

Christie, J. D., and Prentice, M. A. (1978). The relationship between numbers of *Schistosoma mansoni* daughter sporocysts and miracidia. *Ann. Trop. Med. Parasitol.* **72,** 197–198.

Cohen, J. E. (1977). Mathematical models of schistosomiasis. *Ann. Rev. Ecol. Systematics* **8,** 209–233.

Dalton, P. R., and Pole, D. (1978). Water contact patterns in relation to *Schistosoma haematobium* infection. *Bull. WHO* **56,** 417–426.

De Bont, J., Vercruysse, J., Van Aken, D., Southgate, V. R., Rollinson, D., and Moncrieff, C. (1991). The epidemiology of *Schistosoma spindale* Montgomery, 1906 in cattle in Sri Lanka. *Parasitology* **102,** 237–241.

De Jonge, N., Boerman, O.C., and Deelder, A. M. (1989). Time-resolved immunofluorometric assay (TR-IFMA) for the detection of the schistosome circulating anodic antigen. *Trans. R. Soc. Trop. Med. Hyg.* **83,** 659–663.

De Lima e Costa, M. F. F., Rocha, R. S., De Almeida Magalhães, M. H., and Katz, N. (1985). A clinico-epidemiological survey of schistosomiasis mansoni in a hyperendemic area in Minas Gerais State (Comercinho, Brazil). I. Differences in the manifestations of schistosomiasis in the town centre and in the environs. *Trans. R. Soc. Trop. Med. Hyg.* **79,** 539–545.

Dietz, K. (1980). Models for vector-borne parasitic diseases. *Lecture Notes Biomath.* **39,** 264–277.

Doenhoff, M., Musallam, R., Bain, J., and McGregor, A. (1978). Studies on the host-parasite relationship in *Schistosoma mansoni*-infected mice: The immunological dependence of parasite egg excretion. *Immunology* **35,** 771–778.

Etard, J. F., Borel, E., and Segala, C. (1990) *Schistosoma haematobium* infection in Mauritania: Two years of follow-up after a targeted chemotherapy—a life-table approach to the risk of reinfection. *Parasitology* **100,** 399–406.

Farooq, M., and Mallah, M. B. (1966). The behavioural pattern of social and religious water contact activities in the Egypt-49 bilharziasis project area. *Bull. WHO* **35,** 377–387.

Goddard, M. J., and Jordan, P. (1980). On the longevity of *Schistosoma mansoni* in man on St. Lucia, West Indies. *Trans. R. Soc. Trop. Med. Hyg.* **74,** 185–191.

Gryseels, B., and Polderman, A. M. (1991). Morbidity, due to schistosomiasis mansoni, and its control in subsaharan Africa. *Parasitol. Today* **7,** 244–248.

Hagan, P. (1987). The human immune response to schistosome infection. *In* "Biology of Schistosomes from Genes to Latrines" (D. Rollinson and A. J. G. Simpson, eds.), pp. 295–320. Academic Press, London.

Hagan, P., Blumenthal, U. J., Dunn, D., Simpson, A. J. G., and Wilkins, H. A. (1991). Human IgE, IgG4 and resistance to reinfection with *Schistosoma haematobium. Nature (London)* **349,** 243–245.

Jordan, P. (1985). "Schistosomiasis: The St. Lucia Project." Cambridge University Press, Cambridge.

Jordan, P., and Webbe, G. (1982). "Schistosomiasis: Epidemiology, Treatment and Control." Heinemann, London.

Keymer, A. E., and Slater, A. F. G. (1987). Helminth fecundity: density dependence or statistical illusion? *Parasitol. Today* **3,** 56–58.

Klumpp, R. K., and Chu, K. Y. (1977). Ecological studies of *Bulinus rohlfsi,* the intermediate host of *Schistosoma haematobium* in the Volta Lake. *Bull. WHO* **55,** 715–730.

Macdonald, G. (1965). The dynamics of helminth infections, with special reference to schistosomes. *Trans. R. Soc. Trop. Med. Hyg.* **59,** 489–506.

Madsen, H. (1990). Biological methods for the control of freshwater snails. *Parasitol. Today* **6,** 237–241.

May, R. M. (1977). Togetherness among schistosomes: Its effects on the dynamics of the infection. *Math. Biosci.* **35,** 301–343.

McLaren, D. J., and Smithers, S. R. (1987). The immune response to schistosomes in experimental hosts. *In* "Biology of Schistosomes from Genes to Latrines" (D. Rollinson and A. J. G. Simpson, eds.), pp. 233–264. Academic Press, London.

Medley, G. F., and Anderson, R. M. (1985). Density-dependent fecundity in *Schistosoma mansoni* infections in man. *Trans. R. Soc. Trop. Med. Hyg.* **79,** 532–534.

Mott, K. (1987). Schistosomiasis control. *In* "Biology of Schistosomes from Genes to Latrines" (D. Rollinson and A. J. G. Simpson, eds.), pp. 431–450. Academic Press, London.

Nelson, G. S. (1990). Human behaviour and the epidemiology of helminth infections: cultural practices and microepidemiology. *In* "Parasitism and Host Behaviour" (C. J. Barnard and J. M. Behnke, eds.), pp. 234–263. Taylor and Francis, London.

Olveda, R. M., Tiu, E., Fevidal, P., Jr., de Veyra, F., Jr., Icatlo, F. C., Jr., and Domingo, E. O. (1983). Relationship of prevalence and intensity of infection to morbidity in schistosomiasis japonica: A study of three communities in Leyte, Philippines. *Am. J. Trop. Med. Hyg.* **32,** 1312–1321.

Ouma, J. (1988). "The Transmission of *Schistosoma mansoni* in an Endemic Area of Kenya with Special Reference to the Role of Human Defaecation Behaviour and Sanitary Practices." Ph. D. Thesis. University of Liverpool, England.

Pike, E. G. (1987). "Engineering against Schistosomiasis/Bilharzia." Macmillan Press, London.

Pitchford, R. J., Meyling, A. H., Meyling, J., and du Toit, J. F. (1969). Cercarial shedding patterns of various schistosome species under outdoor conditions in the Transvaal. *Ann. Trop. Med. Parasitol.* **63,** 359–371.

Richards, C. S. (1984). Influence of snail age on genetic variations in susceptibility of *Biomphalaria glabrata* for infection with *Schistosoma mansoni. Malacologia* **25,** 493–502.

Rosenfield, P. L., Smith, R. A., and Wolman, M. G. (1977). Development and verification of a schistosomiasis transmission model. *Am. J. Trop. Med. Hyg.* **26,** 505–516.

Shiff, C. J., Evans, A., Yiannakis, C., and Eardley, M. (1975). Seasonal influence on the production of *Schistosoma haematobium* and *S. mansoni* cercariae in Rhodesia. *Int. J. Parasitol.* **5,** 119–123.

Sturrock, B. M. (1967). The effect of infection with *Schistosoma haematobium* on the growth and reproduction rates of *Bulinus (Physopsis) nasutus productus. Ann. Trop. Med. Parasitol.* **61,** 321–325.

Sturrock, R. F. (1973). Field studies on the transmission of *Schistosoma mansoni* and on the bionomics of its intermediate host, *Biomphalaria glabrata,* on St. Lucia, West Indies. *Int. J. Parasitol.* **3,** 175–194.

Tingley, G. A., Butterworth, A. E., Anderson, R. M., Kariuki, H. C., Koech, D., Mugambi, M., Ouma, J. H., Arap Siongok, T. K., and Sturrock, R. F. (1988). Predisposition of humans to infection with *Schistosoma mansoni:* Evidence from the reinfection of individuals following chemotherapy. *Trans. R. Soc. Trop. Med. Hyg.* **82,** 448–452.

Touassem, R., and Theron, A. (1989). *Schistosoma rodhaini:* Dynamics and cercarial production for mono- and plurimiracidial infections of *Biomphalaria glabrata. J. Helminthol.* **63,** 79–83.

Udonsi, J. K. (1990). Human community ecology of urinary schistosomiasis in relation to snail vector bionomics in the Igwun river basin of Nigeria. *Trop. Med. Parasitol.* **41,** 131–135.

Vermund, S. H., Bradley, D. J., and Ruiz-Tiben, E. (1983). Survival of *Schistosoma mansoni* in the human host: Estimates from a community based prospective study in Puerto Rico. *Am. J. Trop. Med. Hyg.* **32,** 1040–1048.

von Lichtenberg, F. (1987). Consequences of infections with schistosomes. *In* "Biology of Schistosomes from Genes to Latrines" (D. Rollinson and A. J. G. Simpson, eds.), pp. 185–232. Academic Press, London.

Wakelin, D. (1988). Helminth infections. *In* "Genetics of Resistance to Bacterial and Parasite Infections" (D. Wakelin and J. M. Blackwell, eds.), pp. 153–225. Taylor and Francis, London.

Warren, K. S. (1973). Regulation of the prevalence and intensity of schistosomiasis in man: immunity or ecology? *J. Infect. Dis.* **127,** 595–609.

Wiemer, C. (1987). Optimal disease control through combined use of preventive and curative measures. *J. Dev. Econ.* **25,** 301–319.

Wilkins, H. A. (1989). Reinfection after treatment of schistosome infections. *Parasitol. Today* **5,** 83–88.

Wilkins, H. A., Goll, P. H., Marshall, T. F. de C., and Moore, P. J. (1984). Dynamics of *Schistosoma haematobium* infection in a Gambian community, III. Acquisition and loss of infection. *Trans. R. Soc. Trop. Med. Hyg.* **78,** 227–232.

Wilkins, H. A., Blumenthal, U. J., Hagan, P., Hayes, R. J., and Tulloch, S. (1987). Resistance to reinfection after treatment of urinary schistosomiasis. *Trans. R. Soc. Trop. Med. Hyg.* **81,** 29–35.

Woolhouse, M. E. J. (1989). On the interpretation of age-prevalence curves for schistosome infections of host snails. *Parasitology* **99,** 47–56.

Woolhouse, M. E. J. (1991). On the application of mathematical models of schistosome transmission dynamics. I. Natural transmission. *Acta Tropica* **49,** 241–270.

Woolhouse, M. E. J. (1992). On the application of mathematical models of schistosome transmission dynamics. II. Control. *Acta Tropica* **50,** 189–204.

Woolhouse, M. E. J., and Chandiwana, S. K. (1990a). Epidemiology of schistosome infections of snails: Taking the theory into the field. *Parasitol. Today* **6,** 65–70.

Woolhouse, M. E. J., and Chandiwana, S. K. (1990b). Temporal patterns in the epidemiology of schistosome infections of snails: A model for field data. *Parasitology* **100,** 247–253.

Woolhouse, M. E. J., and Chandiwana, S. K. (1992). A further model for temporal patterns in the epidemiology of schistosome infections of snails. *Parasitology* **104,** 443–449.

Woolhouse, M. E. J., Chandiwana, S. K., and Bradley, M. (1990). On the distribution of schistosome infections among host snails. *Int. J. Parasitol.* **20,** 325–327.

Woolhouse, M. E. J., Taylor, P., Matanhire, D., and Chandiwana, S. K. (1991a) Acquired immunity and epidemiology of *Schistosoma haematobium. Nature (London)* **351,** 757–759.

Woolhouse, M. E. J., Watts, C. H., and Chandiwana, S. K. (1991b). Heterogeneities in transmission rates and the epidemiology of schistosome infection. *Proc. R. Soc. London B* **245,** 109–114.

World Health Organization (1990a). "World Report on Tropical Diseases." World Health Organization, Geneva.

World Health Organization (1990b). "Health Education in the Control of Schistosomiasis." World Health Organization, Geneva.

Bryan T. Grenfell & Ben M. Bolker

Population Dynamics of Measles

16

I. Introduction

Measles is among the best documented of human diseases in terms of epidemiology and population dynamics. This status arises from a combination of three factors. First, the disease represents a major source of mortality and morbidity in human populations, both in the historical record and currently (Anderson and May, 1991). The Arabic physician Rhazes rated measles as at least as important a disease as smallpox (Black, 1984). Measles is still a major killer in developing countries (killing over 1.5 million children a year), despite the existence of an effective vaccine (McLean and Anderson, 1988a,b; see also Chapter 11). The second special feature of measles (which it shares with many other childhood respiratory viral infections) is the relatively simple natural history of transmission of the etiological agent, measles virus (Fine and Clarkson, 1982a; Black, 1984; Nokes and Anderson, 1988a). Essentially, transmission from infectious to susceptible individuals is direct, via respiratory aerosol. Clinical manifestations are relatively distinct and easily diagnosed; infectees undergo short and well-defined incubation and infectious periods. Immunity to re-infection is subsequently effectively lifelong in immunologically competent individuals who survive the primary infection. The third epidemiological factor, which arises from this simple course of infection, is that infection characteristically occurs in self-extinguishing epidemics, since a wave of infection "removes" individuals from the susceptible to the immune class (Hamer, 1906; Soper, 1929; Bartlett, 1957,1960; Black, 1966; Schenzle, 1984; Anderson and May, 1991). Measles epidemics tend to be episodic in small communities (Cliff and Haggett, 1988), but can show remarkably regular oscillations (Brownlee, 1919; Anderson *et al.*, 1984; Cliff and Haggett, 1988), both seasonally and on longer major epidemic time scales, in urban communities that have a sufficiently large rate of recruitment of susceptible children to make the infection endemic (Fig. 1).

In terms of quantitative research on measles epidemiology, the public health importance and dynamic epidemic behavior of measles has spawned two distinct bodies of work: epidemiological modeling and the study of nonlinear dynamics and chaos. We shall consider each one in turn.

A. Epidemiological Modeling

There is a long and distinguished tradition of attempts to understand the dynamics of measles infection, its persistence in commu-

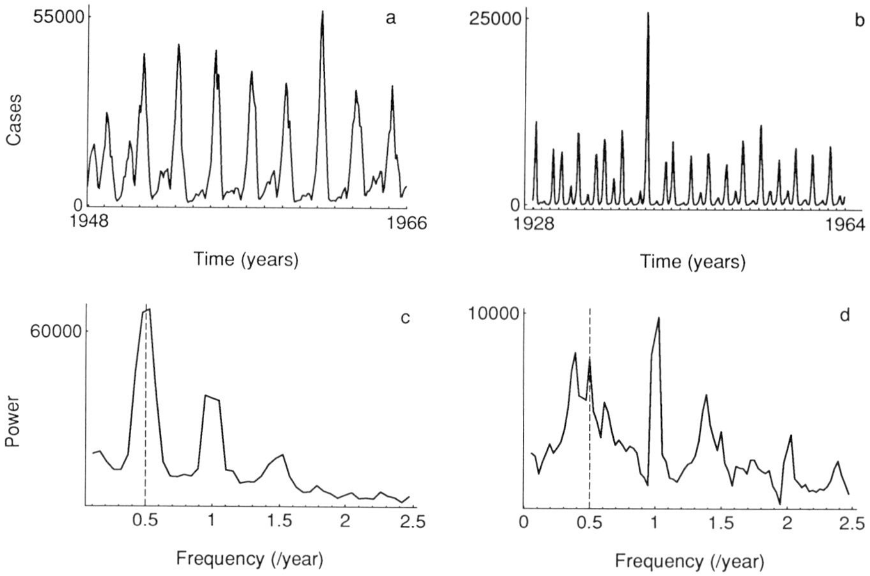

Figure 1 Time series (cases notified) and power spectra for England, Wales, and New York. (a) Weekly case reports for England and Wales, 1948–1968. (b) Monthly case reports for New York City, 1928–1964. (c) Power spectrum for England and Wales data; spectra smoothed with a 3-point running mean (Chatfield, 1975; Priestley, 1982). (d) Power spectrum for New York City, data and method as for c. Spectral analysis of recurring epidemics generates power spectra with peaks that recur at intervals equal to the inverse of the interepidemic period.

nities, and the impact of heterogeneities in transmission with respect to age, season, and spatial organization of the host population (Hamer, 1906; Soper, 1929; Bartlett, 1957,1960; London and Yorke, 1973; Anderson and May, 1985,1991; Dietz and Schenzle, 1985; May, 1986). The advent of an effective vaccine, and its widespread use over the last three decades (Black, 1984), has also generated extensive development of this theory, which addresses the important issue of estimating the critical level of vaccination required to eradicate the infection from a community (Dietz, 1976; Anderson and May, 1982). In parallel with these theoretical developments, the relatively long time series of measles case notifications, which are available for a number of countries, have been analyzed statistically by a number of workers (Brownlee, 1919; Fine and Clarkson, 1982b; Cliff and Haggett, 1988) to assess the relative importance of annual and longer term periodicities in the epidemic pattern (Fig. 1).

B. Nonlinear Dynamics and Chaos

The availability of plausible models and relatively long time series of incidence have also made measles epidemics (which are essentially a manifestation of predator–prey dynamics) of considerable interest to population ecologists (May, 1986). A recent and very stimulating example is the attempt by several groups—notably Schaffer and coworkers (Schaffer, 1985b; Schaffer and Kot, 1985; Kot *et al.*, 1988; Olsen *et al.*, 1988; Olsen and Schaffer, 1990)—to identify fluctuations in measles incidence as manifestations of, for chaotic dynamics. "Chaos" is the general name given to irregular patterns of behavior appearing in systems governed by simple deterministic rules. Chaos

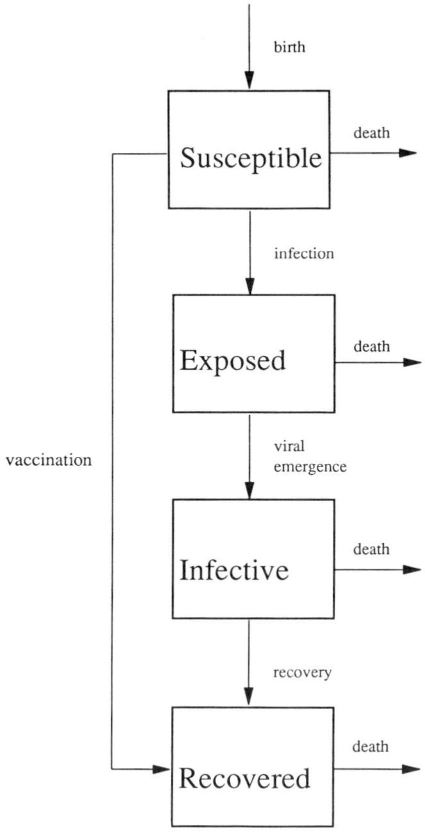

Figure 2 Representation of the compartments and epidemiological transitions in the SEIR (Susceptible/Exposed/Infective/Recovered) model; see text for details.

occurs in models of host–parasite systems as a result of dynamic instability caused by extreme "boom-and-bust" cycles in host and parasite abundance (Schaffer, 1985b; May, 1986). Measles is one of the best candidates for chaotic dynamics in a biological system outside the laboratory. An extensive body of work has identified putative chaotic attractors for measles, based on both observed incidence data (Olsen *et al.,* 1988; Sugihara and May, 1990) and the simplest homogeneous epidemiological model, the SEIR (susceptible/exposed/infectious/recovered) model (Schaffer and Kot, 1985; Rand and Wilson, 1991). However, difficulties do arise with this approach. In addition to the problem that the available measles time series may be simply too short for the statistical methods used to detect chaos (Nychka *et al.,* 1992), the modeling approaches to this problem have tended to ignore more realistic formulations of measles dynamics developed for epidemiological purposes. Currently, this problem is being addressed by analyzing the nonlinear dynamic properties of more realistic epidemiological models (Bolker and Grenfell, 1993).

The following sections review the current status of measles modeling. This subject is vast, so the treatment of most areas necessarily is limited. We begin by discussing the structure and properties of the simple SEIR model, and the effects of biological refinements to improve its epidemiological realism. We then review current work on the nonlinear dynamics of measles, and explore the implications of fusing the epidemiological and nonlinear approaches to measles dynamics. The chapter concludes with a discussion of directions for future work.

II. Basic Model Framework

A. Infection Dynamics

Measles provides one of the most straightforward examples of a directly transmitted microparasitic infection (Anderson and May, 1991; see Chapter 3). The standard pattern of infection for developed countries (Nokes and Anderson, 1988b), which applies effectively to all immunologically competent individuals in the population, is summarized in Fig. 2. After a short period of maternally acquired passive immunity, young children become susceptible to infection. Infective contact with infectious individuals (involving a transfer of virus, predominantly via aerosol) moves individuals into the asymptomatic exposed, or latent, group for a period of 6–9 days. Then individuals progress to become infective for 6–7 days, showing the characteristic rash of measles for the last few days of this period. Finally, immunocompetent individuals who survive infection (and measles is assumed not to be fatal in the simplest models) are effectively immune to subsequent re-infection, and therefore are recovered and effectively removed from the transmission cycle. The main instrument for disease control is the vaccination of susceptible children with a live attentuated measles strain, developed originally in the 1950s (Black, 1984). In terms of the model, vaccination simply moves individuals directly from the susceptible to the recovered class.

B. Compartmental Structure

The essence of this progression is captured at the population level by the well-known SEIR model, which has a long history of study in mathematical epidemiology (Hamer, 1906; Soper, 1929; Kermack and McKendrick, 1927,1932,1933; Dietz and Schenzle, 1985; May, 1986; Anderson and May, 1991). This model is expressed as a set of three nonlinear ordinary differential equations:

$$\frac{dS}{dt} = \mu N(1-p) - (\mu + \beta I)S$$

$$\frac{dE}{dt} = \beta IS - (\mu + \sigma)E \qquad (1)$$

$$\frac{dI}{dt} = \sigma E - (\mu + \gamma)I,$$

and makes the following epidemiological assumptions. S, E, I, and R respectively represent the density of susceptible, exposed, infectious, and recovered individuals in a constant total population of size $N = S + E + I + R$ (given this identity, only the three equations for S, E, and I are necessary to specify the system). Average *per capita* mortality rate due to all causes is μ yr^{-1}, indicating an average life expectancy of $1/\mu$ yr; since infection is assumed not to cause extra mortality, the net birth (μN) and death [$\mu(S + E + I + R)$] rates are equal. In this simplest version of the model (which ignores the effect of passive maternally derived immunity in infants), individuals are assumed to be born susceptible, with vaccination effectively at birth moving a proportion p into the recovered class at birth. The parameters σ and γ (yr^{-1}) are rate parameters controlling the movement of individuals out of the exposed and infectious classes, respectively, such that $1/\sigma$ and $1/\gamma$ are the average incubation and infectious periods. The net infection rate, βIS, is controlled by the infection parameter β (yr^{-1} infective^{-1}) which, in this simplest version of the model, is assumed to be constant. The corresponding infection rate per susceptible (the "force of infection"; $\lambda = \beta I$ yr^{-1}) is, therefore, independent of age, time, spatial structure, or other heterogeneities. The effects of relaxing this homogeneity assumption have provided a major focus for theoretical studies of measles epidemiology in recent years.

The static and dynamic properties of the SEIR model and its various modifications are summarized lucidly by May (1986). Here, we briefly review the properties of the SEIR model, before discussing how epidemiologists have introduced various refinements to improve its explanatory power.

C. Equilibrium Properties of the Model

The equilibrium of the SEIR model (obtained from Eq. 1 when $dS/dt = dE/dt = dI/dt = dR/dt = 0$ at $S = S^*$, $E = E^*$, $I = I^*$, $R = R^*$, E^* and $I^* > 0$) can be used to lay bare many of its essential properties. The crucial quantity is the basic reproduction ratio of infection, R_0—the number of secondary cases caused by the introduction of an infectious individual into a wholly susceptible population. In terms of Eq. 1, at equilibrium

$$R_0 = \frac{N}{S^*}, \tag{2}$$

where

$$S^* = \frac{(\mu + \sigma)(\mu + \gamma)}{\sigma\beta} \tag{3}$$

is the equilibrium density of susceptibles and N is total population size. R_0 must be greater than 1 for the infection to persist in a community. In terms of patterns of measles incidence before the onset of vaccination, the criterion $R_0 > 1$, along with Eqs. 2 and 3, defines a lower population threshold for persistence of infection:

$$N_T = S^*. \tag{4}$$

An analogous dynamic threshold commonly is observed in practice and shall be shown to be crucial to the population dynamics of measles.

Vaccination of a proportion p at birth reduces the effective value of R_0 to

$$R_0' = (1 - p)R_0, \tag{5}$$

so the threshold $R_0' = 1$ leads to the expression

$$p_c = 1 - 1/R_0 = 1 - S^*/N \tag{6}$$

for p_c, the critical vaccination proportion for eradicating the infection. For measles, R_0 is in the range of 12–18, suggesting that p_c is between 92% and 95% (Anderson and May, 1991). The effect on these very approximate calculations of allowing for more biological realism (particularly age-specific heterogeneities in transmission) is discussed subsequently.

D. Dynamic Properties of the Model

For $R_0 > 1$, the SEIR model has a stable positive disease equilibrium that is approached by damped oscillations with approximate period $2\pi[A(1/\sigma + 1/\gamma)]^{1/2}$, where A is the mean age at infection. Typical epidemiological parameters for developed countries before vaccination (when A was 4–5 yrs; May, 1986) generate an "expected" epidemic period of roughly 2 yr, in accord with much of the observed data (Brownlee, 1919; Anderson *et al.,* 1984), although annual and 3-yr periods are seen also (Olsen *et al.,* 1988). The model also indicates an increase in this interval after the onset of vaccination, which again accords roughly with field observations (Anderson *et al.,* 1984). Although these predicted pre- and postvaccination cycle periods are roughly correct for many measles data sets, observed measles time series for large communities show sustained rather than damped oscillations (Fig. 1). Much of measles epidemic theory has been devoted to explaining this recurrent epidemic behavior through refinements of the basic model.

This body of work has aimed at introducing extra biological features into measles models to generate sustained oscillations intermediate between the stability of the continuous-time SEIR model and the unstable neutral cycles of its discrete-time analogs. Since measles is often observed to "fade out" in the troughs between epidemics, the first modification is to allow for the effects of demographic stochasticity (i.e., tracking the fate of individual hosts), replacing the deterministic SEIR model with a discrete stochastic formulation. The seminal work here is by Bartlett (1957,1960,1961,1990), who used a combination of stochastic modeling and data analysis to estimate a lower population threshold for measles persistence of 250 to 500 thousand individuals. This value has also been derived by other workers (Black, 1966; Cliff and Haggett, 1988). However, stochastic effects alone cannot explain the observed pattern of recurrent nonseasonal measles epidemics commonly seen in the cities of developed countries before the vaccination era. The origin of recurrent epidemics has prompted a variety of other modifications to the simple model, most of which are concerned with the inclusion of heterogeneities in transmission.

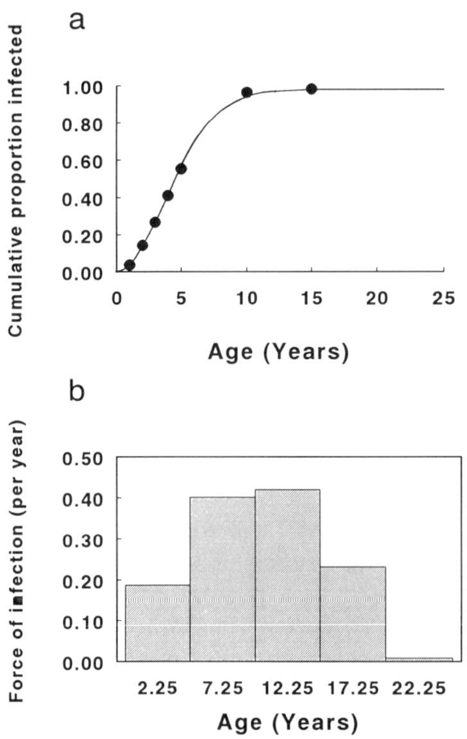

Figure 3 Typical age distribution of measles infections in developed countries before the onset of vaccination. (a) Observed cumulative proportion infected by age in England and Wales. Data from Registrar General's Weekly Reports. (b) Estimated force of infection pattern by age, showing a characteristic peak in school age children. Details of the catalytic model used are given by Grenfell and Anderson (1985). The curve in a shows the fit of this model to the raw data.

III. Modeling Heterogeneities in Measles Transmission

A. Equilibrium Models

1. Age Structure

Since the SEIR model is nonlinear, the simplest way to explore how heterogeneities affect its behavior is by considering the equilibrium properties of the associated models.

Fig. 3a shows a typical cumulative age distribution of infection for measles in developed countries before vaccination (Fine and Clarkson, 1982a; Grenfell and Anderson, 1985). This sigmoid pattern generates estimates of the age-specific force of infection that are emphatically not constant, but rise to a peak in school-age children and subsequently decline in adults (Fig. 3b). The implications of this peak in the force of infection, which reflects higher contact rates between children in school, has been examined by a number of workers (Fine and Clarkson, 1982b; Anderson and May, 1985; Dietz and Schenzle, 1985; Tudor, 1985). The basic theoretical approach is replacing the ordinary differential equations (ODEs) of the SEIR model with partial differential equations (or sets of ODEs), describing the progression of infection through age and time. The equilibrium approach then considers the temporal steady state of the system, leading to a set of ODEs that describes the age distribution of susceptibles, infections, and so on. The general conclusion of this approach is that a peaked force of infection pattern generates a lower estimate of the critical level of vaccination (p_c) than the average infection rate implicitly assumed in the SEIR model (Schenzle, 1984; Anderson and May, 1985), essentially because mass vaccination near birth increases the mean age at infection of unvaccinated individuals, so they are exposed in an older age group with a lower force of infection. This effect is particularly important in other childhood infections, such as rubella and mumps, for which the pattern of morbidity associated with infection peaks in the adult age groups (Anderson *et al.*, 1987; Nokes and Anderson, 1988a).

2. Spatial and Other Heterogeneities

The equilibrium approach has also yielded valuable insights into the impact of spatial heterogeneities and intrinsic (perhaps genetic) differences in infection rate (Bailey, 1975; Anderson and May, 1984; May and Anderson, 1984; Dietz and Schenzle, 1985; May, 1986). The simplest method adds compartments to the basic SEIR model to allow for spatial variations and other differences between groups of individuals.

Probably the most important result to emerge from spatial models of this type is the importance of concentrating limited vaccination efforts on centers of population (May and Anderson, 1984). The dynamic impact of heterogeneities in measles transmission is discussed subsequently.

B. Seasonality and Age Structure in Transmission Dynamics

As shown in the time series analysis of Fig. 1, the most obvious manifestation of heterogeneity in measles transmission dynamics is an annual variation in incidence, generally associated with changes in the contact rate of children, caused by the seasonal pattern of school terms (London and Yorke, 1973; Fine and Clarkson, 1982b). Several workers have modeled the effects of seasonal variations in infection rate (London and Yorke, 1973; Dietz, 1976; Grossman, 1980; Aron and Schwartz, 1984; Schenzle, 1984; Schaffer and Kot, 1985; Aron, 1990). The simplest approach involves replacing the constant infection parameter β of Eq. 1 with a time-varying periodic function, such as

$$\beta(t) = b_0[1 + b_1\cos(2\pi t)], \qquad (7)$$

where b_0 is the average infection rate and b_1 controls the amplitude of variation around it. This formulation has been examined both in the epidemiological literature (London and Yorke, 1973; Dietz, 1976) and in the exploration of nonlinear dynamics in measles incidence (Aron and Schwartz, 1984; Schaffer, 1985a; Schaffer and Kot, 1985; Olsen and Schaffer, 1990; Rand and Wilson, 1991). In terms of the maintenance of recurrent epidemics, seasonal variations in infection rate can interact with the damped epidemics of the SEIR model to produce sustained cycles. An example of this behavior (which can be modeled in terms of resonance for small degrees of seasonal forcing, i.e., for small b_1; Dietz, 1976) is displayed in Fig. 4a. As discussed subsequently, the situation is more complex for large degrees of forcing.

However, an important biological complexity must be addressed first. Specifically, these seasonal variations in transmission are also intimately bound up to the heterogeneities in transmission with age that are illustrated in Fig. 3. In an important paper, Fine and Clarkson (1982a) used measles data for England and Wales to document the sea-

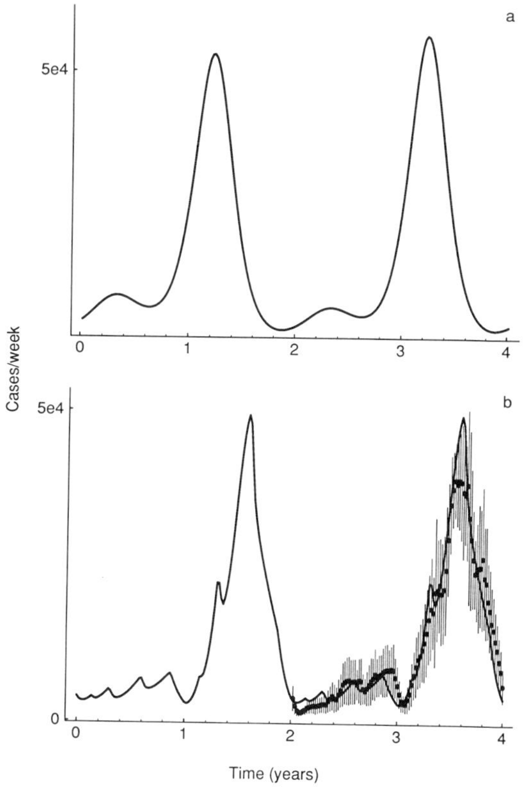

Figure 4 Time series (number of infectives) for two measles models. (a) Sinusoidally forced SEIR model (biennial regime: $c_0 = 1.5 \times 10^{-5}$, $c_1 = 8.5 \times 10^{-6}$). The model is formulated as $\beta = c_0 + c_1[1 + \cos(2\pi t)]$ rather than the more standard form to match the RAS model more closely. (b) RAS model. Contact rates adjusted for best fit (least squares) to England and Wales data. Other parameters are as given by Schenzle (1984). Points represent the mean value of corrected (Fine and Clarkson, 1982a) weekly case reports for England and Wales for a given point in the biennial cycle over the years 1950–1964. Error bars show one standard deviation above and below the mean.

sonal increase in infection rate, particularly of primary school children, associated with the pattern of school terms.

Although introducing age structure alone into the SEIR model does not affect its behavior qualitatively, the combination of age structure and seasonality in transmission has a profound impact. In particular, Schenzle (1984) implemented the qualitative model of Fine and Clarkson in a continuous-time age–cohort formulation that mimics the average prevaccination pattern of measles epidemics in England and Wales very closely (Fig. 4b), as well as the qualitative behavior of the infection after the onset of vaccination. This model (which we shall call the RAS, Realistic Age-Structured, model) is the most sophisticated development to date of the epidemiological approach to measles dynamics. In the next section, the RAS model is used as a point of reference for the nonlinear dynamic approach.

IV. Measles, Chaos, and Biological Realism

Considerable interest has arisen in the last decade in the possibility that measles dynamics may be an ecological candidate for low-dimensional deterministic chaos (Schaffer, 1985; Schaffer and Kot, 1985; Kot *et al.*, 1988; Olsen *et al.*, 1988; Olsen and Schaffer, 1990; Sugihara and May, 1990; Rand and Wilson, 1991). The nonlinear analysis of measles dynamics has been based on both observed incidence time series (Olsen *et al.*, 1988; Sugihara and May, 1990) and data generated from theoretical models (Schaffer, 1985a; Rand and Wilson, 1991; Grenfell, 1992b). Although the main issue in the former case is whether observed time series are sufficiently long to detect the signature of chaos (Nychka *et al.*, 1992), the main theoretical issue is whether the forced SEIR model is sufficiently biologically realistic to represent measles dynamics adequately. This latter issue is addressed in the following sections.

A. Dynamics of the Forced SEIR Model

The dynamical systems literature has spawned a number of approaches for detecting chaos in time series generated from theoretical models. These methods [which are well reviewed by May *et al.* (1992) and Nychka *et al.* (1992)] generally depend on estimating either the dimension of the attractor to which the system's dynamics settle (e.g., zero for a simple equilibrium, 1 for a two-population limit cycle, and fractional for a chaotic attractor), or the tendency for nearby trajectories in phase space (corresponding, for example, to successive minima in the epidemic cycle) to diverge with time in chaotic systems. The characteristic measure of this sensitivity to initial conditions of chaotic systems is the maximum Lyapunov exponent, which is positive if, on average, near trajectories tend to diverge with time and negative if they converge. An important general caveat is that Lyapunov exponents (in addition to all other proposed signatures for chaos) must be interpreted cautiously in the presence of random variability embedded in the nonlinear process.

We illustrate the non-linear properties of the seasonally forced SIR model (a close analog of the SEIR model, but lacking an exposed class) with a series of numerical simulations for a range of values for the amplitude of sinusoidal forcing, controlled by the parameter b_1 in Eq. 7. The results, shown in Fig. 5 as time and I vs S phase space plots, reflect a progression from a simple annual cycle ($b_1 = 0.01$) through a sequence of period doubling as b_1 increases, to irregular chaotic fluctuations at $b_1 = 0.28$, the value proposed by Schaffer and collaborators (Olsen *et al.*, 1988). Although other workers suggest that this degree of seasonal forcing may be unrealistically high (Pool, 1989; Grenfell, 1992b), work on the impact of demographic and environmental noise

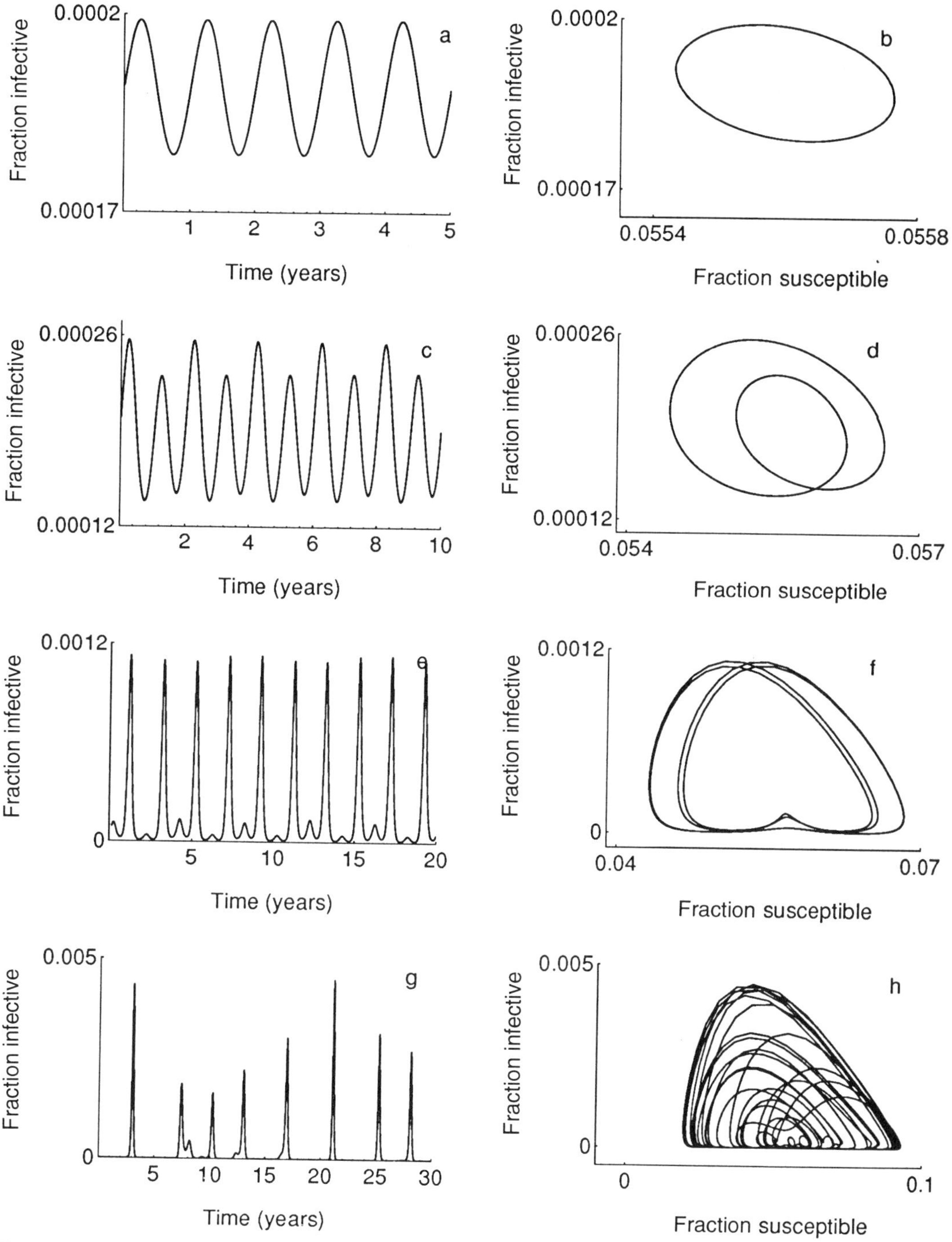

Figure 5 Dynamics of the seasonal SIR model. Time (I vs t) (a,c,e,g) and phase (S vs I) plots (b,d,f,g) for different seasonal amplitudes b_1. Vertical axis shows proportion of population infective; horizontal axis shows time in years. (a,b) Annual cycle ($b_1 = 0.01$); (c,d) biennial cycle ($b_1 = 0.05$); (e,f) 8-yr cycle ($b_1 = 0.267$); (g,h) large amplitude chaos ($b_1 = 0.28$).

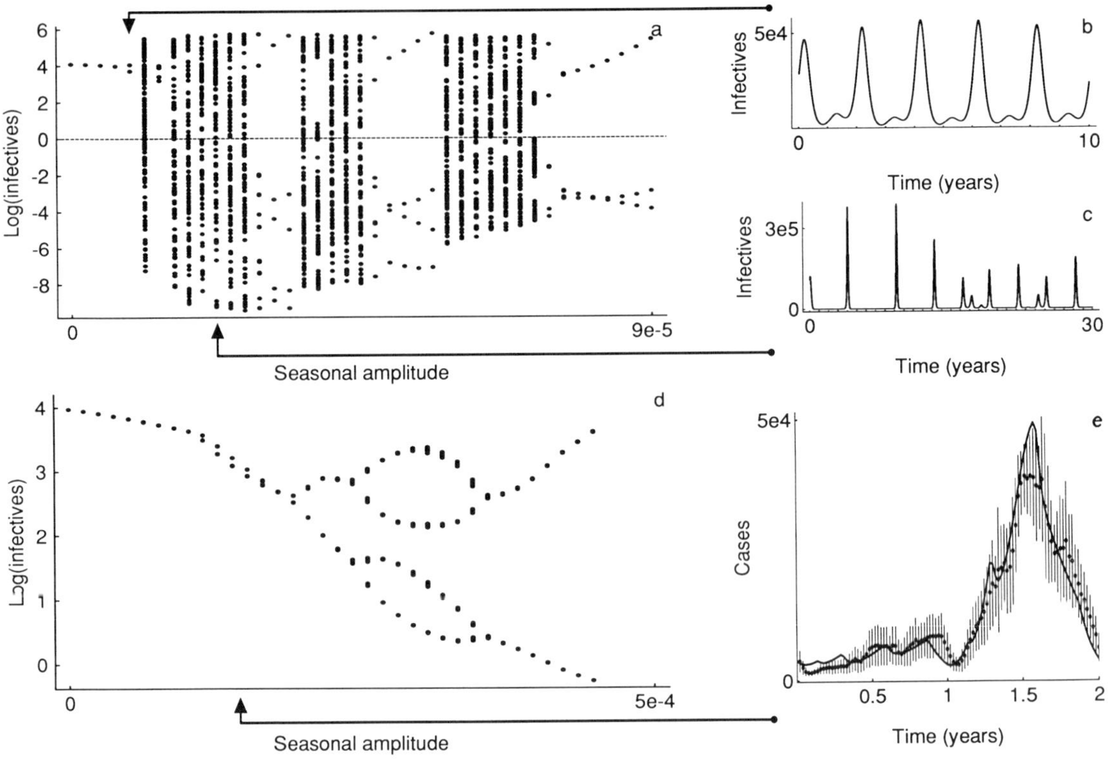

Figure 6 Comparative dynamics of the forced SEIR and RAS models. Bifurcation diagrams show $\log_{10}$ (infectives) in a population of 50 million, sampled annually at the beginning of the epidemiological year (September, near the minimum number of infectives) for 100 yr after a 200-yr transient, for given values of seasonal forcing amplitude. (a) SEIR bifurcation diagram; $\log_{10}$ (infectives) vs c_1. Model as in Fig. 4a. Parameters: $N = 5 \times 10^7$, $\mu = 0.02$, $\gamma = 73.0$, $\sigma = 45.6$, $c_0 = 1.5 \times 10^{-5}$, $c_1 = 0$ to 9.0×10^{-5}. All units yr^{-1} except contact rate (yr^{-1} infective^{-1}). (b) SEIR time series for the biennial regime; $c_1 = 8.5 \times 10^{-6}$. (c) SEIR time series for the chaotic regime; $c_1 = 2.0 \times 10^{-5}$. (d) RAS bifurcation diagram. Parameters as given by Schenzle (1984). Contact rates as in Fig. 4b. Seasonal amplitude is b_2, the additional contact rate among primary school children (effective only during school terms). (e) RAS time series; see Fig. 4b. Arrows indicate the values of the seasonal amplitude parameter in a and d used for simulating the time series in b, c, and e. [See Bolker and Grenfell (1993) for more details.]

indicates that chaos-like behavior can be produced in noisy systems for lower degrees of forcing via the operation of long transients known as "chaotic repellers" (Rand and Wilson, 1991). For the deterministic SEIR model with $b_1 = 0.28$, chaotic dynamics are confirmed by a positive maximum Lyapunov exponent of ≈ 0.45 (Schaffer, 1985a).

Fig. 6a shows a broader view of the dynamics of the model in a standard bifurcation diagram. Bifurcation diagrams are constructed by sampling a system periodically —in this case, recording the number of infectives in the population at the beginning of each year—for a range of values of a particular parameter—in this case, the amplitude of seasonal forcing (b_1). This procedure gives a synopsis of the dynamics for each value of b_1, displaying the eventual dynamic state of the system in terms of the annual pattern of infective abundances. A single point on the infective axis for a given value of b_1 corresponds to a stable equilibrium or an annual cycle at that parameter value; two points correspond to a biennial cycle, and so forth. A large number of points means that the system is continuing to display irregular behavior after long transients, a hallmark of chaos. This picture clearly illustrates the transition to chaos with increasing seasonal forcing of the infection rate. However, this irregular deterministic behavior also is characterized by large amplitude fluctuations with very low minima (to $10^{-14} \times$ population size in Fig. 6a). This unrealistically low figure is the basis of a number of criticisms of the SEIR/chaos result (Pool, 1989; Grenfell, 1992b). In particular, the associated degree of "fade-out" of infection appears to be significantly larger than that observed, even in model cities of one million people [i.e., over twice the population threshold empirically derived by Bartlett (1960), and Black (1966)]. Although spatial heterogeneity appears to reduce the degree of fade-out slightly, fade-out remains a significant problem in terms of modeling the nonlinear dynamics of measles with the SEIR model (Grenfell, 1992b). This situation is explored, in terms of the dynamics of the more realistic RAS model, in the next section.

B. Dynamics of the Realistic Age-Structured Model

Fig. 6d presents a bifurcation diagram for the RAS model, corresponding to the SEIR bifurcation diagram of Fig. 6a. The picture that emerges here is dramatically different. Although the RAS model does undergo some period doubling as seasonal forcing increases, no evidence exists of irregular chaotic fluctuations in the density of infectives. The important biological point apparent in Fig. 6d is that the (interepidemic) infective minima for the RAS model are much higher than in the SEIR model (and therefore more biologically realistic). Bolker and Grenfell (1993) showed that this feature, in turn, leads to a lower propensity for chaos, which is associated in this class of models with divergence of model trajectories at low infective densities.

This qualitative difference between SEIR and RAS model dynamics arises from the dditional population heterogeneity introduced in the more realistic RAS model. Specifically, the relatively small force of infection in the preschool age group (Fig. 3b) allows this segment of the population to act as a reservoir of infection, preventing the violent swings in infective density that are generated in the highly forced SEIR model (Bolker and Grenfell, 1993).

Overall, this result clearly illustrates a general implication for ecological dynamics: introducing extra biological complexity into models sometimes can simplify their behavior, particularly if the introduction of heterogeneity reduces the propensity for extreme dynamics (Grenfell, 1992a). Adding more realism in terms of demographic noise leaves the model essentially intact although, as discussed by Bolker and Grenfell (1993), additional biologically interesting dynamic complexities emerge as human population size decreases. Specifically, whereas stochastic models for large populations (for example, 50 million) exhibit regular biennial epidemics, these models can break down into more complex behaviors with mixtures of 1-, 2-, and 3-yr cycles in populations of 1 million individuals. This tendency raises the issue of spatial dynamics which, as discussed subsequently, appears to be the major direction for future work in this area.

V. Conclusion and Directions for Future Work

This chapter provides a brief summary of theoretical approaches to the epidemiology and population dynamics of measles. Given the huge breadth of this subject, omissions have been inevitable. For example, the very different epidemiological patterns for measles in developing countries (the most important area in an applied sense) has been modeled successfully (McLean and Anderson, 1988a,b), although any implications of seasonal patterns of infection in these systems have not yet been explored. This chapter has concentrated on the situation in de-

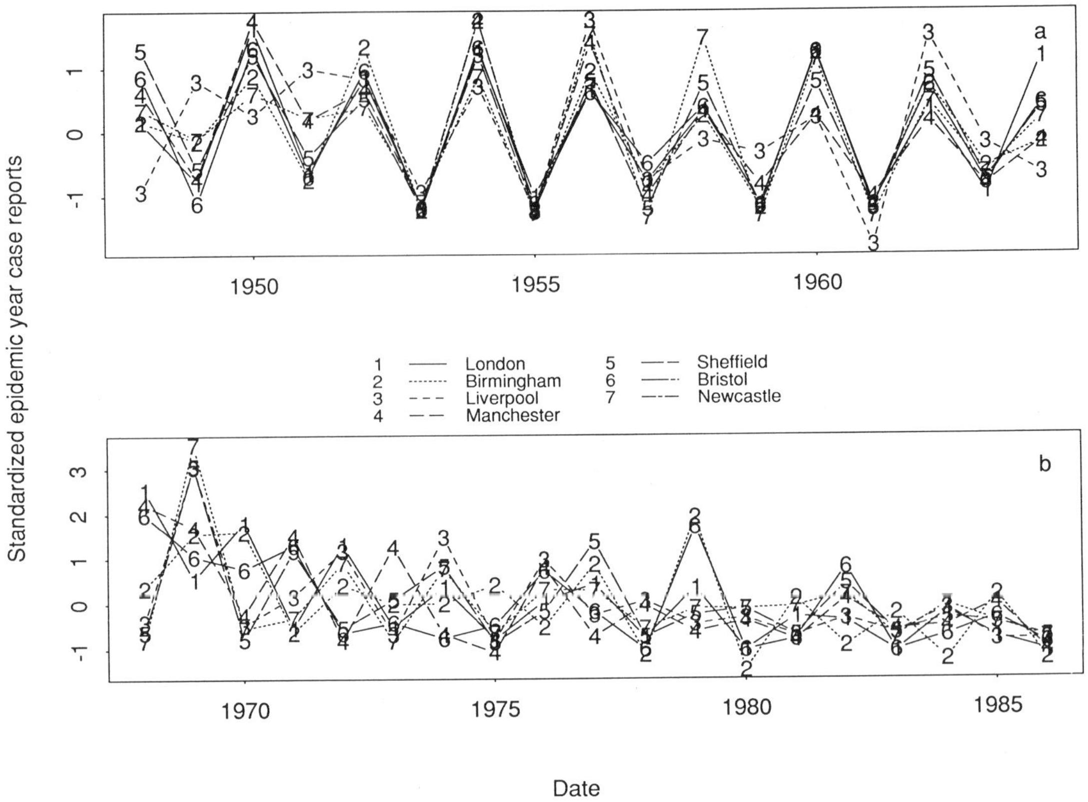

Figure 7 Standardized total measles incidence in epidemiological years (1 Oct–30 Sept) for seven cities in England and Wales. Annual totals for each city were standardized by subtracting the city mean annual incidence, then dividing by the city standard deviation of annual incidence. (a) Prevaccination period, 1948–1966. (b) Vaccination period, 1968–1988.

veloped countries, where the simple SEIR model indicates that relatively strong and predictable seasonality could interact with the propensity for longer term cycles to produce complex chaotic dynamic behavior. An initial evaluation of this result using more realistic epidemiological models indicates that population heterogeneity, in terms of the age distribution of infection, has the potential to simplify this scenario. However, these results interact with the population size (and, therefore, the spatial arrangement) of the host (Bolker and Grenfell, 1993). Spatial dynamics are likely to constitute a major direction for future work in this area.

A. Spatial Heterogeneity and Measles Dynamics

In an interesting discussion of measles dynamics, Olsen *et al.* (1988) underlined the importance of spatial heterogeneity and, in particular, partial spatial coupling between cities in perpetuating the infection. Partial coupling in this context means that susceptible and infective individuals from different places meet often enough to keep disease from fading out entirely in the population, but not so often that the entire population experiences large synchronous epidemics, which also would lead to extinction of the disease. This nonlinear perspective on persistence echoes a large and sophisticated body of work on the spatial geography of measles epidemics (Murray and Cliff, 1975; Cliff and Ord, 1978; Cliff and Haggett, 1988). Although the degree of coupling does not appear to be able to explain completely the persistence of measles in seasonally forced systems, without an appeal to the age effects noted earlier (Grenfell, 1992b), the question of spatial heterogeneity is clearly very important. In particular, more theoretical work is needed to examine the impact of seasonality and age structure on the balance between measles fade-out and coupling in spatial grids with relatively small individual population sizes, corresponding to the spatial arrangement of hosts within and between cities. These questions also can be explored in terms of the available empirical data. For example, Fig. 7 shows observed annual measles notifications for seven cities in England and Wales, before and after the onset of vaccination in 1968. In the prevaccination era (Fig. 7a), epidemics in the cities were relatively in phase, although significant heterogeneities are apparent (Bartlett, 1957), particularly with respect to the Liverpool notifications (which show much higher epidemic minima). The theoretical task is establishing whether England and Wales were coupled sufficiently to be an epidemiological unit before vaccination and the origins (possibly arising from local differences in demography) of systematic departures from the overall pattern. After the onset of vaccination (Fig. 7b), the epidemics in cities were much more irregular in timing, and apparently less in phase with each other. Again, the basis of this irregularity (whether associated with spatial dynamics *per se* or, for example, with regional variations in vaccine uptake) is at the heart of the problem of measles spread and persistence. This issue is also of considerable applied importance since vaccine uptake continues to rise to different extents in different areas.

Overall, recent developments suggest that measles dynamics will continue to present a fascinating challenge for epidemiologists, ecologists, and applied mathematicians. The whole question of the interaction between spatial dynamics and the possibility of chaotic dynamics on some populations scales is likely to be particularly fruitful.

References

Anderson, R. M., and May, R. M. (1982). Directly transmitted infectious diseases: Control by vaccination. *Science* **215,** 1053–1060.

Anderson, R. M., and May, R. M. (1984). Spatial, temporal, and genetic heterogeneity in host populations and the design of immunization programmes. *IMA J. Math. Appl. Med. Biol.* **1,** 233–266.

Anderson, R. M., and May, R. M. (1985). Age-related changes in the rate of disease transmission: Implications for the design of vaccination programmes. *J. Hyg.* **94,** 365–436.

Anderson, R. M., and May, R. M. (1991). "Infectious Diseases of Humans: Dynamics and Control." Oxford University Press, Oxford.

Anderson, R. M., Grenfell, B. T., and May, R. M. (1984). Oscillatory fluctuations in the incidence of infectious disease and the impact of vaccination: Time series analysis. *J. Hyg.* **93,** 587–608.

Anderson, R. M., Crombie, J. A., and Grenfell, B. T. (1987). The epidemiology of mumps in the UK: A preliminary study of virus transmission, herd immunity and the potential impact of immunization. *Epidemiol. Infect.* **99,** 65–84.

Aron, J. L. (1990). Multiple attractors in the response to a vaccination program. *Theoret. Population Biol.* **38,** 58–67.

Aron, J. L., and Schwartz, I. B. (1984). Seasonality and period-doubling bifurcations in an epidemic model. *J. Theoret. Biol.* **110,** 665–679.

Bailey, N. T. J. (1975). "The Mathematical Theory of Infectious Diseases and Its Application," 2d Ed. Griffin, London.

Bartlett, M. S. (1957). Measles periodicity and community size. *J. R. Stat. Soc. A* **120,** 48–70.

Bartlett, M. S. (1960). The critical community size for measles in the U.S. *J. R. Stat. Soc. A* **123,** 37–44.

Bartlett, M. S. (1961). Monte Carlo studies in ecology and epidemiology. *In* "Proceedings of the Fourth Berkeley Symposium on Mathematical Statistics and Probability," Vol. 4, pp. 39–55. University of California Press, Berkeley.

Bartlett, M. S. (1990). Chance or chaos. *J. R. Stat. Soc. A* **153,** 321–347.

Black, F. L. (1966). Measles endemicity in insular populations: Critical community size and its evolutionary implication. *J. Theoret. Biol.* **11,** 207–211.

Black, F. L. (1984). Measles. *In* "Viral Infections of Humans: Epidemiology and Control" (A. S. Evans, ed.), pp. 397–418. Plenum Press, New York.

Bolker, B. M., and Grenfell, B. T. (1993). Chaos and biological complexity in measles dynamics. *Proc. R. Soc. London B* **251**, 75–81.

Brownlee, J. (1919). Periodicities of epidemics of measles in the large towns of Great Britain and Ireland. *Proc. R. Soc. Med.* **12**, 77–117.

Chatfield, C. (1975). "The Analysis of Time Series: Theory and Practice." Chapman and Hall, London.

Cliff, A. D., and Haggett, P. (1988). "Atlas of Disease Distributions: Analytic Approaches to Epidemiological Data." Blackwell Scientific Publications, Oxford.

Cliff, A. D., and Ord, J. K. (1978). Forecasting the progress of an epidemic. *In* "Towards the Dynamic Analysis of Spatial Systems" (R. L. Martin, N. J. Thrift, and R. J. Bennett, eds.), pp. 191–204. Pion, London.

Dietz, K. (1976). The incidence of infectious disease under the influence of seasonal fluctuations. *Lecture Notes Biomath.* **11**, 1–15.

Dietz, K., and Schenzle, D. (1985). Mathematical models for infectious disease statistics. *In* "A Celebration of Statistics" (A. C. Atkinson and S. E. Feinberg, eds.), pp. 167–204. Springer-Verlag, New York.

Fine, P. E. M., and Clarkson, J. A. (1982a). Measles in England and Wales. I. An analysis of factors underlying seasonal patterns. *Int. J. Epidemiol.* **11**, 5–15.

Fine, P. E. M., and Clarkson, J. A. (1982b). Measles in England and Wales. II. The impact of the measles vaccination programme on the distribution of immunity in the population. *Int. J. Epidemiol.* **11**, 15–25.

Grenfell, B. T. (1992a). Parasitism and the dynamics of ungulate grazing systems. *Am. Naturalist* **139**, 907–929.

Grenfell, B. T. (1992b). Chance and chaos in measles dynamics. *J. R. Stat. Soc. B* **54**, 383–398.

Grenfell, B. T., and Anderson, R. M. (1985). The estimation of age-related rates of infection from case notifications and serological data. *J. Hyg.* **95**, 419–436.

Grossman, Z. (1980). Oscillatory phenomena in a model of infectious diseases. *Theoret. Population Biol.* **18**, 204–243.

Hamer, W. H. (1906). Epidemic disease in England—The evidence of variability and of persistency of type. *Lancet* **I**, 569–574.

Kermack, W. O., and McKendrick, A. G. (1927). A contribution to the mathematical theory of epidemics. *Proc. R. Soc. London A* **115**, 700–721.

Kermack, W. O., and McKendrick, A. G. (1932). Contributions to the mathematical theory of epidemics. Part II. The problem of endemicity. *Proc. R. Soc. London A* **138**, 55–83.

Kermack, W. O., and McKendrick, A. G. (1933). Contributions to the mathematical theory of epidemics. Part III. Further studies of the problem of endemicity. *Proc. R. Soc. London A* **141**, 92–122.

Kot, M., Graser, D. J., Truty, G. L., Schaffer, W. M., and Olsen, L. F. (1988). Changing criteria for imposing order. *Ecol. Model.* **43**, 75–110.

London, W. P., and Yorke, J. A. (1973). Recurrent outbreaks of measles, chickenpox and mumps. I. Seasonal variation in contact rates. *Am. J. Epidemiol.* **98**, 453–468.

May, R. M. (1986). Population biology of microparasitic infections. *In* "Biomathematics" (T. G. Hallam and S. A. Levin, eds.), Vol. 17, pp. 405–442. Springer-Verlag, Berlin.

May, R. M., and Anderson, R. M. (1984). Spatial heterogeneity and the design of immunization programs. *Math. Biosci.* **72**, 83–111.

May, R. M., Isham, V., Bolker, B., Renshaw, E., Lawrance, A. J., Spencer, N. M., Smith, L. A., Mayne, A. J., King, G. P., Albrecht, V., Lim, K. S., Dvorak, I., Basterfield, J., Salahuddin, Mollison, D., Chan, K. S., Tong, H., Wolff, R. C. L., Lord, G., Nason, G., Tjostheim, D., Casdagli, M., Smith, R. L., Broomhead, D. S., Huke, J. P., Muldoon, M. R., Nychka, D., Ellner, S., Gallant, A. R., McCaffrey, D., and Cheng, B. (1992). Royal Statistical Society Meeting on Chaos—Discussion. *J. R. Stat. Soc. B* **54**, 451–474.

McLean, A. R., and Anderson, R. M. (1988a). Measles in developing countries. Part I. Epidemiological parameters and patterns. *Epidemiol. Infect.* **100**, 111–133.

McLean, A. R., and Anderson, R. M. (1988b). Measles in developing countries. Part II. The predicted impact of mass vaccination. *Epidemiol. Infect.* **100**, 419–442.

Murray, G. D., and Cliff, A. D. (1975). A stochastic model for measles epidemics in a multi-region setting. *Inst. Br. Geographers* **2**, 158–174.

Nokes, D. J., and Anderson, R. M. (1988a). Measles, mumps, and rubella vaccine: What coverage to block transmission? *Lancet* **2**, 1374.

Nokes, D. J., and Anderson, R. M. (1988b). The use of mathematical models in the epidemiological study of infectious diseases and in the design of mass immunization programmes. *Epidemiol. Infect.* **101**, 1–20.

Nychka, D., Ellner, S., Gallant, A. R., and McCaffrey, D. (1992). Finding chaos in noisy systems. *J. R. Stat. Soc. B* **54**, 399–426.

Olsen, L. F., and Schaffer, W. M. (1990). Chaos versus noisy periodicity: Alternative hypotheses for childhood epidemics. *Science* **249**, 499–504.

Olsen, L. F., Truty, G. L., and Schaffer, W. M. (1988). Oscillations and chaos in epidemics: A nonlinear dynamic study of six childhood diseases in Copenhagen, Denmark. *Theoret. Population Biol.* **33**, 344–370.

Pool, R. (1989). Is it chaos, or is it just noise? *Science* **243,** 25–28.

Priestley, M. B. (1982). "Time Series Analysis and Forecasting," 2d Ed. Academic Press, London.

Rand, D. A., and Wilson, H. (1991). Chaotic stochasticity: A ubiquitous source of unpredictability in epidemics. *Proc. R. Soc. London B* **246,** 179–184.

Schaffer, W. M. (1985a). Can nonlinear dynamics elucidate mechanisms in ecology and epidemiology? *IMA J. Math. Appl. Med. Biol.* **2,** 221–252.

Schaffer, W. M. (1985b). Order and chaos in ecological systems. *Ecology* **66,** 93–106.

Schaffer, W. M., and Kot, M. (1985). Nearly one-dimensional dynamics in an epidemic. *J. Theoret. Biol.* **112,** 403–427.

Schenzle, D. (1984). An age-structured model of pre- and post-vaccination measles transmission. *IMA J. Math. Appl. Med. Biol.* **1,** 169–191.

Soper, M. A. (1929). The interpretation of periodicity in disease prevalence. *J. R. Stat. Soc. A* **92,** 34–61.

Sugihara, G., and May, R. M. (1990). Nonlinear forecasting as a way of distinguishing chaos from measurement error in time series. *Nature (London)* **344,** 734–741.

Tudor, D. W. (1985). An age-dependent epidemic model with application to measles. *Math. Biosci.* **73,** 131–147.

Michael J. Coyne & Gary Smith

Trichostrongylid Parasites of Domestic Ruminants

17

I. Nematodes and Production Losses

Gastrointestinal trichostrongylid nematode infections of domestic ruminants are responsible for millions of dollars of lost production every year because of the death of the animal, decreased meat and wool production, increased time required to reach market or breeding weight, and cost of helminth control (Barger, 1982). Studies of production losses in domesticated and feral sheep showed that mortality in nontreated grazing lambs parasitized with several trichostrongylid nematode parasite species ranged from 10 to 68% compared with 0–10% in treated animals (Barger, 1982; Gulland, 1991); (Fig. 1). Weight gain and fleece weight also are less in nontreated or inadequately treated parasitized grazing lambs (Gordon, 1967; Barger, 1982; Callinan *et al.,* 1982). Young grazing cattle are affected similarly. Weight gain is 18–40% lower than in treated controls in young cattle with a moderate parasite burden (Barger, 1982).

Two of the most important trichostrongylid nematode parasites of ruminants in temperate areas are *Haemonchus contortus* and *Ostertagia ostertagi. Haemonchus contortus* is a blood-feeding gastrointestinal nematode found in the abomasum of sheep, goats, cattle, and a number of nondomesticated ruminant species (Levine, 1980). The parasite is distributed worldwide, and is responsible for considerable morbidity and mortality in sheep raised on pasture. The effects are most severe in young lambs during their first grazing season (Ayalew and Gibbs, 1973; Cornwell, 1975; Boag and Thomas, 1977; Hubert *et al.,* 1979; Brunsdon, 1980; Lyons *et al.,* 1987). The pathological changes seen in sheep infected with *H. contortus* typically include severe normocytic, hypochromic anemia caused by the blood feeding of the larval and adult stages of the parasite, and hemorrhage from the parasite feeding sites. Death occurs after the collapse of the erythropoietic system (Dargie and Allonby, 1975). *Ostertagia ostertagi* is found in the abomasum of cattle and causes severe disease during the first grazing season. Signs include diarrhea, anemia, emaciation, and dependent edema secondary to hypoproteinemia. Death occurs from dehydration, anemia, and hypoproteinemia. Some controversy exists over the exact cause of production losses due to *O. ostertagi,* but work by Xiao and Gibbs (1992a,b,c) is consistent with a previous idea that the principal cause is a disruption of postabsorptive protein metabolism, perhaps related to reduced

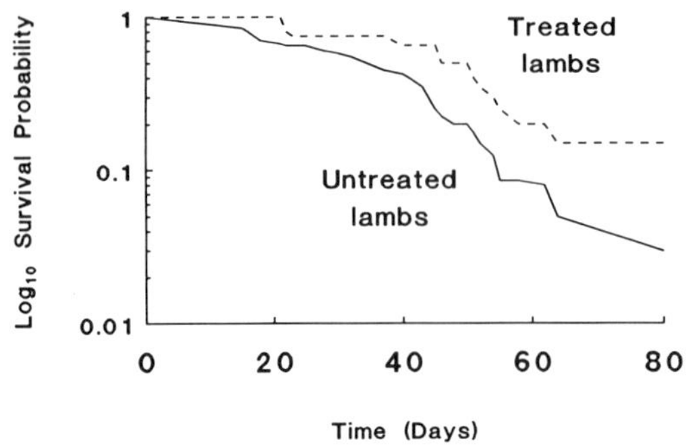

Figure 1 Survivorship analysis of feral sheep on St. Kilda, Scotland, in 1989. The sheep were infected with several species of trichostrongylid nematodes. The principal cause of death was protein-energy malnutrition, exacerbated (in the untreated sheep) by parasitism (Gulland, 1991).

plasma insulin concentrations. The damaging consequences of infection can be fairly subtle, including decreased carcass quality at slaughter manifested in decreased dressing weights and increased capacity for holding water in the rib eye muscle (Xiao and Gibbs, 1992b).

II. Life Cycles

As do all trichostrongylids, *H. contortus* and *O. ostertagi* have direct life cycles (Fig. 2). Third-stage infective larvae on herbage are consumed by the grazing animals. The ingested third-stage larvae exsheath in the rumen and are carried with the ingesta to the abomasum (Rose, 1969; Nicholls *et al.*, 1985) where they molt to the fourth stage and begin to feed. These larvae molt once more to the fifth (adult) stage. On reaching sexual maturity, the worms mate and egg production begins. The period from ingestion of larvae to egg production for *H. contortus* ranges from 15 to 21 days (Veglia, 1915; Silverman *et al.*, 1970); the prepatent period for *O. ostertagi* is ~18 days (Rose, 1969). Eggs are released by the female into the abomasal contents and are carried with the ingesta through the gastrointestinal tract. During the transition from the abomasum to the rectum, the egg undergoes a series of mitotic divisions until the 24-cell morula stage is reached (Veglia, 1915). Additional development within the gastrointestinal tract is precluded by the absence of oxygen (Veglia, 1915; Shorb, 1944). Eggs are passed in the feces in the morula stage. On pasture, embryonation and hatching occurs, and the first-stage larvae are released into the feces. These rhabditiform larvae feed on microorganisms in the feces, undergoing two molts to third-stage filariform larvae. These third-stage larvae do not feed, and are ensheathed in the outer skin retained from the previous molt (Veglia, 1915). The rate of hatching and development from one stage to the next is dependent on the ambient temperature, as well as on the oxygen concentration and moisture content of the feces during development (Veglia, 1915; Shorb, 1943; Dinaburg, 1944; Berberian and Mizelle, 1957; Silverman and Campbell, 1959; Rose, 1963; Jehan and Gupta, 1974). On development, third-stage larvae actively migrate laterally or are carried passively from the feces to the adjacent herbage (Skinner and Todd, 1980), where the larvae actively migrate vertically onto the herbage. This vertical migration is dependent on both the ambient temperature and relative humidity (Buckley, 1940; Rees, 1950).

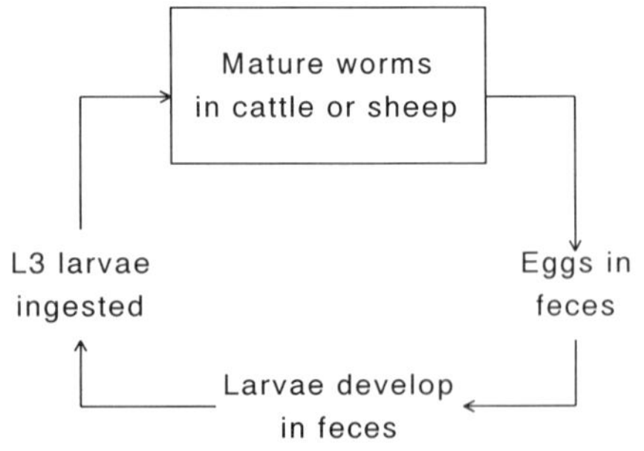

Figure 2 Generalized life cycle for the gastrointestinal trichostrongylid parasites of cattle and sheep. The direct life cycle involves five developmental stages: eggs (in feces); L1, L2, and L3 larvae (on pasture); and L4 larvae and L5 stage worms (in the host).

III. Parasite Control Programs and the Role of Mathematical Models

Modern parasite control programs are based on the prophylactic (rather than therapeutic) use of anthelmintic drugs, to reduce the level of pasture larval contamination and so to minimize the exposure of animals to infective larvae. This goal generally can be accomplished by the strategic administration of anthelmintics to cattle or sheep at the times of year at which the microclimate at the pasture surface is most favorable for the development and survival of the free-living eggs and larvae. Unfortunately, despite this rule of thumb, considerable uncertainty is associated with decisions concerning the precise timing and number of doses of anthelmintic required to achieve the desired impact. This approach does not consist merely of treating the easily recognizable sick animals as in chemotherapeutic strategies. Usually, the drugs must be administered well before the hosts show any signs of parasitism. A plethora of factors must be taken into account including climate, local husbandry practices, and the characteristics of the chosen anthelmintic. A formal mathematical representation of the important processes removes some of the uncertainty because it facilitates a more complete appraisal of the most relevant factors. As a result there is a burgeoning literature on mathematical models intended to assist in the design and evaluation of trichostrongylid control strategies based on anthelmintic drug treatments. A number of models have been developed for *H. contortus* and *O. ostertagi* in particular (Tallis and Leyton, 1969; Gordon *et al.*, 1970; Gettinby *et al.*, 1979; Grenfell *et al.*, 1987; Smith *et al.*, 1987a,b; Smith, 1988,1990a,b; Smith and Guerrero, 1993). The biological basis of two of these models is the subject of this chapter. Most attention is given to a model for *H. contortus* but the model for *O. ostertagi* is referred to for contrast or emphasis.

The following discussion begins with a general survey of the relevant population biology. Most trichostrongylid models assume that the processes that regulate parasite density exert their influence by altering parasite survival and fecundity during the parasitic phase. These processes are regulatory in the sense that they respond negatively to increases in parasite density and vice versa. On the other hand, the abundance of the free-living stages is affected but not regulated by climate because the impact of climate is not a density-dependent phenomenon (see Chapter 5).

IV. Parasitic Phase of the Life Cycle

A. Parasite Fecundity

Parasite fecundity is defined as the number of eggs reaching pasture per female worm per day. In calves experimentally infected with *O. ostertagi,* daily fecal egg counts peak within 2–3 wk following patency and decline exponentially thereafter (Michel, 1969). Little correlation is found between the mature female worm burden and the total daily fecal egg count. The daily fecal egg counts of lambs experimentally infected with *H. contortus* change in a similar fashion (Coyne *et al.*, 1991a) but, unlike calves infected with *O. ostertagi,* the correlation between total daily fecal egg counts and mature female worm burden is very good (Coyne *et al.*, 1991a). The fecundity of *O. ostertagi* is subject to very severe constraints; it is related inversely to the intensity and duration of the infection (Michel, 1969; Smith *et al.*, 1987a). No such constraints have been demonstrated for *H. contortus* (Coyne *et al.*, 1991a,b). In short, the fecundity of *O. ostertagi* is regulated whereas the fecundity of *H. contortus* is not. The average fecundity for *H. contortus* ranges from 4700 to 7000 eggs per female parasite per day irrespective of the intensity or duration of infection (Coyne *et al.*, 1991a,b). Peak fecundity for *O. ostertagi* is about 120 eggs per female parasite per day and soon declines (Smith *et al.*, 1987a).

Changes in the egg output for both parasite species can be represented by

$$\frac{dE_t}{dt} = \lambda\, 0.5P_t, \quad (1)$$

where E_t is the total number of eggs laid since the start of the infection, and P_t is the number of mature worms, and λ is the *per capita* fecundity per female worm per day. The model assumes that the ratio of male to female worms in an average worm burden is 1 : 1. Egg production is unregulated in the case of *H. contortus,* so *per capita* fecundity remains constant throughout the course of the infection. Thus,

$$\lambda = \lambda_0, \quad (2)$$

where λ_0 is potential maximum *per capita* daily fecundity of female worms. For *O. ostertagi,* for which fecundity varies with the intensity and duration of the infection, the change in *per capita* fecundity of worms in a single host is given by

$$\lambda = \lambda_0 e^{-\gamma P_t t}, \quad (3)$$

where γ is a constant that defines the severity of the density-dependent constraint. Smith *et al.* (1987a) argued that Eq. 3 could not be used to describe the mean fecundity of worms in the entire population of hosts because it does not account for the aggregated distribution of parasites that characterizes populations of *O. ostertagi.* These investigators suggested that

$$\lambda = \lambda_0 e^{-\gamma t}\left[1 + \frac{P_t}{Hk}(1 - e^{-\gamma t})\right]^{-(k+1)}, \quad (4)$$

would be a more precise representation of the average *per capita* fecundity of *O. ostertagi* in a population of H hosts (k is the exponent term of the negative binomial probability distribution; see Chapter 4). However, studies by Smith and Guerrero (1993) indicated that the increase in precision is negligible for populations in which k is greater than 1. Since the value of k for trichostrongylid populations in ruminants is typically between 1 and 4, Eq. 3 is used in the most recent model for *O. ostertagi* (Smith, 1990a; Gulland, 1991; Smith and Guerrero, 1993).

B. Parasite Mortality

Parasite abundance in the host is regulated by two processes: decreases in parasite establishment (immune exclusion) and increases in the mortality of established parasites (Smith, 1988,1993). Both these processes are assumed to be manifestations of the hosts' immune response. Barger *et al.* (1985) showed that the proportion of ingested third-stage *H. contortus* larvae that gains access to the abomasum and develops to the fourth-stage decreases with the duration of infection. Jackson *et al.* (1988) found that this immune exclusion process was dose dependent and sensitive to steroids. A simple empirical model (a declining sigmoid) for the decline in the establishment of *H. contortus* in sheep was provided by Smith (1988):

$$\Phi_t = \frac{e^{(i_1 + i_2 t)}}{1 + e^{(i_1 + i_2 t)}}, \quad (5)$$

where Φ_t is the proportion of ingested third-stage larvae that becomes established in the abomasal mucosa, i_1 and i_2 are constants, and t is the duration of infection (Fig. 3a,c). Immune exclusion is assumed to culminate in the death of the third-stage larvae, either directly or as a result of their being unable to complete development. Immune exclusion has been inferred to occur in other trichostrongylid species (Smith and Galligan, 1988); a declining sigmoid function similar to Eq. 5 provides an excellent mimic of the results of trickle infection experiments in which calves are infected with *O. ostertagi* (Fig. 4). Previous work (Grenfell *et al.*, 1987) on this parasite mistakenly assumed that parasite establishment declined in an exponential manner.

In single infection experiments, the mortality of (μ) of established *H. contortus* is described best as an increasing linear func-

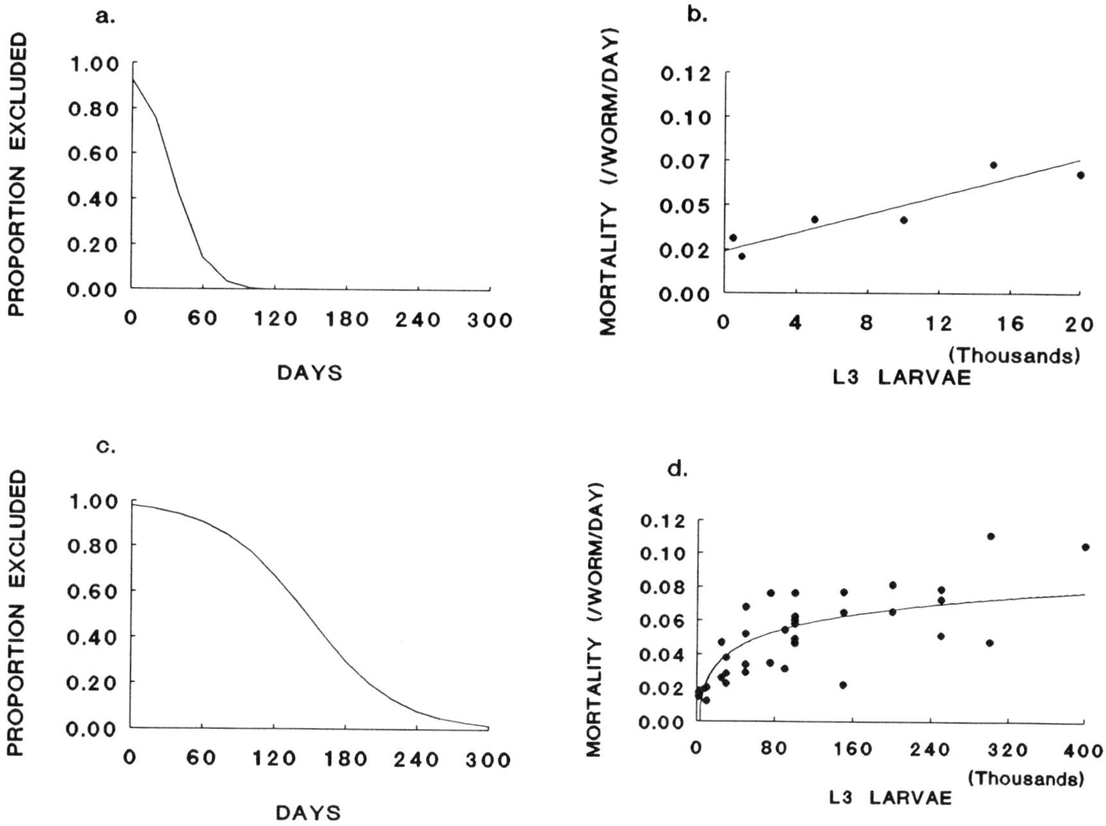

Figure 3 (a) Immune exclusion in *Haemonchus contortus.* (b) Mortality of established *H. contortus.* (c) Immune exclusion in *Ostertagia ostertagi.* (d) Mortality of established *O. ostertagi.* The change in immune exclusion for both parasites was inferred from trickle infection experiments in which hosts were infected with a constant number of larvae each day (Smith, 1988, 1993). The mortality of the established worms was estimated from single infection experiments in which hosts were infected only once and then killed for worm counts at varying intervals after infection. The fitted lines were obtained by standard linear regression and nonlinear regression methods (Smith, 1988, 1990a; Coyne, 1991). Redrawn with permission from Smith (1993).

tion of the magnitude of the initial infection:

$$\mu = p_1 + p_2\Lambda, \tag{6}$$

where p_1 and p_2 are constants, and Λ is the infection dose (Fig. 3b,d; Smith, 1988; Coyne, 1991; Coyne *et al.,* 1991a). To represent mortality of immature and mature fifth-stage worms in natural infections in which animals are infected repeatedly, Smith (1988) found it necessary to assume that the mortality of established worms was a function of the entire experience of third-stage larvae of the host so

$$\mu = p_1 + p_2\Lambda t, \tag{7}$$

where Λt is the total number of third-stage larvae ingested over the interval t. The expression for μ has two components: p_1, the intrinsic mortality rate, and $p_2\Lambda t$, the mortality associated with immune response of the host. That this expression will survive unmodified for very long is unlikely. For example, experimental infection studies with other trichostrongylid species, including *O. ostertagi* (Fig. 3d), suggest that mortality of established worms increases to an

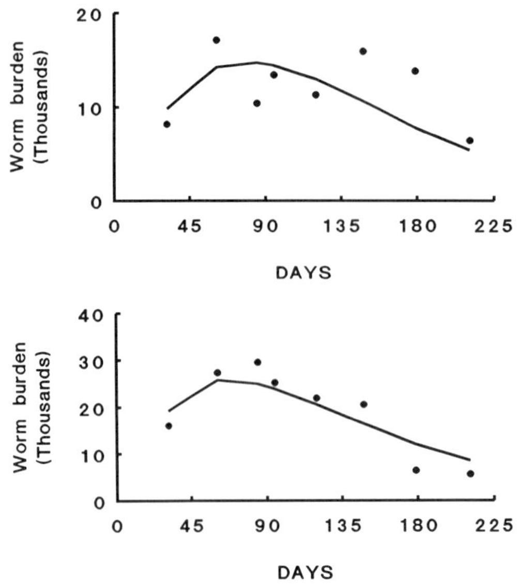

Figure 4 Observed and predicted worm burdens in a trickle infection experiment carried out by Michel (1969). The parasite was *Ostertagia ostertagi*. The solid line was derived independently of the data shown here as a test of the adequacy of a model for the population biology of the parasitic phase of *O. ostertagi* (Smith, 1993). Redrawn with permission from Smith (1993).

asymptotic value as the magnitude of the infecting dose increases (Smith and Galligan, 1988; Smith, 1990a). This property is intuitively reasonable since some minimum time is likely to be required for a host to kill any given parasite. The asymptotic nature of the relationship may be obscured in the case of *H. contortus* by the pathogenicity of the parasite. Equation 7 also must account for the fact that the immunological memory of the host experience of infection is not indefinitely long. For example, in his model for *H. contortus,* Coyne (1991) followed the method suggested by Berding *et al.* (1986,1987) for the inclusion of acquired immunity of limited duration and obtained

$$\mu = p_1 + p_2 \int_{t-\xi}^{t} \Lambda dt, \tag{8}$$

where ξ is the duration of immunity. This expression assumes that larvae ingested more than ξ days prior to the current day no longer influence the mortality of the established worms.

C. Arrested Development

Fourth-stage *H. contortus* and *O. ostertagi* vary in their propensity to enter a phase of arrested development. The proportion of the worm burden consisting of arrested larvae is greatest when the pasture microclimate is unfavorable for the development and survival of the free-living stages (cold winters or hot dry summers). Accordingly, the global patterns of arrested development for both parasites can be resolved into three or four basic types (Connan, 1971; Blitz and Gibbs, 1972; Waller and Thomas, 1975; Smith and Grenfell, 1985). Nevertheless, despite the qualitative predictability of these patterns, predicting the precise *quantitative* pattern from first principles is not yet possible. This problem is important for individuals wanting to write convincing mathematical models for the population biology of the parasite.

An interim solution is measuring temporal variations in the probability (p) that larvae ingested over a given period will enter a phase of arrested development. When incorporated into a model that accurately reflects parasite development and mortality, changes in the probability of entering a phase of arrested development will generate convincing patterns of arrested development.

The probability, p, cannot be measured from data concerning the general pattern of arrested development because such data reflect not only changes in p but also the long-term influence of stage-specific differences in parasite death rates and the mysterious processes that cause arrested larvae to resume development. One consequence of this situation is that grazing hosts can harbor thousands of arrested larvae at a time when newly ingested larvae have no propensity to arrest at all! Data obtained from tracer animals (previously parasite-naive hosts grazed on pasture for a very short pe-

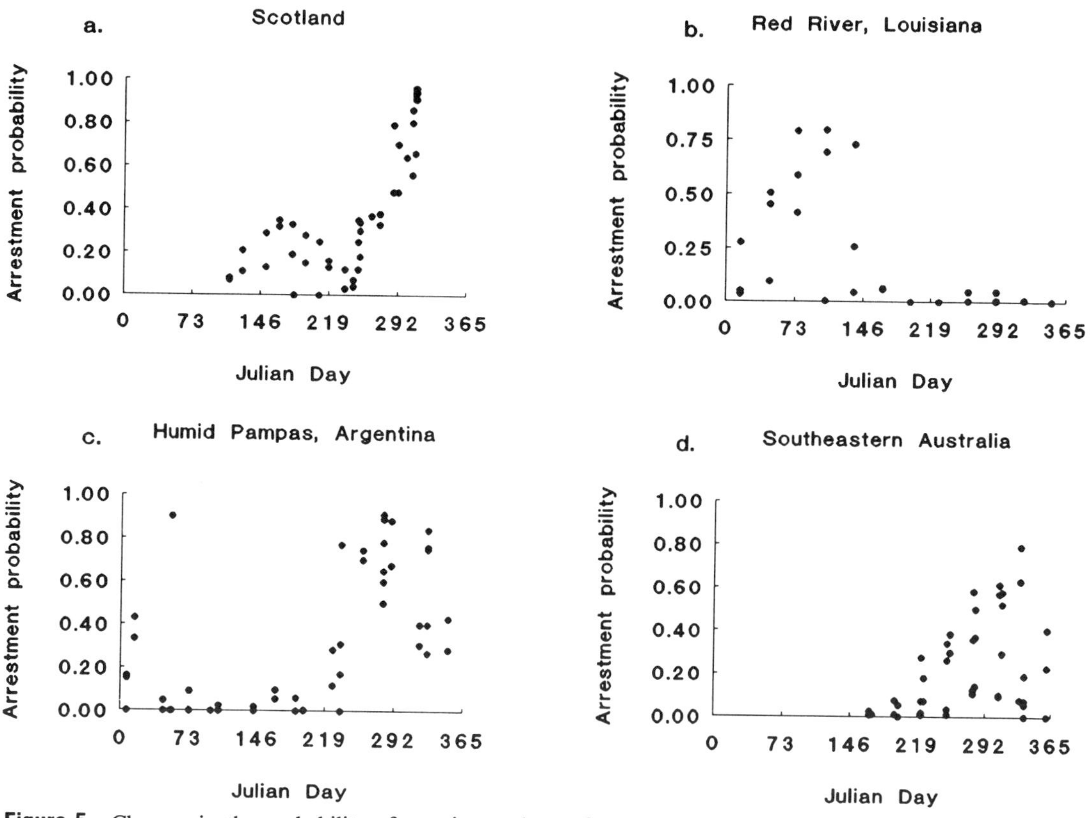

Figure 5 Changes in the probability of entering a phase of arrested development calculated from *Ostertagia ostertagi* worm burdens in tracer calf studies. (a) Data from Anderson *et al.* (1965, 1969) and Armour *et al.* (1969, 1979, 1981). (b) Data from Williams *et al.* (1983). (c) Data from Fiel *et al.* (1990). (d) Data from Smeal *et al.* (1980) and Smeal and Donald (1982). January 1 is Julian Day 1 and December 31 is Julian Day 365.

riod) can be used, however, to estimate p. In such hosts, the short time period of infection enables the researcher to say with confidence that any adult worms found in the tracer animal cannot have undergone an arrested development phase. Such data can also lessen the impact of stage-specific differences in parasite death rates (between the arrested larvae and established worms, for example). The ratio of arrested larvae to total worm burden in tracer hosts provides a simple, intuitively reasonable measure of the average value for p during the grazing period. This method assumes that the mortality of arrested and developing worms is effectively zero. Clearly, this is not the case but, provided the period for which the tracer hosts are grazed is kept relatively short (2 wk is best), the errors introduced by parasite deaths are negligibly small (C. Plyler, G. Smith, and J. Guereto, unpublished observations). Some examples of the patterns obtained for *O. ostertagi* are given in Fig. 5. In northern areas of the northern hemisphere, the highest probability of entering a phase of arrested development occurs in the autumn, prior to cold winter months; in southern areas of the northern hemisphere where winters are mild, and in much of the southern hemisphere, the highest probability of arrest occurs in the spring prior to the dry season. Patterns such as these can be described using simple polynomial functions (Coyne, 1991) and can be used to determine the rate at which newly molted fourth stage larvae move into developing and arrested classes of larvae.

V. Free-Living Stage of the Life Cycle

The demographies of the free-living stages of *H. contortus* and *O. ostertagi* are very similar. Eggs deposited on the pasture hatch in the feces. The first- and second-stage larvae feed on fecal microorganisms. The third-stage larvae move actively or passively out of the feces (translocation); some subsequently are ingested by grazing hosts. Great care must be taken when extrapolating laboratory measurements of larval development and mortality rates to the field. For example, Smith *et al.* (1986) and Young *et al.* (1980) found that the development of the pre-infective stages of *O. ostertagi* proceeded much more slowly in the field than would be predicted from available laboratory studies. Although Smith (1990b) could find no evidence that this was true for *H. contortus,* he did note that the mortality rate of *H. contortus* third-stage larvae in the field was an order of magnitude larger than that measured in the laboratory.

The population biology of the free-living stages of *H. contortus* and *O. ostertagi* has been modeled by a number of workers (Young *et al.,* 1980; Grenfell *et al.,* 1986; Smith *et al.,* 1986; Smith 1990a,b; Coyne and Smith, 1992). The models share a basic architecture that is common to most trichostrongylid models. Development and mortality rates are temperature-dependent (usually modeled by some variant of a logarithmic function). Some models assume that development is only possible given adequate soil-surface moisture conditions; others take account of moisture by increasing the mortality of eggs and larvae in times of drought. Some models also include provision for larval losses from the herbage as a result of active or passive migration into the soil during times of high rainfall (Smith and Grenfell, 1985).

VI. Model for *Haemonchus contortus* in Grazing Lambs

The basic model is a deterministic, compartmental model composed of a series of coupled first-order differential equations, each one representing a particular stage in the life cycle. The model incorporates a constant fecundity, immune exclusion, changes in the mortality of established worms as the host's recent experience of infection changes, temperature-dependent mortality and survival rates for the free-living stages, and the possibility of passive or active downward migration of pasture larvae in the presence of abundant rainfall. Details are given by Coyne (1991). Model behavior was investigated using numerical solution methods.

A. Model for Untreated Lambs

The model was used to simulate changes in the average *H. contortus* burden in lambs during their first grazing season in southeastern Pennsylvania. Results of a field trial performed in the same area of Pennsylvania were used to validate the model (Coyne *et al.,* 1991b). In the simulation, the lambs were born in early January and weaned on April 1 (day 90) at an average weight of 20 kg. After weaning, the lambs were placed on a pasture contaminated with a low level of overwintered *H. contortus* larvae (100 third-stage larvae/kg herbage). The lambs were reared on pasture from April 1 through November 15 when they were housed and given supplemental feed. The stocking density of the lambs on pasture was 12 lambs per hectare. The observed and predicted changes in *H. contortus* worm burdens are illustrated in Fig. 6. The model provided a good fit to the observed daily fecal egg output early in the season but overestimated the counts subsequently. Since the predicted and observed total worm burdens were very similar, the discrepancy indicates that the model overestimated the proportion of ma-

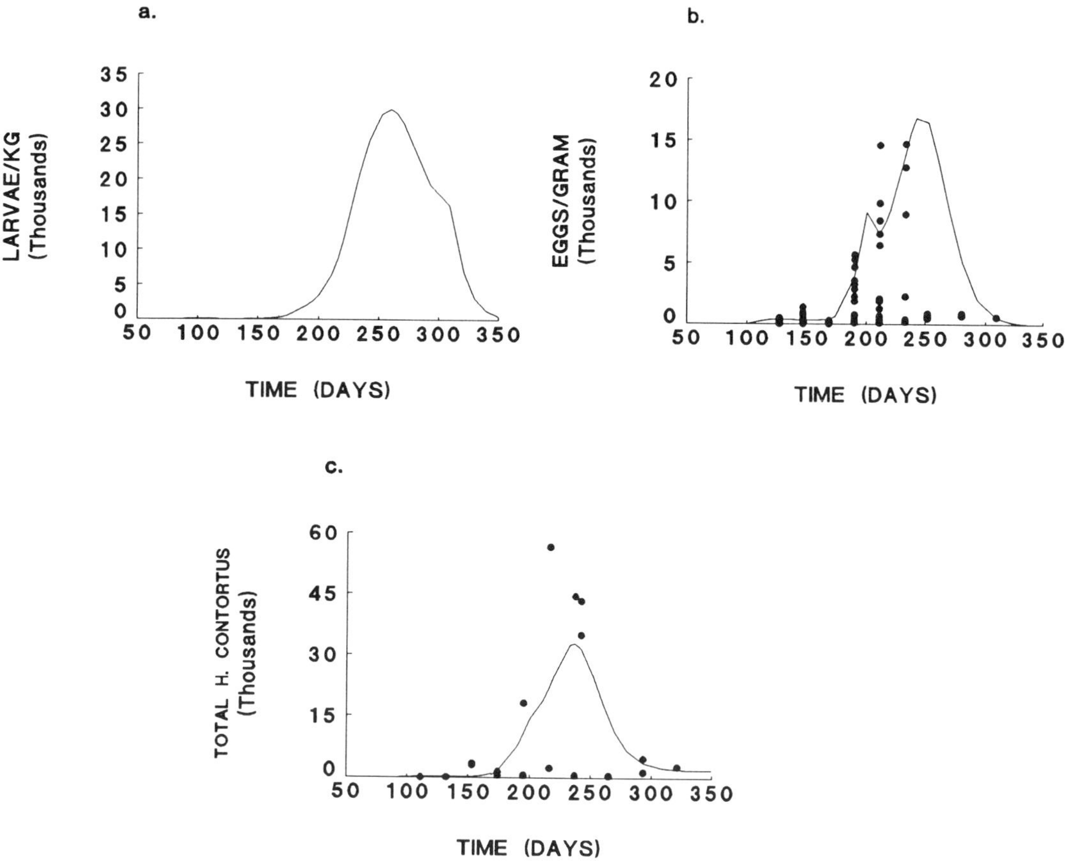

Figure 6 Observed (points) and predicted (solid line) patterns of abundance for *Haemonchus contortus* on a pasture in Southeastern Pennsylvania. (a) Predicted pattern of pasture larval contamination. (b) Fecal egg counts. (c) Mean total worm burden. Data from Coyne (1991) and Coyne *et al.* (1991b).

ture worms in the late summer and early fall.

B. Model for Treated Lambs

The model also was used to evaluate parasite control strategies described in the literature (Brunsdon, 1980; Herd, 1984; Herd *et al.,* 1984). Three strategies were examined: treatment with an anthelmintic (levamisole) in mid-grazing season; treatment with levamisole in mid-July followed by an immediate transfer to a "clean" (parasite-free) pasture; and treatment with levamisole at 3, 6, 9, and 12 wk after turnout onto pasture. The first treatment is a typical therapeutic strategy used by sheep producers, because during the mid-grazing period clinical disease occurs and parasitism is recognized as a problem by the producer. Treating the lambs with levamisole (early to mid-grazing season) and moving them to a pasture free of infective third-stage larvae is a simple prophylactic strategy, limited only by the availability of pasture (Brunsdon, 1980). In this case, susceptible lambs are removed from the pasture prior to the period of heavy larval contamination. Treatment with levamisole before moving eliminates the existing worm burden, and movement to the clean pasture reduces the risk of substantial re-infection in nonim-

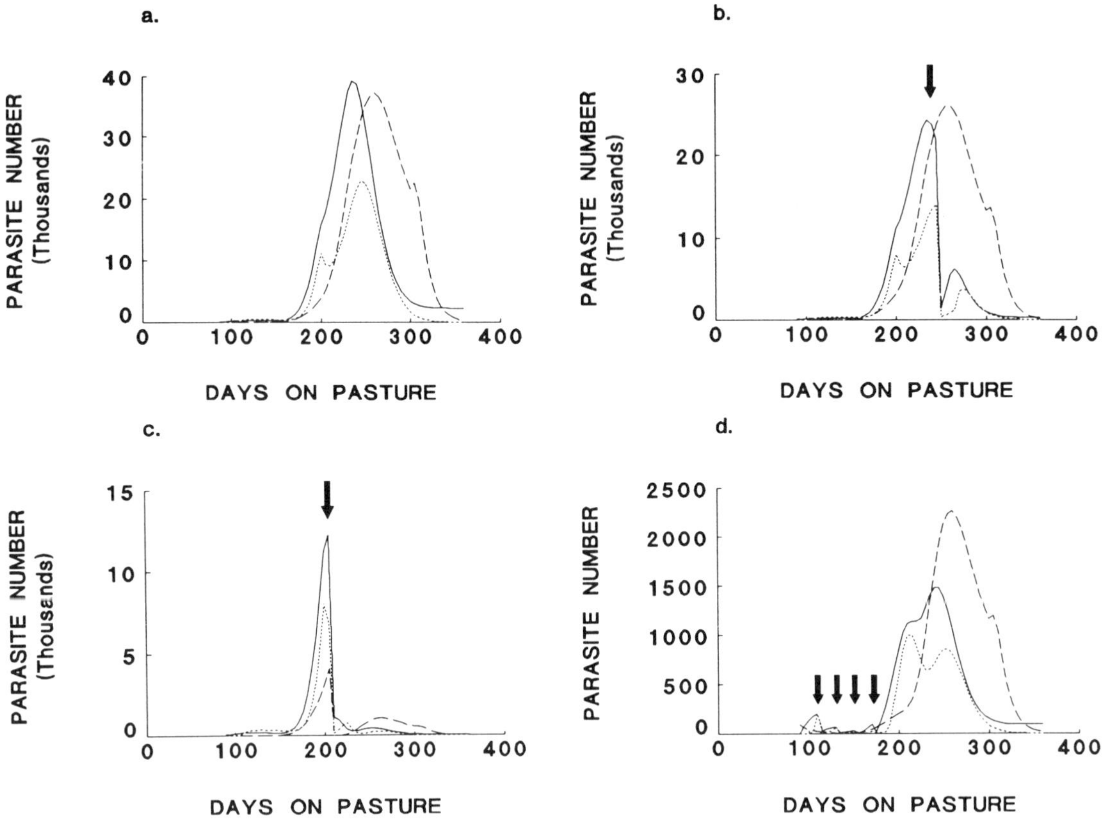

Figure 7 Model behavior under different strategies for control of *Haemonchus contortus* in a pasture containing set-stocked lambs. (a) No anthelmintic treatment. (b) Single therapeutic treatment in mid-summer. (c) Single treatment in late spring followed by immediate transfer to a "clean" pasture. (d) Four early season treatments. Note that the scale of the vertical axis is different in each case. Mean total worm burden, ——; fecal egg counts, · · · ; pasture larval contamination, – – – ; treatment with anthelmintics, arrows (Coyne, 1991).

mune lambs. Multiple treatment of lambs with levamisole early in the grazing period is also a prophylactic strategy because it prevents those larvae ingested early in the grazing season from developing into mature worms and producing eggs. Eliminating the early contamination of pasture with parasite eggs results in a decrease in the number of infective third-stage larvae available to the lambs later in the grazing season. This prophylactic strategy is considered optimal for the control of *H. contortus* (Herd, 1984; Herd *et al.*, 1984).

Model behavior under each strategy is illustrated in Fig. 7. The therapeutic mid-season treatment with levamisole clearly reduced the mean parasite burden. Further, despite the presence of large numbers of infective larvae on the pasture (Fig. 7), the mean burdens never recovered to pretreatment levels, presumably because the lambs had, by the time of treatment, already developed a substantial degree of immunity to additional infection. Nevertheless, in this scenario, treatment was probably too little too late. The lambs already had been subjected to burdens approaching those in the untreated group and would, in the field, have suffered similar production losses. The two prophylactic strategies were much more successful (note the marked change in scale of the vertical axes in Fig. 7). Repeated treatment of lambs early in the grazing season was particularly effective. That Herd *et al.* (1984) obtained comparable results in an experimental evaluation of similar strategies is encouraging. The prophylactic strategies work because they eliminate the large increase in pasture larval contamination that is caused largely by eggs that were deposited early in the grazing season. The dose-and-move strategy often is not feasible because of the requirement for a second "clean" pasture. However, multiple treatments early in the grazing season now are used fairly commonly and are proving very useful. The most effective of the multiple dosing strategies are those in which the interval between successive treatments approximates the prepatent period of the parasite. Additional simulations indicate, for example, that increasing the treatment interval from 21 to 28 days will lead to a twofold increase in the worm burden (Coyne, 1991).

VII. Conclusions

The simple model for the population biology of *H. contortus* outlined in the preceding section is capable of reproducing the patterns of fecal egg output, worm burden, and pasture larval contamination observed in the field. Further, the model mimics the results of established parasite control strategies using anthelmintics. The possibility exists, therefore, that the model now can be used in unfamiliar contexts to assist in the design and evaluation of plausible parasite control strategies.

References

Anderson, N., Armour, J., Jarrett, W. F. H., Jennings, F. W., Ritchie, J. D. S., and Urquhart, G. M. (1965). A field study of parasitic gastritis in cattle. *Vet. Rec.* **77,** 1196–1204.

Anderson, N., Armour, J., Jennings, F. W., Ritchie, J. S. D., and Urquhart, G. M. (1969). The sequential development of naturally occurring ostertagiasis in calves. *Res. Vet. Sci.* **10,** 18–28.

Armour, J., Jennings, F. W., and Urquhart, G. M. (1969). Inhibition of *Ostertagia ostertagi* at the early fourth larval stage. I. The seasonal incidence. *Res. Vet. Sci.* **10,** 232–237.

Armour, J. F., Bairden, K., Duncan, J. L., Jennings, F. W., and Parkins, J. J. (1979). Observations on ostertagiasis in young cattle over two grazing seasons with special reference to plasma pepsinogen levels. *Vet. Rec.* **105,** 500–503.

Armour, J., Bairden, K., Duncan, J. L., Jones, R. M., and Bliss, D. H. (1981). Studies on the control of bovine ostertagiasis using a morantel sustained-release bolus. *Vet. Rec.* **108,** 532–535.

Ayalew, L., and Gibbs, H. C. (1973). Seasonal fluctuations of nematode populations in breeding ewes and lambs. *Can. J. Comp. Med.* **37,** 79–89.

Barger, I. A. (1982). Helminth parasites and animal production. *In* "Biology and Control of Endoparasites" (L. E. A. Symons, A. D. Donald, and J. K. Dineen, eds.), pp. 133–135. Academic Press, Sydney.

Barger, I. A., Le Jambre, L. F., Georgi, J. R., and Davies, H. I. (1985). Regulation of *Haemonchus contortus* populations in sheep exposed to continuous infection. *Int. J. Parasitol.* **15,** 529–533.

Berberian, J. F., and Mizelle, J. D. (1957). Developmental studies on *Haemonchus contortus* Rudolphi (1803). *Am. Midl. Nat.* **57,** 421–439.

Berding, C., Keymer, A. E., Murray, J. D., and Slater, A. F. G. (1986). The population dynamics of acquired immunity to helminth infections. *J. Theoret. Biol.* **122,** 459–471.

Berding, C., Keymer, A. E., Murray, J. D., and Slater, A. F. G. (1987). The population dynamics of acquired immunity to helminth infection: Experimental and natural transmission. *J. Theoret. Biol.* **126,** 167–182.

Blitz, N. M., and Gibbs, H. C. (1972). Studies on the arrested development of *Haemonchus contortus* in sheep. I. The induction of arrested development. *Int. J. Parasitol.* **2,** 5–12.

Boag, B., and Thomas, R. J. (1977). Epidemiological studies on gastro-intestinal nematode parasites of sheep: The seasonal number of generations and succession of species. *Res. Vet. Sci.* **22,** 62–67.

Brunsdon, R. V. (1980). Principles of helminth control. *Vet. Parasitol.* **6,** 185–215.

Buckley, J. J. C. (1940). Observations on the vertical migrations of infective larvae of certain bursate nematodes. *J. Helminthol.* **18,** 173–182.

Callinan, A. P. L., Morley, F. H. W., Arundel, J. H., and White, D. H. (1982). A model of the life cycle of sheep nematodes and the epidemiology of nematodiasis in sheep. *Agric. Systems* **9,** 199–225.

Connan, R. M. (1971). The seasonal incidence of inhibition of development in *Haemonchus contortus*. *Res. Vet. Sci.* **12,** 272–274.

Cornwell, R. L. (1975). The pattern of infection with gastro-intestinal nematodes in young fattening lambs at pasture. *Res. Vet. Sci.* **18,** 1–5.

Coyne, M. J. (1991). "Field and Experimental Studies of *Haemonchus contortus* in Lambs." Ph.D. Thesis. School of Veterinary Medicine, University of Pennsylvania, Philadelphia.

Coyne, M. J., and Smith, G. (1992). The development and mortality of the free-living stages of *Haemonchus contortus* in laboratory culture. *Int. J. Parasitol.* **22,** 641–650.

Coyne, M. J., Smith, G., and Johnstone, C. (1991a). A study of the mortality and fecundity of *Haemonchus contortus* in sheep following experimental infections. *Int. J. Parasitol.* **21,** 847–853.

Coyne, M. J., Smith, G., and Johnstone, C. (1991b). Fecundity of gastrointestinal trichostrongylid nematodes of sheep in the field. *Am. J. Vet. Res.* **52,** 1182–1188.

Dargie, J. D., and Allonby, E. W. (1975). Pathophysiology of single and challenge infections of *Haemonchus contortus* in merino sheep: Studies on red cell kinetics and the 'self-cure' phenomenon. *Int. J. Parasitol.* **5,** 147–157.

Dinaburg, A. G. (1944). Development and survival under outdoor conditions of eggs and larvae of the common ruminant stomach worm, *Haemonchus contortus. J. Agric. Res.* **69,** 421–433.

Fiel, C. A., Steffan, P. E., Almada, A., Ambrustolo, R. R., Entrocasso, C. M., and Bulman, G. M. (1990). Epidemiology of trichostrongyle infection in grazing cattle of the humid pampa (Argentina) with special reference to *Ostertagia ostertagi. In* "Epidemiology of Bovine Nematode Parasites in the Americas" (J. Guerrero and W. D. Leaning, eds.), pp. 15–24. Veterinary Learning Systems, Trenton, New Jersey.

Gettinby, G., Bairden, K., Armour, J., and Benitez-Usher, C. (1979). A prediction model for bovine ostertagiasis. *Vet. Rec.* **105,** 57–59.

Gordon, G., O'Callaghan, M., and Tallis, G. M. (1970). A deterministic model for the life cycle of a class of internal parasites of sheep. *Math. Biosci.* **8,** 209–226.

Gordon, H. McL. (1967). The diagnosis of helminthosis in sheep. *Vet. Med. Rev. (Leverkusen)* **2/3,** 140–168.

Grenfell, B. T., Smith, G., and Anderson, R. M. (1986). Maximum-likelihood estimates of the mortality and migration rates of the infective larvae of *Ostertagia ostertagi* and *Cooperia oncophora. Parasitology* **92,** 643–652.

Grenfell, B. T., Smith, G., and Anderson, R. M. (1987). A mathematical model of the population biology of *Ostertagia ostertagi* in calves and yearlings. *Parasitology* **95,** 389–406.

Gulland, F. M. D. (1991). "The Role of Parasites in the Population Dynamics of Soay Sheep on St. Kilda." Ph.D. Thesis. Girton College, University of Cambridge.

Herd, R. P. (1984). A practical approach to parasite control in sheep in the Northern United States. *Comp. Cont. Ed. Prac. Vet.* **6,** S67–S73.

Herd, R. P., Parker, C. F., and McClure, K. E. (1984). Epidemiologic approach to the control of sheep nematodes. *J. Am. Vet. Med. Assoc.* **184,** 680–687.

Hubert, J., Kerboeuf, D., and Gruner, L. (1979). Study of gastro-intestinal strongylosis in a sheep flock on permanent pasture. 1. Sheep parasitism in 1977. *Ann. Rech. Vét.* **10,** 503–518.

Jackson, F., Miller, H. R. P., Newlands, G. F. J., Wright, S. E., and Hay, L. A. (1988). Immune exclusion of *Haemonchus contortus* larvae in sheep: Dose dependency, steroid sensitivity and persistence of the response. *Res. Vet. Sci.* **44,** 320–323.

Jehan, M., and Gupta, V. (1974). The effects of temperature on the survival and development of the free living stages of twisted wireworm *Haemonchus contortus* Rudolphi 1803 of sheep and other ruminants. *Z. Parasitenk.* **43,** 197–208.

Levine, N. D. (1980). "Nematode Parasites of Domestic Animals and of Man," 2d Ed. Burgess Publishing, Minneapolis, Minnesota.

Lyons, E. T., Drudge, J. H., and Tolliver, S. C. (1987). Epizootiology of internal parasites in lambs and ewes during the periparturient period in Kentucky in 1986. *Proc. Helminthol. Soc. Washington* **54,** 233–236.

Michel, J. F. (1969). Some observations on the worm burdens of calves infected daily with *Ostertagia ostertagi. Parasitology* **59,** 575–595.

Nicholls, C. D., Lee, D. L., and Sharpe, M. J. (1985). Scanning electron microscopy of biopsy specimens removed by a colonoscope from the abomasum of sheep infected with *Haemonchus contortus. Parasitology* **90,** 357–363.

Rees, G. (1950). Observations on the vertical migrations of the third-stage larva of *Haemonchus contortus* (Rud.) on experimental plots of *Lolium perenne* S24, in relation to meteorological and micrometeorological factors. *Parasitology* **40,** 127–143.

Rose, J. H. (1963). Observations on the free-living stages of the stomach worm *Haemonchus contortus. Parasitology* **53,** 469–481.

Rose, J. H. (1969). The development of the parasitic stages of *Ostertagia ostertagi. J. Helminthol.* **43,** 173–184.

Shorb, D. A. (1943). Survival on grass plots of eggs and preinfective larvae of the common sheep stomach worm, *Haemonchus contortus. J. Parasitol.* **29,** 284–289.

Shorb, D. A. (1944). Factors influencing embryonation and survival of eggs of the stomach worm, *Haemonchus contortus. J. Agric. Res.* **69,** 279–287.

Silverman, P. H., and Campbell, J. A. (1959). Studies on parasitic worms of sheep in Scotland. I. Embryonic and larval development of *Haemonchus contortus* at constant conditions. *Parasitology* **49,** 23–38.

Silverman, P. H., Mansfield, M. E., and Scott, H. L. (1970). *Haemonchus contortus* infection in sheep: Effects of various levels of primary infections on nontreated lambs. *Am. J. Vet. Res.* **31,** 841–857.

Skinner, W. D., and Todd, K. S. (1980). Lateral migra-

tion of *Haemonchus contortus* larvae on pasture. *Am. J. Vet. Res.* **41,** 395–398.
Smeal, M. G., and Donald, A. D. (1982). Inhibited development of *Ostertagia ostertagi* in relation to production systems for cattle. *Parasitology* **85,** 21–25.
Smeal, M. G., Fraser, G. C., and Robinson, G. G. (1980). Seasonal changes in the structure of nematode populations of cattle in New South Wales in relation to inhibited larval development. *Austral. Vet. J.* **56,** 80–86.
Smith, G. (1988). The population biology of the parasitic stages of *Haemonchus contortus. Parasitology* **96,** 185–195.
Smith, G. (1990a). The use of computer models in the design of strategic parasite control programs. *In* "Epidemiology of Bovine Nematode Parasites in the Americas" (J. Guerrero and W. D. Leaning, eds.), pp. 107–115. Veterinary Learning Systems, Princeton, New Jersey.
Smith, G. (1990b). The population biology of the free-living phase of *Haemonchus contortus. Parasitology* **101,** 309–316.
Smith, G. (1993). Population biology of the parasitic phase of trichostrongylid nematode parasites of cattle and sheep. *Int. J. Parasitol. (in press).*
Smith, G., and Galligan, D. T. (1988). Mathematical models of the population biology of *Ostertagia ostertagi* and *Teladorsagia circumcincta,* and the economic evaluation of disease control strategies. *Vet. Parasitol.* **27,** 73–83.
Smith, G., and Grenfell, B. T. (1985). The population biology of *Ostertagia ostertagi. Parasitol. Today* **1,** 76–81.
Smith, G., and Guerrero, J. (1993). Mathematical models for the population biology of *Ostertagia ostertagi* and the significance of aggregated parasite distributions. *Vet. Parasitol.* **46,** 243–257.
Smith, G., Grenfell, B. T., and Anderson, R. M. (1986). The development and mortality of the non-infective free-living stages of *Ostertagia ostertagi* in the field and in the laboratory culture. *Parasitology* **92,** 471–482.
Smith, G., Grenfell, B. T., and Anderson, R. M. (1987a). The regulation of *Ostertagia ostertagi* populations in calves: Density-dependent control of fecundity, *Parasitology* **95,** 373–388.
Smith, G., Grenfell, B. T., Anderson, R. M., and Beddington, J. (1987b). Population Biology of *Ostertagia ostertagi* and anthelmintic strategies against ostertagiasis in calves. *Parasitology* **95,** 407–420.
Tallis, G. M., and Leyton, M. K. (1969). Stochastic models of helminthic parasites in the definitive hosts, I. *Math. Biosci.* **4,** 39–48.
Veglia, F. (1915). The anatomy and life history of the *Haemonchus contortus* (Rud.). *Union S. Africa Dep. Agric. Rep. Dir. Vet. Serv. Anim. Indust.* **3–4,** 347–500.
Waller, P. J., and Thomas, R. J. (1975). Field studies on inhibition of *Haemonchus contortus* in sheep. *Parasitology* **71,** 285–291.
Williams, J. C., Knox, J. W., Baumann, B. A., Snider, T. G., Kimball, M. D., and Hoerner, T. J. (1983). Seasonal changes of gastrointestinal nematode populations in yearling beef cattle in Louisiana with emphasis on prevalence of inhibition in *Ostertagia ostertagi. Int. J. Parasitol.* **13,** 133–143.
Xiao, L., and Gibbs, H. C. (1992a). Effects of clinically apparent and subclinical *Ostertagia ostertagi* infections on nitrogen and water metabolism in calves. *Am. J. Vet. Res.* **53,** 2009–2012.
Xiao, L., and Gibbs, H. C. (1992b). Nutritional and pathophysiologic effects of clinically apparent and subclinical infections of *Ostertagia ostertagi* in calves. *Am. J. Vet. Res.* **53,** 2013–2018.
Xiao, L., and Gibbs, H. C. (1992c). Effects of *Ostertagia ostertagi* infection on secretion of metabolic hormones in calves. *Am. J. Vet. Res.* **53,** 2019–2022.
Young, R. R., Nicholson, R. M., Tweedie, R. L., and Schuh, H-J. (1980). Quantitative modelling and prediction of development times of the free-living stages of *Ostertagia ostertagi* under controlled and field conditions. *Parasitology* **81,** 493–505.

M. G. Roberts & M. A. Gemmell

18

Echinococcosis

I. Introduction

Four species of *Echinococcus* are regarded as taxonomically valid. *E. granulosus* (Batsch, 1786), *E. multilocularis* (Leuckart, 1863), *E. oligarthrus* (Diesing, 1863), and *E. vogeli* (Rausch and Bernstein, 1972) (Eckert *et al.*, 1981). The transmission of all species is maintained through a predator–prey relationship. The adult tapeworm parasitizes the small intestine of carnivores, usually Canidae, and sheds proglottids containing eggs. The proglottids pass, sometimes intact with the eggs and sometimes not, with the feces of the host. If the eggs are ingested subsequently by herbivores, they develop into a larval stage (cyst) within the liver or other viscera. The cycle is completed when the carnivore consumes the herbivore, ingesting a mature cyst (Fig. 1). For additional details of the biology of *Echinococcus,* see Thompson (1986).

Echinococcus granulosus is considered to have originated in a wolf–deer life cycle in the holarctic zone of the Northern Hemisphere. After the introduction of animal husbandry (about 10,000 yr ago), the parasite evolved in dogs and sheep as well as in other domesticated animals. Colonization over the past 400 yr has spread this population worldwide. For a host checklist, see Rausch (1986) and Eckert *et al.* (1981). The definitive hosts of *E. multilocularis* are arctic and red foxes *(Alopex lupus, Vulpes vulpes);* the intermediate hosts are rodents of the genera *Microtus, Lemmus,* and *Clethrionomys* (Rausch, 1986; Eckert *et al.*, 1981).

Humans are accidental intermediate hosts for *Echinococcus* species. Cystic hydatid disease is caused by the larval form of *E. granulosus.* Surgery provides a cure in 50–90% of cases, but the disease recurs following treatment in 2–11% of cases (Schantz *et al.*, 1982). Alveolar hydatid disease caused by *E. multilocularis* results in insidious metastases throughout the soft organs. Until recently, the disease was invariably fatal, but progress now is being made with chemotherapy that retards the proliferation of cysts (Schantz *et al.*, 1982; Eckert, 1986; Ammann *et al.*, 1990). Observations of the spread of taeniid eggs by blowflies suggest that, in many instances, hydatid disease may be regarded as foodborne, having its greatest impact on the rural population at the lower end of the socioeconomic scale (Lawson and Gemmell, 1985; Gemmell, 1990).

In Section II, the current epidemiological status of *Echinococcus* species is reviewed.

Parasitic and Infectious Diseases

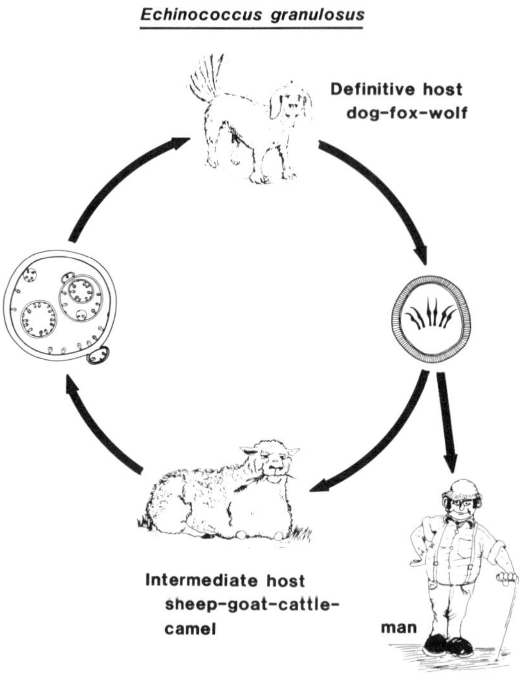

Figure 1 Life cycles of *Echinococcus granulosus* showing the definitive and intermediate hosts and transmission to humans. Adapted from Gemmell *et al.* (1987b) with permission of Elsevier Trends Journal.

Examples are given of situations in which the existence of strains of *E. granulosus* with different host specificities may influence the transmission dynamics of the parasite. The influence of environmental factors on egg survival is mentioned, and the important effect of immunity of the intermediate host to infection with the larval stage is discussed. Examples of sociological factors, such as dog feeding and control practices and animal husbandry, that have been shown to influence transmission of these parasites are discussed. These examples, although not always of primary importance to the global situation, illustrate the variety of interactions between animal hosts, parasites, and humans and the impact on transmission these factors can have.

In Section III, the history of *E. granulosus* control in New Zealand is described in detail. Although the program was successful in eradicating *E. granulosus,* the prevalence of the related parasites *Taenia hydatigena* and *Taenia ovis* did not decrease in sheep. A model is described that was used to explain this result and to determine the epidemiological and biological factors that were the causes of this difference in responses. If this model had been available prior to the start of the control program, it could have been used to assess the program's prospects for success, and as the basis of a benefit–cost analysis to aid in the planning. The use of such models for these purposes is discussed in Section IV.

II. Epidemiology

A. Geographical Distribution

Several reviews of the geographical distributions of *E. granulosus* and *E. multilocularis* have been published in the past three decades (Gemmell, 1960; Schantz and Schwabe, 1969; Williams *et al.,* 1971; Matossian *et al.,* 1977; Eckert *et al.,* 1981; Rausch, 1986; Craig *et al.,* 1991). These reviews clearly demonstrate that *E. granulosus* is a ubiquitous parasite. To date, *E. multilocularis* is confined to the Northern Hemisphere but is spreading south, having reached the north central states of the United States (Schantz, 1991) and the northern islands of Japan (Iida, 1969). *Echinococcus oligarthrus* and *E. vogeli* are found only in Central and South America, and will not be discussed further in this chapter.

In areas in which control campaigns have been mounted against *E. granulosus,* some successes have been had (Fig. 2). Control methods that have been used include the correct disposal of offal, improved dog control, mass or selective dog killing, and chemotherapy to remove tapeworms (Gemmell, 1979).

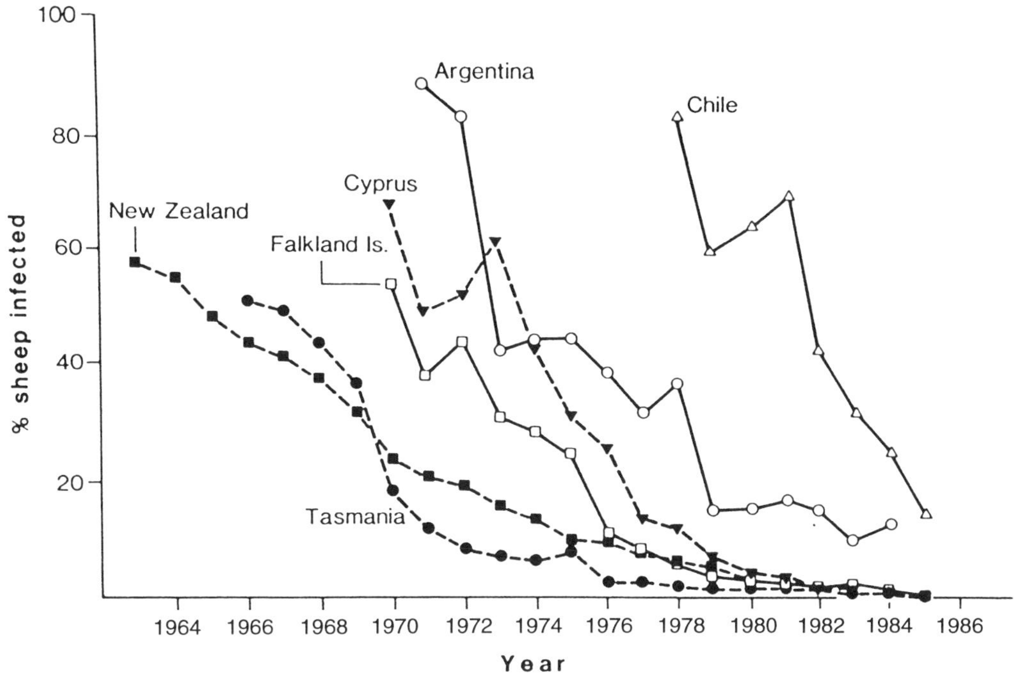

Figure 2 Prevalence of *Echinococcus granulosus* in aged sheep during national control programs in New Zealand, Tasmania, Cyprus, Falkland Islands, Argentina (Neuquen), and Chile (Region XII). Reproduced from Gemmell and Lawson (1986) and Gemmell *et al.* (1987b) with permission of Elsevier Trends Journals.

B. Biological Factors that Influence Transmission

1. Strain Differences

Echinococcus granulosus can exist in domestic, feral, and wildlife predator–prey relationships (Rausch, 1986). Self-insemination by the hermaphrodite adult and polyembryony by the larvae have given rise to clones that sometimes favor specific host assemblages (Smyth and Smyth, 1964; McManus and Bryant, 1986; Thompson, 1986; Thompson and Lymbery, 1988). The practical need to categorize strains lies in their pathogenicity to humans. For example, more than 50% of horses in the United Kingdom have been reported to be infected with *E. granulosus equinus,* but this "horse strain" is not believed to be infective to humans; cystic hydatid disease occurs only in regions in which the "sheep strain" is present (McManus *et al.,* 1989). In this case, the existence of different strains may have a crucial effect on the prevalence of infection in humans.

Strain differences also may shed light on the mechanisms of transmission. In Australia, a well-established cycle of *E. granulosus* exists between dingos and macropods (Durie and Riek, 1952). Thompson and Kumaratilake (1985) suggested that separate domestic and wildlife strains existed in Australia; Hobbs *et al.* (1990) showed that *E. granulosus* originating from different intermediate hosts could be distinguished by virtue of their hook morphology. However, evidence now exists that these two are the same strain. Hope *et al.* (1991) were unable to distinguish between *E. granulosus* from macropod marsupials and sheep from Australia or New Zealand using restriction fragment length polymorphism (RFLP) analysis; direct sequencing of a limited number of loci has failed to detect DNA sequence differences between parasites recovered from British or Australian sheep; macropods, humans, or dingos (Hope *et al.,* 1992). Therefore, foci of *E. granulosus* in Australia appear to be maintained in a wildlife reservoir, and human cases may originate from either domestic or wild animal infections.

2. Environmental Factors

In addition to controlling the blowfly population, which is a major vector for eggs of *Echinococcus* species (Lawson and Gemmell, 1985), environmental factors determine the life-span and infectivity of the free-living egg stage (Gemmell *et al.,* 1987a). For example, *E. granulosus* is uncommon in areas of Australia in which hot dry conditions prevail (Gemmell, 1990). *Echinococcus multilocularis* is endemic in mountainous regions of central Europe, its prevalence in foxes reaching 30–40% in some areas (Delattre *et al.,* 1991). In the latter regions, low temperatures preserve the eggs for a sufficient time to facilitate the infection of rodents and, hence, aid transmission of the parasite.

3. Intrinsic Factors

The factors that are intrinsic to the host–parasite interaction and determine the dynamics of the parasite population are the biotic potential of the parasite and the host's immunity to infection, both innate and acquired (Gemmell *et al.,* 1986a).

We define the biotic potential as the potential number of viable cysts that could be produced per infected dog per day. Table 1 shows that, using estimates for dogs in New Zealand in the late 1950s (Gemmell *et al.,* 1986b,1987a), the biotic potential of *E. granulosus* was approximately 1/100 of that of *T. hydatigena.* The low biotic potential meant that the infection pressure of *E. granulosus* on sheep was insufficient to stimulate a lasting acquired immunity to infection, and accounted for the success of the control program against the parasite (Roberts *et al.,* 1987). In other situations, in which worm burdens in dogs are much higher (e.g., Kenya; Macpherson *et al.,* 1985), the biotic potential would be higher. The impact of biotic potential on the basic reproduction ratio (R_0) is discussed in detail in Section III,C.

Table 1 Biological Parameters of *Echinococcus granulosus* and *Taenia hydatigena* that Determine the Biotic Potential of the Parasites[a]

	E. granulosus	*T. hydatigena*
Mean number of eggs per proglottid	587	38,000
Number of proglottids shed per worm per day	0.0714	1
Mean number of worms per infected dog	202	1
Number of eggs shed from average infected dog per day	8,470	38,000
Proportion of eggs that transform into viable cysts	0.0033	0.071
Potential number of viable cysts per infected dog per day (biotic potential)	28	2,698

[a] Estimates from Gemmell *et al.* (1986b, 1987a).

C. Socioecological Factors That Influence Transmission

The manner in which dogs are fed and controlled has a major influence on the transmission of *E. granulosus,* even in developed countries. In North Wales, for example, farm and gun dogs have been found to be free of the parasite, which is confined to foxhounds (Stallbaumer, 1987). Researchers suggested that a lifestyle that comprises hunting over a large area, and a diet high in uncooked carcasses, makes the foxhound the ideal definitive host. However, in many societies deliberate farming policies promote transmission; for example, in Morocco over 50% of stray dogs were found to be infected, with a high prevalence in sheep and cattle. A questionnaire showed that the human population was ignorant of the parasite life cycle; all farmers admitted to offering hydatid-infected viscera to dogs (Pandey *et al.,* 1988).

In many instances, the lack of knowledge of the parasite life cycle by the local population has been implicated in facilitating transmission. *Echinococcus granulosus* has been recorded over approximately 87% of the territories in China (Craig *et al.,* 1991). In Xinjiang province alone, up to 1000 human cases are treated annually; slaughterhouse records for 1988 showed 71% of sheep and a high proportion of other domestic animals to have hydatid cysts. A local survey of dogs found a prevalence of 58%. A similar survey (with similar results) in the Hutubi County of the Xinjiang/Uygar Autonomous Region also incorporated a questionnaire to assess the villagers' knowledge of the disease (Chi *et al.,* 1990). About half the inhabitants had heard of hydatid disease, but only 18% knew that a tapeworm was the causative agent. Of sheep owners, 76% had seen a hydatid cyst in butchered animals and 77% routinely gave raw offal to their dog.

Echinococcus granulosus is present in Arctic and sub-Arctic regions of North America. The principal definitive host is the wolf; the intermediate hosts are moose and caribou (Schantz, 1991). Of greater threat to human health in these northern regions is *E. multilocularis.* Almost all human cases occur in Eskimo villages, where foxes are trapped and skinned and where greens, roots, berries, and drinking water that become contaminated with fox feces are other potential sources of infection. However, a case study in Alaska found no evidence that trapping and skinning foxes increased the risk of human exposure (Stehr-Green *et al.,* 1988). Patients were more likely to have owned dogs for their entire life, to tether their dogs close to their house, and to have houses built on the tundra rather than on gravel or permanent foundations. Hence, the route to human infection appeared not to be directly from the fox feces but through the domestic dog.

An extensively studied area of Africa is the Turkana district of northwestern Kenya, which has the highest incidence of human hydatid disease in the world (Watson-Jones and Macpherson, 1988). High prevalences of the adult tapeworm are observed in the domestic dog (Macpherson *et al.,* 1985) and the jackal (Macpherson *et al.,* 1983), although prevalence of the larval stage is low in wild herbivores and all domestic livestock except cattle (Macpherson *et al.,* 1983, 1985). The people of this region live in close proximity to their dogs, which are allowed to lick children, scavenge from cooking utensils, and defecate within the home area (Watson-Jones and Macpherson, 1988).

Studies in the Maasai regions of Kenya and northern Tanzania have found the life cycle in domestic animals to be important (Macpherson, 1985). Macpherson *et al.* (1989) speculated on why the incidence of hydatid disease in humans is much lower in the Maasai than in the Turkana region. In both regions, extensive human–dog contact occurs and dogs are fed offal containing cysts. Although the intensity of infection in Maasai dogs is much lower, this difference is compensated for by the cooler climate, which allows eggs to survive longer. These authors concluded that the difference may be the result of the periods of drought in Turkana, during which people and infected dogs are concentrated in famine relief camps.

Therefore, for *E. granulosus,* dog control and feeding practices apparently influence transmission to domestic animals. Risky practices often continue when the dog owner is ignorant of the life cycle of the parasite. For both *E. granulosus* and *E. multilocularis,* the relationship between human and dog influences the risk of transmission of the parasite to humans. Poor socioeconomic conditions often lead directly to conditions that facilitate this transmission.

III. Hydatid Disease in New Zealand

A. Historical Review

Echinococcus granulosus was present in domestic animals in Australia before 1840, and almost certainly was introduced to New Zealand from that source. The first human case of cystic hydatid disease in New Zealand was recorded at Dunedin Hospital in 1862 (Gemmell, 1990). At that time, the cause of the disease was not well understood; in fact, people widely believed that humans contracted the disease from eating undercooked sheep meat. Evidence of the life cycle and modes of transmission became accepted by the mid-1870s and, in 1898, the Department of Agriculture recommended that all dogs should be dosed with extract of areca nut (Gemmell, 1990). In 1937, an amendment to the Dog Registration Act (1908) required local authorities to supply arecoline hydrobromide (the active constituent of the areca nut) to all owners of registered dogs. An addition to the regulations in the Meat Act (1940) made feeding raw offal to dogs illegal. These measures were accompanied by education campaigns by the Departments of Health and Agriculture, but no evidence exists that transmission in animals or humans was modified during this period (Gemmell, 1973).

The Department of Health recorded data on patients with cystic hydatid disease from 1945. The annual number of cases peaked at 123 in 1946 (7.2 per 100,000) and then declined to just 12 cases (0.37 per 100,000) in 1987 (M. G. Roberts and M. A. Gemmell, unpublished data), largely because of the official control program that was initiated in 1959.

B. Experimental and Official Control Programs

The Styx Field trial was initiated in 1943 in an isolated valley of the Maniototo Plain on the South Island of New Zealand (Gemmell, 1968; Gemmell *et al.,* 1986c). The aim of the trial was to determine the effectiveness of a regular anthelmintic treatment program for dogs in controlling infections with *E. granulosus* and *T. hydatigena* in sheep. From 1943 to 1951, dogs were treated with arecoline hydrobromide at 3-mo intervals. The prevalence of larvae in the livers of adult sheep declined from approximately 80% to 40% during this period, but that in lambs remained constant. No differentiation was made between the two parasite species (Gemmell *et al.,* 1986c). The trial was re-introduced in 1958, the chemotherapy being re-inforced by the strong educational component that resulted from the presence of the dog owner during the diagnosis of worms following treatment with arecoline. In 1965, arecoline treatment was replaced with treatments with bunamidine hydrochloride at 4-wk intervals. *Echinococcus granulosus* was eradicated from ewes in the area by 1965, but the prevalence of *T. hydatigena* remained constant. Investigators noted that the numbers of cysts per ewe actually increased (Gemmell, 1968; Gemmell *et al.,* 1986c).

The national control program began with the enactment of the Hydatids Act in 1959. All dogs were required to be treated with arecoline hydrobromide four times annually. Prior to 1958, *T. ovis* had been seen rarely in dogs or sheep in New Zealand, but by 1968 it was being detected in more than 5% of lambs and ewes at meat inspection. Control of this parasite then was included in the Hydatids Act. After 1972, dogs were treated with niclosamide every 6 wk. This drug has no effect on *E. granulosus* (Gemmell and Johnston, 1981). Beginning in

1978, praziquantel, which is effective against both parasites, was used (Gemmell, 1973).

The prevalence of *E. granulosus* in dogs and sheep had been 37% and 89%, respectively, in 1958 (Gemmell, 1990) but by 1990, no dogs were found to be infected and infected sheep were confined to a few farms (National Hydatids Council, 1990; Heath and Kasper, 1991). In contrast, although the detected incidence of *T. ovis* in lambs had been reduced to less than 1%, the prevalence in ewes had not decreased.

Paradoxically, therefore, although the national control program was so successful in eradicating *E. granulosus, T. ovis* appeared as a new problem. A model was developed to explain this difference, and to assess the prospects for control or eradication of *T. ovis*.

C. Analytical Approach

Figure 3 shows the mean numbers of cysts of *E. granulosus* and *T. hydatigena* in sheep of different ages in New Zealand in 1958, prior to the start of the official control program. Clearly, whereas the number of *E. granulosus* in sheep increased with age, the number of *T. hydatigena* did not. Since environmental and behavioral factors had remained relatively constant throughout the life of the sheep, the infection pressure should have remained constant (Roberts *et al.,* 1987). The figure, therefore, demonstrates that sheep had acquired immunity to *T. hydatigena,* but not to *E. granulosus,* at an early age.

Figure 4 demonstrates this principle for a hypothetical parasite. For a low infection pressure, immunity is not acquired, superinfection occurs, and the intensity of infection increases linearly with age. At higher infection pressures, superinfection is prevented and young and old animals have similar numbers of parasites. Indeed, depending on the dynamics of acquired immunity, old animals may have fewer parasites under a high infection pressure than

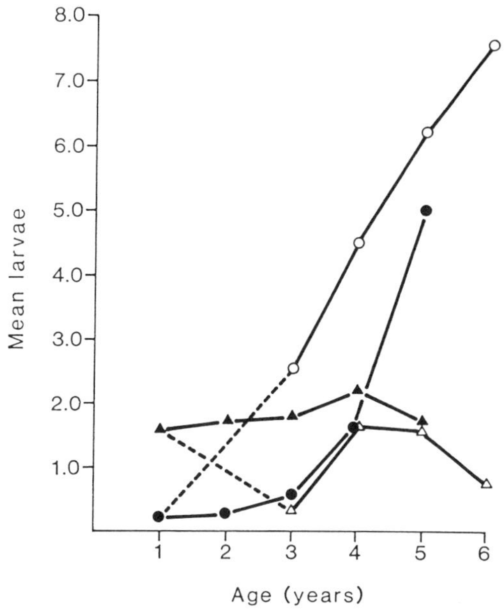

Figure 3 Mean numbers of cysts (larvae) of *Echinococcus granulosus* (●,○) and *Taenia hydatigena* (▲,△) in male and female sheep, respectively, at different ages in New Zealand in 1958. Reproduced from Roberts *et al.* (1987) with permission of Cambridge University Press.

under a low one. For example, if the basic infection pressure were 4 larvae/yr in Fig. 4, and an applied control measure reduced this to 1 larva/yr, although the mean number of cysts in 1-yr-old sheep would be reduced (from A to B in Fig. 4), the mean number in 5-yr-old sheep would increase (from C to D). Hence, a control measure that reduces the infection pressure, for example, dog dosing, could increase parasite numbers in older sheep; if an improvement in meat quality were the objective, such a control measure would be counterproductive. This result was demonstrated in the Styx Field trial, when the larval population of *T. hydatigena* increased in adult sheep following a 4-weekly dog-dosing program with bunamidine hydrochloride in 1964 (Gemmell, 1990).

That sheep can acquire immunity to infection with the larval stages of cestodes

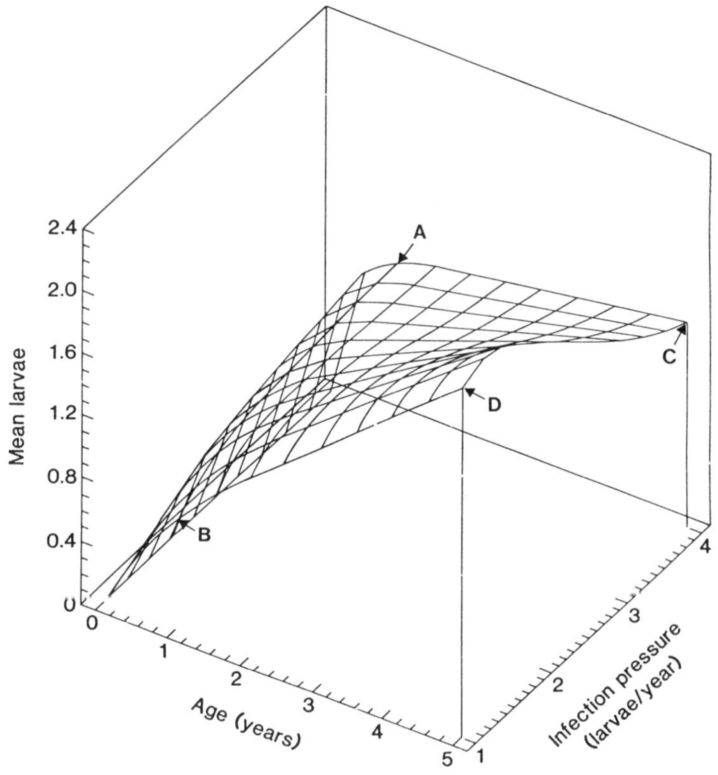

Figure 4 Mean numbers of cysts (larvae) in sheep at different ages and infection pressures derived from the model with hypothetical but realistic parameter values. The points indicated on the surface are: A, age 1 yr, infection pressure 4 larvae/yr; B, age 1 yr, infection pressure 1 larva/yr; C, age 5 yr, infection pressure 4 larvae/yr; D, age 5 yr, infection pressure 1 larva/yr.

(Gemmell and Macnamara, 1972), and that dogs can acquire immunity to infection with *E. granulosus* at very high infection pressures (Gemmell *et al.,* 1986b), is well known. Since no evidence has been found for a "crowding effect," or of parasite-induced host mortality, acquired immunity is the only possible density-dependent constraint in the parasite life cycle. Roberts *et al.* (1986a), however, determined that the infection pressure in New Zealand in 1958 was so low that immunity would not be acquired by either host under natural conditions. Also, no evidence is available of any density-dependent constraint in the life cycles of *T. ovis* or *T. hydatigena,* other than acquired immunity in the intermediate host (Gemmell *et al.,* 1987a). Clearly therefore, Fig. 3 shows that this constraint operated in the case of *T. hydatigena* prior to the introduction of control.

A simple mathematical model that describes the population dynamics of cestodes in domestic life cycles has been described by Roberts *et al.* (1986a,1987). A measure of parasite abundance was termed the basic infection pressure (h), and defined as the mean rate at which an intermediate host animal (sheep) would acquire larvae in the absence of density-dependent constraints. Since acquired immunity in the intermediate host is the only such constraint in the system, h is proportional to the mean number of worms in dogs as well as to the mean density of eggs on the pasture. The dynamics of the variable h are described by the nonlinear integro-differential equation

$$\frac{dh}{dt} = -\mu h + \lambda \int_0^t f(t-\tau)h(\tau)S(\tau)\, d\tau, \quad (1)$$

where $S(t)$ is the proportion of sheep that is susceptible to infection at time t, μ is the rate at which parasites are lost from the life cycle by processes such as the desiccation of eggs or the mortality of adult parasites, and λ is an overall transmission rate. The function $f(t\text{-}\tau)$ is the probability density of the time from the establishment of a cyst (τ) to its being mature and ingested by a dog (t), conditional on this latter event (Roberts *et al.*, 1986a). Hence, the convolution integral introduces a distributed delay that models the transmission delays in the life cycle during the larval stage in the intermediate host.

The proportion of sheep that is susceptible to infection (S) decreases at a rate proportional to the infection pressure (h) and increases as immunity wanes. As acquired immunity is boosted by re-exposure (Gemmell *et al.*, 1987a), the rate of decrease of immunity (γ) is a function of h:

$$\frac{dS}{dt} = -ahS + \gamma(h)(1-S). \quad (2)$$

These two equations form a hybrid model from the system. The integro-differential equation, Eq. 1 is expressed in terms of parasite intensity, whereas the differential equation, Eq. 2 bears more relation to a prevalence framework. These equations are valid only for the domestic situation in which the host populations are constant.

Before considering appropriate parameter values, performing a steady-state analysis of the model is easy. The equations have two equilibria; one at $h = 0$, $S = 1$, at which the parasite is extinct and all sheep are susceptible to infection, and a second at which $S = \mu/\lambda$ and h is the solution of $ah = \gamma(h)(\lambda/\mu - 1)$. Since γ is a decreasing function of h, the second equilibrium is determined uniquely by the equation, but can exist only if the basic reproduction ratio $R_0 = \lambda/\mu$ is greater than 1.

The bifurcation diagram (Fig. 5) shows the possible equilibrium values of h for different values of R_0, calculated for arbitrary but realistic values of $\gamma(h)$, a, μ, and λ. The extinction state ($h = 0$) exists for all values of R_0, and is globally asymptotically stable for $R_0 < 1$ but is unstable for $R_0 > 1$ (Roberts *et al.*, 1987). Therefore, if $R_0 < 1$, the parasite population cannot maintain itself and is doomed to extinction with time. However, if $R_0 > 1$, the conditions are favorable for the parasite population to persist. Under these conditions, if the parasite is absent from a host population and is introduced subsequently it will become established within that population. Roberts *et al.* (1987) showed that the stability of the nonzero equilibrium depends on the precise nature of the function $\gamma(h)$. Insufficient data are available to determine this function at present. The authors coined the terms "endemic state" for the nonzero equilibrium when R_0 is near to but greater than 1 and "hyperendemic state" when R_0 is much greater than 1. A parasite population was said to have extinction, endemic, or hyperendemic status depending on whether $R_0 < 1$, $R_0 \simeq 1$ and $R_0 > 1$, or $R_0 \gg 1$, respectively.

Estimating R_0 with any degree of precision is not easy (see Chapter 3). The equation for the nonzero equilibrium value of h may be rearranged to obtain $R_0 = 1 + ah/\gamma(h)$, which may be interpreted as

$$R_0 = 1 + \frac{\text{mean duration of immunity}}{\text{mean time to immunity}}. \quad (3)$$

If the infection pressure is high, any acquired immunity is life long, so Eq. 3 becomes

$$R_0 = \frac{\begin{array}{c}\text{mean life}\\ \text{expectancy at birth}\end{array}}{\begin{array}{c}\text{mean age at which}\\ \text{immunity acquired}\end{array}}. \quad (4)$$

Using these formulae, Roberts *et al.* (1987)

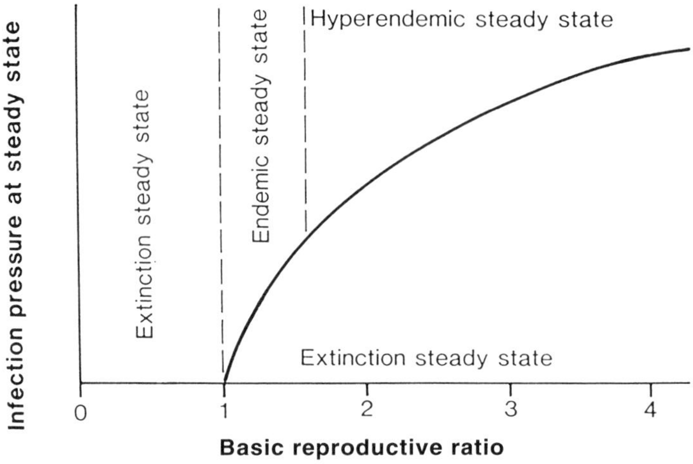

Figure 5 Bifurcation diagram indicating the various model equilibrium infection pressures (h) for different values of basic reproduction ratio (R_0). Reproduced from Roberts *et al.* (1986b) with permission of Elsevier Science Publishers.

estimated the values of R_0 for *E. granulosus* in male and female sheep in New Zealand in 1958 to be not greater than 1.3 and 1.6, respectively. The value of R_0 is higher in female sheep because they have more contact with dogs. Similar estimates for *T. hydatigena* were 2.5 and 4.3; R_0 for *T. ovis* would be comparable. The final piece of information required is the estimate by Roberts *et al.* (1986a) that the dog-dosing program reduced R_0 for *E. granulosus* by 69%. The less-than-perfect result is caused by an estimated 75% capture rate for dogs, 95% efficacy for the treatment, and the small percentage of worms that achieved patency in less than 6 wk. The behavior of these three parasites during the control program now can be explained.

Prior to the official control program in New Zealand, *E. granulosus* was endemic. The education campaign combined with regular dog dosing transformed the parasite population to extinction status (Fig. 5) and inevitable eradication. This eradication will take a long time because sheep act as reservoirs of infection for several years. Prior to control, *T. hydatigena* was hyperendemic. The control program reduced the population to endemic status, and a new equilibrium level was found, old ewes now had more cysts than their lambs. *Taenia ovis* was seen rarely prior to control, but conditions were favorable for it to persist and attain a hyperendemic equilibrium level. The control program then reduced this level to an endemic equilibrium. To eradicate *T. hydatigena* or *T. ovis,* a more effective control effort than that achieved by dog dosing alone is needed.

The results of this research demonstrate that a model does not need to simulate or predict parasite numbers to shed light on parasite population dynamics. Indeed, an earlier simulation model (Harris *et al.*, 1980) had been used to compare potential control strategies for *E. granulosus, T. hydatigena,* and *T. ovis* in New Zealand, but the authors did not analyze the model qualitatively and, hence, missed the broad principles that have been oulined here. A simulation model is essential for a comparative benefit–cost analysis of control strategies, however. Lawson *et al.* (1988) used one based on the model described here to assess economically some proposed strategies for cestode control in New Zealand. Such an analysis for *Echinococcus* species hesitates at the question of the economic value of a human life, and may be of little use in absolute terms. The analysis can, however, compare the costs of strategies that are shown to be equally effective, and thus, can aid the decision makers in their design of control or eradication programs.

A control program is one that maintains the parasite population at an acceptable level and is, therefore, ongoing. In contrast, an eradication program is time limited. Once eradication is achieved, the antiparasite measures cease and the parasite population, although extinct, may return to endemic or hyperendemic status if re-introduced. In other words, the conditions again become favorable for parasite transmission after the control program ceases. Vigilance is needed to ensure that a re-introduction of the parasite does not occur, since this event could restore the population to former levels. Yekutiel (1980) defined four phases of an eradication program: preparatory, attack, consolidation, and maintenance. The preparatory phase in New Zealand was prior to 1959, when education and informal campaigns prepared society for the acceptance of a national control program. The dog-dosing campaign of 1959-1990 was the attack phase, during which all potential hosts were treated with the same measure. The consolidation phase is being entered now; control is targeted at areas where a potential problem still exists. Hopefully the fourth phase will be entered soon; vigilance will be used to maintain an eradication state. A benefit-cost analysis should be used to determine the optimal time for transition between these phases (Lawson *et al.*, 1988).

IV. Future Prospects

The New Zealand experience shows that a mathematical model can be used to explain the dynamics of a domestic *E. granulosus* population and assess its amenability to control measures. Adapting this model for other hosts, and assessing other potential control strategies, is now a simple matter. For example, the development of a vaccine against *T. ovis* is in progress (Johnson *et al.*, 1989). The same technology could, in principle, be used to develop one against *E. granulosus.* Roberts *et al.* (1987) have shown how to include vaccination of the intermediate host in the mathematical model for cestodes, and how to use this model to screen vaccine strategies.

A model for a wildlife *E. granulosus* cycle would be more complicated since the population dynamics of the hosts also must be included. This adjustment has been made for many other diseases and presents no technical problems. Similarly, a situation in which wildlife and domestic cycles coexist and interact presents complications but not mathematical challenges. Obtaining the data required to specify the host population dynamics, estimate R_0, and model control interventions would be a considerable epidemiological challenge. For example, the construction of age-intensity curves such as those in Fig. 3 requires a reliable method of aging a wild animal (Gemmell *et al.*, 1987b). This problem is compounded for *E. multilocularis,* because the rodent intermediate host experiences rapid and seemingly unpredictable population fluctuations (Gipps *et al.*, 1986); even a reliable method of counting cysts has yet to be developed. For this parasite, the sources of density-dependent regulation have not been determined.

Any control or eradication program requires a reliable disease surveillance system. Without this system, researchers cannot know whether a control program is producing the desired result, or when an eradication program should proceed from one phase to the next (Gemmell *et al.*, 1987b; Lawson *et al.*, 1988). Although seemingly self-evident, this aspect of the program often is overlooked or proceeds in a manner that renders the data unreliable.

The final objective of a control program also should be borne in mind. *Echinococcus granulosus* does not cause a considerable loss to agricultural industries, the liver being but a small portion of the price of a carcass. The reason for control of *E. granulosus* and *E. multilocularis* must be prevention of transmission to humans. This process may have ceased effectively long before the parasite has been eradicated (Gemmell, 1990). The mathematical framework provides a method for determining the characteristics of the dynamics of *Echinococcus* populations, and supports a benefit-cost analysis of control options. Hence, an intervention can be assessed for epidemiological and economic merit before it proceeds.

References

Ammann, R. W., Hirsbrunner, R., Cotting, J., Steiger, U., Jacquier, P., and Eckert, J. (1990). Recurrence rate after discontinuation of long-term mebendazole therapy in alveolar echinococcosis. *Am. J. Trop. Med. Hyg.* **43,** 506-515.

Chi, P., Zhang, W., Zhang, Z., Hasyet, M., Liu, F., Ding, Z., Andersen, F. L., Tolley, H. D., and Schantz, P. M. (1990). Cystic echinococcosis in the Xinjiang/Uygur Autonomous Region, People's Republic of China. I. Demographic and epidemiologic data. *Trop. Med. Parasitol* **41,** 157-162.

Craig, P. S., Deshan, L., and Zhaoxun, D. (1991). Hydatid disease in China. *Parasitol. Today* **7,** 46-50.

Delattre, P., Giraudou, P., and Pascal, M. (1991). L'échinococcose alvéolaire. *Recherche* **22,** 294-303.

Durie, P. H., and Riek, R. F. (1952). The role of the dingo and wallaby in the infestation of cattle with hydatids *(Echinococcus granulosus)* (Batsch, 1786) (Rudolphi, 1805) in Queensland. *Austral. Vet. J.* **28,** 249-254.

Eckert, J. (1986). Prospects for treatment of the metacestode stage of *Echinococcus. In* "The Biology of *Echinococcus* and Hydatid Disease" (R. C. A. Thompson, ed.), pp. 250-284. Allen and Unwin, London.

Eckert, J., Gemmell, M. A., Matyas, Z., and Soulsby, E. J. L. (eds.) (1981). "Guidelines for Surveillance, Prevention and Control of Echinococcosis/Hydatidosis," VPH/81.28. World Health Organization, Geneva.

Gemmell, M. A. (1960). Advances in knowledge on the distribution and importance of hydatid disease as world health and economic problems during the decade 1950-1959. *Helminthol. Abstrr.* **29,** 355-369.

Gemmell, M. A. (1968). The Styx Field trial. A study on the application of control measures against hydatid disease caused by *Echinococcus granulosus. Bull. WHO* **39,** 73-100.

Gemmell, M. A. (1973). Hydatid disease in New Zealand. The first 100 years (1883-1972). *In* "Proceedings of a Symposium Presented by the Hamilton Medical Veterinary Association" (R. W. Brown, J. R. Salisbury, and W. E. White, eds.), pp. 54-61. Hamilton, Victoria, Australia.

Gemmell, M. A. (1979). Hydatidosis control: A global view. *Austral. Vet. J.* **55,** 118-125.

Gemmell, M. A. (1990). Australasian contributions to an understanding of the epidemiology and control of hydatid disease caused by *Echinococcus granulosus*—Past, present and future. *Int. J. Parasitol.* **20,** 431-456.

Gemmell, M. A., and Johnston, P. D. (1981). Cestodes. *Antibiot. Chemother.* **30,** 54-114.

Gemmell, M. A., and Lawson, J. R. (1986). Epidemiology and control of hydatid disease. *In* "The Biology of Echinococcus and Hydatid Disease" (Thompson, R. C. A., ed.), pp. 189-216. Allen and Unwin, London.

Gemmell, M. A., and Macnamara, F. N. (1972). Immune response to tissue parasites II. Cestodes. *In* "Immunity to Animal Parasites" (E. J. L. Soulsby, ed.), pp. 235-272. Academic Press, New York.

Gemmell, M. A., Lawson, J. R., and Roberts, M. G. (1986a). Control of echinococcosis/hydatidosis: Present status of world-wide progress. *Bull. WHO* **64,** 333-339.

Gemmell, M. A., Lawson, J. R., and Roberts, M. G. (1986b). Population dynamics in echinococcosis and cysticercosis: Biological parameters of *Echinococcus granulosus* in dogs and sheep. *Parasitology* **92,** 599-620.

Gemmell, M. A., Lawson, J. R., Roberts, M. G., Kerin, B. R., and Mason, C. J. (1986c). Population dynamics in echinococcosis and cysticercosis: Comparison of the response of *Echinococcus granulosus, Taenia hydatigena* and *T. ovis* to control. *Parasitology* **93,** 357-369.

Gemmell, M. A., Lawson, J. R., and Roberts, M. G. (1987a). Population dynamics in echinococcosis and cysticercosis: Evaluation of the biological parameters of *Taenia hydatigena* and *T. ovis* and comparison with those of *Echinococcus granulosus. Parasitology* **94,** 161-180.

Gemmell, M. A., Lawson, J. R., and Roberts, M. G. (1987b). Towards global control of cystic and alveolar hydatid diseases. *Parasitol. Today* **3,** 144-151.

Gipps, J., Alibhai, S., Gurnell, J., and Krebs, C. (1986). A plague of voles: The search for a cure. *New Scientist* **1516,** 48-51.

Harris, R. E., Revfeim, K. J. A., and Heath, D. D. (1980). Simulating strategies for control of *Echinococcus granulosus, Taenia hydatigena* and *T. ovis. J. Hyg.* **84,** 389-404.

Heath, D. D., and Kasper, K. C. (1991). The final

stages of eradication of *Echinococcus granulosus* from New Zealand. *Arch. Hidatidosis* **30,** 123–128.
Hobbs, R. P., Lymbery, A. J., and Thompson, R. C. A. (1990). Rostellar hook morphology of *Echinococcus granulosus* (Batsch, 1786) from natural and experimental Australian hosts, and its implications for strain recognition. *Parasitology* **101,** 273–281.
Hope, M., Bowles, J., and McManus, D. P. (1991). A reconsideration of the *Echinococcus granulosus* strain situation in Australia following RFLP analysis of cystic material *Int. J. Parasitol.* **21,** 471–475.
Hope, M., Bowles, J., Prociv, P., and McManus, D. P. (1992). A genetic comparison of human and wildlife isolates of *Echinococcus granulosus* in Queensland and the public health implications. *Med. J. Australia,* **156,** 27–30.
Iida, H. (1969). Epidemiology of multilocular echinococcosis in Hokkaido, Japan. *In* "Multilocular Echinococcosis in Hokkaido, Japan," pp 1–15. Hokkaido Institute of Public Health, Sapporo.
Johnson, K. S., Harrison, G. B. L., Lightowlers, M. W., O'Hoy, K. L., Cougle, W. G., Dempster, R. P., Lawrence, S. B., Vinton, J. G., Heath, D. D., and Rickard, M. D. (1989). Vaccination against ovine cysticercosis using a defined recombinant antigen. *Nature (London)* **338,** 585–587.
Lawson, J. R., and Gemmell, M. A. (1985). The potential role of blowflies in the transmission of taeniid tapeworm eggs. *Parasitology* **91,** 129–143.
Lawson, J. R., Roberts, M. G., Gemmell, M. A., and Best, S. J. (1988). Population dynamics in echinococcosis and cysticercosis: Economic assessment of control strategies for *Echinococcus granulosus, Taenia ovis* and *T. hydatigena. Parasitology* **97,** 177–191.
Macpherson, C. N. L. (1985). Epidemiology of hydatid disease in Kenya: A study of the domestic intermediate hosts in Masailand. *Trans. R. Soc. Trop. Med. Hyg.* **79,** 209–217.
Macpherson, C. N. L., Karstad, L., Stevenson, P., and Arundel, J. H. (1983). Hydatid disease in the Turkana district of Kenya. III. The significance of wild animals in the transmission of *Echinococcus granulosus,* with particular reference to Turkana and Masailand in Kenya. *Ann. Trop. Med. Parasitol.* **77,** 61–73.
Macpherson, C. N. L., French, C. M., Stevenson, P., Karstad, L., and Arundel, J. H. (1985). Hydatid disease in the Turkana district of Kenya. IV. The prevalence of *Echinococcus granulosus* infections in dogs, and observations on the role of the dog in the lifestyle of the Turkana. *Ann. Trop. Med. Parasitol.* **79,** 51–61.
Macpherson, C. N., Romig, T., Zeyhle, E., Craig, P. S., and Watschinger, H. (1989). Observations on human echinococcosis (hydatidosis) and evaluation of transmission factors in the Maasai of northern Tanzania. *Ann. Trop. Med. Parasitol.* **83,** 489–497.
Matossian, R. M., Rickard, M. D., and Smyth, J. D. (1977). Hydatidosis: A global problem of increasing importance. *Bull. WHO* **55,** 499–507.
McManus, D. P., and Bryant, C. (1986). Biochemistry and physiology of *Echinococcus. In* "The Biology of *Echinococcus* and Hydatid Disease" (R. C. A. Thompson, ed.), pp. 114–142. Allen and Unwin, London.
McManus, D. P., Thompson, R. C. A., and Lymbery, A. J. (1989). Comment on the status of *Echinococcus granulosus* in the UK. *Parasitol. Today* **5,** 365–367.
National Hydatids Council (1990). "Thirtieth Annual Report." Ministry of Agriculture and Fisheries, Wellington, New Zealand.
Pandey, V. S., Ouhelli, H., and Moumen, A. (1988). Epidemiology of hydatidosis/echinococcosis in Ouarzazate, the pre-Saharian region of Morocco. *Ann. Trop. Med. Parasitol.* **82,** 461–470.
Rausch, R. L. (1986). Life cycle patterns and geographic distribution of *Echinococcus* species. *In* "The Biology of *Echinococcus* and Hydatid Disease" (R. C. A. Thompson, ed.), pp. 48–80. Allen and Unwin, London.
Roberts, M. G., Lawson, J. R., and Gemmell, M. A. (1986a). Population dynamics in echinococcosis and cysticercosis: Mathematical model of the life cycle of *Echinococcus granulosus. Parasitology* **92,** 621–641.
Roberts, M. G., Lawson, J. R., and Gemmell, M. A. (1986b). Epidemiology and control of hydatidosis and cysticercosis. *In* "Proceedings of the Pacific Statistical Congress, 1985" (I. S. Francis, B. F. J. Manly, and F. C. Lam, eds.), pp. 460–463. North-Holland, Amsterdam.
Roberts, M. G., Lawson, J. R., and Gemmell, M. A. (1987). Population dynamics in echinococcosis and cysticercosis: Mathematical model of the life cycles of *Taenia hydatigena* and *T. ovis. Parasitology* **94,** 181–197.
Schantz, P. M. (1991). Parasitic zoonoses in perspective. *Int. J. Parasitol.* **21,** 161–170.
Shantz, P. M., and Schwabe, C. W. (1969). Worldwide status of hydatid disease control. *J. Am. Vet. Med. Assoc.* **155,** 2104–2121.
Schantz, P. M., Van den Bossche, H., and Eckert, J. (1982). Chemotherapy for larval echinococcosis in animals and humans. *Z. Parasitenkd.* **67,** 5–26.
Smyth, J. D., and Smyth, M. M. (1964). Natural and experimental hosts of *Echinococcus granulosus* and *E. multilocularis,* with comments on the genetics of speciation in the genus *Echinococcus. Parasitology* **54,** 493–514.
Stallbaumer, M. (1987). The prevalence and epidemiology of cestodes in dogs in Clwyd, Wales. II. Hunting dogs. *Ann. Trop. Med. Parasitol.* **81,** 43–47.
Stehr-Green, J. K., Stehr-Green, P. A., Schantz, P. M., Wilson, J. F., and Lanier, A. (1988). Risk factors for infection with *Echinococcus multilocularis* in Alaska. *Am. J. Trop. Med. Hyg.* **38,** 380–385.

Thompson, R. C. A. (1986). Biology and systematics of *Echinococcus*. *In* "The Biology of *Echinococcus* and Hydatid Disease" (R. C. A. Thompson, ed.), pp. 5–43. Allen and Unwin, London.

Thompson, R. C. A., and Kumaratilake, L. M. (1985). Comparative development of *Echinococcus granulosus* in dingoes *(Canis familiaris dingo)* and domestic dogs *(Canis familiaris familiaris)* with further evidence for the origin of the Australian sylvatic strain. *Int. J. Parasitol.* **15,** 535–542.

Thompson, R. C. A., and Lymbery, A. J. (1988). The nature, extent and significance of variation within the genus *Echinococcus*. *Adv. Parasitol.* **27,** 209–258.

Watson-Jones, D. L., and Macpherson, C. N. L. (1988). Hydatid disease in the Turkana district of Kenya. VI. Man: dog contact and its role in the transmission and control of hydatidosis amongst the Turkana. *Ann. Trop. Med. Parasitol.* **82,** 343–356.

Williams, J. F., López-Adaros, H., and Trejos, A. (1971). Current prevalence and distribution of hydatidosis, with special reference to the Americas. *Am. J. Trop. Med. Hyg.* **20,** 224–236.

Yekutiel, P. (1980). Contributions to Epidemiology and Biostatistics "Eradication of Infectious Diseases: A Critical Study," (M. A. Klingberg, ed.), Vol. 2, Karger, Basel.

Stephen Vail, Gary Smith, & Cynthia Lord

Population Biology of *Ixodes scapularis,* the Vector of Lyme Disease in the Eastern and North Central United States

19

I. Introduction

Lyme borreliosis is a disease that occurs in an unknown proportion of hosts bitten by ixodid ticks infected with the spirochete *Borrelia burgdorferi.* The distribution of Lyme disease in Europe, Asia, and the United States is determined by the distribution of appropriate tick vectors (Ai *et al.,* 1988; Ciesielski *et al.,* 1988; Stanek *et al.,* 1988). Obvious signs of disease have been observed in dogs, horses, cattle, and humans (Carmel and Edwards, 1989). The spirochete is capable of invading multiple organ systems in any of these hosts and may cause a variety of mild to severe musculoskeletal, neurological, and cardiac symptoms (Carmel and Edwards, 1989; Rahn and Malawista, 1991). Ecological epidemiological research, which naturally has focused on ways of preventing transmission from ticks to humans, is especially important because Lyme disease is increasingly refractory to antibiotic treatment in the later stages of infection.

This chapter addresses in detail what little we know about the population biology of the tick vector in the United States, *Ixodes scapularis.* We shall show how mathematical models of tick demography can be useful in making sense of field and laboratory observations of tick development and survival, and then describe some more inclusive models that are designed to elucidate the likely outcome of plausible disease control strategies.

II. Vector Population Biology

Field surveys and epidemiological data suggest that the seasonal incidence of Lyme disease (Schulze *et al.,* 1985; Piesman *et al.,* 1987; Lane *et al.,* 1991), as well as its geographic distribution and epidemic spread (Wallis *et al.,* 1978; Main *et al.,* 1981; Spielman *et al.,* 1985; Anderson, 1989), depend primarily on changes in the abundance of the tick vector. Therefore, understanding the processes that control and regulate tick numbers is important. *Ixodes scapularis* is the vector in eastern North America, where over 85% of reported North American cases of Lyme borreliosis have occurred [Centers for Disease Control (CDC), 1992]. *Ixodes pacificus* is the vector on the West Coast (about 4% of reported cases; CDC, 1992).

In this chapter, we restrict our attention to *I. scapularis.* This is not meant to suggest that *I. scapularis* is necessarily typical of the

Parasitic and Infectious Diseases

Lyme disease vectors. Note that the dynamics of other vectors in other regions may be different, not only quantitatively (e.g., different development rates or seasonal times of occurrence) but also qualitatively. For example, the rapid spread of Lyme disease in the northeastern United States is probably the result of bird-feeding juvenile *I. scapularis* (Spielman *et al.*, 1985; Battaly *et al.*, 1987). In contrast, the principal vector on the West Coast, *I. pacificus,* is seldom seen on birds. Further, generalized spirochetal infections and transovarial transmission of spirochetes are much more common in *I. pacificus* than in *I. scapularis* (Lane and Burgdorfer, 1987; Lane *et al.*, 1991). Such differences are likely to have a profound influence on vector population dynamics and Lyme disease epizootiology.

Current knowledge of the relevant biology of *I. scapularis* falls roughly into two categories: (1) knowledge of distribution, abundance, and host associations obtained from observation in the field and (2) knowledge of development, mortality, and behavior obtained from field and laboratory experiments. A principal goal of the population biologist is linking these two kinds of information. The intention is to explain patterns of abundance in terms of mechanisms that constrain tick survival and development at the individual level. A primary tool is the mathematical model.

Mathematical models are built for some purpose, which the form of the model reflects. Most published models for other tick species are of a sort designed to mimic, as closely as possible, the observed dynamics of abundance for a particular population (Gardiner *et al.*, 1981; Haile and Mount, 1987; Mount and Haile, 1989). Such models have been called "tactical" models in the ecological literature (May, 1974). These models have been successful for several important tick species, but depend on a comprehensive and detailed database quantifying the developmental rates, mortality, and behavior of all stages in the life cycle as functions of temperature, humidity, host density, and so on. Empirical knowledge of *I. scapularis* falls far short of this level of detail. Instead, bits and pieces of the whole picture are available, so models are required as research tools to determine how the pieces fit together, to deduce the consequences under various assumptions, to suggest which assumptions most need to be tested, and to identify the areas in which quantitative data are most sorely lacking. Toward these ends, we employ a variety of models—some very simple, others more complex.

In the sections that follow, current knowledge of *I. scapularis* population biology as it relates to tick and Lyme disease dynamics is reviewed. We will begin with a discussion of observed population dynamics and life-cycle phenomena, proceed to a review of demographic data from field and laboratory experiments, and conclude with a brief description of two attempts to integrate this information into the framework of a mathematical model.

A. *Ixodes scapularis* Life Cycle

The life cycle of *I. scapularis* includes three post embryonic developmental stages: larva, nymph, and adult. Each stage can be subdivided in turn according to three sequential phases of activity: "questing," in which the unfed tick seeks a host and attaches to it; "feeding," in which the attached tick feeds, becomes engorged, and then drops off; and "developing," in which the detached engorged tick develops to emergence as a questing tick of the next stage (or lays eggs, in the case of an adult female). During each substage, a tick has some probability of dying before passing to the next substage (Fig. 1). *Ixodes scapularis* must seek a new host during each of the three main stages of the life cycle.

B. Host Associations

The considerable literature on the host associations of *I. scapularis* has been reviewed by Lane *et al.* (1991). Like all ixo-

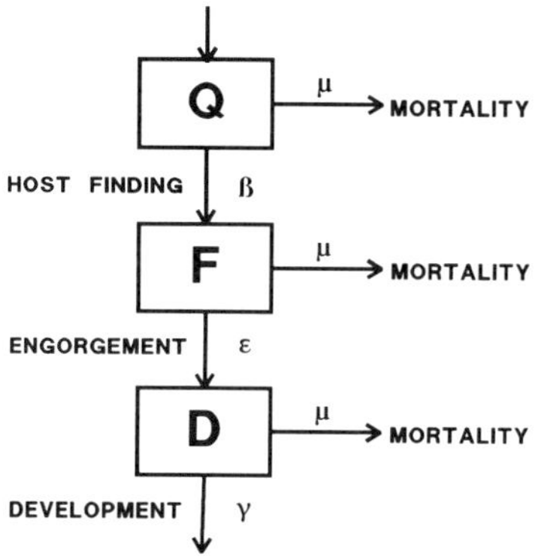

Figure 1 Juvenile and adult ticks pass through the three sequential phases of activity shown here. *Q*, Questing; *F*, feeding; *D*, developing. The Greek letters refer to the instantaneous *per capita* rates used in Eq. 1.

dids of the *Ixodes ricinus* complex, *I. scapularis* is found on many vertebrate hosts (31 mammals and 49 birds), the adults having a somewhat more restricted host range (13 species) than the larvae and nymphs (Lane *et al.*, 1991). Nevertheless, the dynamics of *I. scapularis* and Lyme disease appear to depend largely on the relationships with just two host species: the white-footed mouse *(Peromyscus leucopus)* and the white-tailed deer *(Odocoileus virginianus)*. The former is the most common host species for juveniles (larvae and nymphs) in many habitats, and the most frequently infested (Lane *et al.*, 1991). White-footed mice are also a highly competent reservoir for Lyme spirochetes (Donahue *et al.*, 1987; Lane *et al.*, 1991). Adult *I. scapularis* generally are found on medium-sized and large mammals; among these, *O. virginianus* is the most important. Deer are incompetent as reservoirs for Lyme disease (Telford *et al.*, 1988), but strong circumstantial evidence (Peisman *et al.*, 1979; Main *et al.*, 1981; Spielman *et al.*, 1985; Wilson *et al.*, 1985,1990; Spielman, 1988; Lane *et al.*, 1991), and some experimental evidence (Wilson *et al.*, 1988) suggest that deer populations are a crucial and limiting resource for *I. scapularis*. Note that humans are strictly incidental to the overall dynamics of *I. scapularis* and Lyme disease.

Although birds appear to be relatively unimportant in the dynamics of *I. scapularis* populations in any one location, good evidence exists that they have played an important role in the geographic spread of *I. scapularis* and Lyme disease (Anderson and Magnarelli, 1984; Anderson *et al.*, 1986; Battaly *et al.*, 1987). Both larvae and nymphs have been found on a wide variety of bird species, some of which appear to be competent reservoirs for the spirochetes (Anderson *et al.*, 1986; Lane *et al.*, 1991). Further, the expansion of the range of *I. scapularis* and Lyme disease has occurred primarily along the axes of important migratory flyways. The south-biased asymmetry of the expansion is consistent with the seasonal timing of spring and fall migrations and tick availability (Spielman *et al.*, 1985; Battaly *et al.*, 1987).

C. Population Dynamics

Population dynamics patterns have within-year and between-year components. Seasonal patterns of *I. scapularis* abundance have received the most attention. The basic pattern is a sequence of peaks in the abundance of active nymphs, larvae, and adults through the course of the year (Fig. 2; Carey

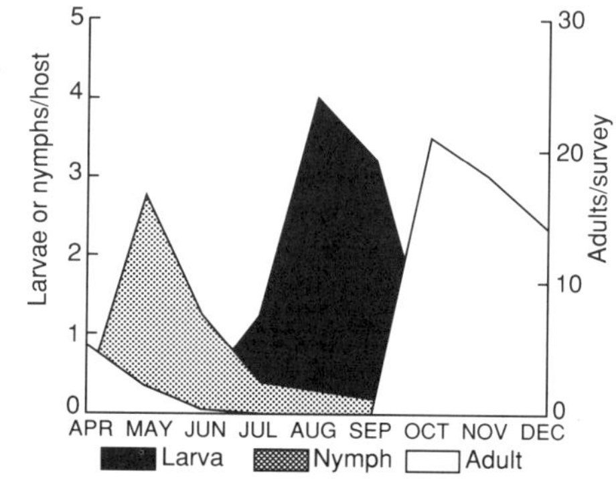

Figure 2 Seasonal changes in the relative abundance of *Ixodes scapularis* larvae, nymphs, and adults in New Jersey. Data from Schulze *et al.* (1985).

et al., 1980; Schulze *et al.*, 1985; Spielman *et al.*, 1985; Wilson and Spielman, 1985; Lane *et al.*, 1991; Siegel *et al.*, 1991). The pattern of adult mortality in field experiments suggests that unfed adults of spring are survivors of a cohort that emerged the preceding fall (Daniels *et al.*, 1989; Yuval and Spielman, 1990). In some locations, a minor peak of larval abundance seems to precede the main one in late summer (Main *et al.*, 1982; Schulze *et al.*, 1985; Main, 1986). Two features of the seasonal pattern are of special interest with respect to Lyme disease epidemiology. The first is the inversion of order of larvae and nymphs in the seasonal sequence of occurrence. Larvae develop into nymphs but, paradoxically, nymphs have an earlier peak in abundance. The surge of nymphal activity detected early in the year is a consequence of the large numbers of unfed nymphs that over-winter and begin questing in the spring. The questing larvae that are detected later in the year have hatched from eggs laid at about the time the overwintered nymphs resume activity. The nymphs that develop from these larvae have little opportunity to seek new hosts before the declining temperatures of autumn, so they are not detected in large numbers until the following spring. Because transovarial transmission of spirochetes is rare in *I. scapularis,* unfed larvae are generally free of disease (Piesman *et al.*, 1986) and must acquire it from infected spirochete-competent hosts if the cycle of infection is to be maintained (Spielman *et al.*, 1985). Because of the high turnover in populations of *P. leucopus* (50–80% over-wintering mortality in several mid-Atlantic and midwestern states; Krohne *et al.*, 1988), these hosts must acquire the infection from spring-feeding nymphs (which acquired the infection as larvae the preceding year) before the late summer emergence of unfed larvae. The second significant feature is the occurrence of the nymphal abundance peak in those summer months during which many people are engaged in outdoor activities (Piesman *et al.*, 1987; Miller *et al.*, 1990; Lane *et al.*, 1991).

Between-year patterns are harder to discern directly. Several studies report tick infestation levels over 2 or more years, but the data are, at best, suggestive of trends. Multiyear studies between 1976 and 1978 on deer (Anderson and Magnarelli, 1980) and between 1975 and 1978 on *P. leucopus* (Piesman and Spielman, 1979) in Connecticut reported continuously increasing infestation levels (mean ticks/individual). Later studies on deer (Main *et al.*, 1981) and *P. leucopus* (Main *et al.*, 1982) in Connecticut, and on *P. leucopus* in Massachusetts (Wilson and Spielman, 1985), reported constant or declining average levels of infestation. These data suggest, but cannot show conclusively, that *I. scapularis* abundance, after growing rapidly during the late 1970s in these locations, has approached a saturation value near which it will be regulated.

Lyme disease case data provide indirect corroboration for this view (e.g., Cartter *et al.*, 1989). Although the data are far from conclusive, the total number of Lyme cases for the United States seems to be leveling off (Fig. 3); to the extent that Lyme case data reflect abundance of the vector, they suggest that tick abundance is being regulated. Regional comparisons are interesting in this regard. Investigators generally accept that the geographic range of *I. scapularis* has expanded from New England, primarily to the south and west (Spielman *et al.*, 1985; Lane *et al.*, 1991). The case data from the state of New York are strongly indicative of a stable equilibrium. The case data from Maryland, in contrast, indicate either continuing exponential growth or the very beginning of saturation.

D. Generation Time, Diapause, and Other Conundrums

No one has followed a cohort of ticks through an entire generation, that is, from egg to egg, under field conditions. There-

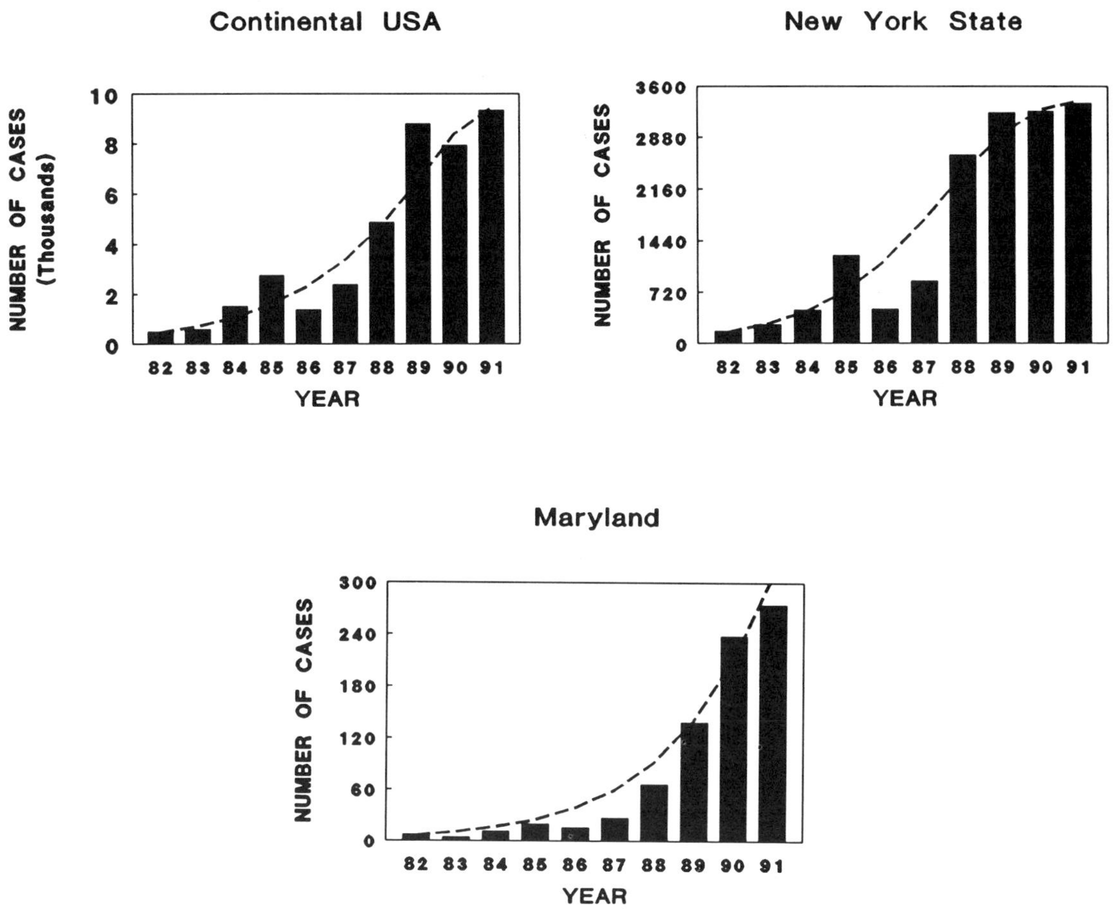

Figure 3 Total number of reported cases of Lyme disease in humans for 1982–1991 in the continental United States, New York, and Maryland. Data from Centers for Disease Control.

fore, our knowledge of the generation time of *I. scapularis* is limited to a reasonable presumption that most individuals complete a generation within 2 yr. Clearly, individuals cannot complete a generation within a single season, because larvae deriving from eggs deposited in the spring first become engorged in late summer, and nymphs generally are not present until the following spring. Spring-feeding nymphs molt to adults in the fall; adult females, after feeding in the fall, oviposit the following spring to complete the canonical 2-yr life cycle (Fig. 4).

The generation time may be extended beyond 2 yr for some individuals, for example, if some unfed larvae or nymphs survive the winter and feed the following summer. Some of the available data on survivorship of larvae and nymphs are suggestive of this possibility (Yuval and Spielman, 1990). The extent to which this occurs, if at all, has an important bearing on the genetic and demographic structure of *I. scapularis* populations and on their long-term dynamics. Insofar as generation time is invariably 2 yr, alternate-year cohorts must be demographically and genetically independent. Clearly,

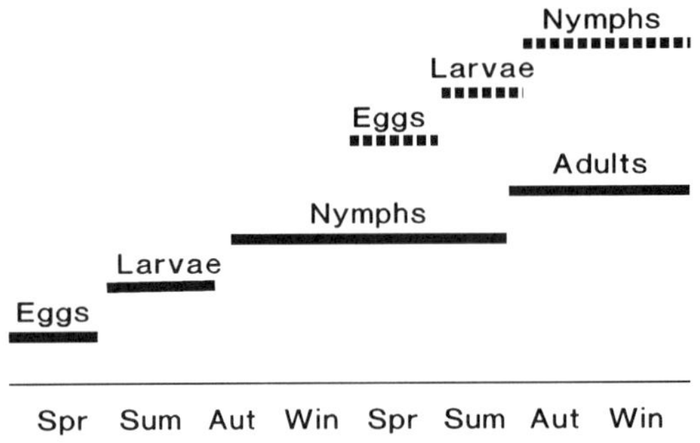

Figure 4 Developmental history (by season) of two consecutive cohorts of ticks.

this independence would have implications for such practical matters as the design of sampling schemes for genetic studies, as well as for experimental tests of control measures. Current general understanding of population regulation, dispersal, and responses to varying conditions also would have to account for such cohort structure. Of special interest, for example, is that both larvae and nymphs are found on dispersing birds (Anderson and Magnarelli, 1984; Anderson *et al.,* 1986; Battaly *et al.,* 1987). If not, and if alternate-year cohorts were substantially independent, outbreaks of a given stage (e.g., adults) would be expected only in alternate years in some locations. This pattern has not been reported for any location. Note that these questions of generation time and independence of cohorts also remain unresolved for the much better studied sheep tick *(I. ricinus),* the European vector of Lyme disease (Donnelly, 1978a,b).

Related to this question of generation time is some uncertainty about the nature and duration of overwintering diapause and quiescence. In general, researchers know that ticks may undergo complicated responses to temperature and photoperiod, including "quiescence" (direct suppression of development or behavior by unfavorable conditions) and true "diapause" (suppression of development or behavior by a "token" stimulus, such as photoperiod, in anticipation of unfavorable conditions) (Belozerov, 1967,1971; Pound and George, 1991). Experimental work (Yuval and Spielman, 1990) establishes that the activity and development of *I. scapularis* depend in some way on seasonal variables, but nothing is known about the nature of this dependence. We do not even know whether it represents "quiescence" or true "diapause."

III. Experimental Demography

The foundation of the population biological approach described in this chapter is a thorough understanding of relevant tick biology at the level of individual development, physiology, and behavior, which does not mean that experiments and models at the population level cannot be useful before such understanding is obtained. Indeed, these models can be extremely valuable in identifying the particular information really needed to answer specific classes of questions. Nevertheless, the goal is to explain population dynamics by appeal to processes at the level of the individual, and such processes are best elucidated by experimentation.

To date, there have been only a handful of quantitative experiments on *I. scapularis* development. Krinsky (1979) published development times for each stage of the life cycle, obtained from laboratory experiments for one specific temperature and level of humidity [27°C and 100% relative humidity (RH)]. This investigator reported that feeding larvae and nymphs required 3–7 days to become fully engorged whereas adult ticks required 8–11 days. Engorged larval ticks required 25–32 days to develop and ecdyse; engorged nymphs required 33–74 days. The interval between ecdysis and oviposition in female ticks was 7–14 days. Eggs required 29–36 days to hatch. Krinksy did not report mortality rates, but the data allow simple estimates for mortality of developing larvae (1.1% per day) and developing nymphs (0.9% per day). Mortality of feeding nymphs was age dependent (S. Vail and G. Smith, unpublished analysis).

Davidar *et al.* (1989) measured larval engorgement success on three host species, including *P. leucopus.* These researchers also compared larval engorgement on hosts subjected to two experimental treatments: (1) no prior exposure to ticks and (2) exposure to cohorts of feeding ticks, 2 and 4 wk previously. A significant negative effect of previous exposure was seen on the number of successfully engorged larvae recovered from *Microtus pennsylvanicus* and guinea pig, but not on the number recovered from *P. leucopus.* A substantial fraction of ticks failed to engorge on all hosts in these experiments (e.g., 30–50% on *P. leucopus*). There was a statistically significant, but very small, negative effect of previous exposure on the weights of engorged larvae recovered from *P. leucopus* and the other hosts, heavier ticks being recovered from naive hosts rather than from previously exposed hosts. The question of how tick populations are regulated remains unanswered.

Daniels *et al.* (1989) estimated the mortality rate of questing adults in laboratory experiments, and compared it with the disappearance rate of marked ticks in the field. We have fitted a number of mortality models to their data and drawn additional conclusions. A constant-mortality model, fitted to laboratory data, gives a mortality rate of approximately 4% per wk (0.6% per day). The actual rate is age dependent and corresponds to a somewhat higher daily rate (6% per wk). The field mortality is constant and substantially higher: 16% per wk (2.3% per day). If we subtract the rate of disappearance in the field from the mortality in laboratory culture, the difference provides the upper bound for an estimate of the host-finding rate for questing adults, that is, 10–12% per wk (1.4–1.7% per day). The best model, fitted to the field data, gives 4.8 wk as the expected duration of the adult questing stage (i.e., average time to either die or find a host), with 8–10 wk as the expected time to find a host.

Yuval and Spielman (1990) performed field and laboratory experiments from which development times for eggs and engorged stages, and mortality rates for questing stages, could be estimated. The measurements are on a monthly time scale and are, therefore, much coarser than those of Krinsky (1979) done in the laboratory. However, to the extent that they can be compared, the data are similar. Yuval and Spielman's results indicate, additionally, that some development times are dependent on an unknown seasonal variable (e.g., temperature or photoperiod). Engorged larvae developed to ecdysis in about 8 wk in May and June, but required only 4 wk in July and August. Larvae feeding in September did not molt until at least the following June. Engorged nymphs molted in 8 wk in May and June, required 4 wk in July, and, if fed in August, did not molt until the following May. Engorged adult females oviposited after 4 wk in the lab (much longer than Krinsky's result), after 8 wk in the field in early summer, and, if fed in November, did not oviposit in the field until the following June. Unfed larvae, caged in the field in September, all survived until the following May; some survived until August. A fitted mortality model (S. Vail and G. Smith, unpublished analysis) yields an expected stage duration (i.e., average time to death) of 308 days. Unfed nymphs survived in the field for 13–15 mo. Unfed adults survived in the field until late June, when all died regardless of whether they first were confined in spring or fall.

C. Lord and K. Curran (unpublished data) performed field experiments from which mortality and development rates of engorged larvae, and mortality rates of questing nymphs, could be estimated. Known numbers of engorged larvae were placed in mesh bags in the field in August, and censused at intervals for engorged larvae and unfed nymphs. A best-fit model (S. Vail and G. Smith, unpublished analysis) yielded the following rate estimates. Engorged larval mortality was 1% per day (expected time to death, 103 days). We assumed a minimum development time of 30 days (based on Krinsky's 1979 data) and a transition rate of 5.6% per day was calculated for the period beyond 30 days (average development time, $30 + 1/.056 = 48$ days). Unfed nymph mortality was 0.6% per day (expected time to death, 180 days). The estimates for engorged larvae are in order-of-magnitude agreement with Krinsky's (1979) laboratory estimates (see previous discussion), but development required twice as much time in the field as in the laboratory and mortality was 20–30% higher in the laboratory (approximately 1.2% vs. 1%).

Each of these experiments has limitations, but provides a foothold in the effort to understand overall population dynamics in terms of individual development and behavior. As a first step toward assessing the current position, expected values of all the stage durations, as estimated from experimental data, can be added and the sum can be compared with the approximately known generation time (2 yr; see previous section on generation time). The estimated stage durations are all as described earlier and can be summed in straightforward fashion for all but the unfed stages. The expected durations of unfed stages depend on mortality and on host finding, and no direct estimates of host-finding rates are available for *I. scapularis.* For unfed adults, the estimate, described earlier, based on comparison of laboratory mortality and field disappearance rates is available. This estimate represents an upper bound of the actual value. The expected time of death provides an upper bound for the expected duration of unfed larval and nymphal stages. For nymphs, two estimates are available that are in sharp disagreement. The sum of stage durations using the smaller value (which takes no account of winter quiescence) is 662 days (1.8 yr). The sum of stage durations using the larger value (which assumes synchronous death of a cohort and, thus, overestimates the probable stage duration) is 902 days (2.47 yr). The mean of these two sums is 2.14 yr, a number that is in close agreement with the consensus generation time of 2 yr.

It is encouraging that, with these few experimental studies, we seem to be on the right track to giving our observations of population dynamics a mechanistic basis. Nevertheless, important gaps remain to be filled. In particular, no direct estimates of host-finding rates are available for any unfed stage. Such estimates are difficult to come by for any tick, and probably will be derived with confidence only by a variety of indirect methods (e.g., Randolph and Steele, 1985; Barnard, 1991; Mount *et al.,* 1991). Also, very little information is available about the dependence of fecundity, mortality, development, and behavior on seasonal variables (e.g., temperature and photoperiod). Such dependence relationships probably contribute to, and may dominate, the seasonal timing of population patterns. Lyme disease epidemiology, in turn, depends critically on the seasonal timing of occurrence of the various stages (Spielman *et al.,* 1985; Piesman *et al.,* 1987). Even less information is available about the dependence of fecundity, mortality, development, or behavior on local tick abundance. The regulation of natural populations of ticks within finite limits necessarily involves such density dependence; understanding the regulation of abundance in natural populations is likely to provide insights into the geographic spread and limits of Lyme disease, and into possible interventions for control.

IV. Models of Components of the Tick Life Cycle

An important use of mathematical models is in helping make sense of the results of field and laboratory experiments. In the context of the population biology of *I. scapularis,* models can be used in this way

1. to obtain concise mathematical descriptions of the available data for possible incorporation into more comprehensive models.

2. to define biologically meaningful parameters in terms of standard fitted models, so data from diverse experimental studies can be compared in a consistent analytical framework.
3. to test hypotheses about the dependence of demographic rates (e.g., development and mortality rates) on such factors as age, temperature, and season.
4. to compare the explanatory power of models that differ in complexity, so we can know how much detail is lost when we simplify.

We begin by formalizing the scheme of Fig. 1 in the following set of generalized differential equations for *I. scapularis* nymphs:

$$\frac{dN_Q}{dt} = \gamma_L L_D - \mu_{NQ} N_Q - \beta_N N_Q$$

$$\frac{dN_F}{dt} = \beta_N N_Q - \mu_{NF} N_F - \epsilon_N N_F \qquad (1)$$

$$\frac{dN_D}{dt} = \epsilon_N N_F - \mu_{ND} N_D - \gamma_N N_D.$$

These equations describe changes in the numbers of questing (N_Q), feeding (N_F), and developing (N_D) nymphs due to mortality (μ_{NQ}, μ_{NF}, μ_{ND}), host finding (β_N), engorgement (ϵ_N), and development (γ_L, γ_N), where the parameters are instantaneous *per capita* rates. L_D is the number of engorged developing larvae. A complete general model for the life cycle of *I. scapularis* would include a similar set of equations for each of the stages (larva, nymph, adult) as well as equations for reproduction and the development of eggs.

In the simplest case, all the parameters would be constants. In reality, these parameters are likely to depend on a whole constellation of variables including temperature, humidity, photoperiod, host density, vegetation type, and so on. In terms of the general model, the problem of understanding *I. scapularis* population dynamics is one of deriving the appropriate mathematical expressions to account for any dependence relationships, and estimating the parameters of those expressions.

The basic approach is cataloguing and analyzing data from various published and unpublished sources (described in the preceding section) that pertain to as many life-cycle processes as the available data allow. We then create models by extracting the appropriate elements from Eq. 1 and substituting various mathematical forms for the generalized parameters. The resulting models are fit to the experimental data using nonlinear regression algorithms (see, for example, Bermann and Weiss, 1978). The fitted models are compared for general conformity to the data, coefficients of determination (% variance explained), mean fractional standard deviations, and expected stage durations.

The method is demonstrated by re-analysis of the data of Daniels *et al.* (1989) on mortality of questing adults. These experiments and the general conclusions are described in the preceding section. We began by fitting two very simple mortality functions to the field and laboratory survivorship data,

$$\frac{dA_Q}{dt} = -\mu_{AQ} A_Q \qquad (2a)$$

$$\frac{dA_Q}{dt} = ae^{bt} A_Q, \qquad (2b)$$

where A_Q is the number of questing adults, μ_{AQ} is a constant *per capita* mortality rate, and the first equation describes exponential decay. The second equation is a simple formulation of age-dependent mortality, in which a and b are constant parameters. The second equation results in a sigmoid, or Type III, survivorship curve (McNaughton and Wolf, 1973). Note that the experimental design of Daniels *et al.* (1989) eliminated all other processes that might have affected the numbers of questing adults (e.g., development of engorged nymphs and host finding by questing adults).

The constant-mortality model explained

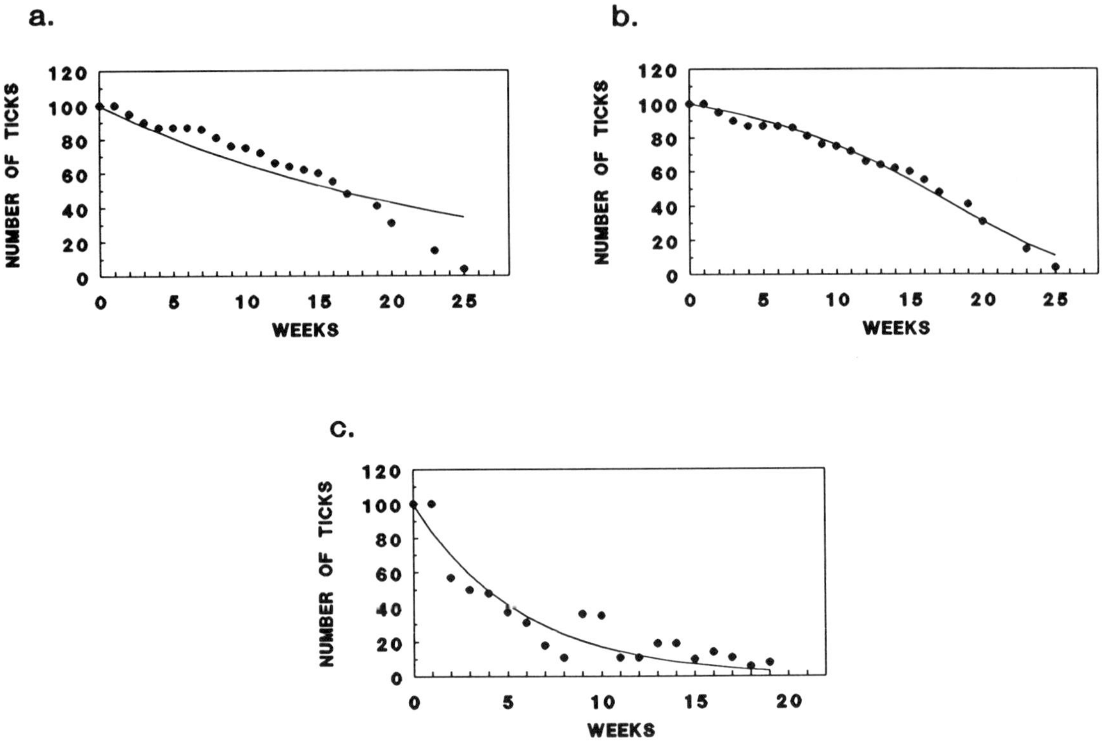

Figure 5 Observed (points) and predicted (solid line) survivorship curves for questing adult ticks. (a) Laboratory data; mortality assumed to be constant. (b) Laboratory data; mortality assumed to be age-dependent. (c) Field data; mortality assumed to be constant. Data from Daniels *et al.* (1989).

83% of the variance in the laboratory data, but did not reproduce the sigmoid shape of the plotted data (Fig. 5a) and substantially overestimated the mean survival time (23.7 wk vs. 15 wk observed). The age-dependent model explained 98% of the laboratory variance (Fig. 5b), reproduced the sigmoid shape of the data, and gave an expected survival time very close to that observed (15.8 wk). This form of age dependence is typical of other organisms whose survival depends principally on stored energy reserves (Smith and Grenfell, 1984).

Field disappearance rates of marked ticks, in contrast, were explained just as well by a fitted constant-mortality model as by an age-dependent model (Fig. 5c). In fact, the best-fit age-dependence parameter (b in Eq. 2) was zero, reducing model 2b to model 2a. The model explained 88% of the variance in the field data and accurately estimated mean time-to-disappearance at 5.75 wk. Mortality in the field was much higher and less age dependent than in the laboratory. We presume that death by starvation is routinely preempted in the field by other sources of mortality (e.g., predation, desiccation) or by successful host finding (see preceding section). Again, this result is standard in helminth parasitology (e.g., Smith *et al.,* 1986), and urges the cautious application of laboratory results to the field situations.

V. More Comprehensive Tick Models

Although we lack a thorough understanding of many details of the *I. scapularis* life cycle, many researchers believe that formulating and analyzing more comprehensive models is valuable nonetheless. Even relatively detail-free models can provide insight into the general mechanisms underlying observed dynamic patterns, and can direct investigators to the empirical information needed most urgently (or to that which can be ignored most confidently). The three models considered here differ from the one in the previous section because each comprises a set of difference rather than differential equations.

A. Model for Questing Nymphs

The first example is the work of Lord (1992) on the within-season dynamics of nymphal *I. scapularis* in the northeastern United States. Lord was concerned with factors controlling the early summer peak in the number of questing nymphs. Infected nymphs are responsible for most human Lyme disease cases (Spielman *et al.,* 1985; Piesman *et al.,* 1987). A peak in nymph abundance during early summer is an important feature of Lyme disease epizootiology because at this time many people are working or playing outdoors. Lord's model is more complicated than the models described in the previous section because it includes all the processes contributing to the abundance of questing nymphs. Following a procedure similar to that described for single-process models, Lord first framed the model as a general equation for nymph abundance at time t (N_t) with unspecified expressions for recruitment (R_t), mortality (M_t), and host finding (H_t):

$$N_{t+1} = R_t - M_t N_t - H_t N_t, \qquad (3)$$

where M_t and H_t are proportions. Various mathematical forms, representing different biological mechanisms, then were substituted for the general parameters R_t, M_t, and H_t. For example, three functional forms for recruitment were derived to represent photoperiodic induction of activity (resulting in a gamma-distributed recruitment rate), a simple developmental delay between larval and nymphal stages (giving a normally distributed recruitment rate), and constant daily recruitment (giving a uniform distribution). Host-finding functions were derived similarly, and mortality was assumed to be approximately constant. Lord then examined the pattern and timing of nymphal activity predicted by each model in this assortment (for a variety of host-finding and mortality rates) and compared the results with patterns observed in the field.

The comparisons indicated that the functional form for recruitment has a profound effect on the nymphal activity pattern, and that the field-observed pattern corresponds best to the recruitment function in which activity is triggered during a relatively narrow window of time, which would happen if diapause were terminated by a change in photoperiod, for example. The temporal pattern of host finding and the magnitude of tick mortality had very little effect on the activity pattern. These findings are significant for the following reasons. First, they indicate that very detailed information on tick mortality or host finding behavior may not be necessary to explain observed patterns of questing tick abundance. This possibility is encouraging. Second, if photoperiod is a cue in terminating nymphal diapause, as the model seems to suggest, this factor will be important in understanding changes in the epidemiology of human infection as *I. scapularis* spreads along the East Coast and midwestern parts of the United States.

B. Models for the Entire Tick Life Cycle

The models that have set out to mimic the entire tick life cycle to this point have done so by sacrificing much of the stage-specific detail seen in Lord's model (Ginsberg, 1988; Porco, 1991; Sandberg *et al.,* 1992). For example, Porco (1991) described a transition matrix model with a 3-mo time step. This "seasonal" view of the life cycle permits a number of simplifying assumptions, principally, that only one feeding stage of any

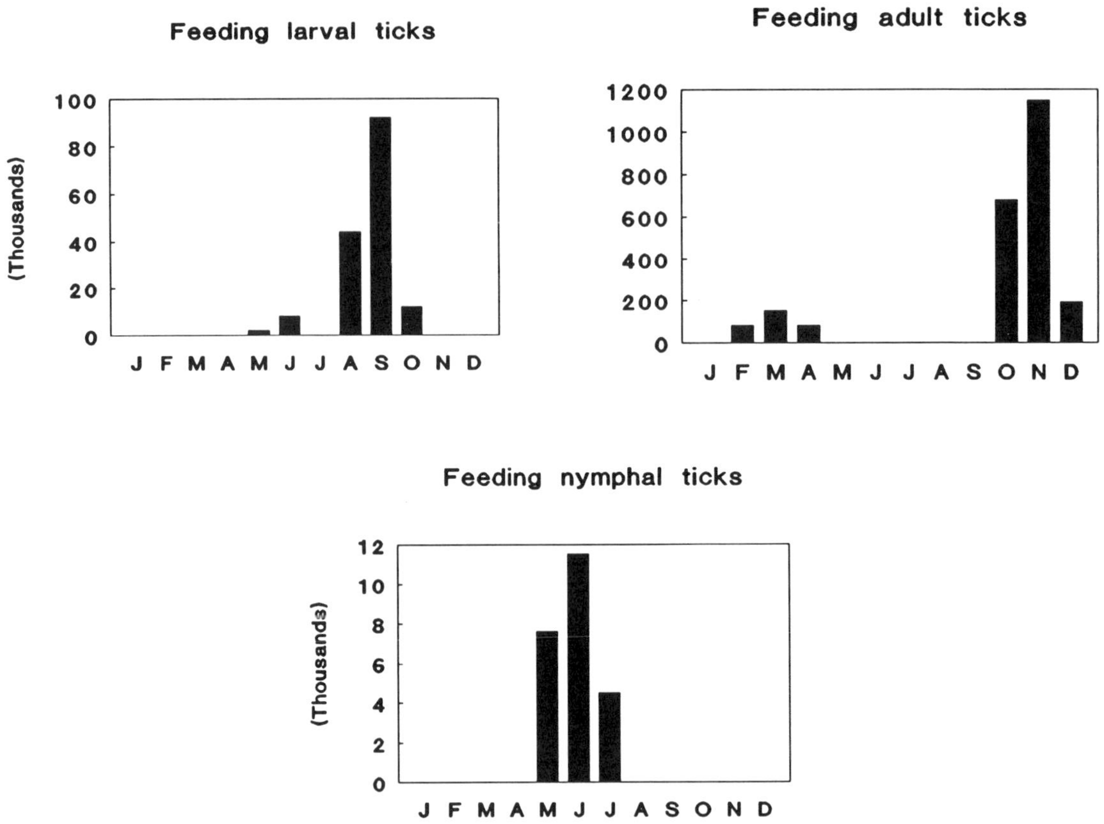

Figure 6 Predicted month by month distribution of feeding ticks based on simulations from the matrix model described by Sandberg *et al.* (1992). Note that the vertical axis is rescaled for each figure. Also note that hosts are parasitized by nymphs (from eggs laid in the previous year) before they are parasitized by larvae (from eggs laid in the current year).

given cohort of ticks will be active at any one time. Porco further assumed that adult *I. scapularis* feed only in the autumn, that temperature has no effect on tick survival or activity, and that host densities were constant, at least within each time step (season). The parameters of the transition matrices were adjusted so the number of tick eggs laid each year was the same. This device was used to maintain tick numbers at some notional equilibrium in the absence of information about the processes that actually regulate tick density. This simplified model architecture permitted model behavior to be examined analytically for some special cases. One interesting conclusion was that moderate reductions in the density of white-tailed deer may decrease or increase the proportion of infected ticks. The precise outcome depends on the fraction of ticks that feeds on the deer and on the functional dependence of tick population size on the number of deer. This ambiguous result is important because it illustrates the counterintuitive point that planned reductions in deer density may have an effect exactly opposite the one desired.

A more detailed transition matrix model was published by Sandberg *et al.* (1992). This model differed from that of Porco because the time step was 1 mo, so 12 transition matrices per year instead of four were analyzed. This change permitted a less limited representation of temporal changes in oviposition and questing behavior. The model was able to reproduce the seasonal pattern of abundance of the various stages quite accurately (Fig. 6). Nevertheless, even this moderate increase in complexity removed any reasonable possibility of examining model behavior analytically. All subsequent studies of this model and its elaborations will depend on repeated numerical simulations. Since the current model also frankly admits the absence of any good information concerning the processes that regulate tick density, tick numbers increase without bound for every successive year included in the simulation. A modification incorporating a putative density-dependent constraint on tick numbers currently is being implemented (T. Auwerbuch, personal communication). This addition will be interesting since density dependence adds considerable complexity to a model of this type, and can have a profound effect on patterns of dynamic behavior (Caswell, 1989).

VI. Conclusions

It should be obvious by now that we know very little about the population biology of *I. scapularis*. By implication, even less is known about the dynamics of Lyme disease. Nevertheless, this chapter has addressed how models can be useful, even in the absence of good information, to guide the direction of inquiries, to provide a quantitative framework for measurement and analysis, and to temper assumptions about methods that are likely to constitute an effective disease control strategy.

References

Ai, C., Wen, Y., Zhang, Y., Wang, S., Qiu, Q., Shi, Z., Li, D., Chen, D., Liu, X., and Zhao, J. (1988). Clinical manifestations and epidemiological characteristics of Lyme disease in Hailin county, Heilongjiang Province, China. *Ann. N. Y. Acad. Sci.* **539,** 302–313.

Anderson, J. F. (1989). Ecology of Lyme disease. *Connecticut Med.* **53,** 343–346.

Anderson, J. F., and Magnarelli, L. A. (1980). Vertebrate host relationships and distribution of ixodid ticks (Acari:Ixodidae) in Connecticut, USA. *J. Med. Entomol.* **17,** 314–323.

Anderson, J. F., and Magnarelli, L. A. (1984). Avian and mammalian hosts for spirochete-infected ticks and insects in a Lyme disease focus in Connecticut. *Yale J. Biol. Med.* **57,** 627–641.

Anderson, J. F., Johnson, R. C., Magnarelli, L. A., and Hyde, L. A. (1986). Involvement of birds in the epidemiology of the Lyme disease agent *Borrelia burgdorferi*. *Infect. Immun.* **51,** 394–396.

Barnard, D. R. (1991). Mechanisms of host–tick contact with special reference to *Amblyomma americanum* (Acari:Ixodidae) in beef cattle forage areas. *J. Med. Entomol.* **28,** 557–564.

Battaly, G. R., Fish, D., and Dowler, R. C. (1987). The seasonal occurrence of *Ixodes dammini* and *Ixodes dentatus* (Acari:Ixodidae) on birds in a Lyme disease endemic area of southeastern New York state. *J. N. Y. Entomol. Soc.* **95,** 461–468.

Belozerov, V. N. (1967). Larval diapause in the tick *Ixodes ricinus* L. and its dependence on external conditions. IV. Interaction between exogenous and endogenous factors in the regulation of the larval diapause. *Entomol. Obozr.* **46,** 447–451.

Belozerov, V. N. (1971). Diapause phenomena in ticks (Ixodoidea) and their regulation. *In* "Proceedings of the 3rd International Congress of Acarology Held in Prague, August 31–September 6, 1971" (M. Daniel and B. Rosicky, eds.), pp. 489–491. Junk, The Hague, Netherlands.

Bermann, M., and Weiss, M. F. (1978). "Simulation, Analysis, and Modeling." DHEW Publication Number (NIH) 78-180. National Institutes of Health, Bethesda, Maryland.

Carey, A. B., Krinsky, W. L., and Main, A. J. (1980). *Ixodes dammini* (Acari:Ixodidae) and associated ixodid ticks in south-central Connecticut, USA. *J. Med. Entomol.* **17,** 89–99.

Carmel, D. K., and Edwards, G. H. (1989). "Lyme Disease in Animals" Fact Sheet 534. University of Maryland Cooperative Extension Service, College Park.

Cartter, M. L., Mshar, P., and Hadler, J. L. (1989). The epidemiology of Lyme disease in Connecticut. *Connecticut Med.* **53,** 320–323.

Caswell, H. (1989). "*Matrix Population Models.*" Sinauer Associates, Sunderland, Massachusetts.

Centers for Disease Control (1992). "Lyme Disease Surveillance Summary," Vols. 3,5. Centers for Disease Control, Atlanta, Georgia.

Cieseilski, C. A., Markowitz, L. E., Horsley, R., Hightower, A. W., Russell, H., and Broome, C. V. (1988). The geographic distribution of Lyme disease in the United States. *Ann. N. Y. Acad. Sci.* **539,** 282–288.

Daniels, T. T., Fish, D., and Falco, R. C. (1989). Seasonal activity and survival of adult *Ixodes dammini* (Acari:Ixodidae) in southern New York state. *J. Med. Entomol.* **26,** 610–614.

Davidar, P., Wilson, M. L., and Ribeiro, J. M. C. (1989). Differential distribution of immature *Ixodes dammini* (Acari:Ixodidae) on rodent hosts. *J. Parasitol.* **75,** 898–904.

Donahue, J. G., Piesman, J., and Spielman, A. (1987). Reservoir competence of white-footed mice for Lyme disease spirochetes. *Am. J. Trop. Med. Hyg.* **36,** 92–96.

Donnelly, J. (1978a). The weather-related ecology of the tick, *Ixodes ricinus* L. *In* "Weather and Parasitic Animal Disease" (T. E. Gibson ed.),Technical Note 159, pp. 97–104. World Meteorological Organization, Geneva.

Donnelly, J. (1978b). The life cycle of *Ixodes ricinus* L. based on recent published findings. *In* "Tick-Borne Diseases and Their Vectors" (J. K. H. Wilde ed.), pp. 56–60. Centre for Tropical Veterinary Medicine, University of Edinburgh.

Gardiner, W. P., Gettinby, G., and Gray, J. S. (1981). Models based on weather for the development phases of the sheep tick, *Ixodes ricinus* L. *Vet. Parasitol.* **9,** 75–86.

Ginsberg, H. S. 1988. A model of the spread of Lyme disease in natural populations. *Ann. N. Y. Acad. Sci.* **539,** 329–380.

Haile, D. G., and Mount, G. A. (1987). Computer simulation of population dynamics of the lone star tick, *Amblyomma americanum* (Acari: Ixodidae). *J. Med. Entomol.* **24,** 356–369.

Krinsky, W. L. (1979). Development of the tick *Ixodes dammini* (Acarina:Ixodidae) in the laboratory. *J. Med. Entomol.* **16,** 354–355.

Krohne, D. T., Merritt, J. F., Vessey, S. H., and Wolff, J. O. (1988). Comparative demography of forest *Peromyscus. Can. J. Zool.* **66,** 2170–2176.

Lane, R. S., and Burgdorfer, W. (1987). Transovarial and transstadial passage of *Borrelia burgdorferi* in the western black-legged tick, *Ixodes pacificus* (Acari:Ixodidae). *Am. J. Trop. Med. Hyg.* **37,** 188–192.

Lane, R. S., Piesman, J., and Burgdorfer, W. (1991). Lyme borreliosis: Relation of its causative agent to its vectors and hosts in North America and Europe. *Annu. Rev. Entomol.* **36,** 587–609.

Lord, C. C. (1992). Nymphal *Ixodes dammini*: Models of the temporal abundance patterns. *Int. J. Parasitol.* **22,** 759–765.

Main, A. J. (1986). Bimodality of seasonal abundance in *Ixodes dammini. J. Med. Entomol.* **23,** 339–340.

Main, A. J., Sprance, H. E., Kloter, K. O., and Brown, S. E. (1981). *Ixodes dammini* (Acari:Ixodidae) on white-tailed deer (*Odocoileus virginianus*) in Connecticut. *J. Med. Entomol.* **18,** 487–492.

Main, A. J., Carey, A. B., Carey, M. G., and Goodwin, R. H. (1982). Immature *Ixodes dammini* (Acari:Ixodidae) on small mammals in Connecticut, USA. *J. Med. Entomol.* **19,** 655–664.

May, R. M. (1974). "Stability and Complexity in Model Ecosystems," 2nd ed. Princeton University Press, Princeton.

McNaughton, S. J., and Wolf, L. L. (1973). "General Ecology." Holt, Rinehart, and Winston, New York.

Miller, G. L., Craven, R. B., Bailey, R. E., and Tsai, T. F. (1990). The epidemiology of Lyme disease in the United States 1987–1988. *Lab. Med.* **21,** 285–289.

Mount, G. A., and Haile, D. G. (1989). Computer simulation of population dynamics of the American dog tick (Acari:Ixodidae). *J. Med. Entomol.* **26,** 60–76.

Mount, G. A., Haile, D. G., Davey, R. B., and Cooksey, L. M. (1991). Computer simulation of *Boophilus* cattle tick (Acari:Ixodidae) population dynamics. *J. Med. Entomol.* **28,** 223–240.

Piesman, J., and Spielman, A. (1979). Host associations and seasonal abundance of immature *Ixodes dammini* in southeastern Massachusetts. *Ann. Entomol. Soc. Am.* **72,** 829–832.

Piesman, J., Spielman, A., Etkind, P., Ruebush, T. K., II, and Juranek, D. D. (1979). Role of deer in the epizootiology of *Babesia microti* in Massachusetts, USA. *J. Med. Entomol.* **15,** 537–540.

Piesman, J., Donahue, J. G., Mather, T. N., and Spielman, A. (1986). Transovarially acquired Lyme disease spirochetes *(Borrelia burgdorferi)* in field-collected larval *Ixodes dammini* (Acari:Ixodidae). *J. Med. Entomol.* **23,** 219.

Piesman, J., Mather, T. N., Dammin, G. J., Telford, S. R. III, Lastavica, C. C., and Spielman, A. (1987). Seasonal variation of transmission risk of Lyme disease and human babesiosis. *Am. J. Epidemiol.* **126,** 1187–1189.

Porco, T. C. (1991). A model of the enzootiology of Lyme disease in the Atlantic North East of the United States. *Nat. Resourc. Model.* **5,** 469–505.

Pound, J. M., and George, J. E. (1991). Selective breeding for photoperiodically induced developmental diapause in laboratory strains of *Amblyomma americanum* (Acari:Ixodidae). *J. Med. Entomol.* **28,** 544–550.

Rahn, D. W., and Malawista, S. E. (1991). Lyme disease: Recommendations for diagnosis and treatment. *Ann. Int. Med.* **114,** 472–481.

Randolph, S. E., and Steele, G. M. (1985). An experimental evaluation of conventional control measures against the sheep tick, *Ixodes ricinus* (L.)(Acari:Ixodidae). II. The dynamics of the tick–host interaction. *Bull. Entomol. Res.* **75,** 501–518.

Sandberg, S., Awerbuch, T. E., and Spielman, A. (1992). A comprehensive multiple matrix model representing the life cycle of the tick that transmits the agent of Lyme disease. *J. Theoret. Biol.* **157,** 203–220.

Schulze, T. L., Bowen, G. S., Lakat, M. F., Parkin, W. E., and Shisler, J. K. (1985). The role of adult *Ixodes dammini* (Acari:Ixodidae) in the transmission of Lyme disease in New Jersey, USA. *J. Med. Entomol.* **22,** 88-93.

Siegel, J. P., Kitron, U., and Bouseman, J. K. (1991). Spatial and temporal distribution of *Ixodes dammini* (Acari:Ixodidae) in a northwestern Illinois state park. *J. Med. Entomol.* **28,** 101-104.

Smith, G., and Grenfell, B. T. (1984). The influence of water temperature and pH on the survival of *Fasciola hepatica* miracidia. *Parasitology* **88,** 97-104.

Smith, G., Grenfell, B. T., and Anderson, R. M. (1986). The development and mortality of the non-infective free-living stages of *Ostertagia ostertagi* in the field and in laboratory culture. *Parasitology* **92,** 471-482.

Spielman, A. (1988). Prospects for suppressing transmission of Lyme disease. *Ann. N. Y. Acad. Sci.* **539,** 212-220.

Spielman, A., Wilson, M. L., Levine, J. F., and Piesman, J. (1985). Ecology of *Ixodes dammini*-borne human babesiosis and Lyme disease. *Ann. Rev. Entomol.* **30,** 439-460.

Stanek, G., Pletschette, M., Flamm, H., Hirschl, A. M., Abere, E., Kristoferitsch, W. and Schmutzhard, E. (1988). European Lyme borreliosis. *Ann. N. Y. Acad. Sci.* **539,** 274-282.

Telford, S. R., III, Mather, T. N., Moore, S. I., Wilson, M. L., and Spielman, A. (1988). Incompetence of deer as reservoirs of the Lyme disease spirochete. *Am. J. Trop. Med. Hyg.* **39,** 105-109.

Wallis, R. C., Brown, S. E., Kloter, K. O., and Main, A. J. (1978). Erythema chronicum migrans and Lyme arthritis: Field study of ticks. *Am. J. Epidemiol.* **108,** 322-327.

Wilson, M. L., and Spielman, A. (1985). Seasonal activity of immature *Ixodes dammini* (Acari:Ixodidae). *J. Med. Entomol.* **22,** 408-414.

Wilson, M. L., Adler, G. H., and Spielman, A. (1985). Correlation between abundance of deer and that of the deer tick, *Ixodes dammini* (Acari:Ixodidae). *Ann. Entomol. Soc. Am.* **78,** 172-176.

Wilson, M. L., Telford, S. R., III, Piesman, J., and Spielman, A. (1988). Reduced abundance of immature *Ixodes dammini* (Acari:Ixodidae) following elimination of deer. *J. Med. Entomol.* **25,** 224-228.

Wilson, M. L., Ducey, A. M., Litwin, T. S., Gavin, T. A., and Spielman, A. (1990). Microgeographic distribution of immature *Ixodes dammini* ticks correlated with that of deer. *Med. Vet. Entomol.* **4,** 151-159.

Yuval, B., and Spielman, A. (1990). Duration and regulation of the development cycle of *Ixodes dammini* (Acari:Ixodidae). *J. Med. Entomol.* **27,** 196-201.

Marilyn E. Scott & Geneviève V. Tanguay

20

Heligmosomoides polygyrus: A Laboratory Model for Direct Life Cycle Nematodes of Humans and Livestock

I. Introduction

Over the past 20 years, the intestinal nematode of mice, *Heligmosomoides polygyrus (Nematospiroides dubius),* has become a common laboratory model. This chapter shows how field observations and mathematical theory have driven experimentation in the laboratory, and how laboratory observations have stimulated field studies. We begin by summarizing some of the more interesting studies on factors that influence three major population processes: parasite reproduction, transmission, and survival. We then review novel studies on specific phenomena of importance to the control of direct life cycle nematode parasites in the real world.

II. Fecundity: A Key to Survival of the Species

A. Parasite Factors that Affect Fecundity

Kerboeuf's early work on *H. polygyrus* was stimulated by concern about the effect of overwintering on infectivity of trichostrongylid nematodes of livestock. In her studies on the effects of age of larvae on the ability of *H. polygyrus* to establish (Kerboeuf, 1978a) and reproduce (Kerboeuf, 1978b), Kerboeuf used a standard infection procedure to establish adult infections in outbred Swiss mice. This researcher detected an unexpected negative relationship between infectivity of larvae and *per capita* egg production of resulting adult females that was independent of the size of the adult population (Fig. 1).

Intrinsic heterogeneity among parasites in their ability to produce eggs enabled Brindley and Dobson (1983) to select lines of parasite on the basis of egg output, thus clearly showing that rate of egg production is a heritable feature. Variation in *per capita* fecundity is usually difficult to quantify, since estimates almost invariably are taken from pools of worms within a single host (Shostak and Scott, 1993). However, it was possible to obtain some idea of interparasite variability using Chang and Wescott's (1972) data on *in utero* egg counts of 6-wk-old worms. The coefficient of variation ranged from 15 to 47%, suggesting high variability among individual worms in the numbers of eggs produced.

Data on the impact of parasite density on *per capita* egg output are confusing (Fig. 2). Slater and Keymer (1986a) and Gregory *et*

Parasitic and Infectious Diseases

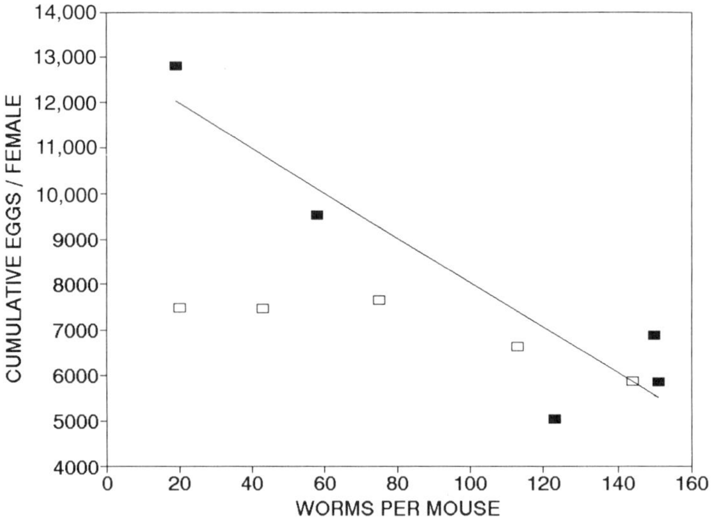

Figure 1 Effect of larval infectivity on fecundity of adult *Heligmosomoides polygyrus*. Male Swiss mice were infected with 200 larvae that had been stored for varying numbers of weeks. This procedure alters the infectivity of the larvae, as seen by the different numbers of worms recovered per mouse (■). As the infectivity of the larvae increased, the cumulative egg output per female worm between 12 and 27 days postinfection decreased ($r = -0.92$; $P < 0.05$). Control mice (□) were infected with sufficient numbers of high infectivity larvae to result in similar worm burdens, thus controlling for possible effects of infection intensity on fecundity. Data from Kerboeuf (1978b).

al. (1990) reported that parasite density had no significant effect on *H. polygyrus* fecundity (Fig. 2a). Using eggs per female per g feces as a measure of egg output, Keymer and Hiorns (1986a) were not able to detect density dependence. However, re-analysis of their data, based on eggs per female per day revealed a significant negative relationship with parasite density (Fig. 2b). Kerboeuf (1978b) showed that cumulative egg output per female worm between 12 and 27 days postinfection (pi) decreased significantly with increasing parasite numbers (Fig. 1). An opposite pattern was observed when data taken from a free-running laboratory colony (Scott, 1988b) were re-analyzed. A significant increase in *per capita* egg production with increasing worm numbers was found (Fig. 2c). Parasites in Scott's study were of mixed ages. Since *per capita* fecundity of adult females is affected by their age (Keymer and Hiorns, 1986b; MacKinnon and Lee, 1988; Gregory *et al.*, 1990; Scott, 1991), increasing during the first 8 wk of infection and then declining, this effect of age on fecundity may account in part for the discrepancy between Scott's data (1988b) and those of others.

B. Host Factors that Affect Fecundity

Since *H. polygyrus* lives in the small intestine, factors that affect the intestine will have an impact on parasite reproduction. Three examples follow. Food intake (or lack thereof) alters both total daily egg production per worm and eggs per g of feces (Keymer and Hiorns, 1986a). Egg output shows a diurnal pattern (Brindley and Dobson, 1982; Kerboeuf and Jacob, 1983), which may relate to host physiology or intrinsic rhythms of the parasite. Parasites living in germ-free or gnotobiotic mice have lower *per capita* egg production than those

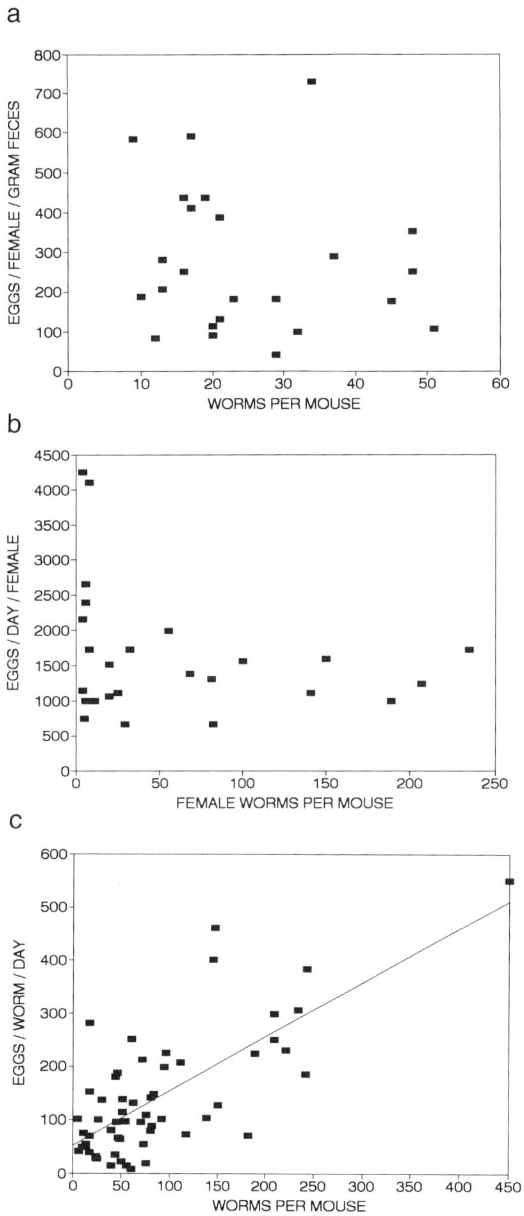

Figure 2 Effect of parasite density on per capita egg production of *Heligmosomoides polygyrus.* (a) CD1 male mice had been exposed to larvae in an indoor arena over a 4-mo period, and had been fed on a 2% protein diet. No significant effect of parasite density was reported ($r = 0.124$; $P > 0.1$). Data from Slater and Keymer (1986a); reprinted with the permission of Cambridge University Press. (b) Male MF1 mice were infected once with varying numbers of larvae. Daily egg production was monitored beginning 17 days postinfection for 6 consecutive days. The average egg production per female worm over this period of time decreased significantly with increasing worm numbers ($t = -2.2$; $df = 123$; $P < 0.05$). Data from Keymer and Hiorns (1986a); reprinted with the permission of Cambridge University Press. (c) A breeding population of mice had been exposed to larvae in an indoor arena over a period of 1 yr. Daily egg production was estimated, and worms were obtained following anthelmintic treatment. Re-analysis of data from Scott (1988b) provided best fit line, $y = a + bx$, where $a = 51.09$ (SE 15.52) and $b = 1.02$ (SE 0.13) ($P < 0.001$). Note that one outlier point was excluded from analysis.

in conventional mice (Chang and Wescott, 1972).

Interestingly, Kerboeuf (1985) demonstrated that mouse density has a significant impact on parasite fecundity. This investigator housed mice at densities of 1, 10, 20, or 30 per cage (20 × 40 cm) and found that *per capita* egg production of *H. polygyrus* was higher when a single mouse was present in each cage (Fig. 3). She also found that fecal output was higher in mice housed at low densities, which may account for the effects on daily egg production of the parasite. This study did not report any effect of mouse density on parasite establishment or survival for a 40-day period. In the calculations of *per capita* fecundity, worm numbers were assumed to be stable over the 40 days of the experiment.

Although Enriquez *et al.* (1988b) found differences in rates of *per capita* egg production between two strains of laboratory mouse (LAF1 and CBA), Scott (1991) reported no difference in *per capita* fecundity between BALB/c and C57BL/6 mice following a primary infection. The effect of mouse strain on *per capita* egg production is more noticeable following a challenge infection, presumably because host immune status has a more dramatic effect on *per capita* rates of egg production and the ability to express effective immunity against *H. polygyrus* differs among mouse strains (Behnke and Robinson, 1985; Cypess *et al.*, 1988; Enriquez *et al.*, 1988a,b,c; Wahid *et*

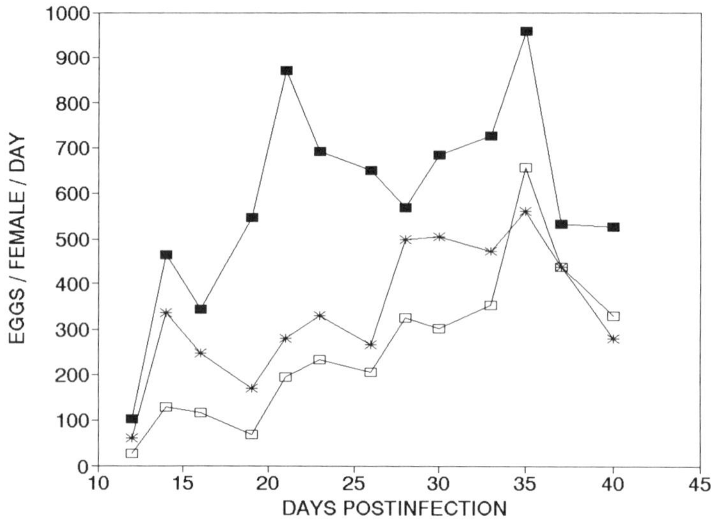

Figure 3 Effect of mouse density on rate of egg production per parasite. One (■), 10 (*), or 30 (□) male CD1 mice were housed together in 20 × 40 cm cages. Each mouse was infected with 150 L_3 and killed 40 days postinfection. The experiment was replicated three times. Fecundity differed significantly with mouse density ($F_{2,78} = 55.8$; $P < 0.001$) and with time ($F_{12,78} = 12.79$; $P < 0.001$) Data from Kerboeuf (1985); reprinted with the permission of Cambridge University Press.

al., 1989). Scott (1991), for example, demonstrated that the median *per capita* egg production 8 wk after challenge infection was reduced fourfold in BALB/c mice and by only one-third in C57BL/6 mice compared with egg output 8 wk after a primary infection. Similarly, Enriquez *et al.* (1988b) showed an average 50% reduction in *per capita* egg production of a challenge infection compared with a primary infection in CBA/J mice. These effects are, in part, related to delayed development of the parasite after a challenge infection (see Monroy and Enriquez, 1992), but also may be related to direct physiological effects of host immunity on the parasite.

Given the clear effect of host immunity on parasite fecundity, and given the effect that host nutritional status has on immunity to *H. polygyrus* (see Slater and Keymer, 1988), one might expect that host nutritional status also would affect *H. polygyrus* fecundity. Studies to date, however, show no effect of severe protein deficiency (Slater and Keymer, 1988) or marginal zinc deficiency (Minkus *et al.*, 1992) on *per capita* egg production.

C. Importance of These Factors

Anyone wishing to formulate a mathematical model for this host–parasite system will need to decide whether the rate of parasite egg production should be made a function of one or more of the previously mentioned factors. Given the wide range of factors shown to influence parasite fecundity, can those that are likely to have the most dramatic effects on *H. polygyrus* population dynamics be identified? Table 1 provides information that suggests that the most dominant source of variability may, in fact, be intrinsic heterogeneity among parasites in their *per capita* fecundity. This feature can be incorporated easily into a theoretical model by defining a distribution for fecundity, rather than using a constant value. A second dominant source of variability appears to be host immune status. As shown in a subsequent section, immunity

Table 1 Magnitude of Effects of Various Parasite and Host Factors on *per capita* Fecundity of *Heligmosomoides polygyrus*

Factor	Effect on fecundity[a]	Reference
Age of larvae at infection	Up to 70% increase in cumulative egg output as larvae aged	Kerboeuf (1978b)
Intrinsic parasite variability	Two-fold difference in egg output between lines bred for high and low fecal egg counts	Brindley and Dobson (1983)
	Five-fold difference in estimated *per capita* egg output at low intensity infections	Keymer and Hiorns (1986a)
	Coefficient of variation of 15–47% in *in utero* numbers of eggs (6 wpi in conventional mice)	Chang and Wescott (1972)
Parasite density	No effect	Slater and Keymer (1986a); Gregory *et al.* (1990); Scott (1991)
	20% decrease in cumulative egg output per parasite over densities ranging from 20 to 144 worms per mouse	Kerboeuf (1978b)
	40–50% decrease in *per capita* egg production per female over female numbers ranging from 1 to 235	Keymer and Hiorns (1986a)
Age of adult	Four-fold difference between peak egg output (6 wpi) and egg output in older worms (22 wpi)	Keymer and Hiorns (1986b)
	Three-fold difference between peak egg output (14 dpi) and egg output in older worms (45 dpi)	Gregory *et al.* (1990)
	Four-fold difference between peak egg output (8 wpi) and egg output in older worms (17 wpi)	Scott (1991)
Food intake	40–50% decrease in eggs per female per day when food is withheld for 24 hr	Keymer and Hiorns (1986a)
Diurnal effects	Three-fold difference in eggs per gram feces over a 24-hr period	Kerboeuf and Jacob (1983)
	Two-fold difference in eggs per gram feces over a 24-hr period	Brindley and Dobson (1982)
Bacterial flora	50–90% decrease in eggs per female per gram feces in germ-free mice relative to conventional mice	Chang and Wescott (1972)
Mouse density	Up to 50% decrease in *per capita* egg production over mouse density ranging from 1 to 30 per cage	Kerboeuf (1985)
Mouse strain	Two-fold difference in peak *per capita* egg production between BALB/c and C57BL/6 mice	Scott (1991)
Host immune status	Four-fold decrease in peak *per capita* egg production in BALB/c mice in challenge infection compared with primary infection	Scott (1991)
	30% decrease in peak *per capita* egg production in C57BL/6 mice in challenge infection compared with primary infection	Scott (1991)
	50% reduction in *per capita* egg production in CBA/J mice in challenge infection compared with primary infection	Enriquez *et al.* (1988b)

[a] wpi, Weeks postinfection; dpi, days postinfection.

must be included in any general model of *H. polygyrus,* for reasons other than effects on egg production. In all but the simplest cases, use of egg production rates for parasites in immune hosts rather than values for parasites in naive hosts is probably more realistic. Parasite age also has an important effect during a primary infection. However, given the major influence of immunity that will appear during more natural infection regimes, in conjunction with the complications introduced by overlapping ages of parasites within a single host, explicit inclusion of parasite age may not be warranted.

III. Transmission: When Parasite Meets Host

In direct life cycle nematodes, successful transmission requires spatial and temporal overlap of infective larvae in the environment and potential hosts, so actual contact between the larvae and the potential host can occur. Also, the larvae must be able to exsheath, penetrate, and establish an infection in the host. Separating these factors and quantifying them independently is difficult, but the following studies reveal the approaches that have been used to date.

A. Overlap of Larvae and Hosts

The spatial distribution of infective larvae on soil, as well as the physical properties of the soil, will be an important determinant of the availability of larvae to mice. No data exist to date on the spatial distribution of *H. polygyrus* larvae in the field, but these larvae are expected to be aggregated in the environment, as are larvae of other nematode parasites of humans and livestock (Donald and Leslie, 1969; Udonsi *et al.*, 1980; Hominick *et al.*, 1987; Boag *et al.*, 1989; Wong and Bundy, 1990).

Quantifying actual contact between larva and host is especially difficult, partly because of the difficulty in quantifying soil contact. This difficulty contrasts with the situation in vector-borne diseases, or in diseases for which transmission is associated with water contact or ingestion of infected intermediate hosts, where points of contact that may lead to transmission are discrete and, therefore, are counted more easily.

Attempts to quantify rates of contact between larvae and mice have involved the use of sentinel or tracer mice, placed for a fixed period of time in a cage containing known numbers of infective larvae on damp peat (Scott, 1991; Tanguay and Scott, 1992). Under such conditions, estimated *per capita* rates of transmission range from 0.004 to 0.024 per larva per mouse per hr (Table 2). Again, note that these estimates assume 100% establishment of ingested larvae and 100% survival to 5 days pi.

Table 2 Uptake of Larvae by Female Mice Exposed Individually to *Heligmosomoides polygyrus*[a,b]

Mouse strain	n	L_4 per mouse (mean ± SE)	Estimate of β[c]
C57BL/6	15	76.5 ± 12.6	0.013
	15	131.9 ± 7.7	0.024
	12	59.8 ± 8.5	0.010
	12	24.7 ± 3.1	0.004
CD1	15	89.7 ± 9.3	0.016
	15	27.9 ± 3.0	0.005
	12	43.8 ± 4.1	0.008
	12	25.0 ± 6.2	0.004

[a] Data from Tanguay and Scott (1992).
[b] Exposure was for 6 hr to 1000 L_3, spread over the surface of damp peat moss in a 26 × 20 cm^2 cage.
[c] The *per capita* transmission rate (β), per larva per host per hour, was estimated using the equation $L_4 = L_{3,0}(1 - \exp^{-\beta t})$, where $L_{3,0}$ is the number of larvae placed on the damp peat, L_4 is the number of fourth stage larvae recovered in the mice 5–6 days after exposure, and t is the duration of exposure to larvae.

B. Establishment of Ingested Larvae

Establishment of larvae, or "larval infectivity," can be quantified by giving mice a known number of larvae by oral gavage, then counting the number of fourth-stage larvae a few days later. Estimates obtained in this way incorporate any mortality of larvae that occurs within the first few days of infection, and are based on the assumption that the proportion of larvae recovered from gavage and from "natural" infections are the same. Larval infectivity has been shown to vary from one larval culture to another (personal observation); researchers rarely report the technique used for determining which larvae to count in calculating their doses. Therefore, care must be taken when comparing different studies.

Table 3 Effect of Mouse Strain, Parasite Dose, and Host Diet on *Heligmosomoides polygyrus* Establishment Following a Primary or Challenge Infection

Mouse strain	*n*	Dose[a]	Time of necropsy[b]	Number of worms (±SE)	Recovery (%)	Reference
CBA	ni[c]	250P	14–28 dpi	156 ± 11	62.4	Parker and Inchley (1990)
BALB/c	ni	250P	14–28 dpi	114 ± 9	45.6	
NIH	ni	250P	14–28 dpi	147 ± 12	58.8	
BALB/c	10	100P	6 dpi	54.3 ± 1.5	54.3	Scott (1991)
C57BL/6	10	100P	6 dpi	62.5 ± 1.7	62.5	
BALB/c	15	100C	28 dpc	18.5 ± 3.0	18.5	
C57BL/6	15	100C	28 dpc	60.8 ± 2.2	60.8	
Refractory	56	100P	21 dpi	58 ± 2	58	Brindley and Dobson (1982)
Liable	56	100P	21 dpi	72 ± 2	72	
Control	50	100P	21 dpi	65 ± 3	65	
CBA/J	6	100P	12 dpi	62	62	Enriquez *et al.* (1988b)
LAF1/J	6	100P	12 dpi	56	56	
CBA/J	6	100C	12 dpc	60	60	
LAF1/J	6	100C	12 dpc	8	8	
MF1	5	10P	14 dpi	7 ± 1	70	Keymer and Hiorns (1986b)
	5	52P	28 dpi	40 ± 8	76.9	
	5	110P	28 dpi	72 ± 6	65.4	
	5	244P	28 dpi	184 ± 4	75.4	
	5	450P	42 dpi	350 ± 30	77.8	
C57BL/10	11	50–100P	14 dpi	36 ± 1.5	48	Robinson *et al.* (1989)
	11	250–300P	14 dpi	206 ± 4	75	
NIH	8	50–100P	14 dpi	49 ± 3	65	
	8	250–300P	14 dpi	245 ± 8	89.2	
BALB/c	10	50–100P	14 dpi	26.8 ± 1.7	35.7	
	10	250–300P	14 dpi	156 ± 6	56.9	
Quackenbush	10	10P	14 dpi		92	Dobson *et al.* (1985)
	10	25P	14 dpi		99	
	10	50P	14 dpi		82	
	10	100P	14 dpi		88	
	10	200P	14 dpi		86	
	10	400P	14 dpi		92	
CD1, 2%[d]	5	154C	14 dpc	140 ± 13	90.9	Slater and Keymer (1988)
CD1, 4%	5	154C	14 dpc	125 ± 17	81.2	
CD1, 8%	5	154C	14 dpc	93 ± 12	60.4	

[a] P, Primary infection; C, challenge infection.
[b] dpi, Days post primary infection; dpc, days post challenge infection.
[c] ni, Not indicated.
[d] Percentage protein provided in diet.

The effect of larval dose on worm establishment is unclear (Table 3). Robinson *et al.* (1989) reported consistently higher proportions of adults following administration of high doses (250–300) of larvae compared with low doses (50–100), whereas Keymer and Hiorns (1986b) reported no effect of larval dose on worm recovery over doses ranging from 10 to 450 infective larvae. Similarly, Dobson *et al.* (1985) found no effect of larval dose (ranging from 10 to 400) on percentage recovery of worms 2 wk pi.

Mouse strain influences larval infectivity somewhat during a primary infection (Brindley and Dobson, 1982; Enriquez *et al.*, 1988b; Parker and Inchley, 1990; Scott, 1991), but much more so after a challenge infection (Enriquez *et al.*, 1988b; Scott, 1991; Table 3). Immunological status has a dramatic effect on larval establishment (Keymer and Tarlton, 1991; Scott, 1991); the degree to which worm establishment is reduced depends on the effectiveness of procedures used to stimulate immunity (Lueker and Hepler, 1975; Enriquez *et al.*, 1988a,c).

To date, data on the effect of nutritional status on establishment of a primary infection suggest no effect of low protein or zinc (Brailsford and Mapes, 1987; Minkus *et al.*, 1992). In contrast, dramatic effects of low protein are seen on worm recovery after a challenge infection. Keymer and Tarlton (1991) reported significantly higher numbers of worms 2 wk post challenge infection in mice fed a deficient diet (3% protein) compared with those fed a sufficient diet (16% protein). Similar results were reported by Slater and Keymer (1988) when an 8% protein diet was compared with diets containing 4% and 2% protein (Table 3). These differences presumably are related to impairment of functional immunity by protein deficiency (Slater and Keymer, 1988).

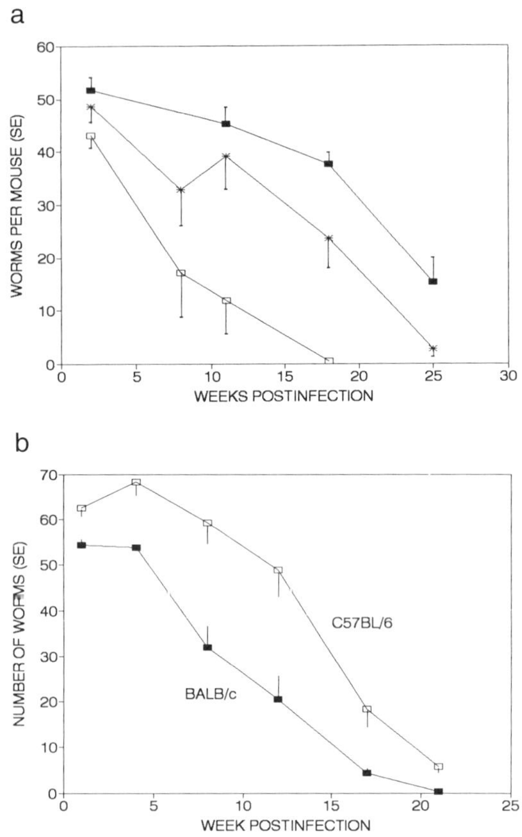

Figure 4 Survival of *Heligmosomoides polygyrus* during a primary infection in different mouse strains. (a) Comparison of worm numbers in female C57BL/10 (■), NIH (*), and BALB/c (□) mice given between 50 and 100 L_3. Data from Robinson *et al.* (1989); reprinted with the permission of Cambridge University Press. (b) Comparison of worm numbers in female C57BL/6 (□) and BALB/c (■) mice given 100 L_3 ($F_{1,78} = 4.47$; $P = 0.038$). Data from Scott (1991); reprinted with the permission of Cambridge University Press.

IV. Survival: Lasting Relationship in the Host

A. Primary Infection

Heligmosomoides polygyrus survives for many months during a primary infection, although survival profiles do differ among mouse strains (Behnke, 1987; Robinson *et al.*, 1989; Scott, 1991; Fig. 4). This chronic infection state has been shown to be related to the production of immunosuppressive factors by adult worms in the lumen of the intestine (see reviews by Behnke, 1987; Monroy and Enriquez, 1992).

Dobson *et al.* (1985) reported that parasite survivorship increased with increasing parasite dose in Quackenbush mice. A similar pattern was observed by Behnke *et al.* (1987) in CFLP mice. This effect of parasite density may be related to a dose-dependent immunosuppression induced by the parasite. In contrast, Keymer and Hiorns (1986b) reported no effect of infection dose on survival of worms in MF1 mice, over doses ranging from 10 to 450 infective larvae.

Differences between male and female worm survival apparently occur within the first 14 days of infection. Keymer and Hiorns (1986b) reported higher death rates per worm per day for males [0.033 (SE 0.003)] than for females [0.021 (SE 0.005)]. Note that this estimation assumes an equal sex ratio in the L_3 population. This assumption has not been tested. Interestingly, however, survival of males overall is higher than that for females (Keymer and Hiorns, 1986b; Scott, 1991). In their estimation of average expected life-spans (11.22 wk for males and 9.92 wk for females), Keymer and Hiorns (1986b) incorporated loss of worms at the time of infection. However, if worms lost at the time of establishment are excluded (approximately 46%), the expected life-span of established worms will be much longer (estimated from their data as 22 and 17 wk for males and females, respectively).

B. Challenge Infection

Although several researchers have published survivorship profiles for *H. polygyrus* after a primary infection, data on the time course of infection following a challenge infection are much more scarce; most researchers report data from only one or two time points after challenge infection. Nevertheless, survival in virtually all strains of mice is reduced relative to a primary infection following a challenge infection. The magnitude of the reduced survival does vary among mouse strains, and the extent of this variation is even more dramatic following a challenge infection than following a primary infection (Behnke and Robinson, 1985; Cypess *et al.*, 1988; Enriquez *et al.*, 1988a,b,c; Wahid *et al.*, 1989; Scott, 1991). These effects have been shown to have a genetic component (see Monroy and Enriquez, 1992), both by correlations between susceptibility/resistance phenotypes and the H2 histocompatibility loci (Wassom *et al.*, 1987) and by selective breeding experiments in which susceptibility or resistance has been shown to be heritable (Brindley *et al.*, 1986; Keymer *et al.*, 1990).

C. Trickle Infection Experiments

All the experiments reported in the previous section were done under tightly controlled artificial conditions. If *H. polygyrus* is to be a useful laboratory model for livestock or human infections, or for *H. polygyrus* in wild-mouse populations, the host and parasite must be allowed to interact in a manner that approaches their interaction in the real world. Also, determining whether the factors shown to influence the biology of the parasite under controlled conditions are relevant under more natural conditions is important.

A first step toward making the system more natural has been the use of a trickle, or repeated, infection procedure. Rarely is a host exposed only once to nematode parasites. Instead, hosts are constantly ingesting infective stages. To simulate this condition, researchers feed low doses of larvae to mice at regular intervals. Four studies have followed worm numbers in mice repeatedly infected with larvae (Kerboeuf, 1982; Keymer and Hiorns, 1986b; Slater and Keymer, 1986b; Scott, 1991).

Kerboeuf (1982) infected mice 9 times with 50 larvae. After 81 days of continuing infection, worm numbers in these mice were compared with worm numbers in mice infected only once with either 50 or 450 larvae. This investigator recovered 59% of the worms from mice infected once with 450 larvae, and 50% of the worms from mice infected once with 50 larvae. In marked contrast, only 8% of the worms were recovered from mice infected 9 times with 50 larvae.

Keymer and Hiorns (1986b) repeatedly infected mice with 50 larvae every 2 wk for 20 wk. At the end of their experiment, these investigators recovered 50 (SE 29) worms per mouse, 9% of the total number of larvae fed to the mice. The pattern of worm num-

Table 4 Estimation of *per capita* Death Rates of *Heligmosomoides polygyrus* during Trickle Infections[a]

Strain	Weekly dose	Total dose	Number of worms (±SE)	Percentage of total	Death rate per day[b]
C57BL/6	10	50	24.3 ± 5.9	48.6	
	10	100	21.8 ± 0.8	21.8	0.06
	50	250	113.5 ± 3.0	45.4	
	50	500	70.3 ± 8.3	14.1	0.08
	100	500	210.2 ± 8.4	42.0	
	100	1000	36.0 ± 3.2	3.6	0.12
BALB/c	10	50	15.7 ± 1.5	31.4	
	10	100	3.0 ± 2.2	3.0	0.15
	50	250	35.0 ± 2.9	14.0	
	50	500	0.8 ± 0.4	0.2	0.40
	100	500	79.7 ± 5.1	15.9	
	100	1000	0.7 ± 0.4	0.07	0.36

[a] Estimates using data from Scott (1991).
[b] Based on a simple immigration–death model where $P_t = (\lambda/\mu)(1 - \exp^{-\mu t})$. Death rate ($\mu$) estimated by nonlinear curve-fitting procedure of SYSTAT using data from 5 and 10 wk of repeated infection.

bers through time suggests that an equilibrium was reached, but at a much lower level than expected from the estimates of rates of mortality of worms (0.02 per day) following a single infection (Keymer and Hiorns, 1986b). Given the observed equilibrium of approximately 58 worms, and an infection rate of 50 per 14 days, the estimated death rate of worms in the trickle infection experiments is 1.16 per 14 days or 0.08 per day.

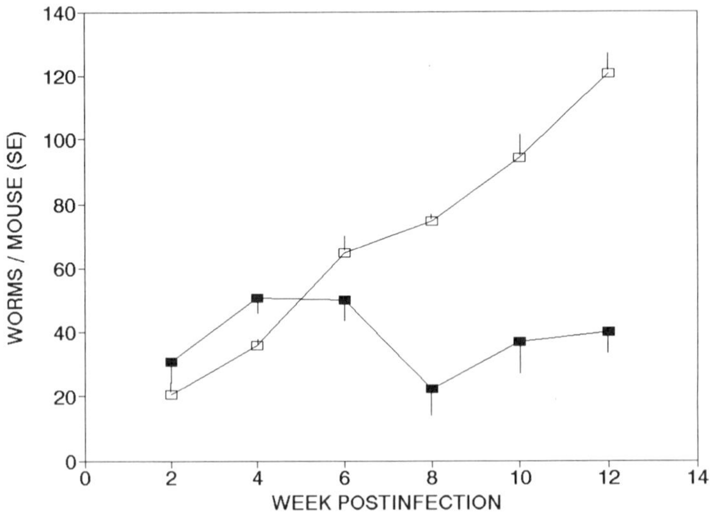

Figure 5 Effect of level of dietary protein on worm numbers following trickle infection of CD1 mice with 40 L_3 every 2 wk. 2% protein diet (□); 8% protein diet (■). Data from Slater and Keymer (1986b); with permission of The Royal Society.

Clearly, in this situation, the establishment and/or survival of worms is lower when mice are infected repeatedly than when mice are given a single infection.

A trickle infection procedure has been used to compare two strains of inbred mice: BALB/c and C57BL/6 (Scott, 1991). Mice were infected weekly with 10, 50, or 100 larvae over a 10-wk period. Estimates of the *per capita* death rate were made, assuming a simple immigration–death process (Table 4). Since the estimates are based on only two points in time, the confidence in the values is limited. Nevertheless, the parasite death rate appears to be higher in BALB/c mice than in C57BL/6 mice, and the death rate appears to increase with increasing rates of infection. This result is consistent with the observation that BALB/c mice are more resistant to *H. polygyrus* than C57BL/6 mice.

Slater and Keymer (1986b) found a similar pattern using outbred CD1 mice infected every 2 wk with 5, 10, 20, or 40 larvae. In addition, these researchers investigated the impact of two levels of dietary protein (2% and 8%) on the worm burdens obtained and found that, when protein was restricted in the diet, worm numbers continued to increase through the experiment (Fig. 5), in contrast to the turnover in worm numbers seen when mice were fed the 8% protein diet. These investigators suggested that this difference was related to impaired immunological response of the protein-restricted mice (Slater and Keymer, 1988).

Box 1 A Free-Running Laboratory Set-Up for *Heligmosomoides polygyrus* in Mouse Populations

In an attempt to provide a more natural environment for both host and parasite, we house populations of mice in a large indoor enclosure. Two different arena dimensions have been used: 1.2×2.4 m^2 (Scott, 1987,1988a,b,1990) and 3.2×0.8 m^2 (Scott, 1991). Populations of mice are introduced into the arena. In some cases, these may be single sex, nonbreeding populations (Scott, 1991); in other cases, mixed-sex breeding populations are used (Scott, 1987,1988a,b, 1989). The arenas contain areas of damp peat moss in which the parasite eggs can hatch, and in which parasite larvae can live and be accessible to the mice. The infection is introduced by infecting a small number of mice by gavage. These mice then pass parasite eggs into the environment with the feces. Other mice become infected when they contact the resulting parasite larvae on the damp peat moss. This system allows the mice to live together as a population, and enables parasite transmission to occur through the host population without the intervention of the researcher. This method also enables the researcher to vary a range of characteristics including host population density and surface area available for transmission.

V. *Heligmosomoides polygyrus* and Current Issues in Parasite Population Biology

Over the past 15 years, much interest has been generated about intestinal worms because of the hypotheses arising from theoretical studies. Researchers have been able to investigate a number of these hypotheses using the *H. polygyrus*–mouse model system. In the following section, some of the advances that have been made are reviewed. In all cases, the ability to test hypotheses has been related fundamentally to the development of a free-running experimental system in which mice live together as a population and in which infection occurs through contact between the mice and the substrate in their colony arena (see Box 1). This system first was reported by Slater and Keymer (1986a), and subsequently has been used extensively by Scott (1987; 1988a,b; 1991).

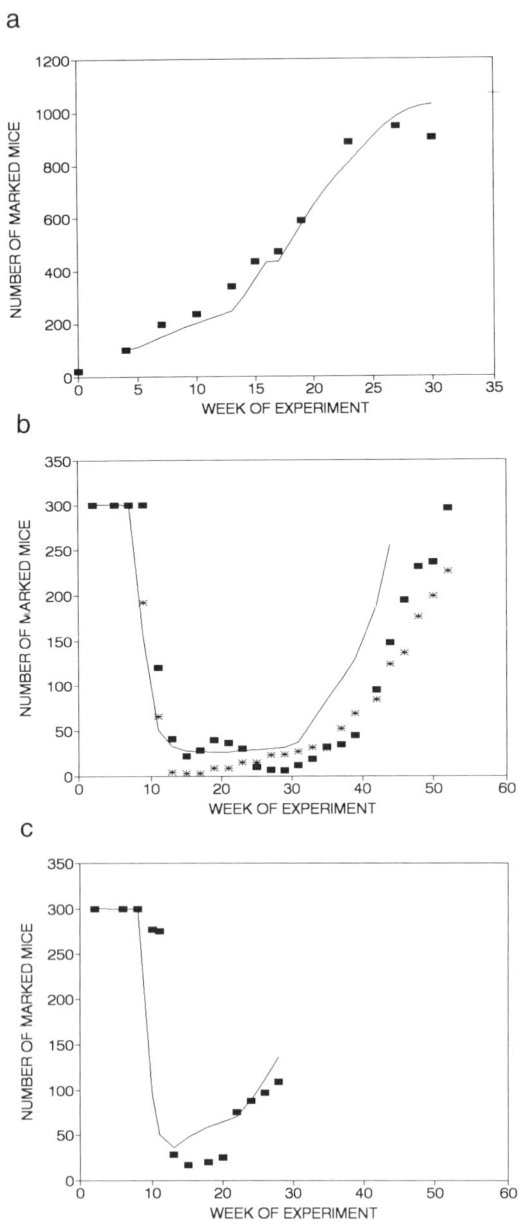

Figure 6 Effect of *Heligmosomoides polygyrus* on population growth of CD1 mice. Symbols represent the numbers of mice (over 1.5 wk of age) transferred to clean arenas at each census. The line represents the prediction of an age-structured Leslie Matrix Model. (a) Uninfected mouse population. (b) Two replicates of mouse population in which *H. polygyrus* was introduced at week 3 and all mice were treated with anthelmintic at week 30. (c) Mouse population in which *H. polygyrus* was introduced at week 3 and the substrate was altered to reduce transmission at week 11. Data from Scott (1990); reprinted with the permission of Cambridge University Press.

A. Parasites as Regulators of Host Population Abundance

It has long been thought that predation and competition provide important regulatory mechanisms in animal populations, yet the role of parasitic infection has been much more controversial. Anderson and May (1978) demonstrated the theoretical potential of macroparasites to regulate the abundance of their host population. To test this hypothesis, Scott (1987) placed breeding populations of outbred CD1 mice into indoor arenas (4 × 8 sq ft). In the absence of *H. polygyrus* infection, mouse numbers increased exponentially and eventually equilibrated at a density of 320 mice/m^2 (Fig. 6a). An age-structured Leslie matrix model provided a good fit to this growth profile (Scott, 1990); the model assumed that first parturition occurs at 9 wk of age, that the sex ratio at birth is 1 : 1, that recruitment into the population is density dependent, and that mouse survival is density independent but age dependent (in male mice).

This uninfected population then was divided into three subpopulations of 300 mice each. Each population was placed in a separate arena. *Heligmosomoides polygyrus* was introduced into the populations, and host and parasite numbers were monitored through time. In all three replicates, significant parasite-induced host mortality led to a decline of over 90% in the host population (Fig. 6b,c). When transmission continued at high levels, the population remained stable at less than 18 mice/m^2 whereas, when transmission rates were reduced (by reducing the proportion of the substrate on which parasite larvae could survive), the mouse population gradually increased in size (Fig. 6c). When mice were treated with anthelmintic drugs (Fig. 6b), the mouse population rapidly increased in size. Scott (1990) reported that the parasites had no significant effect on mouse reproduction, but that infection caused both acute and chronic mortality. This investigator incorporated these effects of parasitism into an age-structured Leslie matrix model that provided a good fit

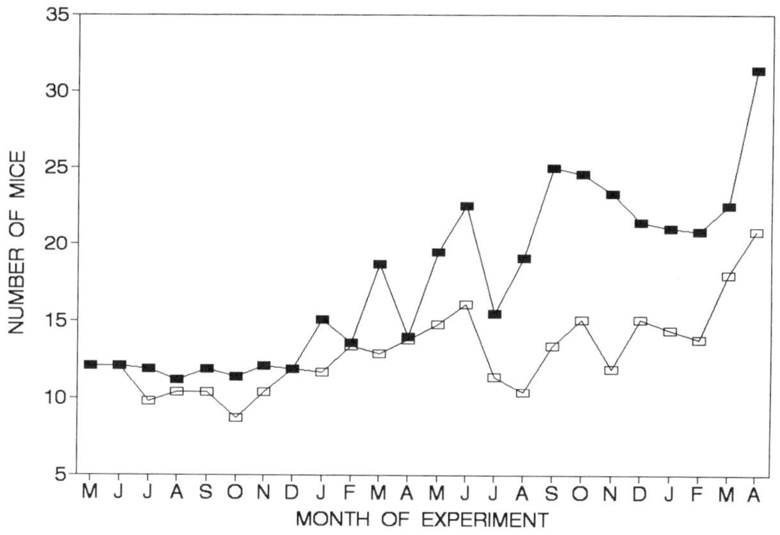

Figure 7 Effect of *Heligmosomoides polygyrus* on population growth of wood mice, *Apodemus sylvaticus,* in an outdoor enclosure. Data represent the average of three replicate populations. Control uninfected populations (■); populations infected with *H. polygyrus* (□). Data from Gregory (1991); reprinted with permission from Blackwell Scientific Publications, Ltd.

to the observed data (Fig. 6b,c).

Gregory (1991) investigated the effect of *H. polygyrus* on wood mouse *(Apodemus sylvaticus)* populations in enclosed outdoor arenas (9 × 12 m) and reported that the presence of the parasite depressed host population growth (Fig. 7) through effects on survival rather than on reproduction.

The extent to which these observations can be extrapolated to the "real world" has been discussed by Gregory (1991) and by Scott (1987, 1990). In both studies, average infection levels were higher than routinely reported in field populations (Elton *et al.,* 1931; Forrester, 1971; Kisielewska and Zubczewska, 1973; Wolton and Flowerdew, 1985; Montgomery and Montgomery, 1990; Gregory, 1992). Also, host population densities were higher than normally reported in the field, even in mouse plagues in Australia (Redhead *et al.,* 1985). Despite these considerations, these studies are important because they demonstrate that intestinal nematodes do have the ability to regulate host population abundance, and because they have taken an experimental approach into a semi-natural setting.

B. Aggregation and Predisposition

1. Relative Importance of Innate and Acquired Resistance and Host Behavior

Nematode parasites characteristically are aggregated in their host population. Much attention has been directed in recent years to understanding why this aggregation occurs. A laboratory study has been undertaken, using *H. polygyrus* in mice, to attempt to quantify the relative contributions of heterogeneity among mice in innate resistance to infection, in acquired resistance to infection, and in behaviors important for parasite transmission (Tanguay and Scott, 1992). To determine whether heterogeneity in innate resistance (presumably of genetic basis) among mice increased the degree of variability in parasite numbers per host, an outbred strain of mouse (CD1) was compared with inbred strains (C57BL/6, C3H/

Table 5 Effect of Innate Resistance, Acquired Resistance, and Host Behavior on Variability in Worm Recovery of *Heligmosomoides polygyrus in Mice*[a,b]

Experiment	Infection	Strain	Exposure	*n*	Mean ± SE	Variance[c]
1	Primary	CD1	Gavage	26	63.7 ± 2.8	0.011
		C57BL/6	Gavage	29	29.2 ± 1.4	0.012
		C3H/HeN	Gavage	30	61.4 ± 3.7	0.040[d]
		CBA/Ca	Gavage	30	50.2 ± 1.7	0.007
	Challenge	CD1	Gavage	30	21.2 ± 4.0[e]	0.533[f]
		C57BL/6	Gavage	30	16.6 ± 1.6[e]	0.208[d,f]
		C3H/HeN	Gavage	30	77.3 ± 2.6	0.007[d,f]
		CBA/Ca	Gavage	30	97.1 ± 2.2	0.003[d,f]
	Primary	CD1	Natural	15	89.7 ± 9.3	0.053[g]
		CD1	Natural	15	27.9 ± 3.0	0.034[g]
		C57BL/6	Natural	15	76.5 ± 12.6	0.185[d,g]
		C57BL/6	Natural	15	131.9 ± 7.7	0.013[d]
	Challenge	CD1	Natural	15	74.6 ± 13.4	0.205[g]
		CD1	Natural	15	33.1 ± 6.5	0.120[g]
		C57BL/6	Natural	15	32.1 ± 9.5	0.263
		C57BL/6	Natural	15	9.7 ± 2.1	0.339[d,f]
2 (Rep 2)	Primary	CD1	Water	11	103.7 ± 6.1	0.010[f]
		CD1	Cortisone	13	96.3 ± 4.6	0.005
		C57BL/6	Water	10	117.9 ± 5.1	0.003[f]
		C57BL/6	Cortisone	8	104.1 ± 6.5	0.007[f]
	Challenge	CD1	Water	12	71.3 ± 7.2	0.045[h]
		CD1	Cortisone	11	81.3 ± 5.3	0.009
		C57BL/6	Water	12	42.3 ± 7.9	0.152[d,h]
		C57BL/7	Cortisone	14	90.0 ± 9.3	0.028[d]

[a] Data from Tanguay and Scott (1992); reprinted with the permission of Cambridge University Press.
[b] Variability due to innate resistance was compared in outbred (CD1) and inbred (C57BL/6, C3H/HeN, and CBA/Ca) mice; variability associated with acquired resistance was compared in mice given a primary or a challenge infection; variability associated with host behavior was compared in mice given a gavage infection or exposed to a natural infection process. In Experiment 2, groups of mice were given cortisone in the drinking water (Cortisone) as an immunosuppressive agent.
[c] Variances were calculated on data standardized to a mean of 100, then normalized with a log-transformation.
[d] Significantly different from outbred CD1 mice ($P < 0.05$).
[e] Significant protection compared with challenge controls (data not shown) ($P < 0.05$).
[f] Significant difference between primary and challenge infection ($P < 0.05$).
[g] Significant difference compared with gavage infection ($P < 0.05$).
[h] Significantly different from mice drinking cortisone ($P < 0.05$).

HeN, CBA) following a controlled primary infection. Interestingly, the results were not consistent. In one case (C3H/HeN), inbred mice had even more variable worm burdens than outbred mice (Table 5).

The role of acquired resistance was investigated by comparing variability in worm burdens between mice given a single infection (no functional acquired resistance) and mice given a challenge infection following an anthelmintic-abbreviated primary infection. Consistently, worm burdens were more variable following the challenge infection (Table 5). This effect was removed when the same protocol was repeated in immunosuppressed mice (Tanguay and Scott, 1992). Gregory *et al.* (1990) examined the role of genetics and immunity as factors generating variability in parasite numbers. These researchers also reported that variability was higher following repeated infections (for which host immunity is functional) than following a single infection.

Investigations into host behavior involved comparing variability in worm numbers between mice given a gavage infection (no host behavior involved in transmission) and mice placed in an infective arena for a period of time. In the latter situation, infection occurred through contact between the mice and larvae present in their environment. Greater variability was expected when mice were exposed to such a "natural" infection process; in many cases this variability occurred (Table 5). However, this result was not as consistent as that obtained with acquired immunity (Tanguay and Scott, 1992).

This study highlighted the complex interplay among the range of factors that determine the degree of aggregation in parasite numbers (see Anderson and Gordon, 1982). Each of these factors in turn is affected by a wide range of conditions. The relative impact of each is likely to vary from day to day and from population to population.

2. Predisposition

The phenomenon of aggregation has raised an important question for individuals interested in control of parasitic nematode infections. Is focusing control efforts on the few heavily infected individuals in the population practical or appropriate? Such individuals are at the highest risk of infection-associated disease, and are passing disproportionate numbers of infective stages into the environment. The difficulty lies in identifying those individuals in the population who are heavily infected, particularly if no other general characteristic is common to each of these individuals. Identifying each one prior to each round of treatment will be very expensive, and may make any control program based on treatment of only heavily infected individuals impractical. Therefore, the issue of whether the same individuals will become heavily re-infected following treatment has attracted attention. Studies in human populations with *Ascaris, Trichuris,* and hookworm have shown repeatedly a significant correlation between pretreatment infection levels and intensities of infection several months after treatment (Schad and Anderson, 1985; Bundy *et al.,* 1987; Haswell-Elkins *et al.,* 1987; Thein-Hlaing *et al.,* 1987; Forrester *et al.,* 1990; see Chapter 14). This phenomenon has been called "predisposition."

Using the *H. polygyrus*-mouse model system, Scott (1988b) demonstrated that predisposition can be detected in this parasite as well, thus reinforcing the importance that this parasite assumes as a laboratory model for studying human gastrointestinal nematodes. During a follow-up study, Scott (1988a) attempted to determine the extent to which the positive correlations persisted after repeated treatments. If community control programs are to take advantage of predisposition as a means of targeting treatment, predisposition must be shown to exist not only from one treatment to the next, but over several rounds of treatment. Data from the *H. polygyrus* system indicated that the correlation between the first pretreatment infection level and subsequent posttreatment infection levels decreased with each successive treatment, to the point that the relationship disappeared (Table 6).

This important laboratory observation identified a potential difficulty for field programs, and prompted a field study by Chan *et al.* (1992) in which a similar protocol was used in Malaysia. Over the course of three repeated treatments, these researchers continued to detect significant correlations when *Ascaris* infection intensities were compared with the original pretreatment data (Table 6). Interestingly, however, virtually all correlations done with *Trichuris* data were reduced after the second round of treatment, although most remained statistically significant (Table 6).

The differences in results obtained with *H. polygyrus* in the laboratory and with *Ascaris* in the field may result from differences in the time scale of the two studies, relative to the life-span of the two host populations. In the mouse study, re-infection was monitored over three intervals of 8 wk.

Table 6 Effect of Repeated Treatment on Ability to Detect Predisposition to Nematode Infections[a]

Population	Re-infection period	Correlation with pretreatment	D.F.	Probability
H. polygyrus				
Young mice	First	0.43	23	0.03
	Second	0.15	22	0.49
	Third	−0.05	22	0.82
Old mice	First	0.53	23	0.006
	Second	0.39	22	0.06
	Third	0.32	21	0.14
Ascaris				
2–6 yr olds	First	0.59	88	<0.0001
	Second	0.59	86	<0.0001
7–12 yr olds	First	0.51	115	<0.0001
	Second	0.58	107	<0.0001
13–17 yr olds	First	0.70	31	<0.0001
	Second	0.56	29	<0.001
18–29 yr olds	First	−0.12	19	>0.05
	Second	0.50	18	<0.05
30–39 yr olds	First	0.53	56	<0.0001
	Second	0.34	58	<0.01
40+ yr olds	First	0.55	30	<0.01
	Second	0.45	27	<0.05
Trichuris				
2–6 yr olds	First	0.50	64	<0.0001
	Second	0.23	71	<0.05
7–12 yr olds	First	0.47	94	<0.0001
	Second	0.29	87	<0.01

[a] Data on *Heligmosomoides polygyrus* from Scott (1988b); reprinted with the permission of Cambridge University Press; data on *Ascaris* and *Trichuris* from Chan *et al.* (1992); reprinted with the permission of Cambridge University Press.

Mice become sexually mature at 6–8 wk of age; by 24 wk of age, females will have had five or six litters. Therefore, significant behavioral and physiological changes would be expected in mice that were 4 wk of age at the beginning of the study. In contrast, the study by Chan *et al.* (1992) covered a time-span of 2 yr in a human community. Although developmental and behavioral changes would be expected over such a period of time in children, the changes may not be as dramatic as those experienced by mice during the duration of the study by Scott (1988a).

3. Host Susceptibility and Resistance

The detection of predisposition again raises the question of the role that genetics may play in determining the susceptibility or resistance of individuals to high intensity infections. Many researchers have shown that immunological responsiveness to *H. polygyrus* infection has a strong genetic component (see review by Monroy and Enriquez, 1992). In all cases, however, these studies have been done using tightly controlled infection conditions. A study by Scott (1991) investigated the extent to which the characterization of mouse susceptibility or resistance under controlled conditions could be used to predict infection levels when mice lived as a population and were exposed to more natural transmission processes.

Two strains of mice were characterized as either susceptible (C57BL/6) or resistant (BALB/c) to *H. polygyrus* infection based on a controlled primary infection (Fig. 4b)

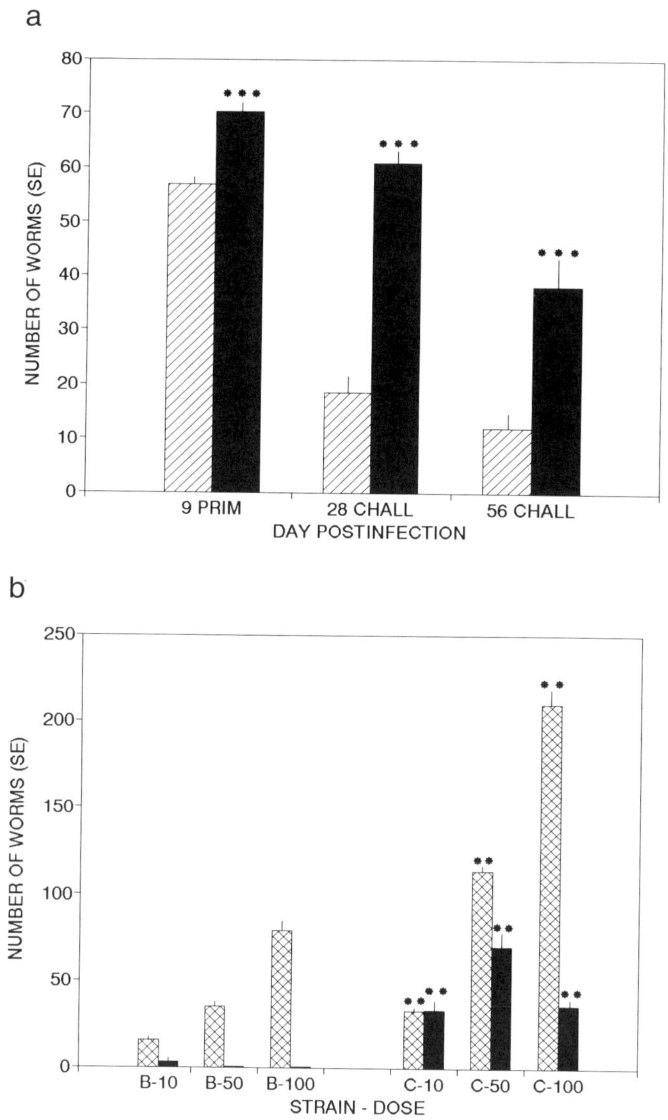

Figure 8 Comparison of "susceptible" C57BL/6 mice and "resistant" BALB/c mice following a variety of infection protocols with *Heligmosomoides polygyrus.* (a) Challenge infection protocol: BALB/c (▨) and C57BL/6 (■) mice were given a primary infection with 100 L_3 that was terminated with anthelmintic treatment 9 days postinfection. Mice were reinfected with 100 L_3. Worm numbers at 28 and 56 days after a challenge (CHALL) infection with 100 L_3 are shown. Control mice were infected only once with 100 L_3 and were killed 9 days postinfection (PRIM). Between strain comparisons: $^{***}P < 0.001$. (b) Trickle infection protocol: worm numbers in BALB/c (B) and C57BL/6 (C) mice infected with 10, 50, or 100 L_3 weekly for 5 (▩) or 10 (■) wk. Between strain comparisons: $^{**}P < 0.01$. Data from Scott (1991); reprinted with the permission of Cambridge University Press.

and a controlled challenge protocol (Fig. 8a). The two strains then were housed together in arenas (see Box 1) in which continuous exposure to parasite larvae in the substrate occurred. The expectation was that the resistant strain (BALB/c) would have fewer parasites than the susceptible strain (C57BL/6). However, in two repli-

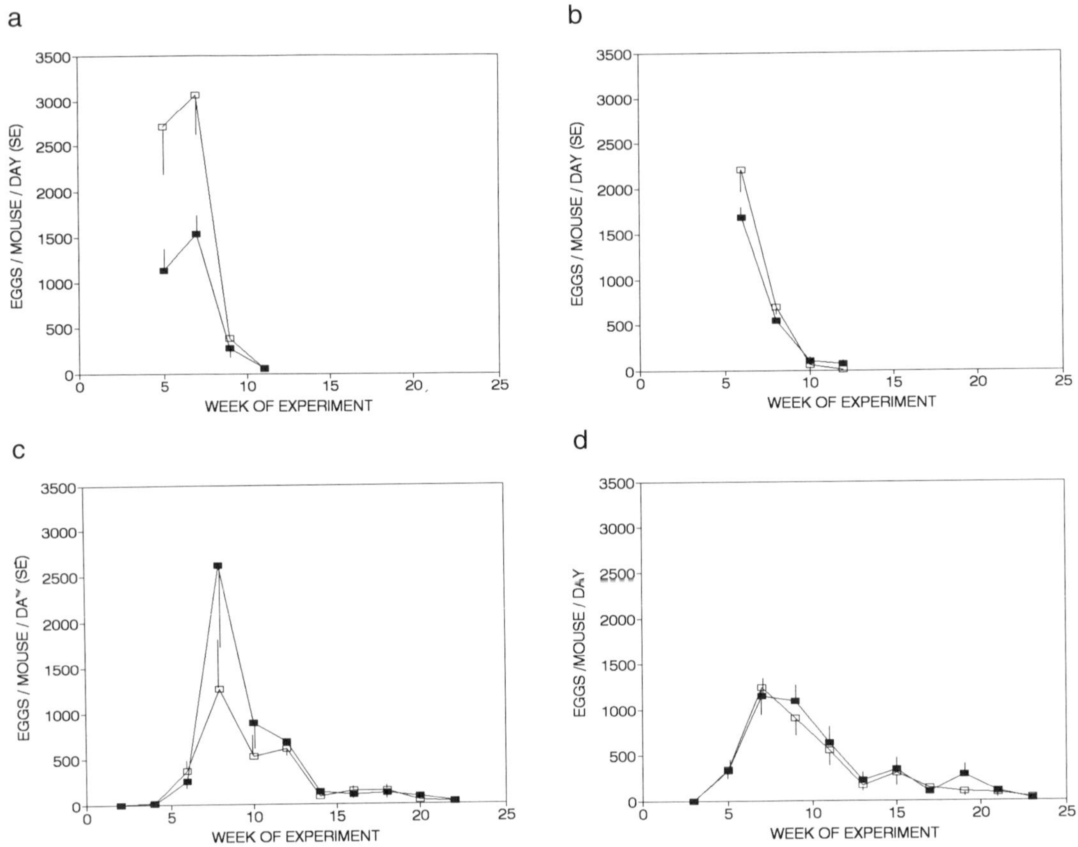

Figure 9 Daily egg production in BALB/c (■) and C57BL/6 (□) mice living together in an indoor arena and exposed to continual transmission of *Heligmosomoides polygyrus*. This experiment was replicated twice; in each replicate two subpopulations of mice were monitored. (a) Replicate 1, Group A (Strain: $F_{1,28} = 13.03$, $P = 0.001$; Time: $F_{3,84} = 37.33$, $P < 0.001$; Time × Strain: $F_{3,84} = 5.54$, $P = 0.002$). (b) Replicate 1, Group B (no significant effects). (c) Replicate 2, Group A (no significant effects). (d) Replicate 2, Group B (no significant effects). Data from Scott (1991); reprinted with the permission of Cambridge University Press.

cates of this experiment, the resulting infection levels in the two strains did not differ (Fig. 9).

A series of follow-up investigations were undertaken to explain the unexpected difference between the results of the controlled infection situation and the free-running system. Two specific hypotheses have been investigated. Scott (1991) suggested that the continuous exposure of mice to infective larvae that occurs in the arena might alter the response of the mice to infection. To test this possibility, both strains were given a trickle infection of 10, 50, or 100 larvae weekly for 10 wk. BALB/c mice had fewer worms than C57BL/6 mice in all comparisons (Fig. 8b), thereby suggesting that the difference between the controlled experiments and the natural infection process was not related to continuous exposure of the mice to parasite larvae.

The second hypothesis was that behavioral differences between the strains might have affected the rate at which they contacted larvae in the peat trays. To test this hypothesis, sentinel mice of both strains were placed in the arenas for a period of 24 hr. Both strains always acquired the same

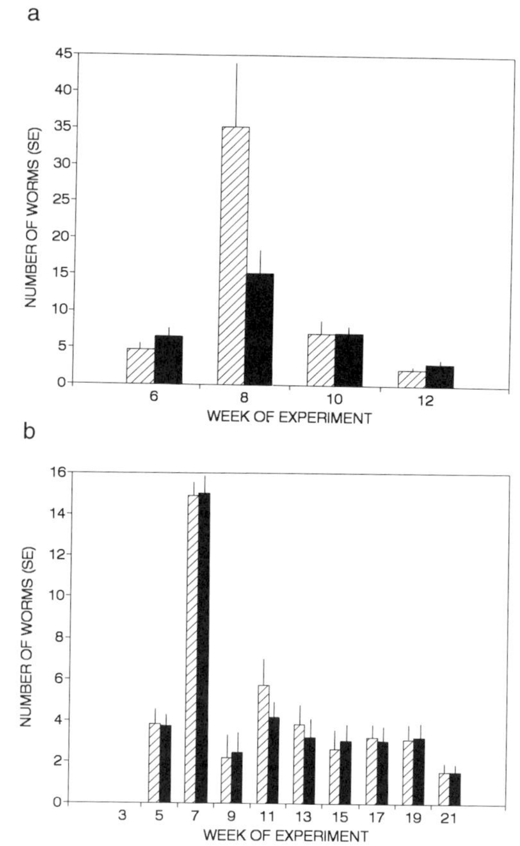

Figure 10 Numbers of *Heligmosomoides polygyrus* L_4 in sentinel BALB/c (▨) and C57BL/6 (■) mice placed into arenas with endemic transmission (see Fig. 9) for 24 hr. (a) Replicate 1. (b) Replicate 2. No significant differences between strains were detected at any time. Data from Scott (1991); reprinted with the permission of Cambridge University Press.

numbers of larvae (Fig. 10), thereby suggesting that the rate of transmission in the arenas was similar for both strains. This conclusion was supported further by preliminary data on mouse behavior. The numbers of mice of each strain that were physically present on infective trays were recorded at 1-min intervals over 25 30-min observation periods. The frequency of contact with infective trays in the arena was found to be similar for both strains of mouse [BALB/c mice: 16.6 (SE 4.4); C57BL/6: 18.4 (SE 4.6)]. Note that neither the sentinel data nor the behavior data provide a conclusive indication that mouse behavior affecting transmission does not differ between strains. The behavior of mice during the first 24 hr after being introduced into a new environment is not necessarily similar to that of mice who have been living in the environment. The preliminary observations of behavior were made only during the day, when the activity of the nocturnal mice is generally low. Also, more than physical contact with infective trays is required to effect transmission.

These studies are being continued to try to arrive at an explanation for these unexpected results. Although caution is exercised in extrapolating beyond this system, the data indicate that susceptibility defined under controlled conditions will not necessarily be an indicator of high infection levels under more natural conditions. As noted earlier regarding studies on aggregation, clearly infection intensities in more natural situations will be determined by a complex mixture of very many factors.

VI. A System to Emulate

What makes the *H. polygyrus*–mouse system so valuable? In part, the degree to which it shares features with the important gastrointestinal nematodes of humans and livestock contributes to its value. However, the more important feature is the ability to create a semi-natural endemic infection in the laboratory, using a host population rather than individual hosts isolated in small cages and using a natural infection process by which the mice continuously contaminate their own environment and re-infect themselves. Similar free-running setups have been developed with parasites of fish (see Scott and Anderson, 1984; McCallum, 1985) and with an acanthocephalan of rats (Monks and Nickol, 1989).

Invariably, development and use of such systems is a long-term proposition. Time is required to develop and test the systems, and each individual experiment takes several months or years to complete. Also, a wide range of background information must be obtained from controlled experiments so the results from the more free-running system can be interpreted. When these background experiments can be tied together in a mathematical framework, the design of experiments for the free-running system is simplified substantially. Despite the requisite time investment, the effort is worthwhile. We encourage other investigators to consider identifying systems that could be adapted similarly.

Acknowledgments

Research in our laboratory and at the Institute of Parasitology is supported by the Natural Sciences and Engineering Research Council of Canada and by the Fonds FCAR pour l'aide à la recherche.

References

Anderson, R. M., and Gordon, D. M. (1982). Processes influencing the distribution of parasite numbers within host populations with special emphasis on parasite-induced host mortalities. *Parasitology* **85,** 373–398.

Anderson, R. M., and May, R. M. (1978). Regulation and stability of host–parasite population interactions. I. Regulatory processes. *J. Anim. Ecol.* **47,** 219–247.

Behnke, J. M. (1987). Evasion of immunity by nematode parasites causing chronic infections. *Adv. Parasitol.* **26,** 1–71.

Behnke, J. M., and Robinson, M. (1985). Genetic control of immunity to *Nematospiroides dubius:* A 9-day anthelmintic abbreviated immunizing regime which separates weak and strong responder strains of mice. *Parasite Immunol.* **7,** 235–253.

Behnke, J. M., Williams, D. J., Hannah, J., and Pritchard, D. I. (1987). Immunological relationships during primary infection with *Heligmosomoides polygyrus (Nematospiroides dubius):* The capacity of adult worms to survive following transplantation to recipient mice. *Parasitology* **95,** 569–581.

Boag, B., Topham, P. B., and Webster, R. (1989). Spatial distribution on pasture of infective larvae of the gastro-intestinal nematode parasites of sheep. *Int. J. Parasitol.* **19,** 681–685.

Brailsford, T. J., and Mapes, C. J. (1987). Comparisons of *Heligmosomoides polygyrus* primary infection in protein-deficient and well-nourished mice. *Parasitology* **95,** 311–321.

Brindley, P. J., and Dobson, C. (1982). *Nematospiroides dubius* in mice selected for liability to infection: Modification of parasite biology through host selection. *Int. J. Parasitol.* **12,** 573–578.

Brindley, P. J., and Dobson, C. (1983). Genetic control of liability to infection with *Nematospiroides dubius* in mice: Direct and correlated responses to selection of mice for faecal parasite egg count. *Parasitology* **87,** 113–127.

Brindley, P. J., He, S., Sitepu, P., Pattie, W. A., and Dobson, C. (1986). Inheritance of immunity in mice to challenge infection with *Nematospiroides dubius. Heredity* **57,** 53–58.

Bundy, D. A. P., Cooper, E. S., Thompson, D. E., Didier, J. M., Anderson, R. M., and Simmons, I. (1987). Predisposition to *Trichuris trichiura* infection in humans. *Epidem. Infect.* **98,** 65–71.

Chan, L., Kan, S. P., and Bundy, D. A. P. (1992). The effect of repeated chemotherapy on age-related predisposition to *Ascaris lumbricoides* and *Trichuris trichiura. Parasitology* **104,** 371–377.

Chang, J., and Wescott, R. B. (1972). Infectivity, fecundity, and survival of *Nematospiroides dubius* in gnotobiotic mice. *Exp. Parasitol.* **32,** 327–334.

Cypess, R. H., Lucia, H. L., Dundsford, H. A., and Enriquez, F. J. (1988). The tissue reactions of mice to infection with *Heligmosomoides polygyrus. J. Helminthol.* **62,** 69–76.

Dobson, C., Sitepu, P., and Brindley, P. J. (1985). Influence of primary infection on the population dynamics of *Nematospiroides dubius* after challenge infections in mice. *Int. J. Parasitol.* **15,** 353–359.

Donald, A. D., and Leslie, R. T. (1969). Population studies on the infective stage of some nematode parasites of sheep. II. The distribution of faecal deposits on fields grazed by sheep. *Parasitology* **59,** 141–157.

Elton, C., Ford, E. B., Baker, J. R., and Gardner, A. D. (1931). The health and parasites of a wild mouse population. *Proc. Zool. Soc. London* **II,** 657–721.

Enriquez, F. J., Cypess, R. H., and Wassom, D. L. (1988a). Influence of immunizing dose and presence or absence of adult worms on the development of resistance to *Nematospiroides dubius* challenge infections in mice. *J. Parasitol.* **74,** 409–414.

Enriquez, F. J., Scarpino, V., Cypess, R. H., and Wassom, D. L. (1988b). *In vivo* and *in vitro* egg production by *Nematospiroides dubius* during primary and challenge infections in resistant and susceptible strains of mice. *J. Parasitol.* **74,** 262–266.

Enriquez, F. J., Zidian, J. L., and Cypess, R. H. (1988c). *Nematospiroides dubius:* Genetic control of immunity to infections of mice. *Exp. Parasitol.* **67,** 12–19.

Forrester, D. J. (1971). *Heligmosomoides polygyrus*

(=*Nematospiroides dubius*) from wild rodents of Northern California: Natural infections, host specificity, and strain characteristics. *J. Parasitol.* **57,** 493–503.

Forrester, J. E., Scott, M. E., Bundy, D. A. P., and Golden, M. H. N. (1990). Predisposition of individuals and families in Mexico to heavy infection with *Ascaris lumbricoides* and *Trichuris trichiura*. *Trans. R. Soc. Trop. Med. Hyg.* **84,** 272–276.

Gregory, R. D. (1991). Parasite epidemiology and host population growth: *Heligmosomoides polygyrus* (Nematoda) in enclosed wood mouse populations. *J. Anim. Ecol.* **60,** 805–821.

Gregory, R. D. (1992). On the interpretation of host-parasite ecology: *Heligmosomoides polygyrus* (Nematoda) in wild wood mouse *(Apodemus sylvaticus)* populations. *J. Zool. London* **226,** 109–121.

Gregory, R. D., Keymer, A. E., and Clarke, J. R. (1990). Genetics, sex and exposure: The ecology of *Heligmosomoides polygyrus* (Nematoda) in the wood mouse. *J. Anim. Ecol.* **59,** 363–378.

Haswell-Elkins, M. R., Elkins, D. B., and Anderson, R. M. (1987). Evidence for predisposition in humans to infection with *Ascaris,* hookworm, *Enterobius* and *Trichuris* in a South Indian fishing community. *Parasitology* **95,** 323–337.

Hominick, W. M., Dean, C. G., and Schad, G. A. (1987). Population biology of hookworms in West Bengal: Analysis of numbers of infective larvae recovered from damp pads applied to the soil surface at defaecation sites. *Trans. R. Soc. Trop. Med. Hyg.* **81,** 978–986.

Kerboeuf, D. (1978a). The effects of time and temperature of storage on the infectivity of third-stage larvae of *Heligmosomoides polygyrus (=Nematospiroides dubius)*. 1. Effects on the development to the adult stage in mice. *Ann. Rech. Vét.* **9,** 153–159.

Kerboeuf, D. (1978b). The effects of time and temperature of storage on the infectivity of third-stage larvae of *Heligmosomoides polygyrus (=Nematospiroides dubius)*. 2. Studies on the fecundity of female worms as a function of the infectivity of the third-stage larvae. *Ann. Rech. Vét.* **9,** 161–168.

Kerboeuf, D. (1982). Egg output of *Heligmosomoides polygyrus (Nematospiroides dubius)* in mice given multiple infections of constant or increasing size. *Ann. Rech. Vét.* **13,** 357–368.

Kerboeuf, D. (1985). The effect of host density on the establishment and fecundity of *Heligmosomoides polygyrus*. *Parasitology* **91,** 177–183.

Kerboeuf, D., and Jacob, C. (1983). Rhythmic egg release by *Heligmosomoides polygyrus (=Nematospiroides dubius),* parasitic Trichostrongylida of the murine gut. *Chronobiologia* **10,** 255–267.

Keymer, A. E., and Hiorns, R. W. (1986a). Faecal egg counts and nematode fecundity: *Heligmosomoides polygyrus* and laboratory mice. *Parasitology* **93,** 189–203.

Keymer, A. E., and Hiorns, R. W. (1986b). *Heligmosomoides polygyrus* (Nematoda): The dynamics of primary and repeated infection in outbred mice. *Proc. R. Soc. London B* **229,** 47–67.

Keymer, A. E., and Tarlton, A. B. (1991). The population dynamics of acquired immunity to *Heligmosomoides polygyrus* in the laboratory mouse: Strain, diet and exposure. *Parasitology* **103,** 121–126.

Keymer, A. E., Tarlton, A. B., Hiorns, R. W., Lawrence, C. E., and Pritchard, D. I. (1990). Immunogenetic correlates of susceptibility to infection with *Heligmosomoides polygyrus* in outbred mice. *Parasitology* **101,** 69–73.

Kisielewska, K., and Zubczewska, Z. (1973). Intestinal helminths as indexes of reproduction dynamics in the host population—Common vole. *Acta Theriol.* **18,** 237–246.

Lueker, D. C., and Hepler, D. I. (1975). Differences in immunity to *Nematospiroides dubius* in inbred and outbred mice. *J. Parasitol.* **61,** 158–159.

MacKinnon, B. M., and Lee, D. L. (1988). Age-related changes in *Heligmosomoides polygyrus* (Nematoda): Neutral lipid content in developing oocysts. *Can. J. Zool.* **66,** 2791–2796.

McCallum, H. I. (1985). Population effects of parasite survival of host death: Experimental studies of the interaction of *Ichthyophthirius multifiliis* in its fish host. *Parasitology* **90,** 529–547.

Minkus, T. M., Koski, K. G., and Scott, M. E. (1992). Marginal zinc deficiency has no effect on primary or challenge infections in mice with *Heligmosomoides polygyrus* (Nematoda). *J. Nutr.* **122,** 570–579.

Monks, S., and Nickol, B. B. (1989). Effect of *Moniliformis moniliformis* density on distribution within the definitive host population *(Rattus norvegicus)*. *Int. J. Parasitol.* **19,** 865–874.

Monroy, F. G., and Enriquez, F. G. (1992). *Heligmosomoides polygyrus:* A model for chronic gastrointestinal helminthiasis. *Parasitol. Today* **8,** 49–54.

Montgomery, S. S. J., and Montgomery, W. I. (1990). Structure, stability and species interactions in helminth communities of wood mice, *Apodemus sylvaticus*. *Int. J. Parasitol.* **20,** 225–242.

Parker, S. J., and Inchley, C. J. (1990). *Heligmosomoides polygyrus:* Influence of infection on lymphocyte subpopulations in mouse mesenteric lymph nodes. *Exp. Parasitol.* **71,** 249–258.

Redhead, T. D., Enright, N., and Newsome, A. E. (1985). Causes and predictions of outbreaks of *Mus musculus* in irrigated and non-irrigated cereal farms. *Acta Zool. Fenn.* **173,** 123–127.

Robinson, M., Wahid, F., Behnke, J. M., and Gilbert, F. S. (1989). Immunological relationships during primary infection with *Heligmosomoides polygyrus (Nematospiroides dubius):* Dose-dependent expulsion of adult worms. *Parasitology* **98,** 115–124.

Schad, G. A., and Anderson, R. M. (1985). Predisposition to hookworm infection in humans. *Science* **228,** 1537–1540.

Scott, M. E. (1987). Regulation of mouse colony abundance by *Heligmosomoides polygyrus. Parasitology* **95,** 111–124.

Scott, M. E. (1988a). Effect of repeated anthelmintic treatment on ability to detect predisposition of mice to *Heligmosomoides polygyrus* and *Aspiculuris tetraptera* (Nematoda) infections. *Parasitology* **97,** 453–458.

Scott, M. E. (1988b). Predisposition of mice to *Heligmosomoides polygyrus* and *Aspiculuris tetraptera* (Nematoda). *Parasitology* **97,** 101–114.

Scott, M. E. (1990). An experimental and theoretical study of the dynamics of a mouse–nematode *(Heligmosomoides polygyrus)* interaction. *Parasitology* **101,** 75–92.

Scott, M. E. (1991). *Heligmosomoides polygyrus* (Nematoda): Susceptible and resistant strains of mice are indistinguishable following natural infection. *Parasitology* **103,** 429–438.

Scott, M. E., and Anderson, R. M. (1984). The population dynamics of *Gyrodactylus bullatarudis* (Monogenea) within laboratory populations of the fish host *Poecilia reticulata. Parasitology* **89,** 159–194.

Shostak, A. W., and Scott, M. E. (1993). Detection of density-dependent growth and fecundity of helminths in natural infections. *Parasitology* **106,** 527–539.

Slater, A. F. G., and Keymer, A. E. (1986a). Epidemiology of *Heligmosomoides polygyrus* in mice: Experiments on natural transmission. *Parasitology* **93,** 177–187.

Slater, A. F. G., and Keymer, A. E. (1986b). *Heligmosomoides polygyrus* (Nematoda): The influence of dietary protein on the dynamics of repeated infection. *Proc. R. Soc. London B* **229,** 69–83.

Slater, A. F. G., and Keymer, A. E. (1988). The influence of protein deficiency on immunity to *Heligmosomoides polygyrus* (Nematoda) in mice. *Parasite Immunol.* **10,** 507–522.

Tanguay, G. V., and Scott, M. E. (1992). Factors generating aggregation of *Heligmosomoides polygyrus* (Nematoda) in laboratory mice. *Parasitology* **104,** 519–529.

Thein-Hlaing, Than-Saw, and Myint-Lwin (1987). Reinfection of people with *Ascaris lumbricoides* following single, 6-month and 12-month interval mass chemotherapy in Okpo village, rural Burma. *Trans. R. Soc. Trop. Med. Hyg.* **81,** 140–146.

Udonsi, J. K., Nwosu, A. B. C., and Anya, A. O. (1980). *Necator americanus:* Population structure, distribution, and fluctuations in population densities of infective larvae in contaminated farmlands. *Z. Parasitenkd* **63,** 251–259.

Wahid, F. N., Robinson, M., and Behnke, J. M. (1989). Immunological relationships during primary infection with *Heligmosomoides polygyrus (Nematospiroides dubius):* Expulsion of adult worms from fast responder syngeneic and hybrid strains of mice. *Parasitology* **98,** 459–469.

Wassom, D. L., Krco, C. J., and David, C. S. (1987). I-E expression and susceptibility to parasite infection. *Immunol. Today* **8,** 39–43.

Wolton, R. J., and Flowerdew, J. R. (1985). Spatial distribution and movements of wood mice, yellow-necked mice and bank voles. *Symp. Zool. Soc. London* **55,** 249–275.

Wong, M., and Bundy, D. A. P. (1990). Quantitative assessment of contamination of soil by the eggs of *Ascaris lumbricoides* and *Trichuris trichiura. Trans. R. Soc. Trop. Med. Hyg.* **84,** 567–570.

Andrew P. Dobson & Peter J. Hudson

21

Population Biology of *Trichostrongylus tenuis* in the Red Grouse, *Lagopus lagopus scoticus*

I. Introduction

The red grouse *(Lagopus Iagopus scoticus)* is a gamebird of economic and conservation importance in the north of England and Scotland (Hudson, 1992). For the last 10 years we have been studying the interaction between red grouse and the trichostrongylid nematode *Trichostrongylus tenuis,* a parasite that lives in the ceca of red grouse. We have undertaken a series of experiments that examines the effects of this parasite on survival and fecundity. These experiments are complimented by analyses of long-term game bag records for red grouse and the development of mathematical models that analyze the long-term dynamics of the interaction between the grouse, the parasite, and other natural enemies of the grouse such as foxes and birds of prey. This chapter summarizes the work that has been done to date.

Red grouse have been utilized as game since the middle of the 19th century. Ecologists have been fascinated by their population biology since the 1930s, when Elton discussed the long-term cyclic patterns of abundance that grouse populations exhibit throughout the north of England and Scotland (Elton, 1942). We have assembled 400 sets of long-term bag data for grouse populations, 70% or more of these exhibit some tendency to cycle (Fig. 1). In this chapter, we discuss the evidence that shows that the interaction between *T. tenuis* and red grouse is the principle mechanism causing the long-term cycles in many red grouse populations. Where this effect occurs, understanding the population biology of the parasite will be crucial to the efficient management of grouse populations.

The chapter is organized as follows. Initially, we describe individual aspects of the interaction between the parasites and their host, and some mathematical models which may be used to examine the population dynamics of the interaction between *T. tenuis* and red grouse. The observed dynamics of grouse and parasite populations are discussed in terms of the properties of this model. The chapter concludes with a brief description of some work on the seasonal dynamics of *T. tenuis* in red grouse.

II. Life Cycle of the Host

The red grouse is a territorial game bird that feeds almost exclusively on ling heather, *Calluna vulgaris.* The distribution

Parasitic and Infectious Diseases

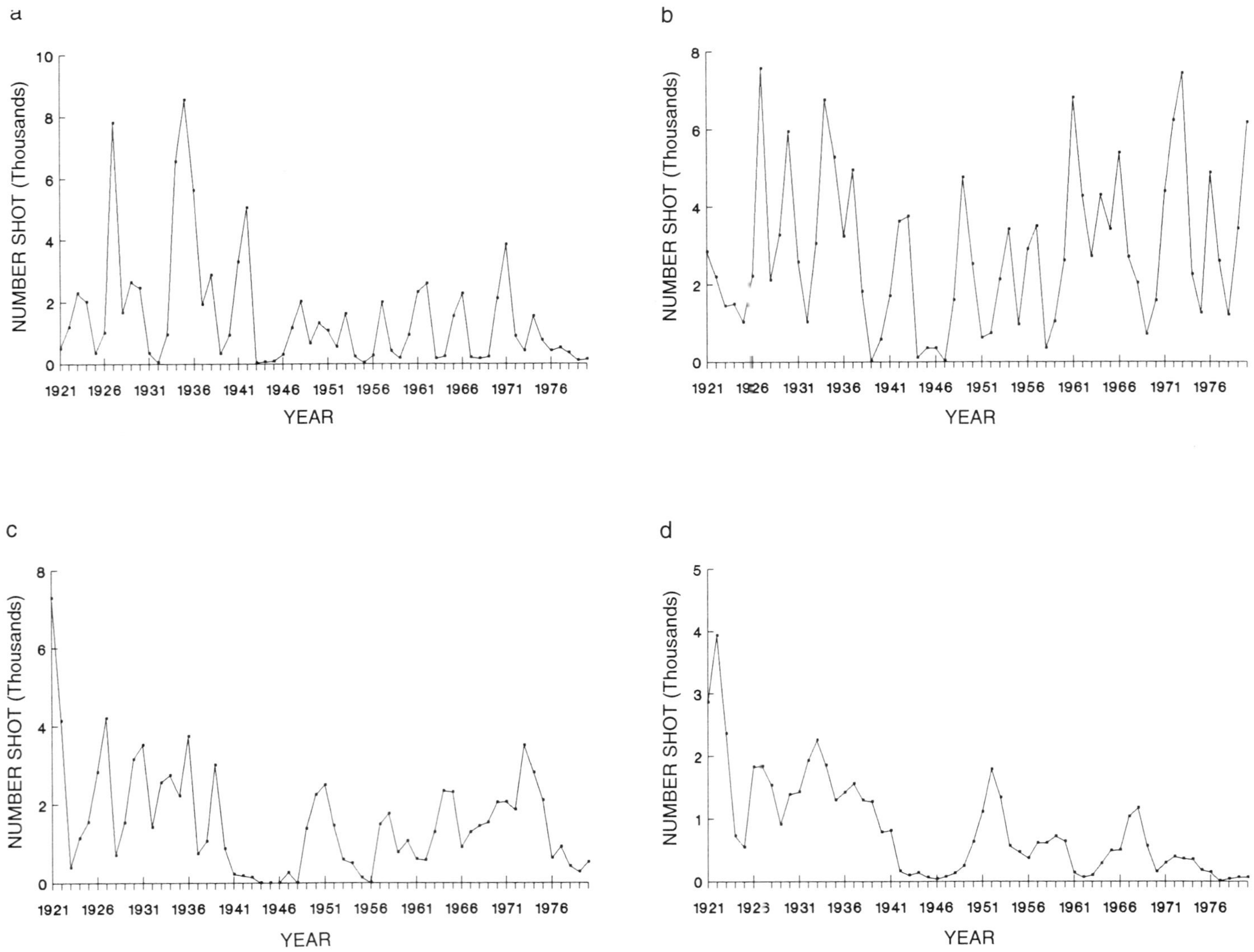

Figure 1 Numbers of grouse shot on four different grouse estates in different parts of Britain. (a) Data from our principle study site at Gunnerside in the north of England. (b) Data from the Dallowgill estate also in the North of England. (c) Data from Castle Grant in Scotland. (d) Data from Ralia in Scotland.

of the bird within the British Isles is limited by the availability of suitable heather-dominant vegetation; the loss of this habitat through overgrazing by ungulates and lagomorphs leads to declines in grouse populations. Much of the heather-dominant moorland in Britain is owned by private sporting estates that manage the heather habitat through a balanced system of grazing and burning. The main predators on grouse estates are the red fox and the carrion crow; these species are controlled to help produce a harvestable surplus of grouse each autumn.

Male grouse establish territories in autumn and spring. Females have started laying by the end of April and commence incubation in early May on an average clutch of 9 eggs. Hatching occurs 22 days later and the chicks leave the nest within the first 1–2 days of life. The hen then takes the chicks to insect-rich areas of bog flush, where they feed on insects for the first 2 weeks of life. By 10 days of age, the chicks are capable of some flight; by 6 weeks of age, they can produce sustained flight over a reasonable distance. The grouse populations in Britain are harvested from 12 August through 10 December usually by driven grouse shooting; on most estates 16 persons will drive the grouse over a line of 9 standing guns. Average daily bags vary from a few birds to 800

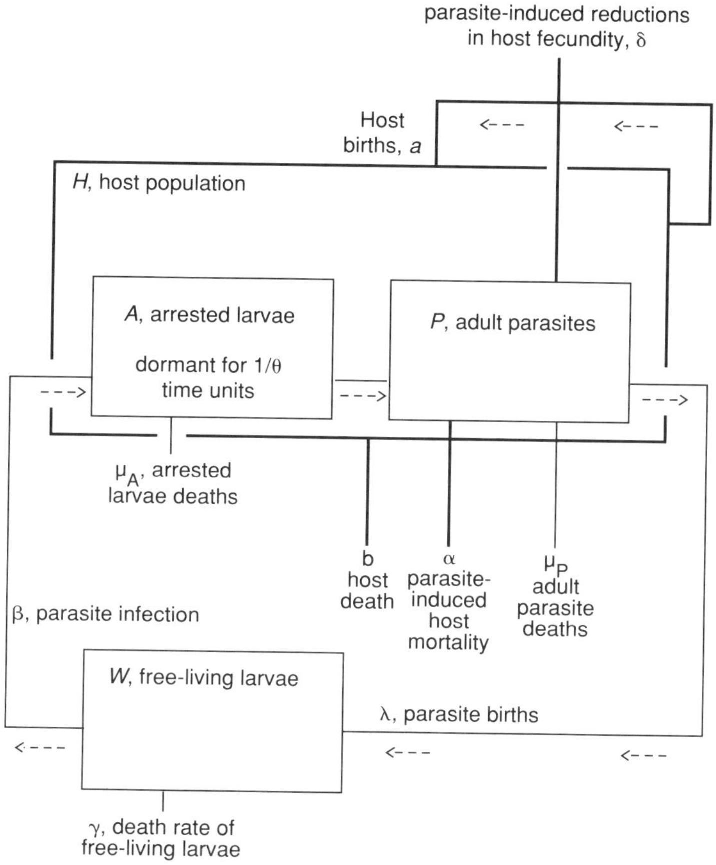

Figure 2 Life cycle of *Trichostrongylus tenuis* in red grouse, represented as a flow chart illustrating the various birth, death, and transmission processes occurring in both the grouse and the parasite. Parameter estimates for each of these rate processes are given in Table 5.

birds, although bags are usually between 150 and 300 birds per day. Most commercial estates harvest more than 27 birds/km^2 within any year.

III. Life Cycle of the Parasite

Trichostrongylus tenuis exhibits a classic monoxenic life cycle (Fig. 2). The adults inhabit the relatively large ceca (70 cm) of red grouse and their eggs pass from the host in the cecal feces. Within the eggs, the worm embryos develop when the temperature exceeds 5°C. The subsequent survival through two larval stages to the infective larvae is dependent on temperature when moisture is adequate. Under optimal conditions, development from egg to the third-stage infective larvae is 7 days, although eggs may remain unhatched for several months. The third-stage infective larvae migrate from the cecal feces to the growing tips of heather, and grouse become infected when they feed on the growing tips of their main food plant, heather (Hudson, 1986a).

Once inside the grouse, the larvae may not develop immediately into an adult stage but molt their outer sheath and cease development in an arrested stage. Arrested larvae are usually found only during the winter months. Although the period of this arrestment and the factors initiating an infective larvae to arrest are not clearly known, these factors are probably temperature dependent. Initial evidence indicates that arrestment is more common in the north of Britain, where temperatures are lower and the winters are longer.

IV. *Trichostrongylus tenuis* in the Red Grouse

A. Parasite Distribution in the Host Population

The prevalence of infection by *T. tenuis* is high: all adult grouse inspected (2739 birds) and 99.2% (2723 birds) of immature grouse were infected with the parasite. The 21 uninfected birds were less than 2 mo old. The statistical distribution of the parasites within the grouse population is invariably aggregated; a large proportion of the parasite population is concentrated in a small proportion of the grouse population (10-yr mean of variance to mean ratio = 1.854, SE = 0.219). Nevertheless, careful monitor-

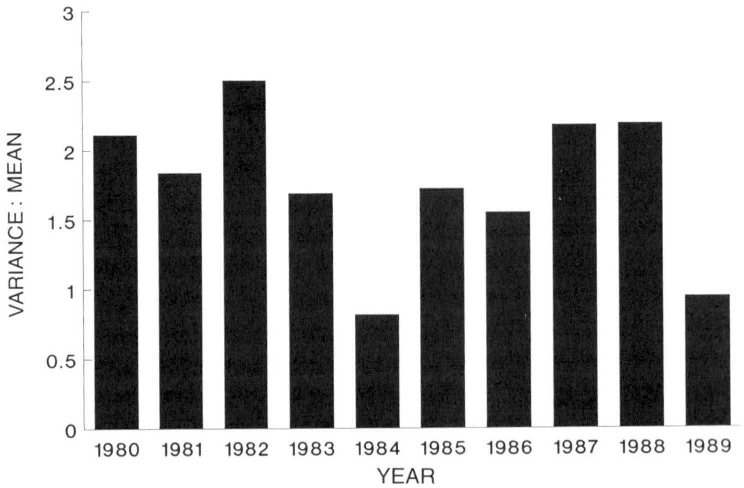

Figure 3 Observed variance to mean ratios of parasite burdens collected from adult grouse on the main study site at Gunnerside.

ing at the main study areas shows that this distribution can vary from year to year (Fig. 3); in some years, the variance to mean ratio can be close to 1, indicating a random distribution of the parasite in the host population (variance to mean ratio less than 1 in 2 of the 10 yr).

Although the overall distribution was aggregated, note that the degree of aggregation is relatively low in comparison with other helminth parasite distributions. Most parasite systems have a much higher level of aggregation, partly because of high prevalence and intensity of infection and partly because the birds have little resistance to the infection.

B. Adult Worm Mortality

Experiments on captive grouse by Wilson (1979) describe the decline in egg production by worms over 72 wk following artificial infection with nematode larvae. Maximum egg production was reached within 6 wk of initial infection and then fell exponentially. If egg production per worm is constant, the observed decline in egg production would indicate a worm survival of ~34% per year. However, subsequent experiments by Shaw and Moss (1989) revealed differences in the rate of egg production between captive and wild grouse, thus making extrapolation between captive and wild birds difficult. This difference in egg production may arise as a consequence of captive grouse having shorter ceca than wild grouse. The experiments also recorded a decrease in egg production with age of infection in captive grouse. Since *per capita* egg production was also found to decrease with age, Wilson's (1979) is probably an underestimate of adult worm survival.

C. Worm Egg Production

Egg concentration in cecal contents was estimated from 422 grouse shot during August and September. Worm eggs were sampled by collecting approximately 1 g cecal contents from the proximal end of one of the two ceca, this being the material most likely to be defecated next. The intensity of worm infection was then determined from the second ceca and egg concentration expressed as eggs per worm per g cecal contents.

The concentration of eggs in cecal contents was 6.05 eggs per worm per g (SE = 1.11) for the 422 individuals sampled. Evidence suggests a slight fall in egg production with intensity of worm infection in old grouse, but this was not a strong relationship and was unlikely to have any direct effects on the dynamics of the system.

The quantity of cecal feces produced per day was determined by weighing feces collected from night roosts, it averaged 18.84 g per bird (SE = 1.06, $N = 52$). Additional cecal feces were rarely found away from night roost sites. Overall worm fecundity can be calculated from the rate of egg production and the average quantity of cecal feces produced as 4×10^4 eggs per worm per yr. Counts of eggs inside the worms by Shaw and Moss (1989) found female fecundity to be 356 eggs per female worm per day. A sex ratio of 1.35 males : 1 female gives an estimate of annual fecundity of 5.5×10^4 eggs per worm per yr.

D. Periods of Infection and Time Delays in Development

Intensity of infection in immature grouse was determined every 2 wk from early July to mid-December in 1982 by collecting samples of young birds on shooting days and by collecting young birds that were killed accidentally by dogs. Serial sampling of grouse indicates two periods of recruitment into the worm population, the first in late summer and the second in late winter. The uptake of worms during the summer is associated with the availability of infective larvae recovered from heather. Serial sampling of grouse revealed an increase in intensity of infection in late winter, when infective larvae could not be recovered from vegetation. The most likely explanation for this recruitment is that larvae during the autumn enter a period of hypobiosis, a period of temporary delayed development that lasts until the following February or March. At the principal study site at Gunnerside, the levels of winter recruitment were greater than those of summer recruitment in all but one year (Hudson *et al.*, 1992).

V. Population Biology of the Red Grouse

Intensive population studies were conducted from July 1979 to July 1989 on an 0.8-km^2 area of managed grouse moor in Swaledale, North Yorkshire. Analysis of the 107 yr of bag record data from the estate produced a damped correlogram with a significant negative coefficient at half the cycle period of 4.7 yr. Total counts of the study area were conducted with the aid of trained pointing dogs. These surveys took place in April to estimate breeding density, and again in July to estimate breeding production when chicks were 7 wk of age. In May of each year, nests of grouse were found using the dogs; clutch size and subsequent number hatching were determined.

A. Key Factor Aanlysis

The population data were analyzed using key factor analysis (Varley *et al.*, 1973). Five causes of female loss were identified:

1. hunting mortality, k_0
2. overwinter loss, k_1
3. inability to lay a maximum clutch, k_2
4. egg mortality, k_3
5. chick loss, k_4

Overwinter loss included losses through natural mortality as well as net loss through emigration and immigration. Losses are expressed as k values:

$$k_i = \log_{10} (N_i/N_{i+1}),$$

where N_i and N_{i+1} are the number of females entering and the number surviving the ith period. Overall loss, K_{TOT}, was calculated as the sum of the losses k_0, k_1, k_2, k_3, and k_4. The key factor that causes year-to-year changes in the population was identified by plotting each loss against total loss and calculating regression coefficients (b). The submortality with the regression coeffi-

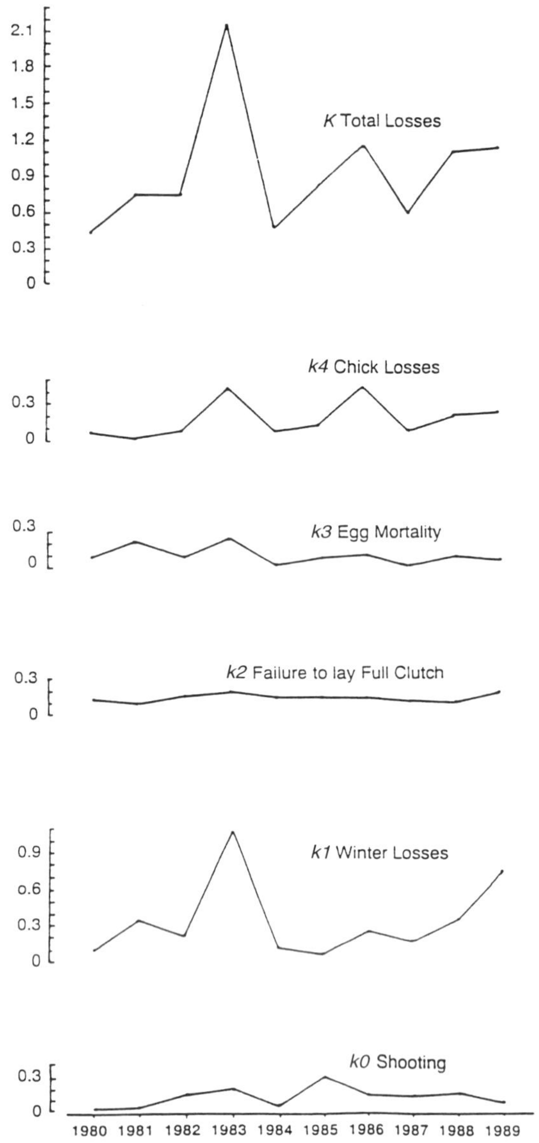

Figure 4 Year to year changes in mortality and losses from the Gunnerside study population. All losses and mortalities are expressed as k values (the logarithm of proportional loss at each stage). The individual k factors are summed to give the overall annual loss, K. Winter losses (k_1) were the key factor. Adapted with permission from Hudson *et al.* (1992).

Table 1 Correlation (r) and Regression Coefficients (b) for Individual Losses and Total Losses from the Gunnerside Grouse Population over 10 Yr[a]

			Losses				
			k_0	k_1	k_2	k_3	k_4
Shooting	k_0	r	—				
Winter	k_1	r	0.11	—			
Clutch	k_2	r	0.37	0.70*	—		
Egg	k_3	r	0.13	0.65*	0.12	—	
Chick	k_4	r	0.38	0.78*	0.63*	0.43	—
Total	k_{TOT}	r	0.38	0.93*	0.75*	0.68*	0.82*
	k_{TOT}	b	0.07	0.59	0.05	0.09	0.20

[a] Significnt coefficients ($P < 0.05$) are indicated with an asterisk; k_1 is considered the key factor with b closest to unity.

cient closest to unity accounts for the greatest year-to-year variation in total loss and is considered the key factor (Fig. 4).

Regression analysis identified overwinter loss (k_1) as the key factor since this factor explained most of the variation in annual loss. This result is, perhaps, not surprising since k_1 covers the greatest time period. However, with the exception of shooting losses (k_0), regression coefficients between total loss and other losses ($k_1 - k_4$) were also significant (Table 1). No significant trends in the k values occurred during the period of study. Only k_1, overwinter loss, was correlated with grouse density immediately prior to the time the loss occurred ($r = 0.651$, $P < 0.05$).

Several of the k values were correlated with each other: overwinter loss (k_1) was correlated positively with each of the breeding losses; failure to lay a maximum clutch (k_2) was correlated positively with chick loss (k_4) (Table 1). These associations between k values may reflect some common factor affecting all losses.

B. Population Losses and Parasite Burdens

Overwinter loss (k_1) increased with both grouse density and the intensity of worm infection in adult grouse ($r = 0.883$, $N = 10$, $P < 0.001$). Partial correlation analysis revealed that overwinter loss was more closely associated with parasite intensity when the effects of density were removed ($r = 0.90$, $P < 0.001$). Both egg mortality (k_3) and chick loss (k_4) increased with the mean intensity of parasite infection (Fig. 5), whereas failure to lay a maximum clutch (k_2) showed only a weak association with intensity of infection. Although such correlations indicate that parasite infection could be a major cause of mortality in grouse, these relationships are not proof and were tested through parasite reduction experiments.

C. Parasite Reduction of Grouse Survival

Grouse were caught, tagged, and treated in spring with antihelmintic levamisole hydrochloride (Nilverm®). Experiments were conducted 0.5 km south of the main study area. Relative survival of treated and untreated birds was measured as the proportion of tagged birds that survived from the time of capture in early spring (February to April) through the breeding season to the shooting season. Most grouse found dead on the study area during field work carried high parasite burdens and died in spring (Hudson *et al.*, 1992); these deaths were classified as overwinter loss (k_1). Worm burdens for treated and untreated grouse were estimated when grouse were shot, after Au-

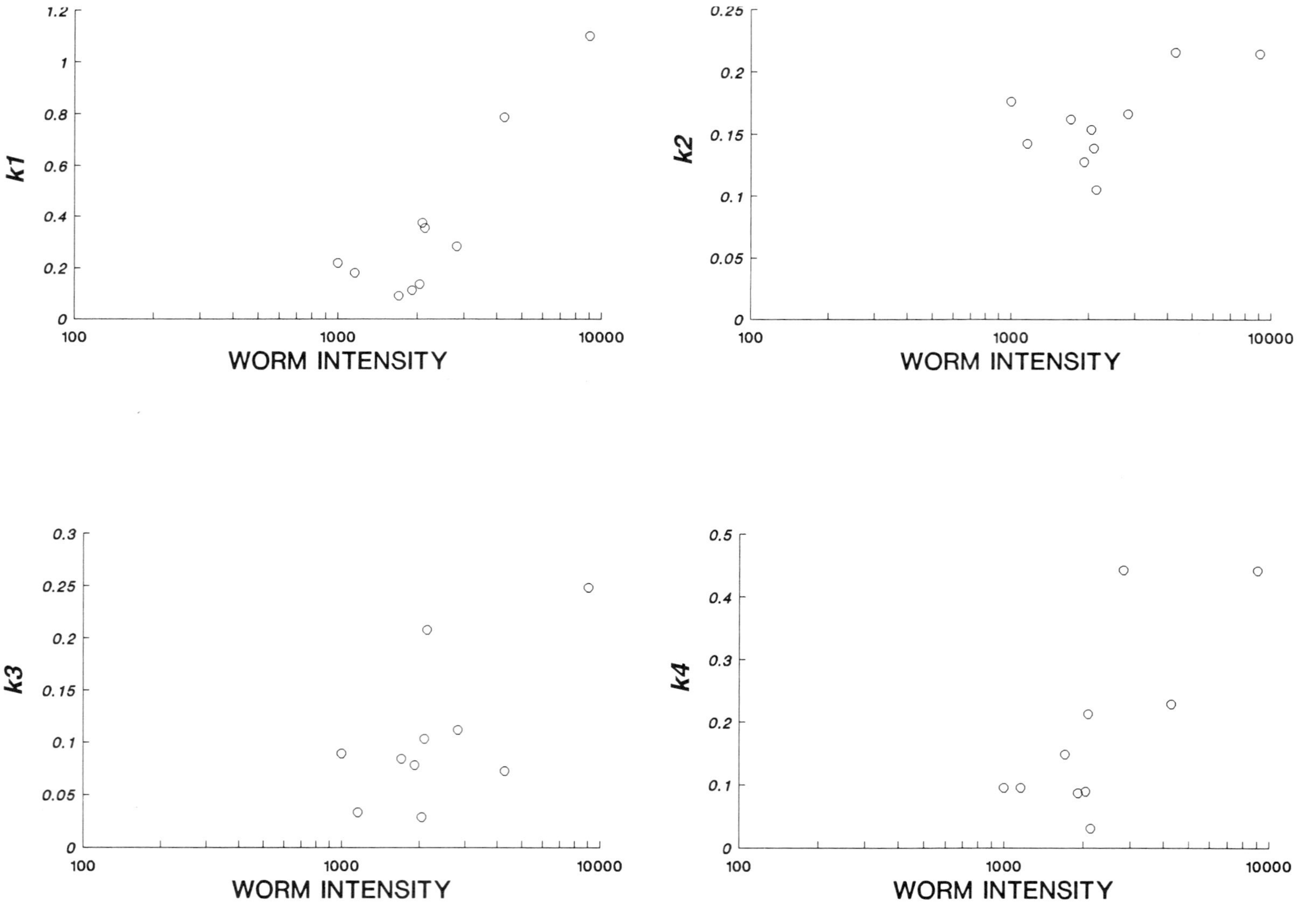

Figure 5 Relationship between mean worm burden (intensity) observed at the Gunnerside study site and four mortality processes operating in each year, each expressed as a k factor (k_1, winter loss; k_2, clutch size; k_3, egg loss; k_4, chick loss). Winter loss, egg loss, and chick loss all show a positive correlation with worm burden ($p < 0.05$). Failure to lay a full clutch (k_2) shows a weak positive association ($0.1 > p > 0.05$). Adapted with permission from Hudson *et al.* (1992).

Table 2 Recovery of Grouse Treated or Untreated with an Anthelminthic and Subsequently Shot in the Following Autumn[a]

Year	Treatment	Number	Number shot	Number not shot	P[b]
1982–1983	Treated	15	7	8	0.17
	Not treated	86	23	63	
1983–1984	Treated	93	38	55	0.048
	Not treated	14	2	12	
1988–1989	Treated	65	5	60	0.104
	Not treated	36	0	36	
Combined P value					<0.05

[a] Overall, treated grouse with reduced worm burdens survived better than grouse with natural levels of infection.
[b] P estimated from Fisher Exact Test.

gust 12 and usually during August and September.

The relative survival of treated birds (from time of treatment to the shooting season) was significantly greater for grouse with reduced worm burdens (Table 2). The data presented in Fig. 5 provide an estimate of the instantaneous mortality rate induced by the parasite of 3×10^{-4} grouse per worm per yr (SE = 5×10^{-5}).

D. Parasite Effects on Grouse Breeding

Female grouse with experimentally reduced parasite burdens consistently produced larger clutches than untreated females; this increase was significant in 4 of 8 yr (Table 3). Hatching success of treated females was significantly greater than that of controls in 3 of 8 yr (Table 3). Overall, brood size was greater in each year of treat-

Table 3 Clutch Size and Hatching Success of Treated and Untreated Female Grouse[a]

Year	Treated females	Control females	Significance[b]
Clutch size			
1982	8.25 + 0.62 (2)	7.93 + 0.36 (14)	NS
1983	8.00 + 0.34 (13)	5.28 + 0.65 (11)	***
1984	8.80 + 0.22 (15)	7.59 + 0.47 (22)	*
1985	9.00 (2)	8.00 + 0.49 (10)	NS
1986	8.67 + 0.33 (3)	7.61 + 0.42 (23)	NS
1987	8.40 + 0.68 (5)	7.90 + 0.38 (10)	NS
1988	10.4 + 0.36 (11)	8.31 + 0.51 (16)	**
1989	8.25 + 0.25 (4)	6.86 + 0.34 (7)	***
Hatching success			
1982	0.97 + 0.02 (8)	0.77 + 0.09 (14)	NS
1983	0.75 + 0.11 (13)	0.38 + 0.13 (11)	**
1984	0.96 + 0.02 (15)	0.92 + 0.24 (22)	NS
1985	0.94 + 0.06 (2)	0.76 + 0.12 (10)	NS
1986	0.96 + 0.04 (3)	0.75 + 0.08 (23)	NS
1987	0.95 + 0.05 (5)	0.91 + 0.03 (10)	NS
1988	0.92 + 0.05 (11)	0.68 + 0.09 (16)	**
1989	1.00 + 0.00 (4)	0.90 + 0.04 (7)	*

[a] In every year, treated birds consistently produced larger clutches with a greater hatching success. Values show mean + SE and sample sizes in parenthesis.
[b] Significance determined using *t* test; *, $P < 0.05$; **, $P < 0.01$; ***, $P < 0.001$; NS, not significant.

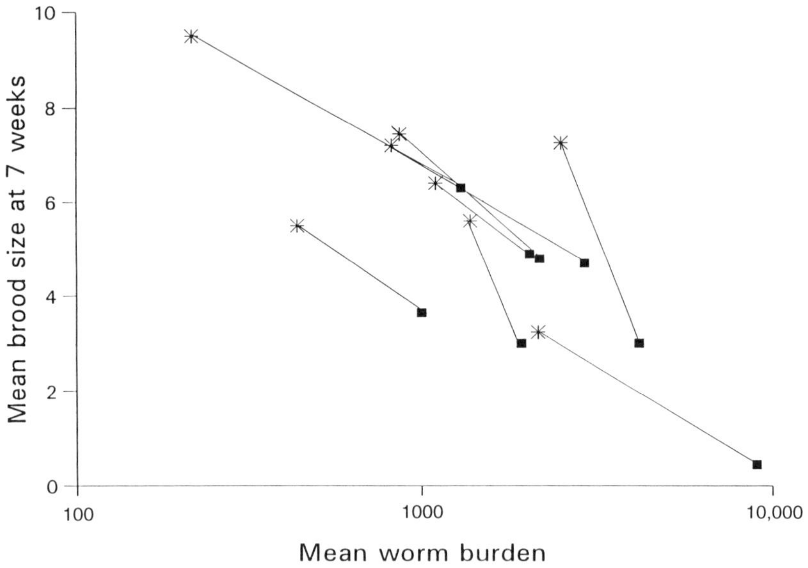

Figure 6 Relationship between reduction in worm burden and the number of chicks surviving until 7 wk. Fecundity of treated birds (*); control birds (with natural worm burdens) (■). The females were treated early in March, approximately 3 mo before their eggs were laid. Adapted with permission from Hudson *et al.* (1992).

ment. These results demonstrate that parasites reduce the breeding production of female red grouse and that differences are consistent between years (Fig. 6). The experimental data provide a series of replicated experiments from which the instantaneous rate of parasite-induced reduction of grouse productivity can be estimated as 5×10^{-4} grouse per worm per yr (SE $= 2 \times 10^{-5}$).

VI. Mathematical Models for *Trichostrongylus tenuis* and Red Grouse

The Anderson and May (1978) model for direct life cycle macroparasites may be adapted to examine the population biology of *T. tenuis* in red grouse. This model has been modified by including additional equations that examine the dynamics of the free-living larval stages and the arrested larval stages. The model therefore consists of four coupled differential equations describing changes in the number of the hosts, H; the free-living larvae, W; the arrested larvae, A; and the adult parasites, P:

$$\frac{dH}{dt} = (a - b)H - (\alpha + \delta)P \quad (1)$$

$$\frac{dW}{dt} = \lambda P - \gamma W - \beta WH \quad (2)$$

$$\frac{dA}{dt} = \sigma\beta WH - (\mu_A + b + \theta)W - \alpha\frac{PA}{H} \quad (3)$$

$$\frac{dP}{dt} = \theta A + (1 - \sigma)\beta WH - (\mu_P + b + \alpha)P - \alpha\frac{P^2}{H}\left(\frac{k+1}{k}\right). \quad (4)$$

The parameters of the model are described in Tables 4 and 5, and correspond to the birth and death rates illustrated earlier in Fig. 2. The model assumes that a proportion of larvae, σ, enter arrestment for a constant period of time $1/\theta$.

Table 4 Notation Used to Denote Various Population Parameters

Parameter	Description
H	Size of host population (numbers of grouse)
W	Numbers of free-living parasite stages (eggs and larvae)
A	Total numbers of arrested larvae
P	Total numbers of adult worms
a	Instantaneous birth rate of grouse (per grouse per unit of time)
b	Instantneous death rate of grouse due to all causes except the parasite (per grouse per unit to time)
α	Instantaneous death rate of host due to the parasite (per parasite per host per unit of time)
δ	Instantaneous reduction in grouse fecundity due to the parasite (per parasite per host per unit of time)
Δ	Density-dependent reduction in grouse fecundity and survival (per host per unit of time)
λ	Instantaneous rate of production of parasite eggs (per parasite per unit of time)
μ_i	Instantaneous death rate of adult μ_P, and arrested, μ_A, parasite stages in grouse (per parasite per unit of time)
γ	Instantaneous death rate of the free-living egg and larval stages (per egg or larva per unit of time)
β	Instantaneous rate of ingestion of parasite infective stages (per larva per host per unit of time)
k	Parameter of the negative binomial distribution which measures inversely the degree of aggregation of parasites within the host population
θ	Rate at which arrested larvae develop into adult worms (per larva per unit of time)
σ	Proportion of larvae that arrest

A. Basic Reproduction Ratio and Critical Threshold Density

Equations 1–4 can be examined when the parasite is first introduced to the host population to produce an expression for the basic reproduction ratio (R_0) of the parasite:

$$R_0 = \frac{T_1}{M_1 M_2}\left[(1-\sigma) + \frac{\sigma\theta}{M_3}\right]. \quad (5)$$

The expression consists of a "birth" term ($T_1 = \beta\lambda H$), the rate of transmission of eggs from birth to establishment in another grouse, and the mortality rates of each stage in the life cycle—adult worms, $M_1 = (\mu_P + b + \alpha)$; free-living larvae, $M_2 = (\gamma + \beta H)$; arrested larvae, $M_3 = (\mu_A + b + \theta)$. Note that, for the fraction of worms that never arrests $(1-\sigma)$, γ/M_1 is the number of eggs laid per worm over the mean life-span of a worm (in the absence of any density-dependent effects) and $\beta H/M_2$ is the proportion of these eggs that develops and survives to be ingested by grouse. For the fraction of parasites that undergoes arrested development (σ), λ/M_1 is the number of eggs laid per worm over the mean life-span of a worm and $\beta H\theta/M_2M_3$ is the proportion of these eggs that develops and survives to be ingested by grouse, and then survives a subsequent period of arrested development to become mature worms.

Essentially, the basic reproduction ratio of the parasite consists of the product of the mean number of new infections established by each female worm and the life expectancies of each stage of the worm life cycle. Here, this ratio consists of the sum of two products: (1) the number of eggs laid over the mature lifetime of an individual parasite that never arrested times the proportion that survived the free-living stage and (2) the number of eggs laid over the mature lifetime of a parasite that passed through arrested development times the proportion that survived both the free-living and arrested stages.

When the basic reproduction ratio equals unity, Eq. 5 can be rearranged to provide an expression for the threshold number of hosts required to continuously sustain an infection of *T. tenuis:*

$$H_T = \frac{\gamma M_1}{\beta\left\{\lambda\left[(1-\alpha) + \frac{\theta\sigma}{M_3}\right] - M_1\right\}}. \quad (6)$$

Figure 7 illustrates the effect of free-living larval life expectancy on the basic reproductive ratio of the parasite and the threshold

Table 5 Population Parameters for *Trichostrongylus tenuis* and Red grouse[a]

Parameter	Symbol	Estimated value (range)	Reference
Grouse fecundity	a	1.8 (0–2)/yr	Hudson (1986a), Hudson *et al.* (1992)
Grouse mortality	b	1.05/yr	Jenkins *et al.* (1963), Hudson (1986a)
Parasite fecundity	λ	11 (9.2–11.5)/yr	Hudson (1986b), Shaw *et al.* (1989), P. Hudson and A. Dobson (unpublished)
Adult worm mortality	μ_P	1.0 (0.8–1.2)/yr	Shaw and Moss (1989), Hudson *et al.* (1992)
Arrested larvae mortality[b]	μ_A	0.5/yr	Gibbs (1986)
Mortality of free-living parasite stages	γ	6.5–13/yr	Watson *et al.* (1988), Shaw *et al.* (1989)
Duration of arrest	$1/\theta$	2–6 mo	Shaw *et al.* (1989)
Parasite pathogenicity	α	3×10^{-4}/worm/yr	Hudson (1986a,b), Hudson *et al.* (1992)
Parasite reduction in host fecundity	δ	5×10^{-4}/worm/yr	Hudson (1986a,b), Hudson *et al.* (1992)
Aggregation of parasites in hosts	k	1.0 (0.5–1.8)	Hudson (1986a,b), Hudson *et al.* (1992)
Transmission rate	β	0.1 ($0.6-1.6 \times 10^1$)/host/yr	This study

[a] The estimates of parasite-induced mortality and reductions in host fecundity are obtained from the slopes of the relationships between parasite burden and host mortality and fecundity given by Hudson *et al.* (1992). The estimates of host mortality and fecundity are obtained by extrapolating these relationships to the point at which parasite burden is zero.

[b] Arrested nematode larvae have very low mortality rates (Gibbs, 1986); by setting this parameter at 0.5, we assume this source of mortality to be negligible.

number of hosts required for the parasite to establish. The figure suggests that, as larval life expectancy decreases, the basic reproduction ratio of the parasite falls below unity and the parasite population is unable to become established. The figure also illustrates the effect of larval arrestment on the magnitude of the threshold number of grouse for establishment. Unless increases in arrestment are compensated for by increases in free-living larval survival, more hosts are required for the parasite to become established in the host population.

B. Equilibrium Dynamics of the Parasite

The factors determining the magnitude of the host and parasite populations can be examined by setting Eqs. 1–4 to equilibrium with $dH/dt = dW/dt = dA/dt = dP/dt = 0$, to give

$$\frac{P*}{H*} = M_P^* = \frac{(a-b)}{(\alpha+\delta)} \tag{7}$$

$$W* = \frac{d'(a-b)}{\beta(\alpha+\delta)} \tag{8}$$

$$H* = \frac{\gamma d' d''}{\beta(\lambda\theta - d'd'')} \tag{9}$$

$$M_A^* = \frac{A*}{H*} = \frac{d'(a-b)}{\theta(\alpha+\delta)}. \tag{10}$$

Here, M_P^* and M_A^* are the mean burdens of adult and arrested larval worms at equilibrium and $d' = \mu_P + \alpha + b + k'(a-b)\alpha/(\alpha+\delta)$, with $k' = (k+1)/k$, whereas $d'' = \mu_A + b + \theta + (a-b)\alpha/(\alpha+\delta)$. These expressions suggest that parasite burdens vary inversely with rates of parasite-induced reductions in host survival and fecundity. The numbers of free-living larvae vary directly

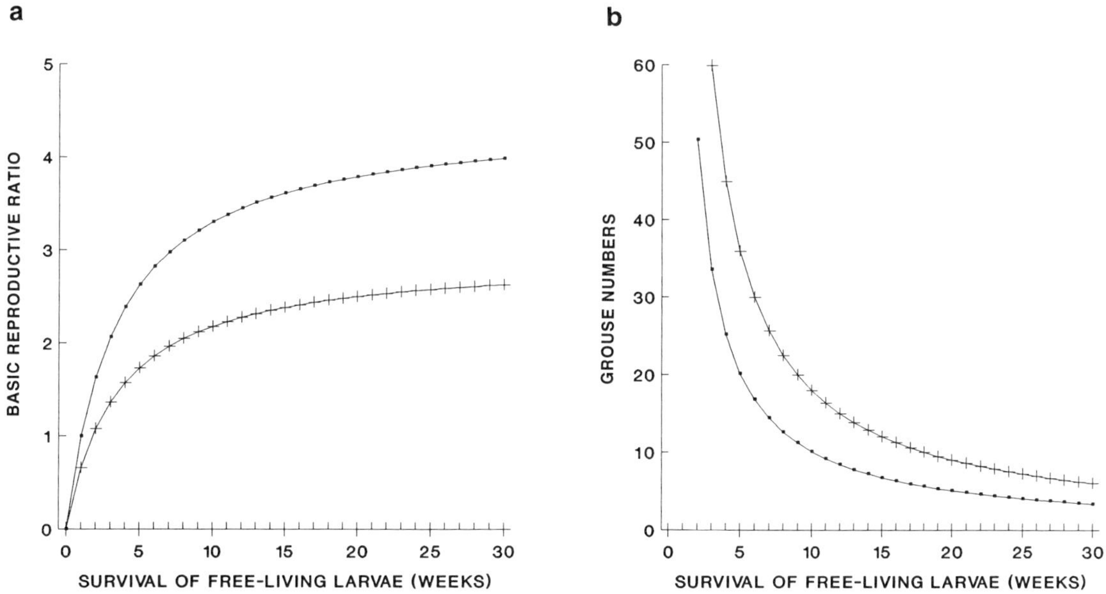

Figure 7 Effects of changes in the life expectancy of the free-living larvae on (a) the basic reproduction ratio of the parasite (R_0) and (b) the threshold host population (H_T) required for parasites to establish. No worms enter hypobiosis (■); all the worms arrest for 4 mo (+) ($\theta = 3$). The other parameter values are $H = 100$, $b = 1.05$, $\alpha = 0.0003$, $\lambda = 10$, $\beta = 0.1$, $\mu_P = 1.2$, $\mu_A = 0.5$. Adapted with permission from Dobson and Hudson (1992).

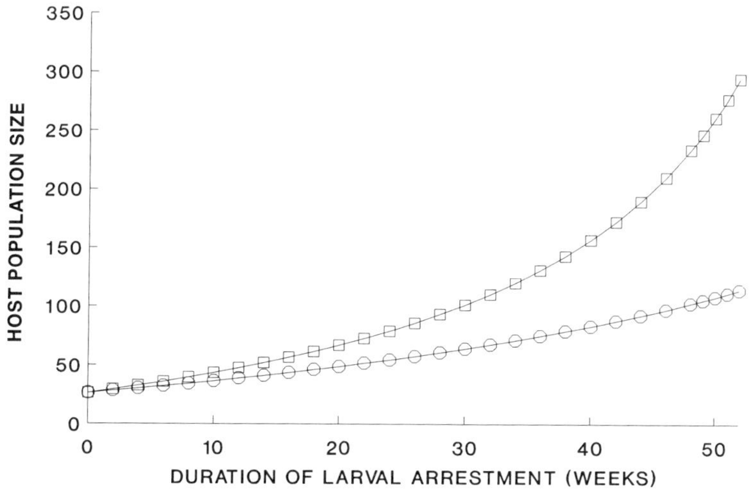

Figure 8 Effect of duration of arrest on the size of the host population at equilibrium. The figure is drawn for two values of arrested larvae mortality rate: $\mu_A = 0.2$ (□) and $\mu_A = 1.0$ (○). The figure assumes that the parasites have no effect on host fecundity. Other parameters are as in Fig. 7. Adapted with permission from Dobson and Hudson (1992).

with mean parasite burden and inversely with their rate of acquisition by the hosts. The expressions for equilibrium host population density closely resemble the expressions for the threshold for establishment (Eq. 6; Fig. 7). In both cases, H^* varies inversely with parasite fecundity (λ) and the rate at which parasite larvae are ingested (β).

Equilibrium host density is higher when the larvae enter hypobiosis (Fig. 8); the size of the equilibrium host population increases with the duration of arrestment. This increase occurs because parasitized hosts live longer if the infective larvae enter an obligatory period of nonpathogenic arrestment before developing into pathogenic adults. Increased host life expectancy will roughly correspond to the period of larval arrest.

C. Transient Dynamics of the Parasite

The transient dynamics of the model are similar to those of the underlying Anderson and May (1978) macroparasite model. Thus, interactions between the parasites and the host will be stable if the parasites are aggregated in their distribution and if the effect of the parasites on host survival exceeds that on host fecundity (see Chapter 5). The data described earlier suggest that neither of these conditions is met in the *T.*

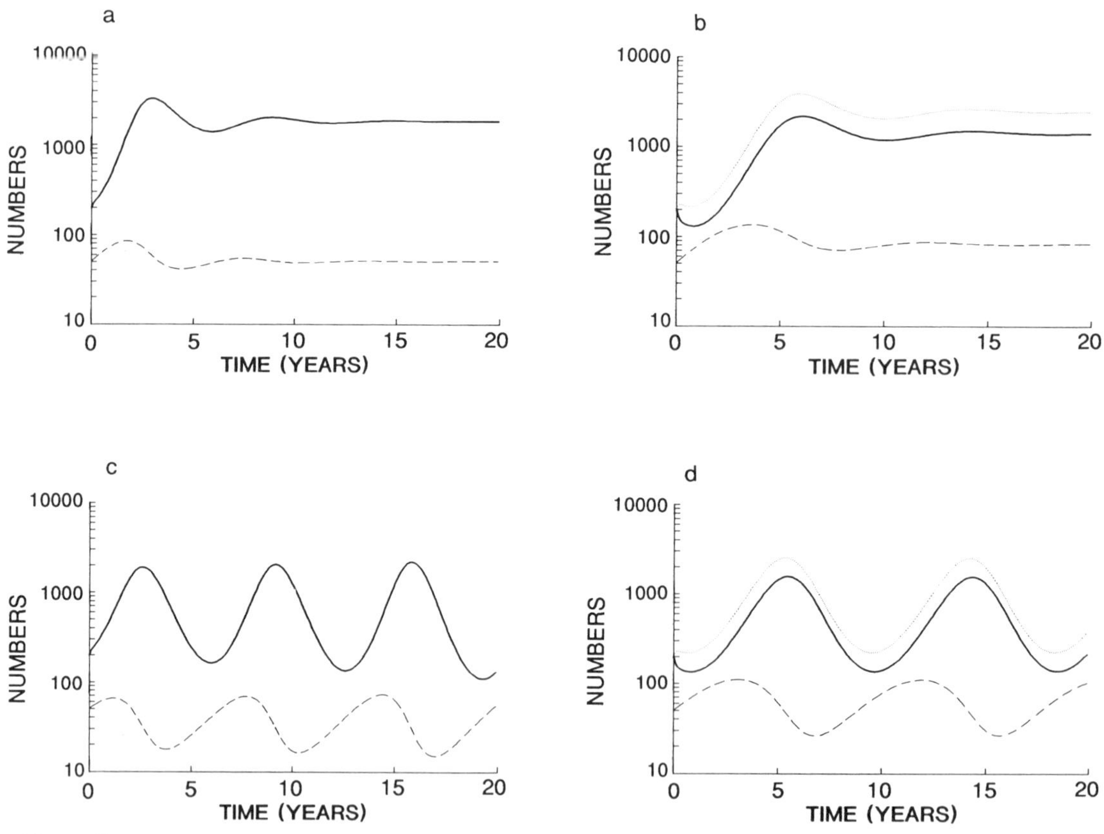

Figure 9 Comparison of the effect of larval arrest and parasite-induced reduction in host fecundity. All the results are depicted for a host population regulated by territoriality. Host population, – – –; total number of adult worms, ——; adult worms plus arrested larvae, · · · . (a,b) Parasite has no effect on host fecundity. (a,c) Parasite with no arrested stages ($\sigma = 0$). (b,d) All the larvae enter arrest for 3 mo ($\sigma = 1$, $\theta = 4.0$). Adapted with permission from Dobson and Hudson (1992).

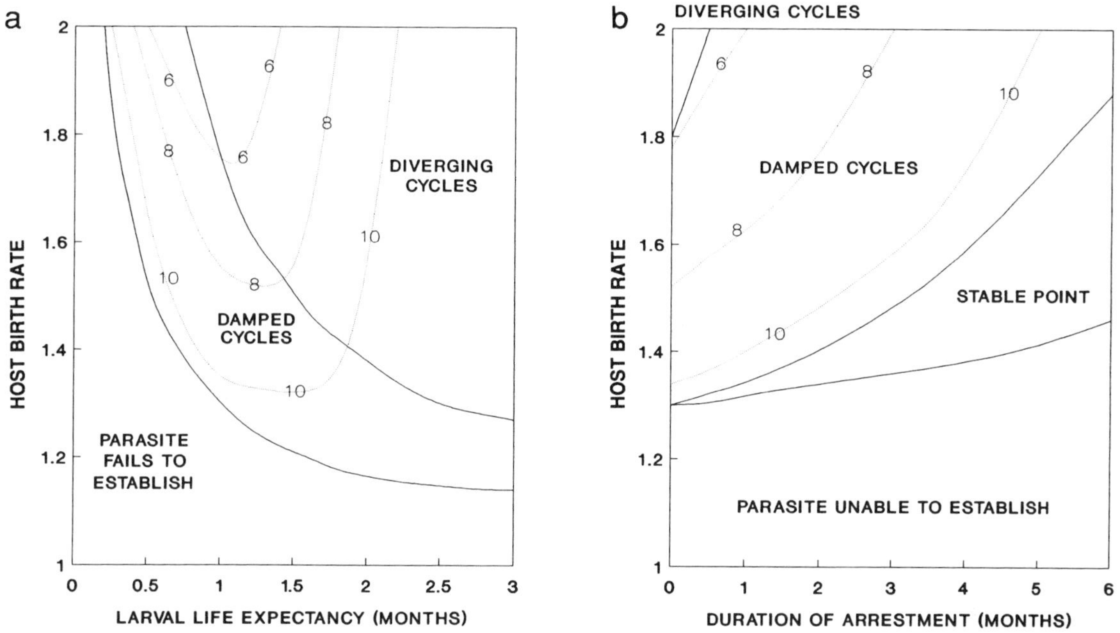

Figure 10 Effects of variations in host fecundity and larval dynamics on the behavior of populations of *Trichostrongylus tenuis* in red grouse. (a) Assume no arrest. (b) Assume constant free-living larval life expectancy of 4 wk ($\gamma = 13$). Different regions of population behavior, ——; expected cycle lengths in regions in which population cycles occur, · · · . All other parameter values are set to the mean values in Table 5. Adapted with permission from Dobson and Hudson (1992).

tenuis–red grouse system. In fact, the parasite has been shown to exhibit greater effects on host fecundity than on survival; and the degree of aggregation of the parasites, although significant, is relatively low. The interaction between these two processes is sufficient to produce damped cycles in the numbers of parasites and hosts (Fig. 9). Increases in the survival of the free-living larval stages of the parasite can lead to expanding cycles of host and parasite abundance, whereas increased duration of arrestment leads to pronounced damping of the cyclic behavior (Fig. 10). The period of the cycles is primarily dependent on host population growth rate, the duration of larval arrestment, and the survival of the free-living larvae. Observations at the principal study site suggest that the observed data roughly correspond to these patterns, with numbers of grouse and parasite burdens cycling over a 5- to 6-yr time interval and peaks of grouse abundance leading peaks in parasite abundance by 1–2 yr (Fig. 11).

D. Seasonal Dynamics of the Parasite

All the analyses presented consider the parasites and hosts to live in a homogeneous environment that remains constant through time. In grouse moors in the north of England and Scotland, pronounced seasonal variation in both temperature and humidity is common, and obviously will affect the survival characteristics of free-living larvae and their propensity to arrest. Furthermore, the birth rate of grouse is highly seasonal; the majority of grouse chicks hatch in the first week of June. This synchronized hatching creates a pulse of new susceptibles that can increase host population size by a factor of ~5. The model can be extended to examine the seasonal dynamics of the system. The following set of eight coupled differen-

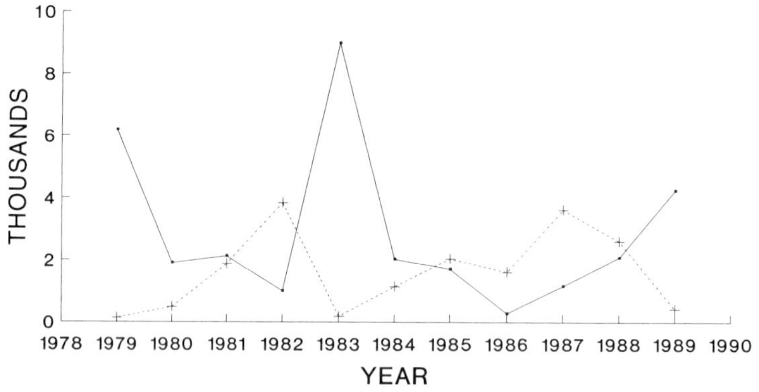

Figure 11 Changes in mean parasite burden per grouse (■) and host population density (birds shot per year) (+) over 14 yr at the Gunnerside. Adapted with permission from Hudson *et al.* (1992).

tial equations describes the dynamics of the adult parasites and free-living and arrested larval stages in the adult grouse and in the young birds born in any year.

Adult grouse:

$$\frac{dH}{dt} = -(b + \alpha P_H)H. \tag{11}$$

Young grouse:

$$\frac{dY}{dt} = -\alpha P_Y - bY. \tag{12}$$

Parasite eggs:

$$\frac{dE}{dt} = \lambda(P_A + P_Y) - (\mu_1 + d_1)E. \tag{13}$$

Infective larvae:

$$\frac{dI}{dt} = d_1 E - [\mu_2 + \beta(H + Y)]I. \tag{14}$$

Arrested larvae in adult grouse:

$$\frac{dA_H}{dt} = \beta HI - (\mu_3 + d_3 + b)A_H - \alpha P_H\left[\frac{A_H}{H}\frac{(k+1)}{k}\right]. \tag{15}$$

Adult worms in adult grouse:

$$\frac{dP_H}{dt} = d_3 A_H - \left[\mu_0 + b + \alpha\frac{P_A}{H}\frac{(k+1)}{k}\right]P_H. \tag{16}$$

Arrested larvae in young grouse:

$$\frac{dA_Y}{dt} = \beta YI - (\mu_3 + d_3 + b)A_Y - \alpha P_Y\left(\frac{A_Y}{Y}\right). \tag{17}$$

Adult worms in young grouse:

$$\frac{dP_Y}{dt} = d_3 A_Y - \left(\mu_0 + b + \alpha\frac{P_Y}{Y}\right)P_Y. \tag{18}$$

Initial number of young grouse:

$$Y_0 = FH - \delta P. \tag{19}$$

The transient dynamics of the host–parasite system can be examined over 1 yr of life, from when the chicks hatch in June until they become adults in the following year. The final conditions at the end of 1 yr can be used to reset the initial conditions for the next year, so the whole exercise can be repeated for as many years as necessary.

The seasonal birth of chicks provides sufficient seasonal forcing to move the damped cycles of the simple model into more complex limit cycles (Fig. 12). The analysis suggests that the system has pronounced seasonal dynamics. In particular, young chicks tend to be infected predominantly during their first few months of life by arrested larvae. These larvae develop into adult worms as they emerge from arrestment. If the chicks survive to become adults, their parasite burdens are dominated by adult worms

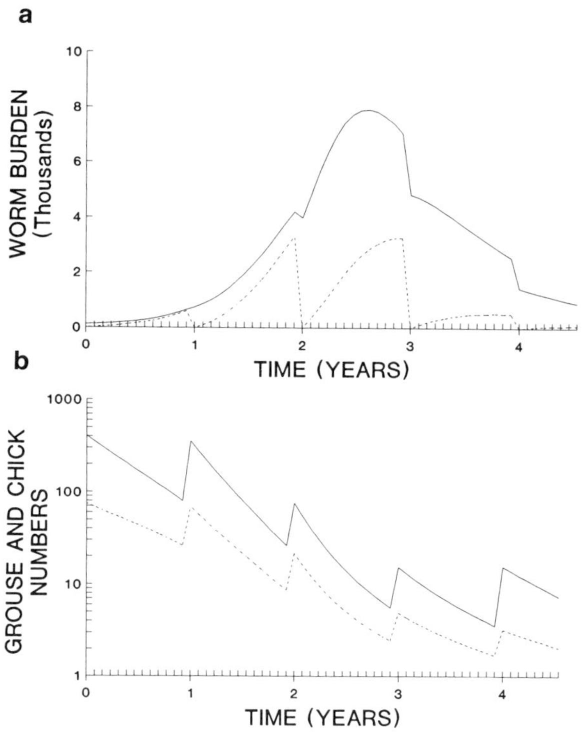

Figure 12 Mean worm burdens in adult and young of the year (- - -) grouse in the model in which host births occur only during the first week of June. (b) Long-term numbers of grouse and chicks in a model for a population in which births are seasonal and occur in the first week of June each year. Adult grouse plus chicks, ——; chicks alone, - - -.

(Fig. 13). This analysis may be pursued further by simple modifications of Eqs. 15–18, which allow the proportion of worms that enters arrestment to vary as a simple annual sine wave or allow the survival of the free-living stages to vary in a similar fashion. Neither of these effects has as dramatic an influence on the long-term dynamics of the host and parasite populations as the seasonal pulse of births created by the annual breeding of the host, although both effects operate to perturb the system sufficiently to help maintain annual cycles (A. P. Dobson and P. J. Hudson, unpublished data).

When propensity to arrest is made a function of time, de-arrestment in the spring is particularly important for young chicks. The seasonal forcing of arrestment in the fall leads to a mass de-arrestment and a precipitous increase in the worm burdens of young birds (Fig. 13). This suggests that the large die-offs of young birds that occur in spring after years in which worm burdens were high are a direct consequence of parasite-induced mortality by de-arresting larvae.

VII. Discussion

All the analyses presented in this chapter suggest that interactions between *T. tenuis* and red grouse are crucial in determining the long-term population dynamics of grouse on upland estates in the north of England and Scotland, as well as in managing grouse populations. The models developed by Anderson and May (1978) and May and Anderson (1978) predicted that, under

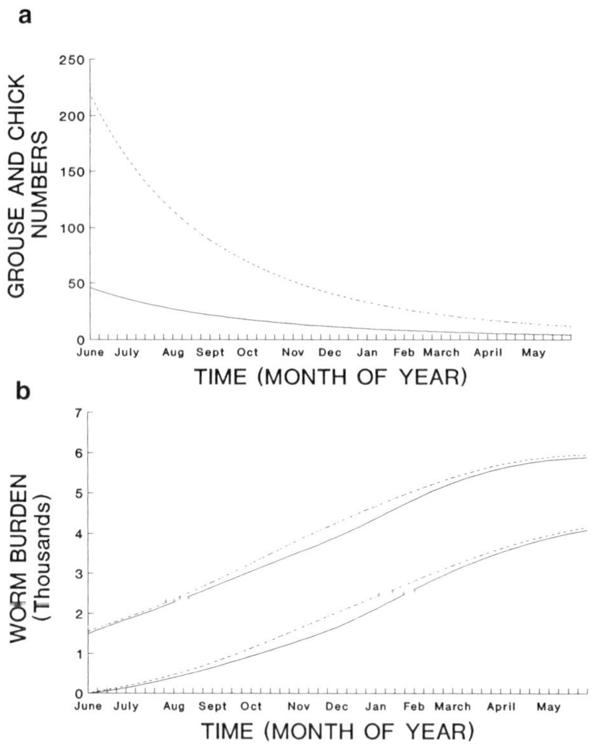

Figure 13 Number of grouse during a year immediately preceding a crash in grouse numbers. Chicks, – – –; grouse, ——. (b) Mean worm burden in adult grouse (upper two lines) and in chicks during their first year of life (lower two lines). Adult worms only, ——; adult worms plus arrested larvae, – – –.

certain conditions, parasitic helminths could cause cyclic fluctuations in the abundance of their host populations. The application of these models to the empirical and experimental data collected on the grouse–*T. tenuis* system suggests that the main interaction causing the system to cycle is the more significant impact of the parasite on host fecundity than on host mortality. The relatively low levels of parasite aggregation observed in the grouse–*T. tenuis* system also increase the tendency of the system to oscillate. The model strongly suggests that the parasite is driving the cycles in grouse abundance that are observed in the long-term study at Gunnerside (Hudson *et al.*, 1992).

An interesting point that emerges from the simulations (Figs. 5,6,9) and the longitudinal study (Fig. 8) is that mean parasite burdens do not peak until after host population density has begun to decline. When combined with the sublethal effects of the parasite on breeding, this result may explain why previous workers (e.g., Jenkins *et al.*, 1963,1967; Wynne-Edwards, 1986) have concluded that the parasites were a consequence rather than a cause of population cycles. Understanding the cause of grouse population cycles has been a major constraint on the efficient management of grouse estates in Britain. Unraveling the role that the parasitic nematode *T. tenuis* plays in these cycles has led to changes in the management of many upland states. Gamekeepers now complement their traditional activities of heather burning and predator control with treatment of grouse to reduce parasite burdens. The drugs can be applied either directly to individual female grouse during the breeding season, or by laying out piles of medicated grit treated with Nilverm®. Grouse soon learn to visit these grit piles, since they need a supply of grit in their gizzards to break down their heather diet. Coating the grit with anthelmintic drugs leads to reductions in worm burdens and is beginning to halt the cycles on moors where grit is in widespread use.

References

Anderson, R. M., and May, R. M. (1978). Regulation and stability of host-parasite population interactions. I. Regulatory processes. *J. Anim. Ecol.* **47,** 219–247.

Dobson, A. P., and Hudson, P. J. (1992). Regulation and stability of a free-living host-parasite system: *Trichostrongylus tenuis* in red grouse. II. Population models. *J. Anim. Ecol.* **61,** 487–498.

Dobson, A. P., and Hudson, P. J. (1993). Seasonal components of the population dynamics of a host parasite system: *Trichostrongylus tenuis* in red grouse. (in press).

Elton, C. S. (1942). "Voles, Mice and Lemmings: Problems in Population Dynamics." Clarendon University Press, Oxford.

Gibbs, H. C. (1986). Hypobiosis in parasitic nematodes—an update. *Adv. Parasitol.* **25,** 129–174.

Hudson, P. J. (1986a). "The Red Grouse. The Biology and Management of a Wild Gamebird." The Game Conservancy Trust, Fordingbridge, Scotland.

Hudson, P. J. (1986b). The effect of a parasitic nematode on the breeding production of red grouse. *J. Anim. Ecol.* **55,** 85–92.

Hudson, P. J. (1992). "Grouse in Space and Time." The Game Conservancy Trust, Fordingbridge, Scotland.

Hudson, P. J., Newborn, D., and Dobson, A. P. (1992). Regulation and stability of a free-living host-parasite system: *Trichostrongylus tenuis* in red grouse. I. Monitoring and parasite reduction experiments. *J. Anim. Ecol.* **61,** 477–486.

Jenkins, D., Watson, A., and Miller, G. R. (1963). Population studies on red grouse, *Lagopus lagopus scoticus* (Lath.) in north-east Scotland. *J. Anim. Ecol.* **32,** 317–376.

Jenkins, D., Watson, A., and Miller, G. R. (1967). Population fluctuations in the red grouse *Lagopus lagopus scoticus*. *J. Anim. Ecol.* **36,** 97–122.

May, R. M., and Anderson, R. M. (1978). Regulation and stability of host–parasite population interactions. II. Destabilizing processes. *J. Anim. Ecol.* **47,** 249–267.

Shaw, J. L., and Moss, R. (1989). The role of parasite fecundity and longevity in the success of *Trichostrongylus tenuis* in low-density red grouse populations. *Parasitology* **99,** 253–258.

Shaw, J. L., Moss, R., and Pike, A. W. (1989). Development and survival of the free-living stages of *Trichostrongylus tenuis,* a cecal parasite of red grouse *Lagopus lagopus scoticus*. *Parasitology* **99,** 105–113.

Varley, G. C., Gradwell, G. R., and Hassell, M. P. (1973). "Insect Population Ecology: An Analytical Approach." Blackwell Scientific Publications, Oxford.

Watson, H., Lee, D. L., and Hudson, P. J. (1988). Primary and secondary infections of the domestic chicken with *Trichostrongylus tenuis* (Nematoda), a parasite of red grouse, with observations on the effect on the caecal mucosa. *Parasitology* **97,** 89–99.

Wilson, G. R. (1979). "The Effects of the Caecal Threadworm *Trichostrongylus tenuis* on Red Grouse." Ph.D. Thesis. University of Aberdeen, Scotland.

Wynne-Edwards, V. C. (1986). "Evolution through Group Selection." Blackwell Scientific Publications, Oxford.

Gerald W. Esch

22

Population Biology of the Diplostomatid Trematode *Uvulifer ambloplitis*

I. Introduction

Uvulifer ambloplitis is a diplostomatid trematode and the causative agent of a disease in centrarchid fishes known as black-spot. This parasite is widespread in North America, having been reported from southern Ontario in Canada to Florida in the United States. The definitive hosts for the parasite are kingfishers, *Mergacyle* spp. Eggs from the adult are shed from the kingfisher in feces; on hatching, miracidia penetrate the pulmonate snail *Helisoma trivolvis*, giving rise to two sporocyst generations. Daughter sporocysts in the hepatopancreas produce cercariae that are released from the snail. The furcocercous cercariae penetrate the flesh of fishes where metacercariae are sequestered in a thick, fibrous capsule of host origin. The cyst is infiltrated by melanocytes, giving it the characteristic black color, hence the name black-spot or black grub.

Most of the research on black-spot disease has been anecdotal, with occasional reference made to its prevalence or density in a local pond or lake (Krull, 1934; Bangham, 1940; McDaniel and Bailey, 1974). Krull (1934) and Hunter and Hunter (1931) reported that heavy infections of the parasite caused weight loss and some mortality of fishes, but these studies were conducted under laboratory conditions using large numbers of cercariae.

Lemly (1979) and Lemly and Dimmick (1982) undertook a series of investigations on the growth and food habits of several species of centrarchid fishes in a number of lakes and ponds in the Piedmont area of North Carolina. During these studies, they observed black-spot in fishes from several locations. At one study site, however, both the prevalence and the density of the parasite were unusually high in juvenile bluegill sunfish, *Lepomis macrochirus*. Moreover, these investigators observed fish with >50 cysts in the fall, but could not find them in the following spring. It was speculated that the disappearance of heavily infected fish could have resulted from parasite-induced host mortality, parasite mortality and resorption of the cyst during the winter, or both.

In view of the potential for such mortality, and because the parasite had not been examined extensively from an ecological perspective, a series of studies was designed (1) to follow the seasonal population biology of the parasite in its snail host, *H. trivolvis*, and in its second intermediate host, the

Parasitic and Infectious Diseases

bluegill sunfish *L. macrochirus;* (2) to test the hypothesis regarding parasite-induced host mortality, using a combination of laboratory and field protocols; and (3) if parasite-induced host mortality in a natural population of fish infected with *U. ambloplitis* could be shown to occur, to identify the nature of those factors that could contribute to the mortality.

For a description of Reed's Pond and the collecting procedures, experimental protocols, and statistical analyses, see Lemly and Esch (1983,1984,a,b,c,1985).

II. Field Observations

A. Snails and Infection

Changes in the densities of the snails in the pond and prevalence of infected snails followed a consistent seasonal pattern over the duration of the study (Fig. 1). Shedding of cercariae from *H. trivolvis* began in May and continued through August in each year. The peak in patent infections (12–13%) coincided with the highest snail densities (95/m^2) in the littoral zone during July. Snail densities were quite low from September to May.

B. Fish and Infection

The seasonal population dynamics of bluegill, *L. macrochirus,* and largemouth bass, *Micropterus salmoides,* were quite consistent throughout the study period. Densities of bluegill remained at 24–26 fish/m^2 from June through September. The age distributions of bluegill in the littoral zone of the pond suggested that they were nearly all young of the year (0+) from May through the next February. Annulus forma-

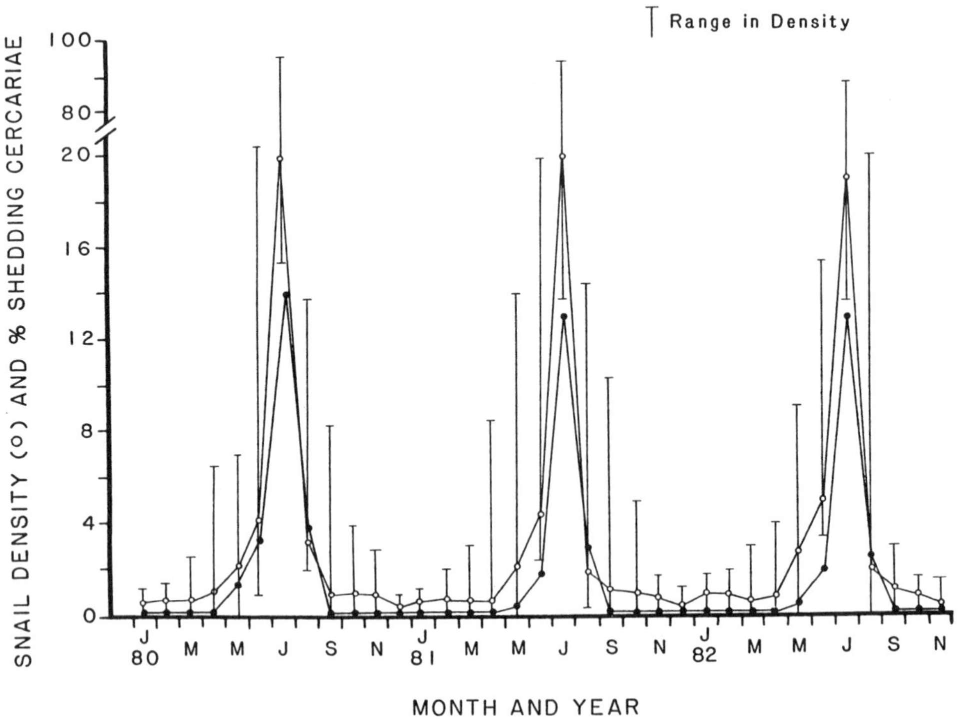

Figure 1 Density (number of individuals/m^2) of the snail *Helisoma trivolvis* in the littoral zone of Reed's Pond (○) and percentage of snails shedding cercariae of *Uvulifer ambloplitis* (●).

tion took place in March. These fish grew beyond 70 mm total length (the largest size class employed in this study) by April.

Live boxes, containing uninfected bluegill, were placed in the littoral zone of the pond each month throughout the year to determine the annual period of parasite recruitment. In each year, recruitment began in May and ended in September (Fig. 2). The peak in parasite acquisition by bluegill in all three size classes examined was in July, corresponding to the time of highest prevalence of shedding snails and to maximum snail densities. The live boxes were scattered in the littoral zone around the pond, but parasite recruitment was similar throughout.

Parasite densities in bluegill measuring 31–50 mm in total length exhibited the same annual patterns of change over a 4-yr sampling period (Fig. 3). In each year, maximum densities occurred in September (up to 269 cysts/fish) and declined over the winter. The patterns of recruitment in the other two size classes (≤30 mm and 51–70 mm) were similar. Prevalence (not shown) of black-spot reached 100% in the two larger size classes of bluegill during the summer months; in the smallest size class (≤30 mm), maximum prevalence was generally ~90%. The minimum prevalence in any size class never declined below 50%, which usually occurred in the months of April and May.

Especially in September, the frequency distributions of *U. ambloplitis* were overdispersed in bluegill (Fig. 4). The variance : mean ratios oscillated over the year, but were always highest in the September–October period. The annual changes in the variance : mean ratios were significant

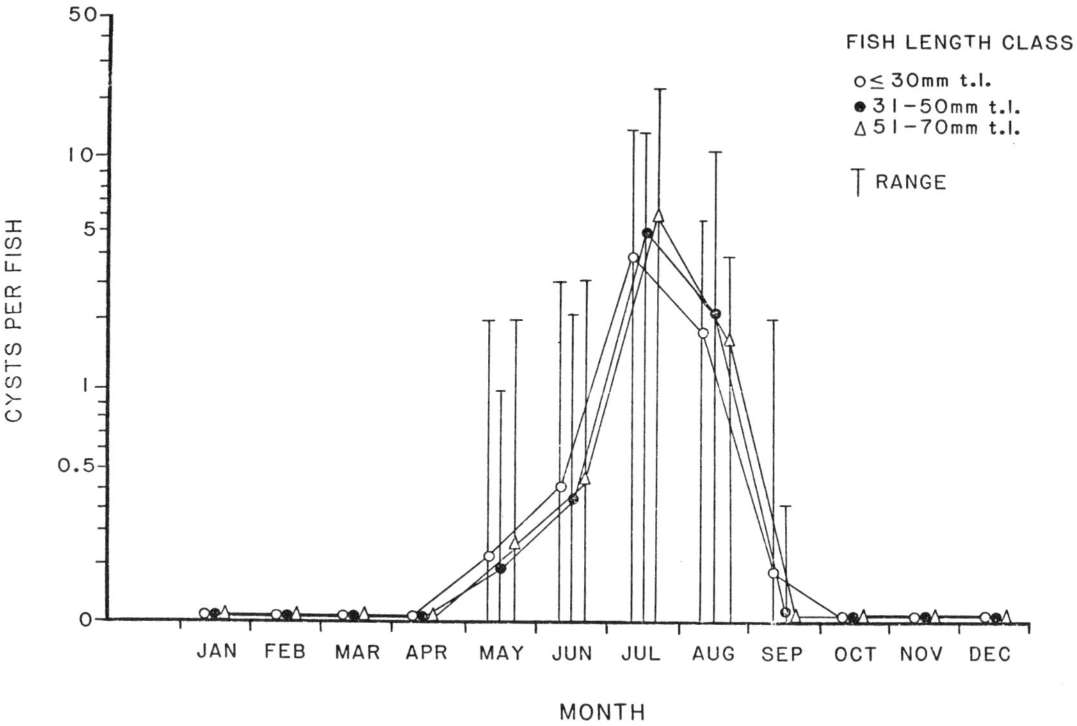

Figure 2 Recruitment of *Uvulifer ambloplitis* by bluegill held in live boxes in the littoral zone of Reed's Pond, January 1, 1981 through December 31, 1981. Fish in the live boxes were replaced with new uninfected individuals on the first day of each month. $n = 10$ fish per data point.

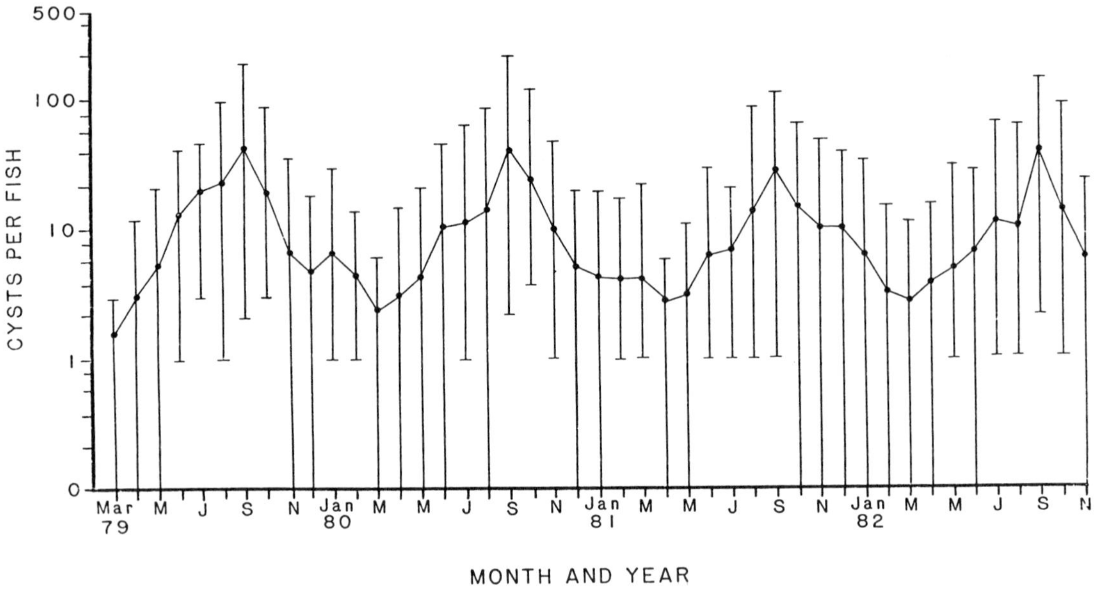

Figure 3 Mean (O) and range (T) of density of parasitism by *Uvulifer ambloplitis* in bluegill from Reed's Pond, March 1979 through November 1982. $n = 10$ fish per data point; all fish were 31–50 mm total length.

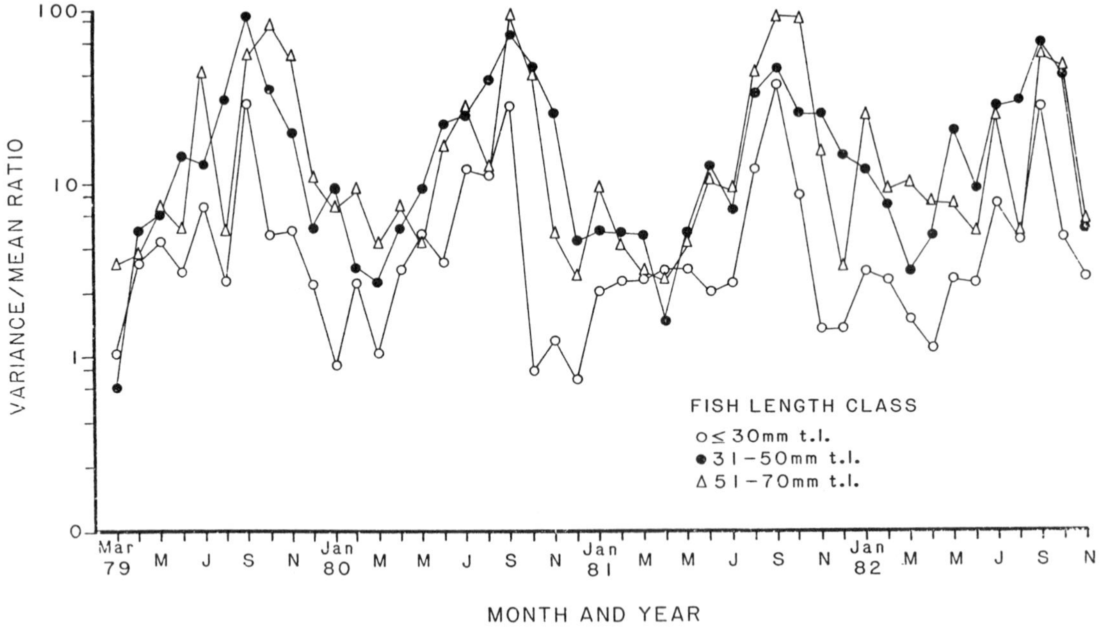

Figure 4 Variance to mean ratios of the density of *Uvulifer ambloplitis* in bluegill from Reed's Pond, March 1979 through November 1982. $n = 10$ fish per data point.

($F_{5,30} = 24.812$, $P < 0.0001$). After peaking in September–October, these ratios declined sharply so by December a 10- to 100-fold decrease had occurred. The observed frequency distributions were described best by the negative binomial model (Fig. 5).

III. Infection and Fish Physiology

A. Field Observations

Body condition, or coefficient (K), is a useful measure for assessing the relative robustness or health of an individual fish (Bagenal, 1978). Provided condition coefficients are computed using data from fish of similar size or age classes, the K value is a useful comparative device at the population level. The coefficient is expressed as

$$K = 100 \times \text{weight (g)/standard length (cm}^3\text{)}.$$

A strong correlation existed between K and the number of cysts within bluegill from the pond (Fig. 6); as cyst numbers increased, especially >50, the K values declined precipitously. Indeed, body conditions were significantly lower in fish with >50 cysts than in bluegill with <25 cysts during each month in which comparisons were made (F values ranged from 4.99 to 9.96 and, in each comparison, $P < 0.05$). Also, a significant negative correlation was found between parasite densities and total body lipid (Fig. 7). Total body lipid changed seasonally, with the lowest concentrations in February and, as would be expected, the highest dur-

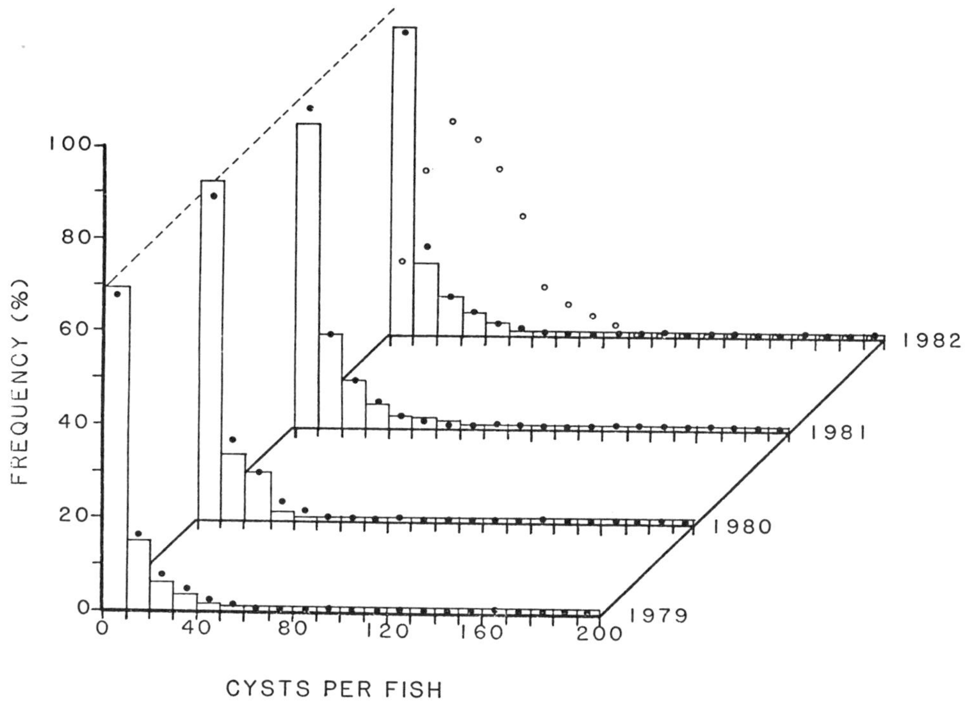

Figure 5 Observed (histograms), Poisson (open circles) and negative binomial expected (closed circles) frequency distributions of *Uvulifer ambloplitis* in bluegill from Reed's Pond, March 1979 through November 1982. $n = 493$ (1979), 566 (1980), 543 (1981), 612 (1982); all fish were <70 mm total length.

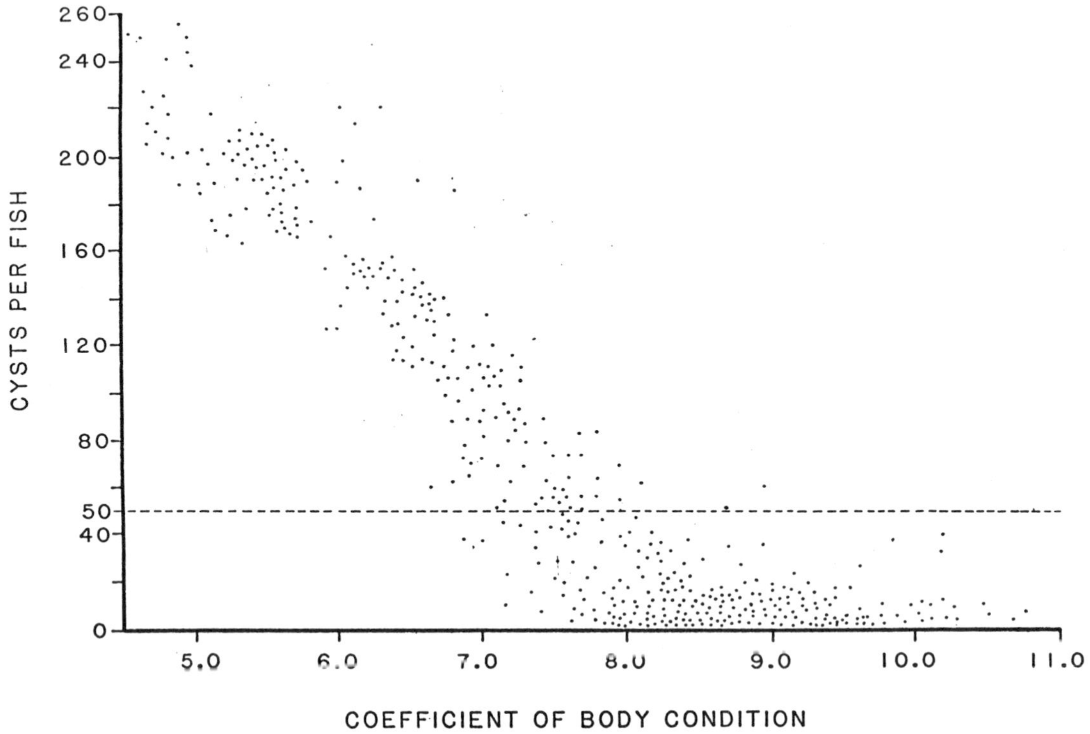

Figure 6 Association between parasite density and body condition of bluegill from Reed's Pond, March 1979 through November 1982. All fish (31–50 mm total length) were infected naturally with *Uvulifer ambloplitis*. $n = 440$, correlation coefficient $(r_s) = -0.933$ $(P < 0.001)$. Dashed line indicates the maximum density observed for bluegill that overwintered successfully in Reed's Pond.

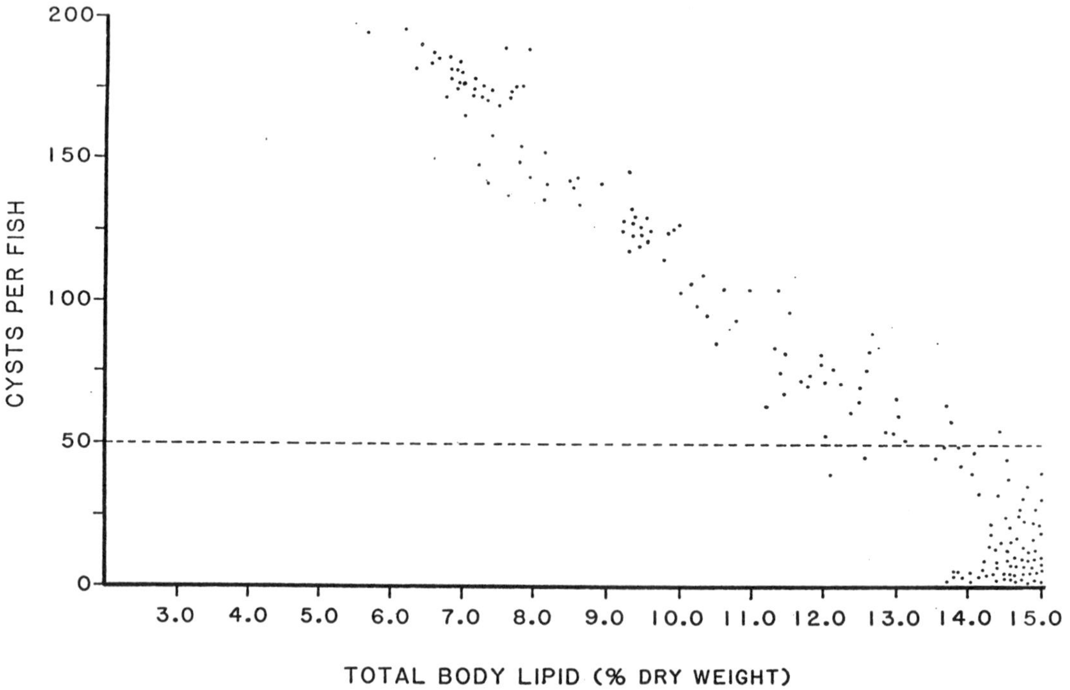

Figure 7 Association between parasite density and total body lipid for bluegill from Reed's Pond, November 1981 through October 1982. All fish (<mm in total length) were infected naturally with *Uvulifer ambloplitis*. $n = 180$, correlation coefficient $(r_s) = -0.963$ $(P < 0.001)$. Dashed line indicates the maximum density observed for bluegill that overwintered successfully in Reed's Pond.

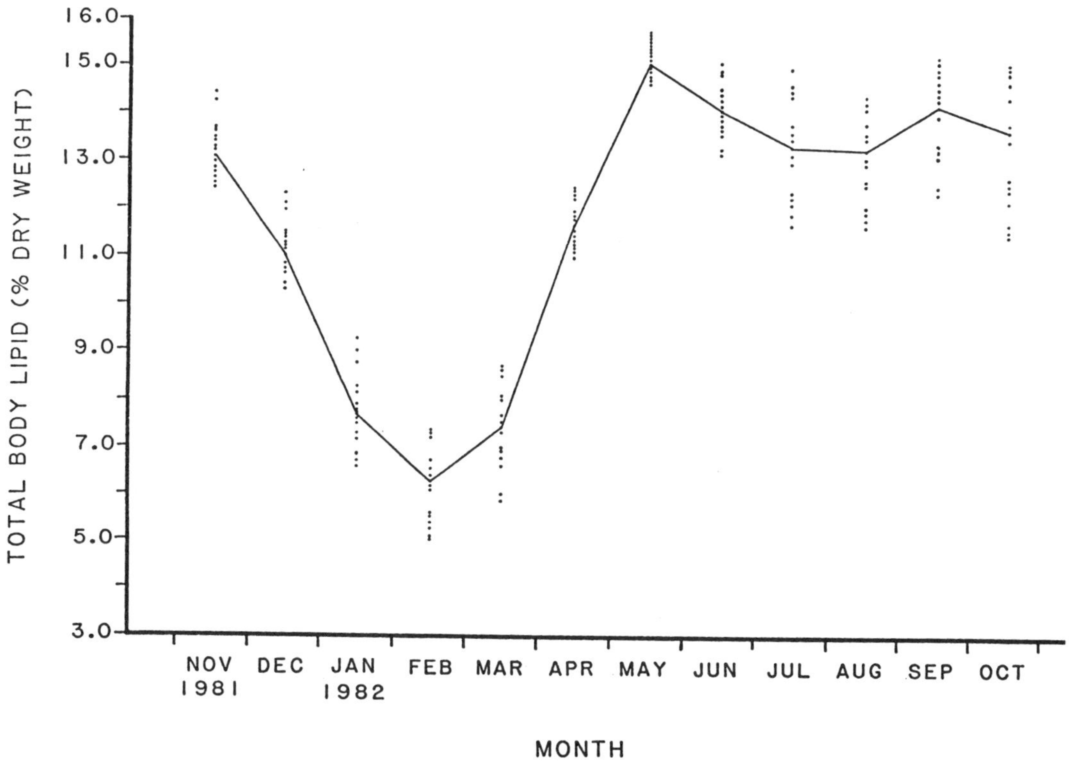

Figure 8 Seasonal variation in total body lipid of bluegill from Reed's Pond, November 1981 through October 1982. All fish were infected lightly (<25 cysts per fish) or uninfected. $n = 15$ fish per month (all <70 mm in total length). Lines connect the mean values.

ing the summer months (Fig. 8). All the fish used to measure seasonal body lipid were lightly infected (<25 cysts/fish). Body conditions of bluegill from the pond were correlated strongly with total body lipid ($r^s = 0.882$; $P < 0.001$).

B. Effect of Temperature on Infected Fish

An experiment was designed to determine the direct effect of changing temperatures on fish with a wide range of metacercariae (0 to >100 cysts per fish). One set of bluegill was placed in live boxes in the littoral zone of the pond and left through the winter. A second group of fish was held in outside aquaria during the winter months. A third set of fish was brought to the laboratory and exposed to light and temperature regimes that simulated winter conditions in the pond. A control group of bluegill was maintained in aquaria in the laboratory, but at a constant temperature of 20°C. In the caged bluegill held in the pond, in the outside aquaria, and in inside aquaria subjected to simulated winter temperatures, almost all fish with <50 cysts survived and those with >50 cysts died. Differences in survival times between heavily (>50 cysts/fish) and lightly (<25 cysts/fish) infected bluegill were also significant in all groups of fish.

C. Effect of Infection on Oxygen Consumption and Lipid Content of Fish

In another experiment, uninfected bluegill were returned to the laboratory. One group was exposed to cercariae and held long enough to give the hosts time to develop a strong tissue reaction. The controls

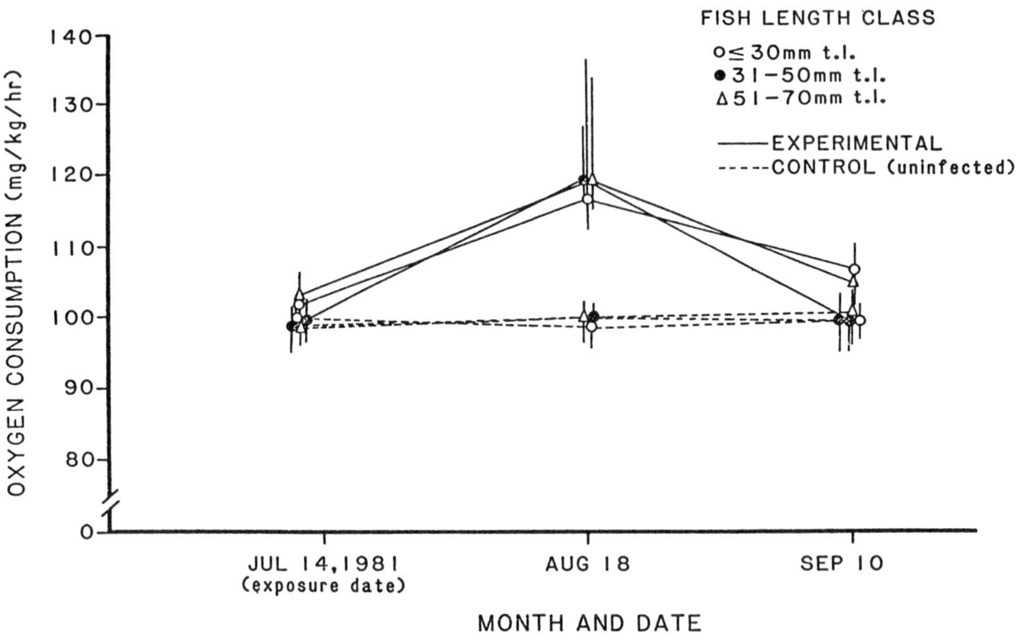

Figure 9 Effects of *Uvulifer ambloplitis* on the oxygen consumption of bluegill from Reed's Pond held at constant temperature (25°C). All fish were uninfected when captured (July 1) and were infected experimentally with cecariae of *U. ambloplitis* on July 14. Resulting artificial infections were 69–167 cysts per fish. The oxygen consumption experiments were conducted prior to exposure to cercariae. $n = 5$ fish per data point. Experimental values were significantly different from controls for each size class of fish on August 18 ($P > 0.001$), but not on September 10.

remained uninfected. Approximately 4 wk after exposure to cercariae, the oxygen consumption was measured in both control and experimental groups. The rates of oxygen consumption were significantly higher in all three size classes of infected fish (Fig. 9), providing clear evidence of the metabolic drain induced by the parasite during the initial phases of infection. When the oxygen consumption was measured approximately 4 wk later, no differences were detected between the experimental and control groups, suggesting that the host response to infection had run its course.

Uninfected fish were brought to the laboratory and starved at 5°C and at 25°C; total body lipid then was measured at the time of death (Fig. 10). In all cases, the total body lipid ranged from 3.0 to 6.0% dry weight at death, which also corresponded to the lipid concentration of pond fishes carrying >50 cysts that had died in the winter months. The relationships between total body lipid, oxygen consumption, and numbers of cysts/fish are illustrated in Fig. 11. As the number of cysts increased, oxygen consumption also increased, and body lipid declined.

IV. Discussion

The changes in densities of *H. trivolvis* in the littoral zone of Reed's Pond suggest that this snail is an annual, reproducing in the spring and early summer, and dying by August. The population during part of the year, therefore, consists of two overlapping cohorts, a situation not unlike that of its congener *Helisoma anceps* in another pond in the Piedmont area of North Carolina (Fernandez and Esch, 1991). The density of *H. anceps* in Reed's Pond is not exceedingly

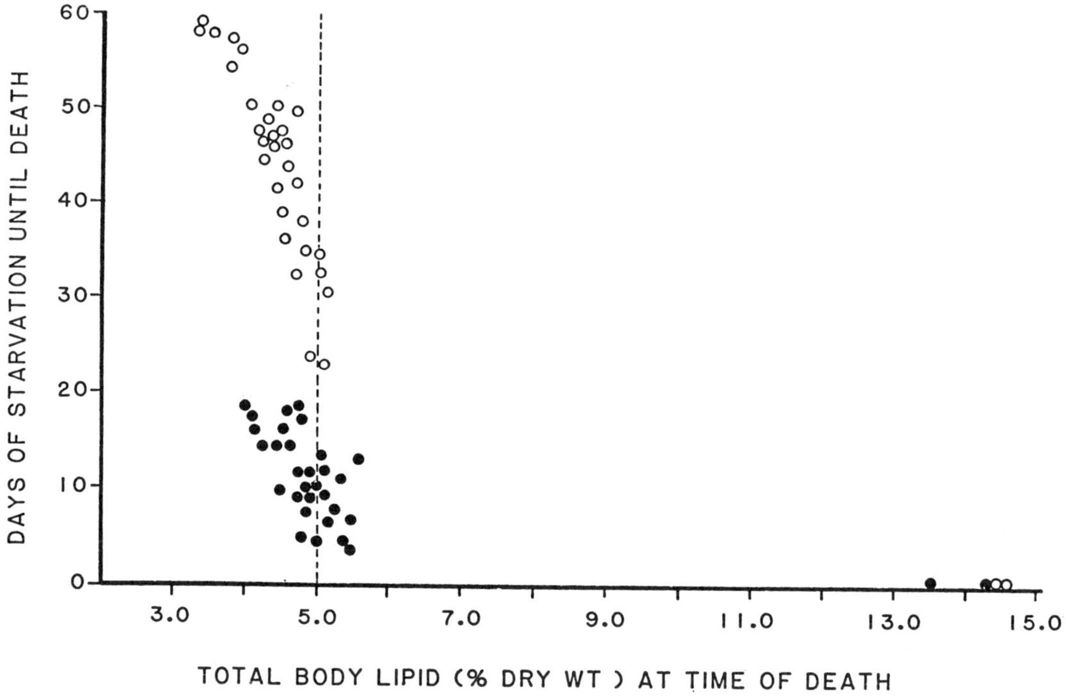

Figure 10 Total lipid present at time of death for uninfected bluegill from Reed's Pond following starvation. $n = 32$ for each water temperature. All fish were <70 mm total length. Correlation coefficient $(r_s) = -0.211$ for 5°C ($P < 0.05$), ○; $r = -0.364$ for 25°C ($P < 0.05$), ●.

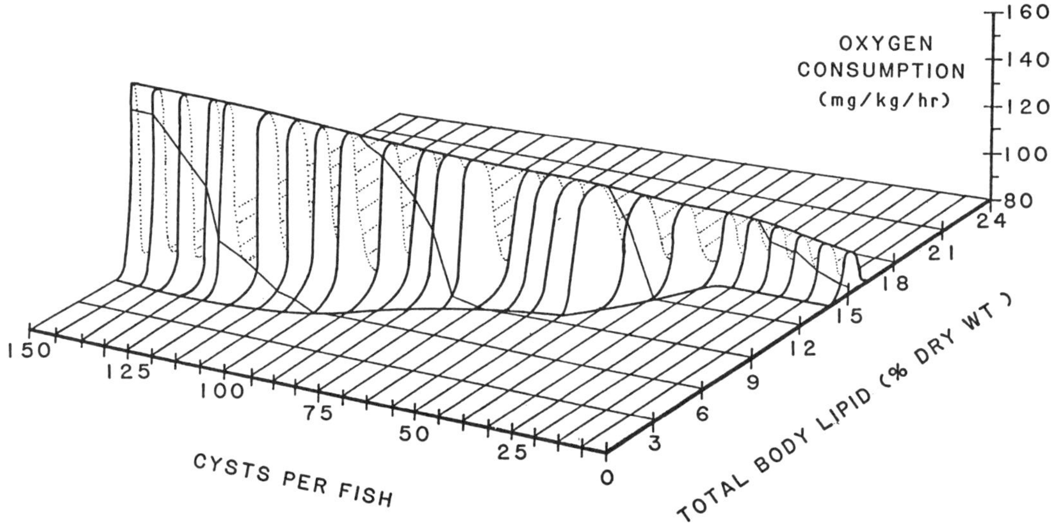

Figure 11 Relationship between parasite density, total body lipid, and oxygen consumption for bluegill from Reed's Pond experimentally infected with cercariae of *Uvulifer ambloplitis* and maintained at constant temperature (25°C). All fish were <70 mm total length.

high relative to densities in other ponds and lakes in the region. Although the prevalence of *U. ambloplitis* in snails is greater than that reported by Cort and Brooks (1928) and Hobgood (1938), it is not particularly high, except in July.

Based on avian life history data from Bent (1940), the highest frequency of foraging by kingfishers in the pond corresponds to the increasing prevalence of the parasite and densities of the snail from May through July. When young kingfishers fledge, they disperse. Foraging pressure by adult kingfishers on bluegill in the pond then subsides, reducing the introduction of parasite eggs into the pond by kingfishers and, in part, accounting for the decline in parasite prevalence in snails after July. A dilution effect created by reproduction of snails from the spring cohort, as well as by natural senescence and snail death, also may contribute to the overall decline in parasite prevalence after the July peak.

The prevalence and density of *U. ambloplitis* in juvenile bluegill in Reed's Pond were both higher than reported by Dechtiar (1972) and Cone and Anderson (1977) and, indeed, are atypical for other sites in the Piedmont region of North Carolina. Two factors were suggested to contribute to this observation. First, kingfishers congregated at the pond during the spring and early summer, corresponding to the time at which snail densities were peaking and when cercariae production was highest. Second, the highest densities of juvenile bluegill occurred in the littoral zone during July. Thus, all three hosts of the parasite simultaneously occurred in high numbers, maximizing the opportunity for parasite transmission.

As for several other larval trematodes in fishes (Pennycuick, 1971; MacKenzie and Liversidge, 1975; Lester, 1977; Rau and Gordon, 1978; Gordon and Rau, 1982), the frequency distribution of black grubs in juvenile bluegill was overdispersed and could be described adequately by the negative binomial model. The variance: mean ratio steadily increased from spring to September, and then declined from 10- to 100-fold by December. Similar declines in variance: mean ratios have been described in other host–parasite systems (Crofton, 1971; Pennycuick, 1971; Lester, 1977; Gordon and Rau, 1982). In these cases, the drop was attributed to the mortality of heavily infected hosts. However, as stated by Lemly and Esch (1984b),

> it should be noted that all of these studies were based on indirect estimates of mortality, subject to variables which would be impossible to assess *a posteriori* (e.g., parasite natality, parasite mortality, expulsion of parasites, emigration or immigration of hosts, and birth of new hosts). Conclusions drawn from these studies are tenuous at best.

In this study, no bluegill movement into or out of the pond occurred during the winter months and no reproduction by the parasite or any of the hosts occurred during the same period, so these factors can be eliminated as possible mechanisms for distorting the variance: mean ratios from September into winter. Three other possible explanations were considered for the sharp decrease in the variance: mean ratios. First, the parasite could have been eliminated from the fish and the cysts resorbed as the result of a strong host response. Second, disproportionate predation by largemouth bass, kingfishers, or both, on heavily infected bluegill in the pond could have taken place. Third, parasite-induced host mortality could have occurred.

The first alternative was eliminated from further consideration after counting metacercariae in caged fish in September and again after overwintering. Although a small percentage of the metacercariae lost their viability over the winter months, none of the cysts were resorbed by the hosts. Efforts to study the feeding behavior of kingfishers were largely unsuccessful. During 87 observation days spent at the pond during late fall and winter, kingfishers were sighted on only 12 occasions, none of which involved foraging. Moreover, no evidence was generated to suggest that black-spot caused abnormal or unusual swimming behavior among infected fishes, thereby increasing their vulnerability to predation. Extensive laboratory and field observations did not reveal any indications of "side-swimming" or surfacing by heavily infected bluegill as occurs in other fish responding to parasitism (Brassard *et al.*, 1982). Stomach analysis of largemouth bass in the pond indicated that they were not discriminately preying on heavily infected juvenile bluegill. Instead, the data generated from the overwintering studies clearly demonstrated that heavily infected bluegill (>50 cysts/fish) did not survive winter conditions and that this factor alone accounted for the sharp decline in the variance : mean ratios from September and through December.

By plotting survivorship data from the live-box studies against cysts/host (Fig. 12), it was "possible to predict the percentage of hosts expected to survive through the winter in Reed's Pond, given a particular level of parasitism" (Lemly and Esch, 1984c). The curves suggest that fish with <20 cysts would be able to survive the winter. Among fishes with 20–50 cysts, however, the survivorship declined from 90% to zero. The curves thus indicate that median survivorship occurred in fish with 38 cysts but, more importantly, the entire range in survivorship was only 30 parasites (from 20 cysts/host to 50 cysts/hosts). "An analysis of the experimental data has, in this case, provided an accurate model of host–parasite interactions and regulation in the field" (Lemly and Esch, 1984c).

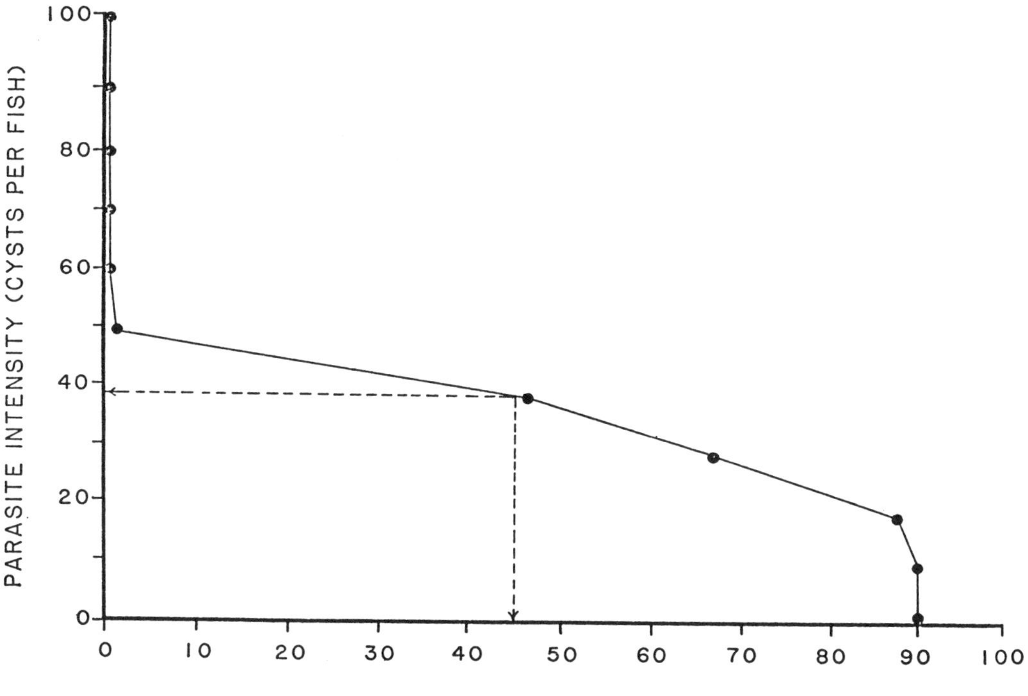

Figure 12 Effects of parasitism by *Uvulifer ambloplitis* on overwintering success in bluegill from Reed's Pond. Data are for fish exposed to naturally declining water temperatures or for fish held in aquaria indoors and subjected to progressively declining temperatures (simulating the onset of winter). $n = 402$. All fish were <70 mm total length. Dashed line indicates the parasite density expected to produce a 50% reduction in the number of bluegill overwintering successfully in the field.

Based on the proportion of fish carrying more than 50 cysts and on the fish population density, Lemly and Esch (1984c) estimated that 10–20% of the young-of-the-year bluegill were eliminated from the pond by the effects of the parasite each year. Nevertheless, the overall bluegill population in the pond appeared to be robust from the standpoint of density, body condition, and growth rate. Therefore, the parasite was suggested to act as a check on overcrowding by cropping excess reproduction that otherwise might lead to stunting or a population decline. Such imbalances commonly are observed when bass–bluegill combinations occur in small ponds and lakes in the southeastern United States.

The mechanism involved in causing parasite-induced host mortality was related to two factors in the infection process by *U. ambloplitis.* The literature (Davis, 1936; Lewert and Lee, 1954a,b; Erasmus, 1960) reveals that migration of strigeid cercariae within infected fish produces severe tissue damage. Cellular destruction, internal bleeding, intense inflammation, and localized edema occur. These reactions are accompanied by elevated metabolic activity, including increased oxygen consumption. Shortly after penetration, *U. ambloplitis* becomes sequestered after it secretes a cyst wall around itself. However, the host also responds by forming a thick fibrous capsule around the parasite (Hunter and Hamilton, 1941). The process of mobilizing fibrocytes and creating the cyst wall lasts 3–5 wk; melanocytes also infiltrate the cyst wall, creating the dark pigmentation associated with the black grub.

The metabolic drain associated with the tissue damage and cyst formation in heavily infected fish dramatically affects stored lipid in the infected fish. The observed correlation between total body lipid and temperature-dependent host mortality also is related to bluegill feeding behavior during the winter months. Many smaller fishes, including centrarchids, are almost completely reliant on stored energy reserves to survive the winter months (Oliver, 1977; Oliver *et al.,* 1979; Salamon, 1979; Pierce *et al.,* 1980; Toneys and Coble, 1980). Dependence on stored energy reserves in juvenile centrarchids during the winter was confirmed in this study through observations on the seasonal changes in total body lipid. If the fish were in poor body condition at the onset of winter, they did not survive. Lemly and Esch (1984c) concluded that heavy infection by *U. ambloplitis,* accompanied by the severe tissue reaction, the heightened metabolic rate, the depletion of stored energy reserves, and the absence of feeding by juvenile bluegill during the winter months, contribute to host mortality. More specifically, however, severe lipid depletion probably generates an increase in protein catabolism. Thus, after lipids are virtually depleted, the normal intermediary catabolic balance between carbohydrates, lipids, and proteins is shifted so increased protein catabolism produces an accumulation of ammonia. Acidosis and ketonemia would result from increased lipid catabolism (Love, 1970). This combination of physiological changes in the host directly causes the heavily infected fish to die.

Note that the prevalence and density of *U. ambloplitis* metacercariae were not as high in juvenile largemouth bass in Reed's Pond as in bluegill. This observation is consistent with those made by Lemly (1983) and Lemly and Esch (1985) for other ponds and lakes in the area. Since young of the year of both species occupy the same areas of the littoral zone, expecting that individuals of each species would be exposed to similar numbers of cercariae is reasonable. The differences between species suggest that the bluegill is a more suitable host for blackspot, and that the capacity to become established in bass is reduced, possibly because of adverse physiological conditions within the flesh of the fish. Several authors (Mettrick and Podesta, 1974; Holmes, 1976; Nickol, 1979; Cook *et al.*, 1980) have pointed out that "differences in the microenvironment of host tissues are well known to influence the invasion and establishment of parasites" (Lemly and Esch, 1984b).

The significance of the Lemly and Esch studies (1983,1984a,b,c,1985) is threefold. First, most of the previous evidence implicating parasites as the cause of host mortality has been based on extrapolations from laboratory-based studies or on comparisons of theoretical and observed frequency distributions (Pennycuick, 1971; Kakonge, 1972; Lester, 1977; Rau and Gordon, 1978; Brassard *et al.*, 1982; Gordon and Rau, 1982). The investigations by Lemly and Esch provided unequivocal evidence for mortality in a natural population of fish caused by high densities of a helminth parasite. Second, the studies were instructive because they actually described a mechanism for the induction of host mortality by the parasite; moreover, these observations were obtained through a combination of laboratory and field experimentation and observation. Third, evidence presented clearly demonstrates that 10-20% of the bluegill young of the year are eliminated from Reed's Pond by the parasite on an annual basis, as well as that 70-80% of the parasites infecting juvenile bluegill are eliminated each year because of the overdispersed frequency distribution of the parasite.

On the other hand, in a survey of 43 other ponds, streams, and lakes in the Piedmont region of North Carolina, the prevalence and density of black-spot in *L. macrochirus* were always significantly lower than in Reed's Pond and, apparently, not high enough to cause mortality. Lemly and Esch (1985) concluded that mortality induced by this parasite in Reed's Pond was, therefore, exceptional. Based on these studies, and a review of the literature, these investigators asserted that parasite-induced mortality and, therefore, regulation is probably is not a widespread phenomenon under natural conditions. When parasitism and host mortality can be linked, they almost always are associated with parasite-induced behavioral changes that increase host vulnerability to predation, thereby facilitating parasite transfer to another host in its life cycle (Dence, 1958; Arme and Owen, 1968; Wilson, 1971; Brasard *et al.*, 1982).

Acknowledgments

All the results and discussion in this chapter are derived ultimately from the doctoral dissertation of Lemly (1983) and from a subsequent series of papers by Lemly and Esch (1983,1984a,b,c,1985). The *Journal of Parasitology* kindly permitted the reproduction of all the figures appearing in this paper.

References

Arme, C., and Owen, R. W. (1968). Occurrence and pathology of *Ligula intestinalis* infections in British fishes. *J. Parasitol.* **54**, 272-280.

Bagenal, T. B. (1978). "Methods for Assessment of Fish Production in Fresh Waters." IBP Handbook No. 3. Blackwell Scientific Publishers, London.

Bangham, R. V. (1940). Parasites of freshwater fish of southern Florida. *Proc. Florida Acad. Sci.* **5**, 289-307.

Bent, A. C. (1940). "Life Histories of North American Cuckoos, Goat-Suckers, Hummingbirds, and Their Allies." Smithsonian Inst. Bull. No. 176. Smithsonian Institute, Washington, D.C.

Brassard, P., Rau, M. E., and Curtis, M. A. (1982). Parasite-induced susceptibility to predation in diplostomiasis. *Parasitology* **85**, 495-501.

Cone, D. K., and Anderson, R. C. (1977). Parasites of pumpkinseed (*Lepomis gibbosus* L.) from Ryan Lake, Algonquin Park, Ontario. *Can. J. Zool.* **55**, 1410-1423.

Cook, C. B., Pappas, P. W., and Rudolph, E. D. (eds.) (1980). "Cellular Interactions in Symbiosis and Parasitism," Ohio State University Press, Columbus.

Cort, W. W., and Brooks, S. T. (1928). Studies on the holostome cercariae from Douglas Lake, Michigan. *Trans. Am. Microscop. Soc.* **47**, 179-221.

Crofton, H. D. (1971). A quantitative approach to parasitism. *Parasitology* **62**, 179-193.

Davis, D. J. (1936). Pathological studies on the penetration of the cercaria of the strigeid trematode *Diplostomum flexicaudum*. *J. Parasitol.* **22**, 329-338.

Dechtiar, A. O. (1972). Parasites of fish from the Lake of the Woods, Ontario. *J. Fish. Res. Board Can.* **29**, 275-283.

Dence, W. A. (1958). Studies on *Ligula* infected common shiners (*Notropis cornutus frontalis* Agassiz) in the Adirondacks. *J. Parasitol.* **44,** 334–338.

Erasmus, D. A. (1960). The migration of *Cercaria X* Baylis (Strigeida) within the fish intermediate host. *Parasitology* **49,** 173–190.

Fernandez, J. C., and Esch, G. W. (1991). Guild structure of larval trematodes in the snail *Helisoma anceps:* Patterns and processes at the individual host level. *J. Parasitol.* **77,** 528–539.

Gordon, D. M., and Rau, M. E. (1982). Possible evidence for mortality induced by the parasite *Apatemon gracilis* in a population of brook sticklebacks *(Culaea inconstans). Parasitology* **84,** 41–47.

Hobgood, J. O. (1938). The metacercaria of *Cercaria flexicorpa* Collins. *Trans. Am. Microscop. Soc.* **57,** 158–164.

Holmes, J. C. (1976). Host selection and its consequences. *In* "Aspects of Parasitology" (C. R. Kennedy, ed.), pp. 21–39. North Holland, Amsterdam.

Hunter, G. W., and Hamilton, J. M. (1941). Studies on host–parasite reactions to larval parasites. IV. The cyst of *Uvulifer ambloplitis* (Hughes). *Trans. Am. Micro. Sc. Soc.* **60,** 498–507.

Hunter, G. W., and Hunter, W. S. (1931). "Studies on the Fish Parasites in the St. Lawrence Watershed." Supplement to the 20th Annual Report, New York State Conservation Department, Biological Survey No. V, St. Lawrence Watershed, 1930.

Kakonge, S. A. K. (1972). "The Ecology of Some Metazoan Parasites of, and Their Effect on, Small Stream Fishes and Fry. Ph.D. Thesis. University of Waterloo, Canada.

Krull, W. H. (1934). *Cercaria bessiae* (Cort and Brooks, 1928), an injurious parasite of fish. *Copeia* **1934,** 69–73.

Lemly, A. D. (1979). "A Study of the Plankton in the Littoral Zone, and Age and Growth of Centarchids in Four Piedmont, North Carolina, Lakes." M.S. Thesis. Wake Forest University, Winston-Salem, North Carolina.

Lemly, A. D. (1983). "Ecology of *Uvulifer ambloplitis* (Trematoda: Strigeidae) in a Population of Bluegill Sunfishes, *Lepomis macrochirus* (Centrarchidae)." Ph.D. Thesis. Wake Forest University, Winston-Salem, North Carolina.

Lemly, A. D., and Dimmick, J. F. (1982). Growth of young-of-the-year and yearling centrarchids in relation to zooplankton in the littoral zone of lakes. *Copeia* **1982,** 305–321.

Lemly, A. D., and Esch, G. W. (1983). Differential survival of metacercariae of *Uvulifer ambloplitis* (Hughes, 1927) in juvenile centrarchids. *J. Parasitol.* **69,** 746–749.

Lemly, A. D., and Esch, G. W. (1984a). Population biology of the trematode *Uvulifer ambloplitis* (Hughes, 1927) in the snail intermediate host *Helisoma trivolvis. J. Parasitol.* **70,** 461–465.

Lemly, A. D., and Esch, G. W. (1984b). Population biology of the trematode *Uvulifer ambloplitis* (Hughes, 1927) in juvenile bluegill sunfish, *Lepomis macrochirus,* and largemouth bass, *Micropterus salmoides. J. Parasitol.* **70,** 466–474.

Lemly, A. D., and Esch, G. W. (1984c). Effects of the trematode *Uvulifer ambloplitis* on juvenile bluegill sunfish, *Lepomis macrochirus:* Ecological implications. *J. Parasitol.* **70,** 475–492.

Lemly, A. D., and Esch, G. W. (1985). Black-spot caused by *Uvulifer ambloplitis* among juvenile centrarchids in the Piedmont area of North Carolina. *Proc. Helminthol. Soc. Washington, D.C.* **52,** 30–35.

Lester, R. J. G. (1977). An estimate of the mortality in a population of *Perca flavescens* owing to the trematode *Diplostomum adamsi. Can. J. Zool.* **55,** 288–292.

Lewert, R. M., and Lee, C. L. (1954a). Studies on the passage of helminth larvae through host tissues. I. Histochemical studies on the extracellular changes caused by penetrating larvae. *J. Infect. Dis.* **95,** 13–35.

Lewert, R. M., and Lee, C. L. (1954b). Studies on the passage of helminth larvae through host tissues. II. Enzymatic activity of larvae *in vivo* and *in vitro. J. Infect. Dis.* **95,** 36–51.

Love, R. M. (1970). "The Chemical Biology of Fishes." Academic Press, New York.

MacKenzie, K., and Liversidge, J. M. (1975). Some aspects of the biology of the cercaria and metacercaria of *Stephanostomum baccatum* (Nicoll, 1907) (Manter, 1934) (Digenea: Acanthocolpidae). *J. Fish Biol.* **7,** 247–256.

McDaniel, J. S., and Bailey, H. H. (1974). Seasonal population dynamics of some helminth parasites of centrarchid fishes. *Southwest. Nat.* **18,** 403–416.

Mettrick, D. F., and Podesta, R. B. (1974). Ecological and physiological aspects of helminth–host interactions in the mammalian gastrointestinal canal. *Adv. Parasitol.* **12,** 183–278.

Nickol, B. B. (ed.), (1979). "Host–Parasite Interfaces." Academic Press, New York.

Oliver, J. D. (1977). "Overwinter Mortality of Smallmouth Bass *(Micropterus dolomieui)* Young-of-the-Year in Relation to Their Size, Environmental Temperature, and Parasite Loading." M.S. Thesis, University of Toronto, Canada.

Oliver, J. D., Holeton, G. F., and Chua, K. E. (1979). Overwinter mortality of fingerling smallmouth bass in relation to size, relative energy stores, and environmental temperature. *Trans. Am. Fish. Soc.* **108,** 130–136.

Pennycuick, L. (1971). Quantitative effects of there species of parasites on a population of three-spined sticklebacks, *Gasterosteus aculeatus*. *J. Zool.* **165,** 143–162.

Pierce, R. J., Wissing, T. E., Jaworski, J. G., Givens, R. N., and Megrey, B. A. (1980). Energy storage and utilization patterns of gizzard shad in Action Lake, Ohio. *Trans. Am. Fish. Soc.* **109,** 611–616.

Rau, M. E., and Gordon, D. M. (1978). The frequency distribution of tetracotyles of *Apatemon gracilis pellucidus* (Yamaguti, 1933) in stickleback *Culea inconstans* (Kirtland) populations of homogeneous age and size structure. *J. Fish Dis.* **1,** 259–263.

Salamon, K. J. (1979). "Physiological, Anatomical, and Ecological Aspects of Overwintering in Largemouth Bass, *Micropterus salmoides* Lacapede." Ph.D. Thesis. Fordham University, New York.

Toneys, M. L., and Coble, D. W. (1980). Mortality, hematocrit, osmolality, electrolyte regulation, and fat depletion of young-of-the-year freshwater fishes under simulated winter conditions. *Can. J. Fish. Aquat. Sci.* **37,** 225–232.

Wilson, R. S. (1971). The decline of a roach *Rutilus rutilus* (L.) population in Chew Valley Lake. *J. Fish Biol.* **3,** 129–137.

Frank Fenner

Myxomatosis

23

I. Introduction

Myxoma virus is a member of the genus *Leporipoxvirus,* poxviruses that produce benign fibromas in certain rabbits of the genus *Sylvilagus* found in the Americas (Fig. 1) but a very severe generalized disease in European rabbits *(Oryctolagus cuniculus)* (Fig. 2a,b). Myxomatosis caused by strains of myxoma virus that occur naturally in *Sylvilagus brasiliensis* in Brazil came into prominence when these strains were used for the biological control of wild European rabbits in Australia (1950) and France (1952). Detailed accounts of the Australian and European experiences can be found in Fenner and Ratcliffe (1965) and Fenner and Ross (1994). Only the Australian experience is described in this chapter.

II. History of Myxomatosis in Australia

A. European Rabbits in Australia

Although tame rabbits had been introduced earlier, wild European rabbits did not become established in Australia until 1859. Within a few years, these animals had spread so extensively over the southern half of the continent that they became the major agricultural pest in the settled areas and an important cause of erosion in the semi-arid interior. Although its use for rabbit control had been suggested as early as 1918, myxomatosis was not introduced into Australian wild rabbits successfully as an epizootic disease until 1950, after testing to insure that the virus would not produce disease in other Australian wild or domestic animals.

Aragao (1943) showed that myxoma virus could be transferred mechanically from one *Sylvilagus* rabbit to another, or from a *Sylvilagus* to an *Oryctolagus* rabbit, by mosquitos biting through skin lesions. Mechanical spread between *Oryctolagus* rabbits by mosquitos and other arthropod vectors was confirmed by Australian investigators in 1944. In many parts of Europe, the rabbit flea *(Spilopsyllus cuniculi)* is an important vector. In 1966, this insect was introduced into Australia to help spread myxomatosis where mosquitos were not effective vectors. Under conditions of close contact between *Oryctolagus* rabbits, myxomatosis can be transmitted by contact or by aerosols.

B. Initial Impact of Myxomatosis

The initial releases of myxoma virus were carried out by inoculating a number of captured rabbits with myxoma virus and releas-

Parasitic and Infectious Diseases

Figure 1 Lesion (the Marshall–Regnery fibroma) produced by the Californian strain of myxoma virus in the Californian brush rabbit *(Sylvilagus bachmani)*. (Courtesy Dr. D. Regnery, Stanford University.)

ing them at several sites in northern Victoria, between May and September 1950. Close observation of the rabbit populations in the release sites showed that, although the virus spread to uninfected rabbits and caused considerable mortality, it appeared to die out. However, in December 1950, the first indications were received of the development of a spectacular epizootic, an epizootic with a scale and speed of spread that was almost without parallel in the history of infectious diseases. Within 1 wk of a report that myxomatosis was occurring among rabbits some miles from the nearest release site, the disease had been reported along the frontage of the River Murray (Fig. 3). During the latter half of December, the disease had spread to various points along the northern tributaries of the Murray. By mid-February 1951, the disease had appeared in practically all parts of a huge area (about 1000 miles from south to north and 1100 miles from east to west) that was to be affected before the advent of winter (Fig. 3). When high mortalities (over 90%) occurred, they often were achieved within 3–4 wk of the date of first observation of the disease in the area. Because of the rainfall distribution in 1950, cases in the southern part of the

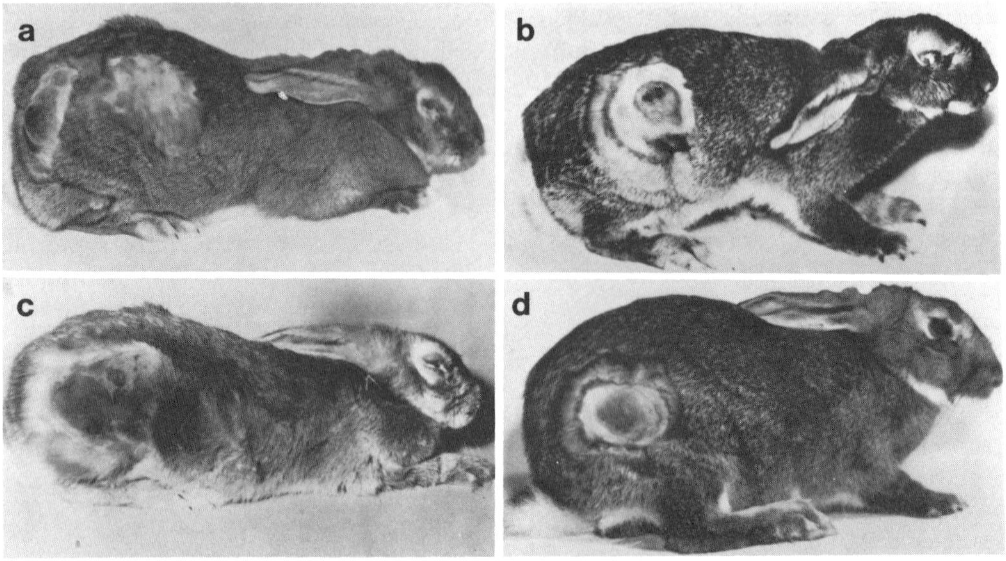

Figure 2 Laboratory rabbits infected with various strains of myxoma virus. (a) Lausanne strain (Grade I virulence), 10 days after infection. (b) Standard laboratory strain (Grade I virulence), 10 days after infection. (c) Australian field strain of Grade III virulence, 21 days after infection. (d) Australian field strain of Grade V virulence, 16 days after infection.

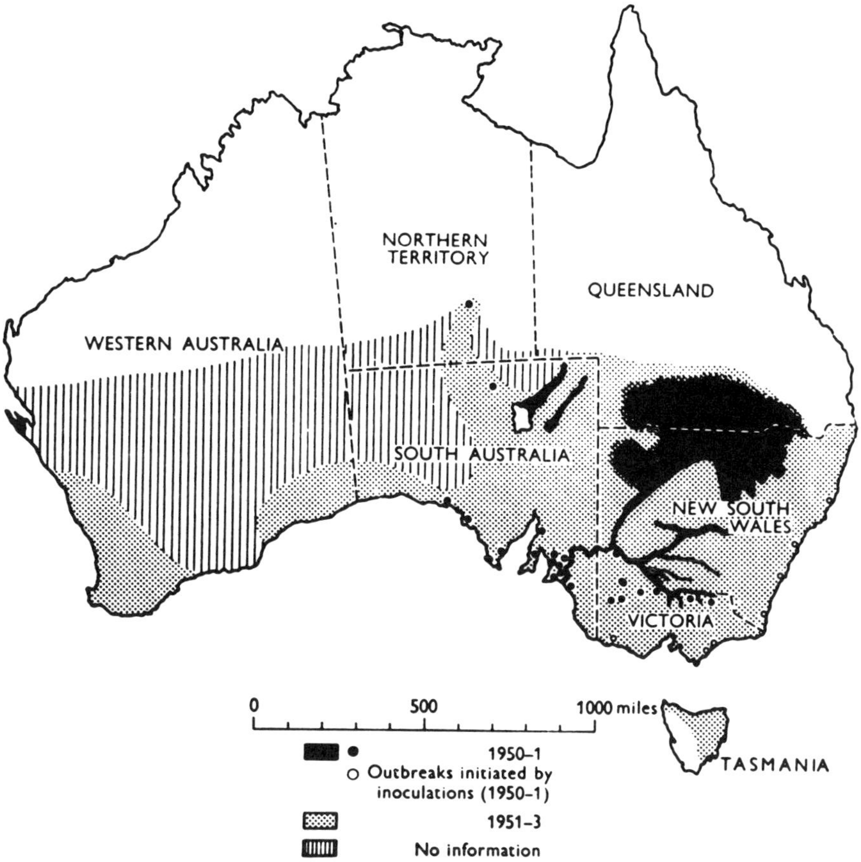

Figure 3 Spread of myxomatosis in Australia during the 3 yr after its escape, late in 1950, from a test site in northern Victoria. The major spread early in 1951 (black) marks out the Murray–Darling river system of southeastern Australia. In the south, the disease was confined to the vicinity of the streams; in the north, heavy rainfall created good conditions for mosquito breeding and the disease spread along and between the streams. Inoculation campaigns with virulent virus were undertaken widely each summer during the 1950s, in all the rabbit-infested agricultural and pastoral lands in southern Australia. Aided by the natural spread of virus that had overwintered, these inoculation campaigns rendered myxomatosis coextensive with the rabbit by 1954. Reproduced with permission from Fenner and Ratcliffe (1965).

area were restricted to the river systems, but the disease spread across the whole countryside in the north. Taking advantage of this new method of rabbit control, extensive inoculation campaigns were carried out by pest control authorities and farmers in the following spring and each spring for many years thereafter. During 1951–1953, with seasonal conditions favoring mosquito breeding, myxomatosis spread over all of eastern Australia where *Oryctolagus* occurred. The disease was introduced by inoculation campaigns and spread in many parts of South Australia and the southwest corner of Western Australia.

III. Opportunity for Research

The 1950–1951 epizootic set the scene for intensive studies of the natural history of the disease and its evolution by scientists of the Wildlife Survey Section of the Commonwealth Scientific and Industrial Research Organization, and well as the Department of Microbiology of the John Curtin School of Medical Research at the Australian National University.

A. Case-Fatality Rates in Naive Populations

In order to study the epidemiology in more detail, a trial site was chosen at Lake Urana in southern New South Wales, where myxomatosis had not occurred in the 1950–1951 outbreak. Two sections of the sandy lake margin were selected for routine counts, providing an index of the rabbit population dynamics. The essential findings from the larger of the two trial sites are presented in Fig. 4. The population count on a standard walk in September 1951, made in the evening when rabbits were grazing, averaged 5000. During that month and the next, some hundreds of rabbits were captured, inoculated with virulent SS myxoma virus (Fig. 2b), and released. The first natural cases were observed in October; during November, a catastrophic fall in the standard count from 5000 rabbits to 50 occurred. Serologic tests showed that 75% of the survivors had escaped infection; thus, the case fatality rate was 99.8%. From casual observations, this case was not exceptional since throughout 1950–1951 the epizootic level of mortality was similar.

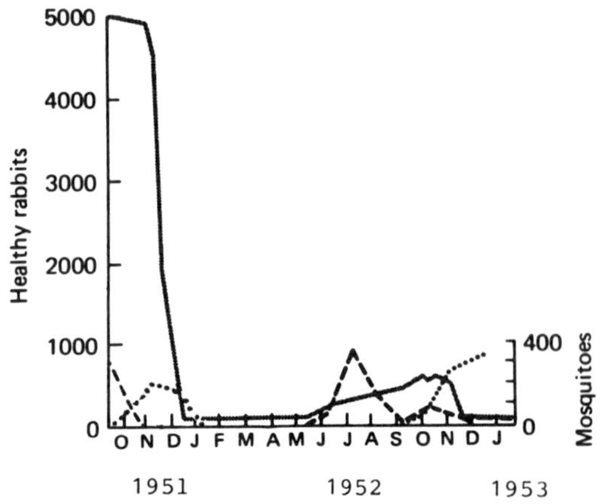

Figure 4 Initial impact of myxomatosis on Australian wild rabbits. Data from a trial site in southern New South Wales, October 1951 to January 1953. Some hundreds of rabbits were captured and inoculated with Grade I virus (Fig. 2b) in October 1951. The solid line indicates the population count of healthy rabbits on a standard transect. The case fatality rates were calculated as 99.8% in the 1951–1952 outbreak and 90% in the 1952–1953 epizootic caused by virus of Grade III virulence (Fig. 2c). The timing of the epizootics suggests that *Anopheles annulipes* mosquitos (···) were, and aedine mosquitos (– – –) were not, efficient vectors. Only the former seek rabbits as a source of a blood meal. Modified with permission from Myers *et al.* (1954).

B. Emergence of Less Virulent Strains

Observations were continued at Lake Urana throughout the winter. In June, the population was augmented by an influx of rabbits from the lake bed, parts of which were flooded. With the emergence of young rabbits from their burrows beginning in July, the population reached 550 in October 1952. The disease probably persisted at the site throughout the winter, since diseased rabbits were seen at least once every month from October 1951 to December 1952, except for June, August, and September. In October, 50 rabbits were captured and inoculated with virulent virus but, before this virus could spread, a sharp epizootic occurred, after which the population was reduced to 60 animals, all of which appeared to be survivors of one or, in a few cases, both epizootics. The case fatality rate in this second epizootic was calculated to be 90%. Eight strains of virus were recovered from pools of mosquitos, all of which were less virulent than the strain used for the inoculations. In test animals, these strains produced a less severe disease (Fig. 2c) with a longer survival time. In fact, a few of the inoculated laboratory rabbits survived the infection.

This field experiment has been described in some detail because it sets the scene for most of the investigations described in this chapter of evolutionary changes in the virus, the rabbit, and the disease that resulted from this host–parasite interaction.

IV. Example of Co-evolution

A. Basis for Natural Selection of Viral Strains

Analysis of the data that had been obtained in the field and from concurrent laboratory experimentation resulted in the postulate that the basis for selection of the virus strains that would become dominant in wild rabbits was high transmissibility. Since the most important mechanism of transmission by far in Australia at this time was mechanical transfer by mosquitos, tests were made of the comparative efficiency of transmission of strains of high, somewhat reduced, and greatly reduced virulence. The efficiency of transmission by mosquitoes that probed through skin lesions at various times after infection was compared with the titer of virus in the skin lesions. Researchers found that the virulent and moderately virulent strains were at about the same concentration in the skin. However, transmission occurred for only ~4 days before host death for the highly virulent strain compared with 16–25 days for the moderately virulent strains (Fig. 2c). Transmitting the "neuromyxoma" strain, a laboratory variant that was extremely attenuated, was almost impossible and the skin lesions of those attenuated strains that were somewhat more virulent than neuromyxoma (Fig. 2d) healed too rapidly for transmission to be possible for more than about 15 days. Applying this concept to the epizootics at the trial site, investigators argued that, especially during the winter months when mosquitos were rare, a strong selection would exist for somewhat but not greatly less virulent strains that would reduce host mortality and thus increase the period of transmissibility through mosquitos.

A few years later this hypothesis was tested in the course of an experiment designed to examine the behavior in the field of the very virulent Lausanne strain of virus (Fig. 2a) that had been used to introduce myxomatosis into France in 1952. Early in the transmission season, a large number of rabbits was captured, inoculated with the Lausanne strain, and released (Fig. 5). During this process, a strain of reduced virulence was isolated from a naturally infected rabbit. A method was devised for testing whether further isolates were the Lausanne strain or the local enzootic strain. A sharp epizootic occurred, during the early part of which the majority of isolates were Lausanne strain; however, during the latter part of the outbreak, only 3 of 30 isolates were the Lausanne strain. Only the somewhat attenuated strains survived until the next season.

B. Changes in the Virulence of the Virus over Time

Using a small number of strains of differing virulence, researchers found that they could "grade" field isolates according to virulence by determining the survival time of groups of 6 laboratory rabbits inoculated intradermally with small doses of virus. On the basis of these survival times and the correlated case fatality rates, field isolates were allocated to one of five (later six) "virulence grades," I being the most virulent. Over the ensuing years, hundreds of strains of myxoma virus recovered from naturally infected rabbits were tested (Table 1). The strains designated "Grade III" (Fig. 2C) were found to become the most common strains almost immediately, and remained so for 30 yr after myxoma virus first was introduced, until 1981 when testing was discontinued.

C. Innate Resistance of Rabbits

With the dominance of strains of virus that allowed the survival of some 10% of infected rabbits, researchers could determine whether this very strong selection would lead to changes in the innate resist-

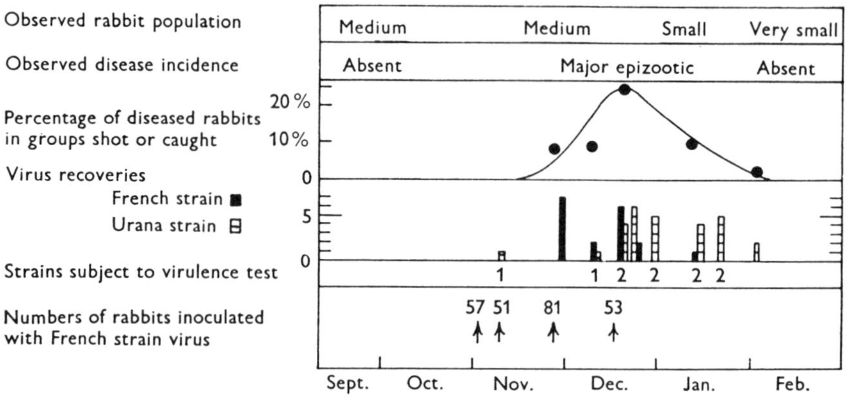

Figure 5 Competition between viruses of Grade I and Grade III virulence. In November and December large numbers of rabbits were captured and inoculated with the Lausanne (French) strain of Grade I virulence (see Fig. 2a). During this process, a single strain of Grade III virulence (Urana) was recovered from a naturally infected rabbit (the first isolate to be subject to a virluence test). A sharp epizootic occurred in December–January. Using a skin test that distinguished between the two strains (French and Grade III enzootic Australian), 45 additional isolates from naturally infected rabbits were tested. The initial outbreak was mainly caused by the French Grade I virus introduced by the inoculations, but during the latter part of the outbreak almost all isolates were of the Australian virus of Grade III virulence. Modified with permission from Fenner *et al.* (1957).

ance of rabbits. This possibility was tested with wild rabbits that were the progeny of animals that had survived infection, and by breeding experiments with laboratory rabbits, each generation being challenged with a Grade III or, later, a Grade I virus.

Over a period of 7 yr, young wild rabbits were captured before the onset of the myxomatosis season from a site where epizootics occurred each year. These animals were raised in isolation until they were ~4 mo old, when they were bled to see whether

Table 1 Classification of the Virulence of Strains of Myxoma Virus Recovered from the Field in Australia between 1951 and 1981, Determined on the Basis of Survival Times[a]

	Classification				
Virulence grade	I	II	III	IV	V
Case fatality rate (%)	>99	95–99	70–95	50–70	<50
Mean survival time (days)	9–13	14–16	17–28	29–50	—
Years					
1950–1951 (1)[b]	>99[c]				
1952–1955 (60)	13.3	20.0	53.3	13.3	0
1955–1958 (432)	0.7	5.3	54.6	24.1	15.5
1959–1963 (449)	1.7	11.1	60.6	21.8	4.7
1964–1966 (306)	0.7	0.3	63.7	34.0	1.3
1967–1969 (229)	0	0	62.4	35.8	1.7
1970–1974 (174)	0.6	4.6	74.1	20.7	0
1975–1981 (212)	1.9	3.3	67.0	27.8	0

[a] Data from Fenner and Ross (1994).
[b] Numbers in parentheses indicate number of samples tested. Data given are in percentages.
[c] Only one field strain was tested, but the field observations in the early outbreaks (see Fig. 2) justify this extrapolation.

Table 2 Evolution of Genetic Resistance to Myxomatosis[a,b]

Number of epizootics to which the population has been exposed	Case fatality rate (%)	Clinical signs in challenged rabbits (%)		
		Severe[c]	Moderate	Mild
0	90	93	5	2
2	88	95	5	0
3	80	93	5	2
4	50	61	26	12
5	53	75	14	11
7	30	54	16	30

[a] Data from Fenner and Ross (1994).
[b] Case fatality rates in groups of wild rabbits collected from an area that experienced annual epizootics of myxomatosis and that, when challenged, lacked both passively and actively acquired antibodies, after inoculation with a standard small dose of myxoma virus of grade III virulence.
[c] Including fatalities.

their sera contained antibody against myxoma virus. These animals and a number of laboratory rabbits then were challenged with small doses of Grade III strain (Fig. 2c) from a single batch of myxoma virus prepared and frozen. The results are indicated in Table 2. The data provided a clear indication that selection for resistance to the myxoma virus had occurred, since both the case fatality rate and the severity of clinical conditions declined over the 7 yr. Subsequently, investigators found that rabbits also became resistant to Grade I virus (Table 3). Breeding experiments in laboratory rabbits (Fig. 6) showed a similar rapid acquisition of increased resistance to myxomatosis, so after several generations of selection rabbits showed resistance to the highly virulent Grade I strain.

It was assumed that, since the influence of maternal antibody had been excluded, the increased resistance was the result of selection for genetically more resistant rabbits. This event undoubtedly is partly responsible for resistance, but experiments with laboratory and wild rabbits show that a substantial nongenetic component in this resistance exists as well. Because of lowered fertility in their "selected" line of rabbits, Sobey and Conolly (1986) had recourse to cross-breed with "unselected" female laboratory rabbits. Comparisons of the survival times of the progeny of unselected does revealed that these progeny survived longer than expected if they had been mated to "selected" bucks during the previous 7 mo, even when the

Table 3 Case Fatality Rates of Strains of Virus of Grade I Virulence[a]

Strain	Case fatality rate (%)	
	Standard laboratory[b]	Lausanne[c]
Selection experiments[d]		
grade 0	99	
grade 4	88	
grade 6.5	79	
Gippsland wild rabbits[e]		
1961–1966	94	
1967–1971	90	
1972–1975	85	~100
1975–1981	79	~100
Mallee wild rabbits[f]		
1961–1966	68	
1967–1971	66	
1972–1975	67	~100
1976–1981	60	98

[a] Data from Fenner and Ross (1994).
[b] Used for introduction into Australia in 1950.
[c] Used for introduction into France in 1952.
[d] See Fig. 4 for explanation of grades.
[e] Gippsland in a region in which epizootics of myxomatosis are infrequent.
[f] The Mallee is a region that experiences annual outbreaks of myxomatosis.

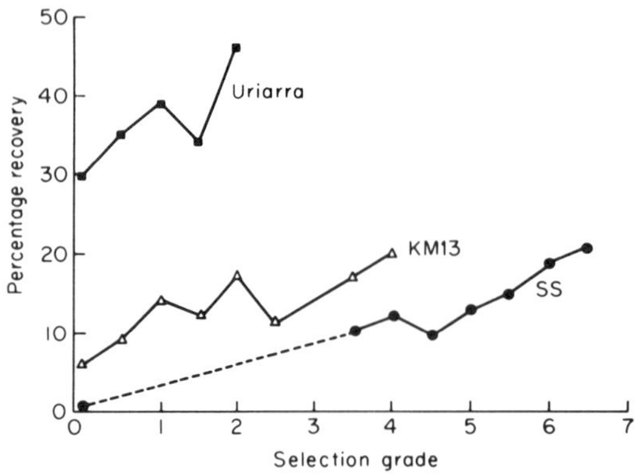

Figure 6 Results of breeding from laboratory rabbits that survived inoculation with various strains of myxoma virus (Sobey, 1969). "Selection grade" indicates the parentage of the rabbits under test: 0.5, one parent had survived infection with myxoma virus; 1, both parents had recovered; 2, both parents and both grandparents had recovered; and so on. From selection grades 0 to 2, a Grade IV (Uriarra) and a Grade III (KM13) virus were used; from selection grades 2 to 4, a Grade III virus was used; from selection grade 3.5, a selection line was begun with the SS virus, of Grade I virulence, a strain that was uniformally lethal for unselected laboratory rabbits. Reproduced with permission from Fenner *et al.* (1974).

"selected" buck had not sired the offspring! Williams and Moore (1991) reported a similar "paternal immunity" in wild rabbits, and suggested that this effect may be responsible for a substantial proportion of the innate resistance previously ascribed to genetic resistance. The mechanism is unknown.

D. Environmental Effects on the Response to Myxomatosis

Researchers have known that elevated temperatures greatly reduce the lethality of myxomatosis. Marshall (1959; Table 4) demonstrated this effect clearly, using a regimen of environmental temperature exposures that mimicked summer and winter field conditions in parts of Australia. Subsequently, field observations confirmed that the "temperature effect" plays an important role in some seasons and some districts.

E. Interplay between Rabbit Resistance and Virus Virulence

If natural selection for virulence were based on transmissibility, investigators might expect that the development of resistance in the rabbit populations would select for somewhat more virulent strains of myxoma virus. To a limited extent, this selection has occurred (Table 5). When the results of virulence tests on field isolates from various parts of Victoria were analyzed according to whether they were obtained from the Mallee region (where resistance was high; Table 3) or elsewhere in the state (where resistance was lower), investigators found that from 1970 onward substantially more Grade I and Grade II strains were found among samples obtained from the Mallee region.

V. Conclusions

Myxomatosis first was recognized as an "emerging virus" in 1896, when laboratory rabbits in Montevideo were struck down by an unknown disease. Over 50 years later, the myxoma virus was introduced deliberately into the wild European rabbit populations of Australia and Europe. Each introduction caused enormous mortalities. Because of the economic importance of the rabbit as a pest in Australia, detailed studies have been carried out there illustrating the evolution of an infectious disease and its host and viral components. Natural selection for the virus is based on transmissibility. Since the dominant mode of transmission of myxoma virus in Australia is mechanical carriage by mosquitos that bit rabbits, selection, especially for a virus that would overwinter, favors strains that would allow infected rabbits to survive with virus-containing lesions for a prolonged time. The highly virulent strain that was introduced repeatedly killed rabbits too quickly to allow it to overwinter, although it spread efficiently when mosquitos were numerous. Highly attenuated variants produced lesions that contained fewer virus particles and healed rapidly. Selection favored strains of intermediate virulence.

Table 4 Myxomatosis in Rabbits after Inoculation with a Virus of Grade III Virulence[a,b]

Inoculum	Result	Cold room (0–4°C)	Mild room (15–22°C)	Hot room (37–39°C)
Myxoma virus	Mortality	36/39	23/35	2/23[c]
	Clinical signs	Severe, progressive generalized disease	Variable	Mild, with early demarkation of lesions
	Viremia	High and prolonged	Moderate and prolonged	Low and transient
Controls	Mortality	0/36	—	0/31

[a] Data from Marshall (1959).
[b] Rabbits were held at room temperature (15–22°C) for 8 hr each day, and at low (0–4°C) or high (37–39°C) temperatures for the remaining 16 hr.
[c] Seven deaths not due to myxomatosis omitted (five deaths in control group).

The early emergence and subsequent dominance of viral strains that allowed >10% of rabbits to survive provided an opportunity for the selection of rabbits of greater innate resistance. Observations showed that this selection occurred rapidly at first, then reached a plateau. Part of this resistance was genetic, but some observations suggest that an unknown "paternal immunity" factor may have played a part. Environmental effects, notably temperature, played a major role in the fate of infected rabbits, since high ambient temperatures favor recovery and low temperatures enhance the severity of the disease.

Finally, as rabbits with higher innate re-

Table 5 Frequency Distribution of the Virulence of Strains of Myxoma Virus Recovered from the Field in the Mallee Region and Elsewhere in Victoria between 1951 and 1981, Calculated on the Basis of Survival Times[a]

	Distribution				
Virulence grade	I	II	III	IV	V
Case fatality rate (%)	>99	95–99	70–95	50–70	<50
Mean survival time (days)	9–13	14–16	17–28	29–50	—
Years of collection in Mallee region[b]					
1959–1963 (70)[c]	0	43	57.1	34.3	4.3
1964–1966 (51)	2.0	0	64.7	31.3	2.0
1967–1959 (51)	0	0	68.1	31.9	0
1970–1974 (102)	1.0	6.9	77.5	14.7	0
1975–1981 (121)	3.0	5.8	67.8	23.4	0
Years of collection in Victoria except for Mallee region[b]					
1959–1963 (379)[c]	2.1	12.4	61.2	19.5	4.7
1964–1966 (255)	0.4	0.4	63.5	34.5	1.2
1967–1969 (198)	0	0	61.6	36.4	2.0
1970–1974 (72)	0	1.4	69.4	29.2	0
1975–1981 (91)	0	0	65.8	34.2	0

[a] Data from Fenner and Ross (1994).
[b] Enzootics occurred almost every year in the Mallee region, but less frequently elsewhere in Victoria. In consequence, the genetic resistance of rabbits was much higher in the Mallee region (see Table 3).
[c] Number in parentheses indicate number of samples tested. Data given are in percentages.

sistance became more common, selection that was dependent on transmissibility led to the persistence of a higher proportion of more virulent strains in populations of wild rabbits.

References

Aragao, H. B. (1943). O virus do mixoma do coelho do mato *(Sylvilagus minenses),* sua transmissao pelos *Aedes scapularis* e *aegypti. Mem. Inst. Oswaldo Cruz.* **38,** 93–99.

Fenner, F., and Ratcliffe, F. N. (1965). "Myxomatosis." Cambridge University Press, Cambridge.

Fenner, F., and Ross, J. (1994). Myxomatosis. *In* "The European Rabbit. The History and Biology of a Successful Colonizer" (H. V. Thompson and C. King, eds.), pp. 205–239. Oxford University Press, Oxford.

Fenner, F., Poole, W. E., Marshall, I. D., and Dyce, A. L. (1957). Studies in the epidemiology of infectious myxomatosis of rabbits. VI. The experimental introduction of the European strain of myxoma virus into Australian wild rabbit populations. *J. Hyg.* **55,** 192–206.

Fenner, F., McCauslan, B. R., Mims, C. A., Sambrook, J., and White, D. O. (1974). "The Biology of Animal Viruses," 2d Ed. Academic Press, New York.

Marshall, I. D. (1959). The influence of ambient temperature on the course of myxomatosis in rabbits. *J. Hyg.* **57,** 484–497.

Myers, K., Marshall, I. D., and Fenner, F. (1954). Studies in the epidemiology of infectious myxomatosis of rabbits. III. Observations on two succeeding epizootics in Australian wild rabbits on the Riverine plain of south-eastern Australia. *J. Hyg.* **52,** 337–360.

Sobey, W. R. (1969). Selection for genetic resistance to myxomatosis in domestic rabbits *(Oryctolagus cuniculus). J. Hyg.* **67,** 743–754.

Sobey, W. R., and Conolly, D. (1986). Myxomatosis: Non-genetic aspects of resistance to myxomatosis in rabbits, *Oryctolagus cuniculus. Austral. Wildl. Res.* **13,** 177–187.

Williams, C. K., and Moore, R. J. (1991). Inheritance of acquired immunity to myxomatosis. *Austral. J. Zool.* **39,** 307–311.

Michael D. B. Burt

24 The Sealworm Situation

I. Introduction

This chapter addresses a real Canadian problem—a problem that costs the East Coast fisheries alone millions of dollars annually and a problem that is growing worse year by year. The crux of the problem is the presence of larval nematodes in the flesh of commercially important fishes. These nematodes are mostly the third larval stage (L_3) of an anisakid worm *Pseudoterranova decipiens* (Krabbe, 1878). The cost involved relates largely to the removal of the larvae from fish fillets prior to sale of the fish. In earlier times, Canada's most important commercial fish species was cod. Among individuals in the fishing industry, this anisakid worm was widely known as "codworm". The current vernacular name "sealworm" has replaced the older term to reflect the definitive host of the worm (seals), as well as the fact that larval stages are by no means restricted to cod, since they are found in many demersal fish species including plaice, halibut, turbot, and other flatfish as well as in cod, haddock, and other gadoids.

Much effort has been spent over the last 40 years, largely by scientists working for the Department of Fisheries and Oceans (DFO) in Canada, in addressing the sealworm problem. Through the efforts of J. E. Stewart (Director, Biological Sciences Branch, Scotia-Fundy Region, DFO, Halifax), two international workshops were held, one in 1987 and the other in 1988, at which information relating to the sealworm was pooled by scientists from Canada and Western Europe. By sharing all this knowledge, researchers hoped that methods of attacking the problem might emerge. In addition, investigators hoped that a predictive model of the population dynamics of the nematode could be developed. The proceedings of these workshops were published in the *Bulletin* series of DFO (*Can. Bull. Fish. Aquat. Sci.,* **222**) in 1990. Although the hoped-for predictive model could not be developed, the gaps in current knowledge were identified clearly, as were those areas in which more detailed information was required. The life cycle of *P. decipiens* is not simple; several hosts can be involved between the hatching of the free-living larva from the egg and eventual maturation of the adult in the seal. As discussed in a later section, various crustaceans can serve as the first (or only?) intermediate host and various fishes can serve as paratenic or transport hosts. As also discussed later, at least three sibling species and perhaps two variants of *P. decipiens* ap-

Parasitic and Infectious Diseases

pear to exist, with differences in their protein profiles and their abilities to infect different crustaceans.

The ever-increasing severity of the sealworm problem is linked clearly to the growth of populations of seals in which these worms mature and reproduce. The three species of seals involved on the East Coast are the gray seal *(Halichoerus grypus),* the harbor seal *(Phoca vitulina),* and the harp seal *(Pagophilus groenlandicus).* Of these, the most important definitive host is the gray seal because of its increasing abundance and the large number of *P. decipiens* that usually are found in its stomach. Although the population of harp seals is growing rapidly because of severe restrictions placed on harvesting this animal, most animals are found north of the Gulf of St. Lawrence, where the low water temperatures may not be suitable for the development of the early stages of the sealworm. Harbor seals, like harp seals, do not carry nearly as many adult worms as gray seals and are far less numerous than gray seals in eastern Canada.

II. Historical Perspective

A. Name

The worm known as the sealworm first was described as *Ascaris decipiens* by Krabbe in 1878. The genus *Porrocaecum* was erected by Railliet and Henry (1912) and included the species *decipiens.* Johnston and Mawson (1945) restricted the genus *Porrocaecum* to those species with birds as the definitive host and reinstated the genus *Terranova* of Leiper and Atkinson (1914) for other worms previously included in the genus *Porrocaecum.* Karokhin (1946) referred specifically to *Terranova decipiens* (Krabbe, 1878) and accepted the restriction of *Porrocaecum* indicated by Johnston and Mawson (1945). Dollfus (1953) preferred to consider *Terranova* to be a subgenus of *Porrocaecum;* the name, accordingly, became *Porrocaecum (Terranova) decipiens* (Krabbe, 1878).

Myers (1959) erected the genus *Phocanema* to contain the single species *decipiens* but this did not meet with wide acceptance. Until 1982, the sealworm was referred to in the literature variously as *Porrocaecum decipiens, Terranova decipiens,* and *Phocanema decipiens.* Gibson and Colin (1982) transferred the species to the genus *Pseudoterranova* of Mozgovia, in Skrjabin *et al.* (1951); Gibson (1983) rendered the genus *Phocanema* a synonym of *Pseudoterranova,* a treatment that was widely accepted.

B. Definitive Hosts

Sealworm has been reported in the stomachs of gray seals, harbor seals, and harp seals from the Gulf of St. Lawrence by Scott (1950) and Fisher (1950,1951) (see Fig. 1). Later, Scott and Fisher (1958) identified *P. decipiens* in a large number of the same three seal species on the Atlantic Coast of Canada (127 gray seals, 318 harbor seals, and 812 harp seals).

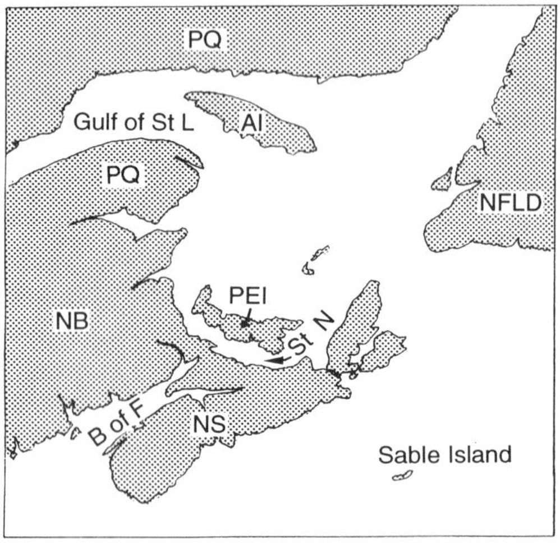

Figure 1 Map of the northwest Atlantic region. AI, Anticosti Island; B of F, Bay of Fundy; Gulf of St L, Gulf of St. Lawrence; NB, New Brunswick; NFLD, Newfoundland; NS, Nova Scotia; St N, Northumberland Strait; PEI, Prince Edward Island; PQ, Province of Quebec.

In the early 1950s, few data on population sizes of gray and harbor seals were available, although anecdotal evidence indicates that the gray seal population in the northwest Atlantic may have numbered about 15,000 in the mid-1950s. In 1976, Mansfield and Beck (1977) suggested a population size of 24,000 of age 1+ for gray seals in eastern Canada. This population increased to 44,000–48,000 age 1+ by 1980 (Zwanenberg *et al.*, 1981). During the period 1962–1989, the Sable Island population had increased about 20-fold (Zwanenburg and Bowen, 1990) (from 2,500 to 50,000) and the Gulf of St. Lawrence population had increased 2- to 5-fold to between 30,000 and 60,000. [If this number represents a 5-fold increase of the 1962 population to 30,000, an earlier Gulf population as low as 6,000 is indicated; if only a 2-fold increase to 60,000 occurred, the 1962 Gulf population could have been as high as 30,000.] During the period 1977–1989, pup production on Sable Island alone increased exponentially from 2,181 to 9,712 pups at an annual rate of 12.6% (Stobo and Zwanenburg, 1990).

Population sizes of harbor seals are less well documented than those of gray seals and are censused by counting the total number of seals hauled out at particular locations (Thompson *et al.*, 1990). Fisher (1949) estimated a population size of 10,000–15,000 for the Maritime Provinces excluding Sable Island. By 1973, the population size for the same area was estimated to be only 5,453; for Eastern Canada (excluding Labrador and the Arctic), the population was thought to be 12,700 (Boulva and McLaren, 1979). Pup production of the Sable Island breeding colony increased about 2-fold from 1978 to 1989 (313 to 521) according to Thompson *et al.* (1990), but population trends elsewhere are not known.

Scott and Fisher (1958) provided information on infection rates of gray seals with *P. decipiens* for the period 1949–1956. The average number of adult worms varied from 48 (Sable Island) to 187 (Miramichi Estuary) with an overall average for the northwest Atlantic (based on 52 seals) of about 136 worms per seal. Mansfield and Beck (1977) reported higher mean intensities of 102 adult female worms (about 204 adults in total) for gray seals. Intensities of 8 adult female worms in harbor seals and 1.5 adult females in harp seals were similar to those recorded by Scott and Fisher (1958).

McClelland (1980a) provided information on infection rates of gray seals with *P. decipiens* for the period 1975–1978. The mean intensity varied from 1 worm (Amet Island, Nova Scotia) to 483 worms (September–October, Ecum Secum, Nova Scotia), with marked seasonal differences. The overall average for the northwest Atlantic was 194 worms (based on 44 seals). Previously unpublished information, provided by McClelland in Brattey *et al.* (1990) and Stobo *et al.* (1990) for 1983–1987 shows a range in average numbers of adult *P. decipiens* from 69 to 510 per gray seal, with an overall average of 208 adults per seal. Again, marked differences in seasonal infection rates were observed; seals sampled in January and February had fewer adults than those sampled at other times of the year.

For the northwest Atlantic, based on 104 harbor seals examined between 1949 and 1956, the number of adult worms per seal averaged about 13 with a range of 6–23 according to Scott and Fisher (1958). One report by McClelland (1980a) indicated an average of 2 adults per seal in a small sample of 5 seals from Ecum Secum in 1975–1978. Later reports by Stobo *et al.* (1990) and unpublished information of McClelland in Brattey *et al.* (1990) for the period 1983–1987 showed an overall average of about 42 adult worms per harbor seal. More detailed information on local infection rates for gray seals, harbor seals, and harp seals is given by Brattey *et al.* (1990).

Although local populations of harbor seals may be large enough to contribute in some measure to the sealworm problem in some areas, in view of the relatively low numbers of adult *P. decipiens* in these seals as well as the relatively low numbers of harbor seals in relation to gray seals, harbor seals probably have little impact on the sealworm problem in Eastern Canada.

C. Fish Hosts

Early surveys of larval *P. decipiens* infections in fish centered almost entirely on cod, which were known to have larval nematodes in their flesh. During the period 1947–1953, infection rates varied considerably by location; the greatest abundance of *P. decipiens* larvae (3.11) in fillets of cod $\geq$41 cm was in the southwest Gulf of St. Lawrence and the lowest abundance (0.01–0.02) in fillets of cod of the same size was in the Grand Bank, offshore Labrador, and northeast Newfoundland (Templeman *et al.*, 1957). Prevalences also varied considerably according to Scott and Martin (1957) for the period 1946–1956; 89–92% of the fillets of cod from the southern Gulf of St. Lawrence, 66% of the fillets of cod from Passamaquoddy Bay, and only 5–7% of the fillets of cod from Sable Island and offshore Scotian banks were infected.

More recent reports show that many different fish species are infected with *P. decipiens* and that larval sealworm abundances seem to be increasing in fish populations throughout the North Atlantic (McClelland *et al.*, 1990a). For example, for fish collected from Sable Island Bank from May 1, 1986, to April 31, 1987, almost all cod over 50 cm and almost all sea raven and American plaice over 30 cm were infected with *P. decipiens* larvae (McClelland *et al.*, 1990b). The densities of these infections, measured in terms of numbers of worms per kg fish fillet, varied in relation to the size and species of the fish: average densities of infection for cod ranged from about 14 to nearly 43 per kg fish; for sea raven from about 64 to over 168; and for American plaice from about 13 to over 290. Even higher infections have been recorded since then, particularly for fish taken near Sable Island but also in those found in the Gulf of St. Lawrence, the Breton and Scotian Shelves, and the Lower Bay of Fundy and northwestern Gulf of Maine (G. McClelland, *personal communication*).

D. Life Cycle

A good and detailed account of the life cycle is provided by McClelland *et al.* (1990a). Adult *P. decipiens* mature and lay eggs in the stomach of their seal host. These eggs are shed from the seal with its feces. Development of the fertilized eggs is inversely proportional to temperature; little or no development occurs at temperatures below 0°C, although eggs kept at −0.5°C for up to 3 mo hatched normally when the temperature was raised (Brattey, 1990). Prolonged incubation at cold temperatures ($\leq$1.0°C) may result in mortalities of over 70% (McClelland, 1982). The eggs can complete embryonation and hatch in fresh water as well as in brackish water and sea water (Burt *et al.*, 1990a).

The larvae that hatch have molted within the egg and are assumed here to be at the L_2 stage; the L_1 cuticle is retained as a sheath. [Note: Opinion is divided on whether one or two molts occur within the egg. If two molts occur, the newly hatched larva would be at the L_3 stage.] These larvae are highly active between temperatures of 10 and 22°C, but become progressively more sluggish as the temperature drops from 10°C to 5°C, and show little or no movement at temperatures below 5°C. The upper lethal temperature is about 25°C, at which rapid hatching (5 days) of some eggs occurs, with subsequent death of the L_2 larvae within a few hours (Burt *et al.*, 1990a).

The newly hatched larvae of *P. decipiens* are infective, either directly or indirectly, to a wide array of different intermediate hosts including some 18 species of copepod crustaceans, some 17 species of malacostracan crustaceans, 2 polychaete annelids, and 1 gastropod mollusk as listed by McClelland *et al.* (1990a). Several additional crustaceans also serve as hosts (G. McClelland, personal communication). Experimental infections by Jarecka *et al.* (1988) indicated that although several (12) species of crustaceans could become infected, others could not be infected experimentally, indicating at least some degree of host specificity. Further, the source of the larvae appeared to play an important role. For example, the larval progeny of some worms from gray seal could infect juvenile *Gammarus oceanicus* directly whereas progeny of worms from harbor seal could not. Other differences were noted between the experimental infections attempted by McClelland (1980b) and those of Jarecka *et al.* (1988), suggesting the possibility of sibling species of *P. decipiens* with different life cycles, which will be discussed subsequently.

Some evidence exists that the molt from the L_2 larva to the L_3 larva occurs in the macroinvertebrate host (Jarecka *et al.*, 1988), but this evidence is not conclusive. Fish become infected either by eating infected macroinvertebrates or by eating other infected fish. In both cases, the larvae migrate through the wall of the alimentary tract into the coelomic cavity prior to penetrating the flesh, where they may or may not become encapsulated within material of host origin containing lipofuschin (Ramakrishna and Burt, 1991). Larvae can infect up to 4 fish hosts in series, but in no case could the same L_3 larva be passaged experimentally into a fifth fish host (Burt *et al.*, 1990b).

Little or no immunity is developed by infected fish since they can recruit new infections continuously. The tissue reaction to new infections in already infected fish resembles closely the initial tissue reaction to the first infective worm (Ramakrishna and Burt, 1991).

Seals become infected by eating infected fish or, in the case of recently weaned pups, by eating infected macroinvertebrates (G. McClelland, personal communication). Of the 63 fish species recorded as carrying sealworm larvae (McClelland *et al.*, 1990b), benthic feeders are infected most heavily. In the study by McClelland *et al.* (1990b) of fish from waters around Sable Island, highest overall densities of infection (numbers of worms per kg fish) were found in sea raven (45), American plaice (34), windowpane (23), longhorn sculpin (22), ocean pout (14), monkfish (12), cod (10), cusk (10), yellowtail flounder (6), and gray sole (3). If fish are grouped into separate size classes, then American plaice measuring 21–30 cm had the highest density of infection (73) followed by sea raven $\leq$30 cm (62), windowpane $\leq$20 cm (28), longhorn sculpin 21–30 cm (23), ocean pout 18–50 cm (22), monkfish 51–60 cm (17), cod 31–40 cm (14), cusk 61–88 cm (10), yellowtail founders 21–30 cm (7), and gray sole 31–40 cm (3).

In culinary terms, these frequencies translate into about 3 worms per 4 oz serving of American plaice, about 1 worm per 4 oz serving of cod, and about 0.5 worms (!) per 4 oz serving of yellowtail flounder, if the worms are not detected and removed during processing.

III. Nature of the Problem

Before being sold to consumers, fillets of those fish species known to be infected with *P. decipiens* are examined routinely using candling tables, which are glass-topped benches with lights underneath that reveal the presence of worms inside the fillet. Detected worms usually are removed using the tip of a thin filleting knife. Several variations in lighting exist; other detection methods have been and are being investigated (Power, 1958,1961; Odense, 1978). Currently, however, the only method employed is that just described. The cost of candling fillets can be great and the efficiency is low; many of the worms, depending on the thickness of the fillets are missed routinely (Homans and MacFarlane, 1957; Power, 1961). Larger worms that may have been in the flesh for a long period are usually brown and readily seen; smaller and unpigmented worms are much more difficult to see and, accordingly, are missed frequently. When fish are parasitized heavily, more candlers must be employed, their numbers frequently exceeding the numbers of filleters (see Templeman, 1990). In addition to the extra cost of paying candlers, heavily parasitized fillets look ragged and torn and must be sold at reduced prices or converted to less profitable products such as fish sticks or fish meal. Malouf (1986) estimated the actual cost to the fishing industry in Atlantic Canada at about $26 million annually; other anecdotal estimates for Atlantic Canada are as high as $60–100 million annually, but these costs may include those related to other worms such as *Anisakis simplex* (G. McClelland, personal communication).

Several attempts have been made to improve the efficiency of candling. Modifications have involved laser, ultrasound, and X-ray technology in various combinations. Although the detection *per se* can be improved vastly, the systems are very costly and actual removal of the worms still must be done manually. As stated earlier, no system other than candling is currently in commercial use.

Several surveys conducted within the last decade by McClelland *et al.* (1983a,b,1985, 1987) and Brattey *et al.* (1990) have determined that *P. decipiens* larvae have become increasingly more numerous in ground fish from southern Newfoundland, the Gulf of St. Lawrence, Scotian Shelf, and Breton Shelf over the past 30 years. With the extra cost involved in worm removal, the loss in revenue from wasted fish, the depressed price of "wormy" fish, and the ever-increasing abundance of sealworms because of the increasing abundance of their definitive hosts, the sealworm problem is one of increasing economic importance to the Atlantic fishing industry.

The growth of seal populations exacerbates the problem for the fishing industry: not only do the seals increase the abundance of sealworm, but they consume some 4% of their own weight in food per day (e.g., 5.6 kg cod/day to 3.1 kg capelin/day) (Benoit and Bowen, 1990) to maintain their caloric input, thus playing a significant role in the overall reduction of fish in the northwest Atlantic. Even with the assumption that not all fish eaten are of commercial size or even commercially important, simple calculations of numbers of harp seals (3 million), gray seals (>100,000), and harbor seals (~30,000) consuming about 4 kg fish/day translates into about 12.5 thousand tons of fish consumed per day which, even at conservative estimates, provides a "quota" of well over 4 million tons of fish and shellfish per year for the seals in the northwest Atlantic. This amount is more than 5 times the total landed catch of fish and shellfish in 1991 from the northwest Atlantic and more than 6 times the total for 1992 (1991: 715,817 finfish, 83,512 shellfish; 1992: 549,266 finfish; 96,808 shellfish). Other studies (see Benoit and Bowen, 1990) clearly show that the great majority of fish consumed by seals (>90%) is well below commercial size. Hence, seals do not compete directly with commercial fisheries. By consuming such large quantities of juvenile fish, however, the seals not only may affect "ground fish recruitment" (in fishery terms) but also may come into direct competition with larger, commercially important fish species such as cod.

IV. Population Dynamics

The development of a model to show the population dynamics of the sealworm must consider the biotic potential of the species and the mortality expected at each stage in its life cycle. The biotic potential is determined solely by the fecundity of the adult worms in their definitive seal hosts since no asexual reproduction occurs at any stage of the life cycle. During the life cycle, mortality can occur at the egg stage because not all eggs released from gravid females will hatch. Once hatched, additional mortality can occur at the L_2 stage, which has a finite life-span; if not ingested by an appropriate host during this time, the L_2 larvae die.

Once the surviving L_2 larvae are in their first host, additional mortality can occur if this host, a copepod for example, is not eaten by a macroinvertebrate such as a larger crustacean, annelid, or mollusk or if the larvae succumb to host responses. Further losses can occur if those infected macroinvertebrates are not eaten by a fish host. Significant additional mortality probably occurs at the fish stage, since not all infected fish are eaten by seals and since some larvae succumb to responses of the fish host. Finally, some evidence suggests that, within the seal host, not all L_3 larvae ingested by seals actually mature.

A. Population Processes

Looking at each of these stages in greater detail will provide a better idea of the complexities involved, as well as of the incredible diversity of hosts that can carry sealworm larvae at different stages in the life cycle.

At the egg stage, development to the L_2 with subsequent hatching is dependent on appropriate fertilization within the female worm and on temperature after leaving the seal. In general, the lower the temperature (down to 0°C), the slower the development. At 4°C, for example, development of eggs to hatching requires ~60 days (Burt *et al.*, 1990a) whereas at 1.7°C this process requires over 120 days (Brattey, 1990). At the other extreme, at 15°C development to hatching requires about 10 days, whereas at 22°C the process requires about 8 days (Burt *et al.*, 1990a). Eggs stored at −0.5°C for up to 3 mo were still viable and hatched at 12°C. Since ambient temperatures in the northwest Atlantic range from about −2°C to about 15°C (Brattey, 1990), clearly hatching times in different locations and at different times of the year vary considerably. This difference can, of course, influence significantly the numbers of L_2 larvae available to the next host, which itself may have seasonal variation with respect to life cycle stages and sensitivity to infection.

At the newly hatched L_2 stage, worms can survive for varying times depending on temperature. In warm water, at 22°C, newly hatched L_2 larvae are active for about 24 hr, become less and less active over the next 2 days, and are moribund or dead by day 4. At 14°C, newly hatched L_2 larvae are active for about 12 days, become slower over the next 2 days, and eventually are moribund or dead by day 21. At 7°C, newly hatched larvae are active for about 28 days, slow down over the next 14 days or so, and are moribund or dead by day 58 (Burt *et al.*, 1990a). At 4–5°C, newly hatched larvae can persist for 120–140 days and remain infective to copepods for 111 days (McClelland, 1982). These data clearly indicate that, as for eggs, the life-span and, hence, the infectivity of the larvae is variable and temperature dependent. As shown by McClelland (1982), the spectrum of susceptible copepod species narrows at lower temperatures.

Within the crustacean first intermediate host, growth of the larvae probably is limited by the space available. Small copepod hosts such as *Harpacticus* spp. and *Thisbe* spp. die within a few days of being infected with sealworm larvae; these infected copepods are eaten by larger invertebrates. Fish are unlikely to become infected until the larvae reach ~2 mm in length (McClelland *et al.*, 1990a). If the first copepod hosts are not eaten by macroinvertebrates, the larvae will not develop further and the life cycle will not be completed. The frequency of consumption of the infected first intermediate hosts by macroinvertebrates is not known; also lacking are data on the frequency of predation on infected macroinvertebrates by fishes. Macroinvertebrates usually live for more than 1 yr; copepods, however, survive only 3–5 wk at 5°C and even shorter periods at higher temperatures (McClelland, 1982). This distribution could keep the larvae available, in macroinvertebrates, to fish over an extended time period. How long the larvae can survive in their various hosts is not known. Clearly this information is essential for the development of a predictive model of the population dynamics.

Within the fish host, the larvae occupy the visceral mesenteries and flesh, where they continue to grow and may become encapsulated. In the sea raven, a more rapid reaction to sealworm larvae appears relative to cod, since the proportion of heavily pigmented and encapsulated worms is much greater in sea raven than in cod (personal observation). The cellular response seems to be lacking in some fish such as smelt and plaice (G. McClelland, personal communication). The actual time required, in any fish species, to render the parasite inactive and unable to develop in the seal host also is not known. Again, this information is required to construct a predictive model.

Within the seal host, the L_3 larvae molt to the L_4 stage and eventually to the adult worm. Experimental infections of captive seals in the laboratory by McClelland (1980a) indicated that at least 50% of the larvae eaten by the seal established themselves in the stomach. In view of possible damage to the worms used in the infection plus stress on the captive hosts, this proportion is probably low compared with natural infection in the wild, especially since gray seals have been found to be infected with very large numbers of worms, over 20,000 (including *Contracaecum osculatum*) in one case (McClelland, 1980b). *In vitro* work showed the time for the L_3 larvae to become sexually mature and start producing eggs to be about 21 days (Likely and Burt, 1989). Adult females, *in vitro,* can survive for about 8 wk although egg production starts to decline after about 6 wk (Likely and Burt, 1989). *In vivo* experiments by McClelland (1980a) indicated loss of adults after about 6 wk in captive seals, but stress factors may be involved in these results. The sex ratio of L_3 larvae is about 1 : 1, as determined by *in vitro* work (Likely and Burt, 1989) as well as by observations by various workers looking at adult worms in seals (McClelland, 1980a; Stobo *et al.,* 1990).

As indicated elsewhere, differences exist in the proportion of mature adults to other stages at different times of the year, suggesting seasonal changes in feeding or some intrinsic density-dependent mechanism. The possibility of some influence of host hormones was investigated initially by Oluoch-Munda (1992), but no firm conclusion reached. Work along these lines continues.

The number of eggs produced by adult females is dependent on the size of the female. McClelland (1980b) showed that the larger worms found in gray seals had a mean fecundity of 360,000 eggs/female and that the smaller worms found in harbor seals had a mean fecundity of 156,000 eggs/female. *In vitro,* worms grow as large as those that occur naturally in gray seals and typically can produce over 500,000 eggs/female over the extended *in vitro* life-span. The proportion of eggs that eventually hatches is about 85% and is independent of temperature within the range 4–22°C (Burt *et al.,* 1990a).

B. Development of a Predictive Model

The difficulties of developing a predictive model of the population dynamics are obvious in light of all the unknowns and the variable interrelationships presented in the previous sections.

If the sealworm population was stable, the difference between the number of eggs produced and the number of mature adults would be equal to the sum of all the mortalities at each stage. Unfortunately, various surveys, including those of McClelland *et al.* (1983a,b,1985,1987,1990b), demonstrate that the numbers of sealworm in fish are increasing. Also, populations of the definitive hosts are increasing (gray seals, for example, increasing from about 15,000 in 1955 to over 100,000 in 1987). With increasing numbers of definitive hosts will be a concomitant increase in the numbers of sealworm eggs produced but, assuming similar levels of infection with adult worms in the definitive hosts, data are insufficient at present to be able to predict the level of reduction of L_3 larvae in fish consequent to a cull of seals. Occam's razor dictates that reducing the gray seal herd will reduce the number of adult worms, which in turn will reduce the numbers of eggs produced. The unknown, however, is the extent to which mortalities occur at the different stages of the life cycle. For example, *if* 50% of the L_2 larvae die because of insufficient numbers of first intermediate hosts, *in theory* reducing the number of eggs produced by 50% by culling 50% of seals simply might allow all the resulting L_2 larvae to infect intermediate hosts, resulting in no change in the numbers of L_3 larvae in the flesh of fishes.

To develop a predictive model, not only the mortality rates at each stage of the life cycle but also data on host availability and on intrinsic self-regulation by the sealworm itself are required. Seasonal variation of worms in seals is described by Stobo *et al.* (1990) for both adult and juvenile seals, particularly in the proportion of sexually mature adult worms to juvenile worms. In mature seals and in juvenile seals, the ratios of sexually mature worms to total worm burdens were highest in July and August, subsequently dropping to low values in January. Whether some intrinsic density-dependent mechanism is at work in the seal stomach or whether the changes are caused solely by host effects such as diet (Stobo *et al.,* 1990) or hormones (Oluoch-Munda, 1992) is not fully understood.

One of the most challenging aspects of studying the population dynamics of sealworms is the apparent wide diversity of potential intermediate hosts, both at the crustacean (invertebrate) level and in fishes. Clearly copepods are not essential as the first intermediate host (although this may be the usual route) since the L_2 larvae have been shown to infect juvenile amphipods and larval decapods directly (Jarecka *et al.,* 1988; McClelland, 1990). Further, that the larvae must be in fish before they can be infective to seals is by no means certain. If the L_3 larvae are big enough in the crustacean or other macroinvertebrate host (e.g., crab/mysid, amphipod) and this host is eaten by a seal, the larvae could develop to maturity in the seal, never having passed through a fish. Such considerations are discussed at length by McClelland (1990). Unpublished data (G. McClelland, *personal communication*) indicate that recently weaned gray seal pups on Sable Island feed primarily on invertebrates and nevertheless accumulate significant numbers of sealworms.

V. Possible Solutions

A. "Aspirin" Approach

Treating the symptom rather than the cause is no more effective in fisheries than in medicine! Taking an aspirin for a headache may alleviate the headache; removing the worms from fish allows them to be sold. Thus, there is clearly some usefulness in alleviating headaches; and, there is some usefulness in candling fillets. However, what is really important is to stop the cause of the headache; and to prevent the fish from being so heavily infected with larval sealworm in the first place.

The fishing industry has spent enormous effort and money in trying to improve the candling efficiency. However, a concerted effort is needed to reduce the number of worms, that is, in attacking the cause of the problem. As indicated earlier, laser, ultrasound, and X-ray technologies are being harnessed to produce better detection methods. Although one of these methods eventually may provide better detection, the increasing numbers of worms renders this approach less useful, since worm removal to any extent will result in ragged fillets, thereby reducing the market value of the fillet.

B. Reduction of the Natural Abundance of Worms

An approach of reducing the natural abundance of worms addresses the actual problem rather than the symptom. Several ways to accomplish this goal can be identified: (1) reduction of adult worms by culling seals; (2) reduction of adult worms through seal birth control; (3) reduction of adult worms through seal chemotherapy; (4) reduction of adult worms through seal immunization; (5) reduction of adult worms using a biological competitor in seals; and (6) reduction of larval worms by intervening at earlier stages in the life cycle.

1. Culling of Seals

In view of the link between the increase in numbers of seals, especially gray seals in which adult *P. decipiens* mature and reproduce, and abundance of larval sealworm in fish, this approach is probably the most likely to achieve the end result of fewer *P. decipiens* in fish. The increase in the seal population is linked anecdotally to the decrease in predatory sharks through long-line fishery. If insufficient "natural" predators are left to control the seal populations, then appropriate government department(s) must control and manage these populations through appropriate reductions. However, to embark on a seal cull of the dimensions required without a good predictive model that would provide some expectation of worm reduction based on numbers of seals culled is neither good science nor good management. Accordingly, knowing the reduction in worm abundance that might be expected with, for example, a 50% cull of the seal population is important. Until such a predictive model is produced, culling seals can hardly be recommended.

2. Birth Control in Seals

As expressed earlier, knowing the level to which reduction should be taken is important. The birth control approach certainly is more socially and politically acceptable among that segment of the population that abhors any kind of "cull" and, accordingly, may be more acceptable to politicians who want to be re-elected. However, even this approach should be adopted only after a predictive model, however rough, is developed. Of course a feasible method must exist and such a method must be able to be implemented. For gray seals pupping on Sable Island and for some other island or ice-breeding populations, delivery may be possible, but for the rest of the gray seal population, delivery may not be feasible even if an efficacious procedure could be developed.

3. Chemotherapy in Seals

Although the idea of ridding seals of their parasites through anthelmintics such as ivermectin has gathered some favor, several problems arise. (1) Even if all worms present in the stomach could be eliminated by way of a drug, this event would not prevent recruitment of new infections as soon as the drug effects wear off. (2) Ivermectin itself appears to be efficacious against sealworm only when administered orally, an approach that would not be feasible in the field. (3) The problem of delivery to the seals remains. For the Sable Island population, treating most, if not all, of the pups each year is theoretically possible but the other half of the gray seal population that is spread over a wide area and, therefore, is more or less untreatable remains.

4. Immunization of Seals

As for chemotherapy, a vaccine remains to be developed and, as for birth control chemicals and chemotherapy, delivery of a vaccine to a significant proportion of the seal population is a major impediment. The fact that continual recruitment of larval worms by seals, through infected fish, occurs indicates little if any development of natural immunity (McClelland, 1980b). This result is not surprising since the worms remain largely in the stomach and do not penetrate deeply into the mucosal lining (McClelland, 1980b). In gray seals, little host reaction occurs, although in harbor seals the stomach lining often is raised in hyperplastic-type swellings where *P. decipiens* are, or have been, attached (McClelland, 1980b; personal observations).

5. Biological Competitor in Seals

Many years ago, Berland (1963) suggested, based on observations of presumably displaced *Phocascaris* spp. nematodes in the intestine, that a related anisakine, *Contracaecum osculatum,* was a successful competitor of other anisakines in the stomach of harp seals. This suggestion was reiterated by McClelland *et al.* (1985). Although difficult and very costly to test experimentally *in vivo,* analyses of natural infections of *C. osculatum* in seals indicate no interspecific competition (G. McClelland, personal communication). Any competitive interaction should be easy to test experimentally using an *in vitro* system. After some 14 mo of trial and error, an *in vitro* system was developed for *P. decipiens* (Likely and Burt, 1989); a modification that suited *C. osculatum* (as well as *P. decipiens*) was further developed based on this culture method (Likely and Burt, 1992). Both *P. decipiens* and *C. osculatum* were grown successfully in single-species culture with production of viable eggs that produced L_2 larvae capable of infecting copepods. When grown together, however, the numbers of eggs produced by the gravid *P. decipiens* were reduced greatly (to as little as 10% of the normal number in single-species culture in one experiment). The nature of the inhibition is not known, although current research is now focused on elucidating this interaction. Once this phenomenon is better understood, competition with *C. osculatum* could represent an excellent means of long-term reduction in the *P. decipiens* population. *Contracaecum osculatum* is already present in some areas. For such an approach to be effective, areas with high concentrations of seals would have to be "seeded" with developing eggs of *C. osculatum.* A prime location for such an approach might be Sable Island. When hatched, the L_2 larvae would infect the copepod host, which in turn would be eaten by fish and then be passaged to the seal. In the stomach of the seal, *C. osculatum* adults mature, presumably to the detriment of *P. decipiens* that also might be present.

Surprisingly, infection of ground fish with *C. osculatum* on Sable Bank is rare (McClelland *et al.,* 1990b; G. McClelland, unpublished data), despite the presence of migrant seals from the Gulf of St. Lawrence with heavy infections of *C. osculatum.* Although these data do not support the competitive interaction found *in vitro* with adult worms, factors in the life cycle of *C. osculatum* may explain the findings presented. Work done by Choudhury (1989) indicates that only one crustacean host is required and that development of *C. osculatum* then proceeds directly in its fish host. Research also has shown that if newly hatched larvae of *C. osculatum* are injected directly into the body cavity of rainbow trout, they can develop there, whereas *P. decipiens* injected similarly cannot develop (Smith *et al.,* 1990). When *C. osculatum* naturally infects its fish host, the L_3 larvae usually remain within the body cavity where they become encapsulated on the cecae or mesenteries. They do not, accordingly, pose any problem for the fishing industry since the worms are discarded with the viscera when the fish are gutted.

6. Attacking Earlier Worm Life-Cycle Stages

As evident from the discussion of the life cycle, the diversity of both invertebrate and fish species that can play a role in the life cycle of *P. decipiens* is extensive, making identification of a vulnerable stage that might be sensitive to human interactions difficult. However, although the life cycle of *P. decipiens, sensu lato,* may appear to have a great diversity of host species, *P. decipiens* is believed to be a complex of sibling species based on differences in soluble protein profiles (Appleton and Burt, 1991) and infectivity to invertebrates. Each of the sibling species may have narrower host range, particularly of crustacean hosts, which could account for the differences found in the experimental infections by Jarecka *et al.* (1988) and by McClelland (1990).

Identification of the sibling species is

currently under investigation in this laboratory (M. D. B. Burt). Larval decapods *(Hyas araneus, Hyas coarctatus, Cancer borrealis)* were found to become readily infected experimentally with the L_2 larvae of harbor seal worms. Trying to infect these species of crabs with larvae from gray seal worms will be interesting. Repeated attempts to infect *G. oceanicus* with the L_2 larvae of harbor seal worms all failed, yet these crustaceans are highly successful hosts for larvae from gray seal worms. If the natural hosts in the life cycles that are responsible for the transmission of most larvae to fish can be determined, consideration could be given, if feasible, to selective trapping of such hosts and their removal from areas of high seal densities. Given the wide diversity of suitable experimental hosts, such a remedy is probably impractical currently. The essential first steps, however, are determining how many sibling species are present and how to identify them microscopically rather than biochemically. Subsequently, researchers must determine the life cycles of these species as well as the specific crustacean hosts involved.

VI. Related Species Contributing to the Sealworm Problem

A related anisakid, *Anisakis simplex,* also has been increasing in abundance in recent years. Unlike the sealworm, this species matures in the stomach of whales and other cetaceans. However, like the sealworm, the L_3 larvae commonly are found in various fishes in which they are referred to as the sashimi or sushi worm, named after raw seafood preparations through which human infections are incurred. In herring, especially in the northeast Atlantic, these anisakids are known as "herring worm;" in low numbers (<15), they seem to be restricted to the body cavity of the herring (McGladdery, 1986). As the numbers increase in herring, however, the worms also penetrate the flesh where they become established (McGladdery, 1986). The proportion of *A. simplex* in the flesh varies with size and species of host but may remain more or less constant in a given host species (McClelland *et al.,* 1990b). Previously, this species was thought to be essentially a parasite of pelagic fishes such as herring and mackerel; more recently, however, the worm has been found in various benthic fishes with high prevalence, especially in larger fish in which the prevalence of *A. simplex* is 100% in cusk and pollack over 61 cm, in cod over 66 cm, and in silver hake over 51 cm. Haddock and white hake over 61 cm show prevalences of 81% and 78% infection with *A. simplex,* respectively (McClelland *et al.,* 1990b). The L_3 larvae in these benthic dwelling fishes, even in low numbers, will become established in the flesh where they can be confused with *P. decipiens* unless examined carefully. Like the larvae of *P. decipiens,* larvae of *A. simplex* in the flesh of the host pose a problem initially to the fish processor and ultimately to the consumer.

A third species, *Contracaecum osculatum,* also is found in large (≥ 51 cm) cod and white hake (McClelland *et al.,* 1990b), but only rarely penetrates the body wall to establish in the flesh; usually, the worm is found encapsulated on the mesenteries in the body cavity of the host and, as such, poses no problem to the processor or the consumer because it is removed during the gutting process.

VII. Aquaculture Implications

Anisakid worms have been reported in various wild salmonid fishes including Atlantic salmon (*A. simplex:* Beverley-Burton *et al.,* 1977; Beverley-Burton and Pippy, 1978; personal observations; *P. decipiens:* personal observations), Arctic char (*Anisakis*-type larvae: Polyanskii, 1966; Hicks and Threlfall, 1973) and brook char (*Anisakis*-type larvae: Hicks and Threlfall, 1973; *P. decipiens:* McClelland *et al.,* 1990b). Atlantic salmon, *Salmo salar,* are infected easily experimentally with both *A. simplex* and *P. decipiens* L_3 larvae removed from other fish such as herring and cod, respectively. The L_3 larvae establish themselves in the flesh of Atlantic salmon and, because of the pigmentation of the flesh, are not easily seen. Since salmon usually is not eaten as fillets that have been candled, the presence of worms could pose a problem for consumers if the fish were smoked cold or eaten as sashimi (raw).

The chances of aquacultured fish being infected are low if artificial or processed feed is used. If untreated fresh fish offal is used as a food source, the chances of parasitic infections increases in direct proportion to the amount of fresh fish or offal consumed. With the increasing abundance of both *P. decipiens* and *A. simplex* in wild fish populations, an increasing abundance of infected macroinvertebrates, many of which could easily enter sea cages accidentally and be eaten by the caged Atlantic salmon or Arctic char, also is evident. This behavior already has resulted in caged salmon being infected with the acanthocephalan *Echinorhynchus gadi* (G. McClelland, personal communication) and, anecdotally, with sealworm. Sea cages around Grand Manan or in the Lower Bay of Fundy are prone to the impact of both infected gray and infected harbor seals. The presence of parasitic worms in cultured fish would affect the sales of such fish adversely and preventive methods should be developed. To that end, investigators are in the process of developing a vaccine that will prevent Atlantic salmon from being infected with *P. decipiens.*

Ramakrishna and Burt (1991) showed that little if any natural immunity develops in fish, which can and do recruit worms continuously. The tissue reaction to each new infection is similar to that to the initial infection. However, if fish are injected with crude extract of *P. decipiens* before being challenged with a live worm, the response to the challenge infection involves an influx of lymphocytes to the site of the lesion, a reaction not seen in fish previously infected with whole L_3 worms and a reaction typical of a delayed-type hypersensitivity reaction (Ramakrishna *et al.,* 1993). Preliminary work has shown that the number of challenge worms that become established after prior injection with crude extract of *P. decipiens* L_3 larvae in phosphate-buffered saline (PBS) was reduced greatly from about 85% in the PBS controls to about 15% (M. D. B. Burt, B. M. MacKinnan, N. Ramakrishna, and C. G. Likely, 1992, unpublished data). Should infection of caged Atlantic salmon with anisakids become a problem, based on results to date, the prognosis for development of a vaccine is good.

VIII. Conclusion

In the final analysis, reducing the abundance of sealworms in the northwest Atlantic (and elsewhere) is not easy. *Pseudoterranova decipiens, sensu lato,* appears to have a natural competitor in *C. osculatum.* This related anisakid, however, has not made much headway in the Sable Bank area. Although seals with heavy infections of *C. osculatum* have come into the Sable Bank region from the Gulf of St. Lawrence, fish in the Sable Bank area are only lightly infected with this worm. The crustacean host for *C. osculatum* may be absent or scarce in this region, preventing any increase of worms in the fish population, although significant numbers of eggs may be produced. Note that, off the East Coast of Scotland, harbor seals are heavily infected with *C. osculatum* as are fish in the same region and *P. decipiens,* the dominant worm there in the 1960s, has almost disappeared. Whether this effect is caused by the competitive interaction reported in this chapter or by some other factor(s) is not known.

Although assuming that the sealworm problem will resolve itself is tempting, this possibility is highly unlikely without even more serious depletions of marine fish stocks, which might force a reduction in the seal predators. Thus, researchers must continue to try to find a solution. The most obvious solution is reducing the size of the gray seal population, a solution that would not be popular and one that should not be employed until the effects on reduction in worm abundance achieved by any particular level of culling can be predicted with some degree of certainty. The development of a predictive model of the population dynamics of the sealworm remains, therefore, a high priority and should be recognized as such by federal and provincial governments.

Acknowledgments

The author gratefully acknowledges the help of C. G. Likely in growing the worms and B. M. MacKinnon in writing the first draft, and the valuable advice of G. McClelland. The author also gratefully acknowledges the facilities of HMSC, where much of the work reported was done, the technical assistance of F. Purton, and the dedicated help of I. Johnston and M. Morton in typing the chapter. Financial support has been provided by the Natural Sciences and Engineering Research Council of Canada, Fisheries and Oceans, Canada, the New Brunswick Department of Commerce and Technology, the New Brunswick Salmon Growers Association, the Atlantic Canada Opportunities Agency, and the Department of Industry, Science and Technology, Canada.

References

Appleton, T. E., and Burt, M. D. B. (1991). Biochemical characterization of third-stage larval sealworm, *Pseudoterranova decipiens* (Nematoda: Anisakidae), in Canadian Atlantic waters using isoelectric focusing of soluble proteins. *Can. J. Fish. Aquat. Sci.* **48,** 1800–1803.

Benoit, D., and Bowen, W. D. (1990). Summer diet of grey seals *(Halichoerus grypus)* at Anticosti Island, Gulf of St. Lawrence, Canada. *Can. Bull. Fish. Aquat. Sci.* **222,** 227–242.

Berland, B. (1963). *Phocascaris cystophorae* sp. nov. (Nematoda) from the hooded seal with an emendation of the genus. *Acta Univ. Bergensis* **17,** 1–21.

Beverley-Burton, M., and Pippy, J. H. C. (1978). Distribution, prevalence and mean numbers of larval *Anisakis simplex* (Nematoda: Ascaridoidea) in Atlantic salmon, *Salmo salar,* and their use as biological indicators of host stocks. *Environ. Biol. Fishes* **3,** 211–222.

Beverley-Burton, M., Nyman, O. L., and Pippy, J. H. C. (1977). The morphology and some observations on the population genetics of *Anisakis simplex* larvae (Nematoda: Ascaridata) from fishes of the North Atlantic. *J. Fish. Res. Board Can.* **34,** 105–112.

Boulva, J., and McLaren, I. A. (1979). Biology of the harbour seal, *Phoca vitulina,* in eastern Canada. *Bull. Fish. Res. Board Can.* **200,** 24 pp.

Brattey, J. (1990). Effect of temperature on egg hatching in three ascaridoid nematode species from seals. *Can. Bull. Fish. Aquat. Sci.* **222,** 27–39.

Brattey, J., Stobo, W. T. (Rapporteurs), Bjørge, A., Burt, M. D. B., des Clers, S., Fanning, L. P., Hauksson, E., Jarecka, L., Landry, T., Mansfield, A., McClelland, G., Mohn, R., Möller, H., Myers, R. A., Ni, I. H., Pálsson, J., Smith, J. W., Thompson, D., and Wootten, R. (1990). Group report 2. Infection of definitive hosts. *Can. Bull. Fish. Aquat. Sci.* **222,** 139–145.

Burt, M. D. B., Smith, J. W., Jarecka, A., Pike, A. W., Wootten, R., and McClelland, G. (1990a). *Pseudoterranova decipiens* (Nematoda: Ascaridoidea): Time of development to hatching of larvae at different temperatures and salinities. *Can. Bull. Fish. Aquat. Sci.* **222,** 41–45.

Burt, M. D. B., Campbell, J. D., Likely, C. G., and Smith, J. W. (1990b). Serial passage of larval *Pseudoterranova decipiens* (Nematoda: Ascaridoidea) in fish. *Can. J. Fish. Aquat. Sci.* **47,** 693–695.

Choudhury, A. (1989). "Aspects of the Early Life History of *Pseudoterranova decipiens* (Krabbe, 1878) and *Contracaecum osculatum* (Rudolphi, 1802) (Nematoda: Ascaroidea: Anisakidae)." *M.Sc.* Thesis. University of New Brunswick, Fredericton, N. B. Canada.

Dollfus, R. P. (1953). Aperçu général sur l'histoire naturelle des parasites animaux de la morue Atlanto-Arctique, *Gadus callarias* L. *Encyclop. Biol.* **43,** 1–428.

Fisher, H. D. (1949). Harbour seals. *Fish. Res. Board Can. Annu. Rep. Atl. Biol. Sta. St. Andrews* **1949,** 104–106. (Appendix # 72)

Fisher, H. D. (1950). Incidence of cod-worms in seal stomachs. *Fish. Res. Board Can. Annu. Rep. Atl. Biol. Sta. St. Andrews* **1950,** 113–116. (Appendix # 68)

Fisher, H. D. (1951). Incidence of cod-worms in seal stomachs. *Fish. Res. Board Can. Annu. Rep. Atl. Biol. Sta. St. Andrews* **1951,** 114–118. (Appendix # 71)

Gibson, D. I. (1983). The systematics of ascaridoid nematodes—A current assessment. *In* "Concepts in Nematode Systematics" (A. R. Stone, H. M. Platt, and L. F. Khalil, eds.), pp. 321–338. Academic Press, New York.

Gibson, D. I., and Colin, J. A. (1982). The *Terranova* enigma. *Proc. Br. Soc. Parasitol.* **85,** XXXVI–XXVIII.

Hicks, F. J., and Threlfall, W. (1973). Metazoan parasites of salmonids and coregonids from coastal Labrador. *J. Fish. Biol.* **5,** 399–415.

Homans, R. E. S., and MacFarlane, A. S. (1957). "Preliminary Report on the Occurrence of Codworms in Flounders of the Maritimes and on the Efficiency of Present Candling Methods in Fish Plants." Inspection Branch, Department of Fisheries, Maritime Area, Halifax, Nova Scotia.

Jarecka, L., Choudhury, A., and Burt, M. D. B. (1988). On the life cycle of *Pseudoterranova decipiens:* Experimental infections of micro and macroinvertebrates. *Bull. Can. Soc. Zool.* **19,** 32.

Johnston, T. H., and Mawson, P. M. (1945). Parasitic nematodes. *Rep. British, Australian, and New Zealand Antarctic Res. Exped. B* **5,** 73–159.

Karokhin, V. I. (1946). Two new species of *Porrocaecum* in birds of prey in Siberia. *In* "Collected Papers on Helminthology Dedicated by his Pupils to K. I. Skryabin in His 40th Year of Scientific, Educational, and Administrative Achievement" (V. P. Pod'yanolskaya, ed.), pp. 132–141. Izdat. Akad. Nauk, Moscow. (In Russian)

Krabbe, H. (1878). Saelerness og Tandhvalernes Spolorme. Overs. Kl. *Danske Videnskab. Selskab. Forh.* **1,** 43–51.

Leiper, R. T., and Atkinson, E. L. (1914). Helminthes of the British Antarctic expedition, 1910–1913. *Proc. Zool. Soc. London* **1,** 222–226.

Likely, C. G., and Burt, M. D. B. (1989). Cultivation of *Pseudoterranova decipiens* (sealworm) from third-stage larvae to egg-laying adults *in vitro. Can. J. Fish. Aquat. Sci.* **46,** 1095–1096.

Likely, C. G., and Burt, M. D. B. (1992). *In vitro* cultivation of *Contracaecum osculatum* (Nematoda: Anisakidae) from third-stage larvae to egg-laying adults. *Can. J. Fish. Aquat. Sci.* **49,** 347–348.

Malouf, A. H. (1986). "Report of the Royal Commission on Seals and Sealing in Canada," Vol. 3. Ottawa, Canada.

Mansfield, A. W., and Beck, B. (1977). The grey seal in eastern Canada. *Environ. Can. Fish. Marine Serv. Tech. Rep.* **704.**

McClelland, G. (1980a). *Phocanema decipiens:* Growth, reproduction, and survival in seals. *Exp. Parasitol.* **49,** 175–187.

McClelland, G. (1980b). *Phocanema decipiens:* Pathology in seals. *Exp. Parasitol.* **49,** 405–419.

McClelland, G. (1982). *Phocanema decipiens* (Nematoda: Anisakinae): Experimental infections in marine copepods. *Can. J. Zool.* **60,** 502–509.

McClelland, G. (1990). Larval sealworm *(Pseudoterranova decipiens)* infections in benthic macrofauna. *Can. Bull. Fish. Aquat. Sci.* **222,** 47–65.

McClelland, G., Misra, R. K., and Marcogliese, D. J. (1983a). Variations in abundance of larval anisakines, sealworm *(Phocanema decipiens)* and related species in cod and flatfish from the southern Gulf of St. Lawrence (4T) and the Breton Shelf (4 Vn). *Can. Tech. Rep. Fish. Aquat. Sci.* **1210.**

McClelland, G., Misra, R. K., and Marcolgiese, D. J. (1983b). Variations in abundance of larval anisakines, sealworm *(Phocanema decipiens)* and related species in Scotian Shelf (4Vs and 4W) cod and flatfish. *Can. Tech. Rep. Fish. Aquat. Sci.* **1202.**

McClelland, G., Misra, R. K., and Martell, D. J. (1985). Variations in abundance of larval anisakines, sealworm *(Pseudoterranova decipiens)* and related species, in eastern Canadian cod and flatfish. *Can. Tech. Rep. Fish. Aquat. Sci.* **1392.**

McClelland, G., Misra, R. K., and Martell, D. J. (1987). Temporal and geographical variations in abundance of larval sealworm, *Pseudoterranova (Phocanema) decipiens,* in the fillets of American plaice *(Hippoglossoides platessoides)* in eastern Canada: 1985–1986 surveys. *Can. Tech. Rep. Fish. Aquat. Sci.* **1513.**

McClelland, G. (Rapporteur), Bjørge, A., Brattey, J., Burt, M., des Clers, S., Fanning, P., Hare, G., Jarecka, L., Landry, T., Margolis, L., McGladdery, S., Misra, R., Mohn, R., Möller, H., Pálsson, J., Smith, J., Stobo, W., and Wootten, R. (1990a). Group Report 1. Hatching and infection of intermediate hosts. *Can. Bull. Fish. Aquat. Sci.* **222,** 17–25.

McClelland, G., Misra, R. K., and Martell, D. J. (1990b). Larval anisakine nematodes in various fish species from Sable Island Bank and vicinity. *Can. Bull. Fish. Aquat. Sci.* **222,** 83–118.

McGladdery, S. E. (1986). *Anisakis simplex* (Nematoda: Anisakidae) infection of the musculature and body cavity of Atlantic herring *(Clupea harengus harengus). Can. J. Fish. Aquat. Sci.* **43,** 1312–1317.

Myers, B. J. (1959). *Phocanema,* a new genus for the anisakid nematode of seals. *Can. J. Zool.* **37,** 459–465.

Odense, P. H. (1978). Some aspects of the codworm problem. *Fish. Marine Serv. Indust. Rep.* **106.**

Oluoch-Munda, M. A. O. (1992). "Effect of Oxytocin on Growth and Reproductive Tract Development of *Pseudoterranova decipiens* (Krabbe, 1878) *in Vitro.*" M.Sc. Thesis. University of New Brunswick, Fredericton, N. B., Canada.

Polyanskii, Y. I. (1966). Parasites of the fish of the Barents Sea. *Trudy Zool. Inst. Leningr.* **19,** 1966, 1–58. *Israel Prog for Sci. Translations* (Translated from Russian)

Power, H. E. (1958). The effect of various lighting conditions on the efficiency of "candling" cod fillets for detection of parasites. *J. Fish. Res. Board Can.* **15,** 537–542.

Power, H. E. (1961). Slicing of fillets as an aid in detection and removal of codworms from Atlantic cod fillets. *J. Fish. Res. Board Can.* **18,** 137–140.

Railliet, A., and Henry, A. (1912). Quelques nématodes parasites des reptiles. *Bull. Soc. Pathol. Expt.* **5,** 251–259.

Ramakrishna, N. R., and Burt, M. D. B. (1991). Tissue

response of fish to invasion by larval *Pseudoterranova decipiens* (Nematoda: Anisakidae). *Can. J. Fish. Aquat. Sci.* **48,** 1623–1628.

Ramakrishna, N. R., Burt, M. D. B., and MacKinnon, B. M. (1993). Cell-mediated immune response of rainbow trout *(Oncorhynchus mykiss)* to larval *Pseudoterranova decipiens* (Nematoda: Ascaridoidea) following sensitization to live sealworm, sealworm extract, and nonhomologous extracts. *Can. J. Fish. Aquat. Sci.* **50,** 60–65.

Scott, D. M. (1950). A preliminary report on the codworm investigation. *Fish. Res. Board Can. Atl. Prog. Rep.* **48,** 10–12.

Scott, D. M., and Fisher, H. D. (1958). Incidence of the ascarid *Porrocaecum decipiens* in the stomachs of three species of seals along the southern Canadian Atlantic mainland. *J. Fish. Res. Board Can.* **15,** 495–516.

Scott, D. M., and Martin, W. R. (1957). Variation in the incidence of larval nematodes in Atlantic cod fillets along the southern Canadian mainland. *J. Fish. Res. Board Can.* **14,** 975–996.

Skrjabin, K. I., Shikhobalova, N. P., and Mozgovia, A. A. (1951). Oxyuroids and ascarids. pp. 1–631. *Opred. Parazit. Nematod.* **2,** (in Russian)

Smith, J. W., Elarifi, A. E., Wootten, R., Pike, A. W., and Burt, M. D. B. (1990). Experimental infection of rainbow trout, *Oncorhynchus mykiss,* with *Contracaecum osculatum* (Rudolphi, 1802) and *Pseudoterranova decipiens* (Krabbe, 1878) (Nematoda: Ascaridoidea). *Can. J. Fish. Aquat. Sci.* **47,** 2293–2296.

Stobo, W. T., and Zwanenburg, K. C. T. (1990). Grey seal *(Halichoerus grypus)* pup production on Sable Island and estimates of recent production in the northwest Atlantic. *Can. Bull. Fish. Aquat. Sci.* **222,** 171–184.

Stobo, W. T., Beck, B., and Fanning, L. P. (1990). Seasonal sealworm *(Pseudoterranova decipiens)* abundance in grey seals *(Halichoerus grypus). Can. Bull. Fish. Aquat. Sci.* **222,** 147–162.

Templeman, W. (1990). Historical background to the sealworm problem in eastern Canadian waters. *Can. Bull. Fish. Aquat. Sci.* **222,** 1–16.

Templeman, W. H., Squires, H. J., and Fleming, A. M. (1957). Nematodes in the fillets of cod and other fishes in Newfoundland and neighbouring areas. *J. Fish. Res. Board Can.* **14,** 831–897.

Thompson, D., Mansfield, A. W. (Rapporteurs), Beck, B., Bjørge, A., Bowen, D., Hammill, M., Hauksson, E., Myers, R., Ni, I. H., and Zwanenburg, K. (1990). Group report 3. Seal ecology. *Can. Bull. Fish. Aquat. Sci.* **222,** 163–170.

Zwanenburg, K., Beck, B., and Smith, S. J. (1981). Eastern Canadian grey seal *(Halichoerus grypus)* research report and 1980 stock assessment. *Can. Atl. Fish. Sci. Adv. Comm. Res. Doc.* **81.**

Zwanenburg, K., and Bowen, W. D. (1990). Population trends of the grey seal *(Halichoerus grypus)* in Eastern Canada. *Can. Bull. Fish. Aquat. Sci.* **222,** 185–197.

Author Index

SCIENTIFIC NAME INDEX

SUBJECT INDEX